WILDLIFE
LAND
and
PEOPLE

CARLETON LIBRARY SERIES

The Carleton Library Series publishes books about Canadian economics, geography, history, politics, public policy, society and culture, and related topics, in the form of leading new scholarship and reprints of classics in these fields. The series is funded by Carleton University, published by McGill-Queen's University Press, and is under the guidance of the Carleton Library Series Editorial Board, which consists of faculty members of Carleton University. Suggestions and proposals for manuscripts and new editions of classic works are welcome and may be directed to the Carleton Library Series Editorial Board c/o the Library, Carleton University, Ottawa K1S 5B6, at cls@carleton.ca, or on the web at www.carleton.ca/cls.

218 *The Ordinary People of Essex*
Environment, Culture, and Economy on the Frontier of Upper Canada
John Clarke

219 *So Vast and Various*
Interpreting Canada's Regions in the Nineteenth and Twentieth Centuries
Edited by John Warkentin

220 *Industrial Organization in Canada*
Empirical Evidence and Policy Challenges
Edited by Zhiqi Chen and Marc Duhamel

221 *Surveyors of Empire*
Samuel Holland, J.F.W. Des Barres, and the Making of The Atlantic Neptune
Stephen J. Hornsby

222 *Peopling the North American City*
Montreal, 1840–1900
Sherry Olson and Patricia Thornton

223 *Interregional Migration and Public Policy in Canada*
An Empirical Study
Kathleen M. Day and Stanley L. Winer

224 *How Schools Worked*
Public Education in English Canada, 1900–1940
R.D. Gidney and W.P.J. Millar

225 A Two-Edged Sword
The Navy as an Instrument of Canadian Foreign Policy
Nicholas Tracy

226 *The Illustrated History of Canada*
25th Anniversary Edition
Edited by Craig Brown

227 *In Duty Bound*
Men, Women, and the State in Upper Canada, 1783–1841
J.K. Johnson

228 *Asleep at the Switch*
The Political Economy of Federal Research and Development Policy since 1960
Bruce Smardon

229 *And We Go On*
Will R. Bird
Introduction and Afterword by David Williams

230 *The Great War as I Saw It*
Frederick George Scott
Introduction by Mark G. McGowan

231 *The Canadian Oral History Reader*
Edited by Kristina R. Llewellyn, Alexander Freund, and Nolan Reilly

232 *Lives in Transition*
Longitudinal Analysis from Historical Sources
Edited by Peter Baskerville and Kris Inwood

233 *W.A. Mackintosh*
The Life of a Canadian Economist
Hugh Grant

234 *Green-lite*
Complexity in Fifty Years of Canadian Environmental Policy, Governance, and Democracy
G. Bruce Doern, Graeme Auld, and Christopher Stoney

235 *Canadian Expeditionary Force, 1914–1919*
Official History of the Canadian Army in the First World War
G.W.L. Nicholson
Introduction by Mark Osborne Humphries

236 *Trade, Industrial Policy, and International Competition, Second Edition*
Richard G. Harris
Introduction by David A. Wolfe

237 *An Undisciplined Economist*
Robert G. Evans on Health Economics, Health Care Policy, and Population Health
Edited by Morris L. Barer, Greg L. Stoddart, Kimberlyn M. McGrail, and Chris B. McLeod

238 *Wildlife, Land, and People*
A Century of Change in Prairie Canada
Donald G. Wetherell

WILDLIFE LAND *and* PEOPLE

A Century of Change in Prairie Canada

Donald G. Wetherell

Carleton Library Series 238

McGill-Queen's University Press   Montreal & Kingston · London · Chicago

ISBN 978-0-7735-4791-9 (cloth)
ISBN 978-0-7735-9988-8 (ePDF)
ISBN 978-0-7735-9989-5 (ePUB)

Legal deposit third quarter 2016
Bibliothèque nationale du Québec

Printed in Canada on acid-free paper that is 100% ancient forest free (100% post-consumer recycled), processed chlorine free

This book has been published with the help of a grant from the Canadian Federation for the Humanities and Social Sciences, through the Awards to Scholarly Publications Program, using funds provided by the Social Sciences and Humanities Research Council of Canada. Funding was also received from the McLean Foundation.

McGill-Queen's University Press acknowledges the support of the Canada Council for the Arts for our publishing program. We also acknowledge the financial support of the Government of Canada through the Canada Book Fund for our publishing activities.

Library and Archives Canada Cataloguing in Publication

Wetherell, Donald G. (Donald Grant), 1949–, author
Wildlife, land, and people : a century of change in Prairie Canada / Donald G. Wetherell.

(Carleton library series ; 238)
Includes bibliographical references and index.
Issued in print and electronic formats.
ISBN 978-0-7735-4791-9 (hardback).–ISBN 978-0-7735-9988-8 (PDF).
–ISBN 978-0-7735-9989-5 (ePUB)

1. Human-animal relationships–Prairie Provinces–History.
2. Nature–Effect of human beings on–Prairie Provinces–History.
3. Animal ecology–Prairie Provinces–History. 4. Human ecology–Prairie Provinces–History. 5. Prairie Provinces–History. I. Title.
II. Series: Carleton library series ; 238

QL85.W47 2016 591.72'7 C2016-903687-1
C2016-903688-X

For
Irene Kmet

Contents

Figures

Abbreviations

In the Text

CWS	Canadian Wildlife Service
DUC	Ducks Unlimited (Canada)
HBC	Hudson's Bay Company
NRTA	Natural Resources Transfer Agreement
NWMP	North-West Mounted Police
NWT	North-West Territories or Northwest Territories
PFRA	Prairie Farm Rehabilitation Administration
RCMP	Royal Canadian Mounted Police

In the Notes and Figures

AMNH	American Museum of Natural History (Rare Book Collection)
CCA	City of Calgary Archives
CCA, CCDF	City of Calgary Archives, City Clerk's and Department Fonds
CCA, PRDF	City of Calgary Archives, Parks and Recreation Department Fonds
CEA	City of Edmonton Archives
DA	Department of Agriculture
DAI	Department of Agriculture and Immigration
DI	Department of the Interior
DIA	Department of Indian Affairs
DLF	Department of Lands and Forests
DMNR	Department of Mines and Natural Resources
DNR	Department of Natural Resources
GA	Glenbow Museum and Archives
LAC	Library and Archives Canada
MA	Manitoba Archives
MLL	Manitoba Legislature Library
PAA	Provincial Archives of Alberta
QUA	Queen's University Archives
SAB	Saskatchewan Archives Board (Regina)
UAA	University of Alberta Archives

Acknowledgments

The history of the encounter of humans and wild animals in prairie Canada touches in many ways on social, cultural, economic, and political life. And since historians are inevitably shaped by their times, I have grappled in writing this history with the spirit of my own age, which is so different in some ways – but so similar in others – to those dealt with in this book. Attitudes familiar to many people in 1900 about managing natural resources such as wildlife for economic gain or to meet social needs remain. And some contemporary attitudes about human responsibility for wild animals and the natural environment or about the character and lives of wild animals would not seem strange to some people a century ago. Yet, we also now live in a world with different priorities and expectations about the human encounter with wild animals. Such things as animal rights, the dominance of the human species in nature, and the relationship among different species are now thought about in very different terms. And while worrying about the health of natural systems and the survival of certain species has a venerable pedigree, it is now conjoined with a growing belief that an important and irreversible divide has been, or is about to be crossed. Growing global environmental changes, from escalating extinction of wild species (and many domestic ones) to heightened levels of greenhouse gases in the atmosphere, suggest to some observers that the human impact on the planet has become part of geological time and that we have left the Holocene and are now living in what some are calling the Anthropocene.

But other things endure, notably the assistance of colleagues and friends that a historian enjoys when writing about a topic of such scope. I have benefited over the years from the knowledge and insight of many, especially Peter Clancy, Bill Davies, Hugh Dempsey, Max Foran, Doug Francis, Don Smith, Angela Specht, and Jim Taylor. And as a model of coexistence with wild animals, Heather MacEwan Foran is an inspiration.

Thanks are due to Heather Mayan for her research assistance and to David Relkoff for his work on official and newspaper records in Winnipeg. So too, Doug Cass at the Glenbow Archives in Calgary and archivists at the Manitoba, Saskatchewan, and Alberta provincial archives, at Library and Archives Canada in Ottawa, and, among others, at Queen's University Archives in Kingston provided much appreciated guidance and assistance. Mark Abley at McGill-Queen's University Press helped and encouraged in many ways to guide this book through the editorial process. I am grateful for his patience and sincerity, and that he so generously gave his experience and talent to help bring this book to publication. Two anonymous reviewers also offered useful suggestions. Thanks are also due to Judith Turnbull, who copy-edited the manuscript with dispatch, sensitivity, and skill, and to Bill Nelson, who expertly drew the map. The excerpt from Lorna Crozier's "Buffalo Stone" is from her book *A Saving Grace* and I thank her for permission to quote it. Valuable too was leave time from the University of Calgary and Athabasca University, as was a Standard Research Grant from the Social Sciences and Humanities Research Council. The latter was crucial and without it, this work could not have been completed. This book has also been published with the help of a grant from the Federation for the Humanities and Social Sciences, through the Awards to Scholarly Publications Program, using funds provided by the Social Sciences and Humanities Research Council of Canada. The McLean Foundation also generously provided financial support for the publication. But my greatest debt is to my life and work partner, Irene Kmet. My gratitude to her goes beyond her patience, her unerring judgment and advice, her inspiration, and her research assistance and editing skills to all that we have shared, and will continue to share, together.

Preface

> [Historians] need to reckon with animals to comprehend the full power and permeable boundaries of our humanity. Erasing the differences between people and beasts distorts history, but so does overlooking the ties of exploitation and violence, curiosity and affection that bind all living creatures to a common past.
>
> Jon Coleman[1]

The Blackfoot say that when Napi created the earth and humans, he needed the help of animals. The world was then covered with water and he persuaded first a duck, then an otter, then a badger to dive below the water to find what was there. Others say it was a loon and a beaver. But whichever animals dove into the vast waters, they came up empty-handed. Then Napi sent down a muskrat. The muskrat stayed below a very long time. When it finally resurfaced – some say nearly dead from exhaustion – it held a lump of mud in its paws. From this mud Napi created the earth. Then he used a bit of this earth to create the first humans and he taught them to hunt and about the right way to live. His task completed, Napi disappeared into the mountains.[2]

Another account of the creation of the world is the Biblical one in Genesis. There it is said that God created the earth in six days by issuing eight commands. In the first four days He commanded the appearance of light, the skies, the earth, sun, moon, stars, seas, and plants. On the fifth day He commanded the appearance of fish and birds. And on the sixth day, He created wild animals, reptiles, livestock, and, as the culminating command, a man and a woman. In this account of the origin of the world, God created it through His Word. The only actor was God, who ended His creation by ordering humans to be "fruitful, and multiply, and fill the earth, and subdue it." He also commanded them to worship Him, whose immanence had created and would continue to fill the material world.

In 1859 Charles Darwin published *The Origin of Species*. He proposed that all life was shaped by the ongoing adaptation of living things to their environment. Others had speculated and worked for decades on such ideas, and Darwin built upon their thinking and findings and his own extensive research. While he did not say how the earth had originated, his theory confirmed an understanding that life arose not from a single great design. Evolution revealed only its own logic; it happened differently in different places because of different contexts and needs. This evidence created troubling personal questions for Darwin because the God of Genesis did not square with the evidence of life's immense past, extraordinary variety, and inexorable force.

These are widely different ways of understanding the world, but like many other accounts, they tend to posit that animals were on the earth before humans and that, with the arrival of humans, there evolved dynamic historical connections between animals and people. This book looks at one part of this history – the relationships that people have had in their everyday lives with wild animals on the Canadian prairies after 1870, the year that the vast and diverse lands that are now Manitoba, Saskatchewan, and Alberta became part of Canada. Wild animals are important in this history; they are a fundamental part of our natural heritage, and the values ascribed to them are essential references to cultural and social life. Literary scholar Susan Kollin has observed that the historic connections between humans and wild animals are part of a "larger history" that cannot be separated from culture. As she phrases it, "the social cannot be so easily left behind." This larger history contains many – often contradictory – streams in the cultural, political, economic, and social connections between people and wild animals in the same landscape. The same thing that historian Alan MacEachern says about environmental history in general can be said about the relationships between people and wild animals: they provide an opportunity to appreciate better "the place of nature in human affairs, and thus ultimately of human affairs itself." Or, as historian Ted Steinberg puts it, study of the history of the human place in the environment is not merely a preface to social and political history but a critical dimension of the working of power in society and the implications of human agency for the world.[3]

In these terms, this book deals with the relationships people had with wild animals in social, political, and economic terms in everyday life in prairie Canada until the early 1960s. An old divide between those who saw animals as sentient and thinking creatures and those who saw them in more mechanical terms was evident in various degrees and expressions in this history. Wild animals have, for the most part, been losers in these relationships – the very existence of some species was deliberately challenged and most other wild animals lost out because leaving room for them was not as important as what could be gained by pushing them aside. While revealing fundamental personal choices and formative social attitudes about the world and about Euro-Canadian settlement of the prairies, the judgment that wild creatures did not merit the lost productivity that their presence on the land would lead to was also a product of the region's integration into Canadian society and the national and international economies after 1870. Aboriginal economies and the fur trade had depended on wild animals, but the new prairie farm economy had no such long-term needs. Indeed, its success was predicated upon changes in regional fauna and flora. This was seen – often eagerly but sometimes reluctantly – as one of the inevitable costs of development. In part this reflected that most Euro-Canadian settlers did not see wild animals as having intrinsic values, nor did they see any personal gain to be derived from accommodating the region's existing natural systems. This anthropocentric attitude was often expressed in the context of a vague knowledge about animal behaviour and through a popular classification of wild animals as friends or foes. Friends offered economic, social, or intellectual benefits, while foes created or threatened economic loss or social or personal harm. In an environment of ongoing change, relationships contingent on social and economic needs that varied by circumstance, place, and time were inherently unstable. Only with a broadening popular acceptance by the 1960s that all animals (including humans) were part of interconnected ecological systems did the dichotomy of friends and foes begin to erode under the weight of growing evidence that all elements in nature had intrinsic value.

Views about the implications of the human–animal interface were made especially clear in the farm sector, the dominant economic activity and a critical point of connection in the relationships between people and wild animals on the prairies. The defence of farming as the foundation for the region's future was a default position in government, business, and popular rhetoric. Farmers had the political and social power to shape public policy on wildlife and to influence views on such matters as predator control and the protection of insectivorous birds. Nonetheless, it is difficult to generalize about a farm "community" made up of hundreds of thousands of individuals of different economic status, ethnicity, and personal histories and grappling with different sub-regional environments. Different farm sectors also had different relationships with various species. Wolves, for example, were considered by ranchers to be foes, coyotes were of concern to mixed farmers but only occasionally to ranchers, and gophers were an obsession for many grain farmers. Insectivorous birds were considered friends by mixed and grain farmers, but waterfowl, cranes, skunks, weasels, and other wild animals were their foes. None of

these animals were of much economic concern to ranchers. But while different agrarian sectors had different relationships with particular species, all were united by an overarching utilitarian view that judged wild animals as allies or adversaries in a struggle for resources. This view had a profound impact on some species; it led, for example, to the extermination of wolves on the southern plains. However, an equally important impact of agriculture was the change that it brought to the land. Whatever the number of wild animals killed, the greatest overall change in animal populations and distribution came because of habitat change. Clearing and breaking the land and dedicating every available acre of land to production devastated the habitat of some species while inadvertently creating new niches for others.

Relationships with wildlife were, of course, framed more broadly than by a single economic sector. Wild game and fish provided food for early settlers and continued to meet subsistence needs for many Aboriginal people and settlers in marginal and fringe areas. Historian George Colpitts has shown that such subsistence use was an important element in attitudes about wild game, and he also notes that the region's wealth of game animals was seen as evidence of the bounty of the land, a doctrine that nicely served the needs of real estate, tourism, and other promoters.[4] Various economic sectors and social groups were also points of connection with wild animals, either singly or collectively. Wild fur-bearing animals were central to the trapping industry (although it gradually lost market share to domesticated fur-bearers), and several species of fish were the basis for the commercial fishery. Certain animals and fish served recreational and commercial functions as commodities for sport hunting and fishing and the development of a tourist sector. Some of these activities, such as sport hunting, further demonstrate how a single connection with wild animals could have very broad implications. Like the fish and game associations that lobbied on its behalf, sport hunting was male dominated and helped form gender identity through a particular expression of masculinity. It also revealed urban and rural tensions and differences about the use and value of game. Sport hunters, for example, sometimes came into conflict with farmers about access to land and damage to property. They also reoriented views about some wild animals by changing public definitions of their value. In the late 1800s, for example, crows were thought to be beneficial for agriculture and were thus considered friends. However, by the end of the First World War, lobbying by sport hunters had turned policy makers and public opinion towards the view that crows threatened game birds and were vermin that should be eliminated.

Given that all natural resources were owned by the Crown, the state shaped the public relationship with wildlife in many important respects. Almost from the beginning of Euro-Canadian settlement, worries about depletion of big game and fur-bearing animals led governments to legislate the seasons when hunting was allowed and the ways that certain animals could be killed and used. These rules were often difficult, if not impossible, to enforce, but this legislation about wildlife was important well beyond its enforcement for it created and shaped standards for public encounters with wildlife and asserted the state's legal authority over all wild animals. This legislation also validated certain patterns of behaviour towards wild animals, explicitly and implicitly promoted assumptions about the value of individual species, and sustained particular social and political relationships with them, including their treatment as natural resources to be exploited and managed for long-term productivity. Reflecting the working of power in prairie society, the state almost never interfered with private property rights in its regulation of wildlife, instead asserting the view that all citizens had equal rights to use wildlife on

Crown land. To a large extent, this favoured the interests of the sport hunting and fishing fraternity and Euro-Canadian farm settlers. This left Aboriginal peoples on the margins. While many Aboriginal peoples had different traditions that explained how humans and animals related to one another, this had little influence beyond their own communities. And in the case of First Nations peoples in particular, the rejection of treaty hunting rights by Euro-Canadians showed where the power lay and demonstrated that officially sanctioned relationships with wildlife formed another element in the Euro-Canadian goal of assimilating Aboriginal peoples.

Despite their centrality, utilitarian objectives such as those that served the interests of farmers or sport hunters were only one aspect of people's relationships with wildlife on the prairies. Keeping wild animals as pets, watching them in national parks or zoos, visiting museums, and participating in natural history outings and meetings often validated human dominance and use of the natural world. But it is equally evident that, for some people, such activities reflected their curiosity and fascination with wild animals and that watching, studying, and interacting with them revealed the magic of life and provided a connection to the world. So, too, some people found that wild animals helped them define what was unique and exceptional about being human. And still others drew symbolic meaning from animal lives and found personal, spiritual, and emotional strength from their encounter with wild animals, sometimes contending that their lives provided models that could inspire and improve human conduct.

These emotional, intellectual, and social relationships were not static and were continually reshaped by changing circumstances. Bison, for example, only became nostalgic symbols of the prairie past when they had been confined to zoos and parks and no longer challenged Euro-Canadian agricultural settlement. Their transformation in the public imagination in the late nineteenth and early twentieth centuries from barriers to settlement to a symbol of a lost and finer world reflected a growing attitude that the appreciation of nature and wild animals was one way to enhance human well-being. An appreciation of what were thought of as "wild" spaces and an assumption that nature was a repository of truth and a source of human well-being gained increasing acceptance. These changing attitudes implicitly endorsed a view that the current generation had an obligation to future ones and that people should leave the natural world no worse than when they had come into it. Such shifting attitudes demonstrate that people's relationships with wild animals were not watertight or self-contained. As historian Tina Loo illustrates in her book *States of Nature*, the human–animal encounter in terms of wildlife conservation was not linear, as it occurred within particular contexts that were contingent, variable over time, and shaped by class, personal experience, and belief. It was thus not surprising that these relationships often reflected contradictory and inconsistent approaches and views. For instance, the aesthetic value of wild animals was often lauded as part of a blunter ambition to conserve economically valuable resources. And some sport hunters expressed admiration and respect for game animals and credited them with sentience and cognitive abilities (sometimes including reason) at the very time that they celebrated the animals' death at their own hands. Certain animals, moreover, were conserved in national parks, and threatened species were preserved in parks specially created for them because of commercially inspired motives as well as a belief in a social and moral duty to preserve animals crucial for defining the nation's natural heritage. And further illustrating this point, wildlife managers organized the killing of predatory animals to meet the needs of hunters, farmers, and others but equally claimed that it would rejuvenate and restore what was believed to be a balance of nature that humans had upset.

Such contradictory and inconsistent approaches to wild animals – what Terry Glavin in his book *Waiting for the Macaws* calls "idiosyncrasies" – have been common around the world.[5] These inconsistencies and contradictions sometimes grew from nothing more than self-interest, self-justification, or a limited knowledge about wild animals and natural systems. They also arose from cultural and social attitudes such as the common presumption that people had the right to reshape the world to serve their personal and collective needs. So, too, the common bloody-minded treatment of wild animals was often accepted as socially normative behaviour. Technological change – from new farm equipment and field practices to the proliferation of automobiles and the growing popularity of photography – sometimes also sharpened or refocused contradictions in the human–animal encounter in both form and practice. But these contradictions become more understandable when we recognize that people usually interacted with individual species, not with an abstract entity called "wildlife." On the prairies, these contradictions were reinforced by the friend and foe duality and also developed from personal experience of the diverse impacts that wild animals had on human needs. Less concretely, some species were thought to be more sociable or more intelligent than others, some animals were thought to be handsomer or cuter than others, and emotive values such as morality, courage, cruelty, and treachery were sometimes ascribed to particular animals. All of these factors contributed to the range of ways that humans encountered wild animals. Such encounters were therefore rarely separate from people's vocations, economic ambitions, and personal, recreational, and social needs. While this was a messy relationship, it was one grounded in the diversity of human experience – one that had a lasting and profound impact on the wildlife of the region.

Postmodern linguistic approaches of the past several decades have led some scholars to contend that nature can only be understood as text that is constructed by language and culture. This includes the knowledge produced by science. Animals and nature can therefore be known only as forms of social production through the stories we tell and the representations we construct about them. This view rubs against a strong tradition in environmental history where the interactions between culture and a nature that exists apart from human society form a dynamic interplay. In this latter view, nature exists, whatever we think about it. In this connection, environmental historian William Cronon has observed that "nonhuman creatures are by and large part of the world that we have not made" and have been "radically silenced in the wake of the linguistic turn." He notes that while his own work has profited from approaches in which nature can be understood as being constituted through culture, he also insists on recognizing the reality of a natural world and its "silent fellow beings." Pointedly, he observes that "without some form of realism to assign some degree of autonomy to these non-human beings, it is not clear that they can be part of the 'conversation' at all."[6] Indeed, as Cronon concludes, ignoring the reality of animal lives and judging nothing beyond human perception and attribution as important shallows out the human experience of being-in-the-world. The American writer Barbara

Ehrenreich describes this well when she tells of an adolescent experience. She recalls that while she was staring at a grove of trees, suddenly "something peeled off the visible world, taking with it all meaning, inference, association, and words." As she realized, "when you take away all human attributions – the words, the names of species, the wisps of remembered tree-related poetry, the fables of photosynthesis and capillary action – that when you take all this away, *there is still something left*."[7]

SOME NOTES ABOUT TERMINOLOGY AND ORGANIZATION

Terminology in animal studies has changed over time, particularly since the 1960s. Some of this changed terminology is old wine in new bottles, but it, both implicitly and explicitly, has become part of the discourse about human–animal relationships. David Fennell, for example, notes in his book *Tourism and Animal Ethics* that contemporary attitudes towards animals include what is called "misothery" (negative feelings towards animals, including hatred, disgust, and fear), "theriophily" (a belief that animals provide models of good behaviour and are superior in some ways to humans), and "biophilia" (an instinctive bond between humans and other living systems). He further argues that contemporary criteria used to assign value to animals are informed by the continuing attraction of the notion that there is a great chain of being as well as by concepts such as "speciesism" (a term first used in 1971 to expresses the attitude that humans have rights that other species do not), "anthropomorphism" (ascribing human traits to non-humans), and "anthropocentrism" (the notion that human needs are central).[8] These concepts have led in some cases to new terminology to describe human–animal relationships, such as "non-human animal" instead of "animal" and "human animal companions" instead of "pets." I have tended overall to retain terms such as "pets" and "animals." In the latter case, I have done so not out of a rejection of the premise of the more contemporary term (humans are animals after all, whatever uniqueness they possess) but because it, like the term "pet," remains popularly recognizable, serviceable, and understood. Historically, however, some customary terms have had variable meaning. The term "wildlife," for example, has often been used as a synonym for game mammals and birds, but it has also been understood more broadly. Gordon Hewitt, in his 1921 book *The Conservation of the Wild Life of Canada*, defined "wild life" as including "the game, non-game, and fur-bearing animals, the game and non-game birds, the fishes and other members of our fauna."[9] I use the term in Hewitt's sense of it, and employ the term "wild animals" in an inclusive sense as including all mammals, birds, and fish living in the wild, whether native or not. I use the term "game" to mean wild animals hunted for sport, recognizing it as a legal term that changed as new animals were included in game legislation and the public relationships that such legislation created. I tend to avoid the term "wilderness" because it has different definitions over time, depending on personal and cultural circumstances.[10] While "nature" can be seen as a constructed concept, it is generally used here to indicate the environments and natural systems that lie outside of formal human society. I follow current usage in employing the term "Aboriginal people" to include Inuit, First Nations, and Métis. The term "First Nations" describes Aboriginal peoples in Canada who are neither Métis nor Inuit. For the prairie context the term is typically used here to refer to people with treaty rights. The term "Métis" refers to members of a people of mixed Aboriginal and European ancestry. I use current names such as Banff National Park instead of Rocky Mountains National Park and terms such as "national park" instead of "dominion park." When identification is certain, current common animal names that are accepted as authoritative by experts in the field are used, such as "whooping crane" instead of "white crane." I have avoided the use of historical terms that are still popularly used but are ambiguous. Examples of this include "prairie chicken," which was sometimes loosely used for various members of the grouse family but usually meant sharp-tailed grouse. "Jumping deer" is another vague term, variously applied at some points to both white-tailed and mule deer, while the term "prairie wolf" was used for both coyotes and wolves. I use the term "bison" rather than "buffalo" and "pronghorn" instead of "pronghorn antelope" or "antelope" because they are more accurate. Similarly, while the term "pickerel" was used widely on the prairies when referring to walleye, I use the latter term because there are no true pickerel native to the region.[11]

The time period covered in this book is a more straightforward matter. It begins in 1870, when the beginning of Canadian control initiated a new economy and society in the prairie west. For the study of wildlife, this was important because it began a process in which the natural world of prairie Canada was remade in a remarkably short time. The next three decades brought a steadily expanding Euro-Canadian population with priorities and ambitions that differed from those that had been typical in the region before. After 1870 wildlife, too, began to be treated as a publicly owned natural resource to be managed by the state. Manitoba, fearing the depletion of game and fur-bearing animals, passed the first game legislation in the region in 1876, while the North-West Territories passed an ordinance to conserve bison in 1877 and more comprehensive game legislation in 1883. An increasing array of legislation regulating wildlife followed, including that passed by the newly created provinces of Alberta and Saskatchewan in 1905. The region's entry into Canada in 1870 therefore marks not only the beginning of the influx of settlers and their impact on the land, but the start of more focused and wide-ranging efforts by the state to regulate and conserve natural resources, including wildlife.

By the late nineteenth century, the view that wildlife should serve the economic, social, and personal needs and goals of incoming settlers was entrenched in private and public life. Many of the assumptions and concerns that emerged at this time remained the basis for public relationships with wildlife in the prairie region in the following decades, and new ways of thinking only began to gain traction in broader public thinking and in state policy after the Second World War. The late 1940s and especially the 1950s were instrumental in preparing the way for the changes that took hold in the coming years.[12] During the 1960s the older duality in which wild animals were ranked, classified, and treated as friends or foes broke down under increasingly popular and officially accepted ecological approaches: mammals, birds, fish, insects, reptiles, and plants were all valued as elements in an interconnected web whose individual elements could not be removed – and neither could new elements be added to it – without bringing changes to an ecosystem. This was a profound change in thinking. As historian Donald Worster notes, there had for centuries been examples of such ecological thinking, but it was not until after the Second World War, when concepts had been more formally articulated, that these ideas began to find their way into popular approaches and policies and nature came to be seen as more than the sum of its parts. This was also an instrumental time because prairie society began to urbanize and develop a somewhat wider economy, and to see itself as something more than just a rural and agrarian culture. Broader environmental concerns also developed, and accumulating unease about the impact of humans on the earth coalesced into a more integrated critique. Even though concerns about soil erosion and environmental degradation in the 1930s had been sidelined by the war and largely resolved in the public mind by land rehabilitation efforts, the memory of the astonishing scale of devastation lingered as evidence of people's negative impact on the planet. The atomic age was even more problematic, and by the early 1950s questions were being raised about the long-term contamination of the environment. Added to this was apprehension about the consequences of the postwar chemical revolution and its effects on water, land, and all life forms.[13] Concern about these matters was reinforced by the growing public acceptance that ecosystems, with many local variations, consisted of interlocked and interdependent elements. The coming together of these apprehensions, along with greater acceptance of ecological principles for understanding natural systems, made the 1960s a watershed era in public thinking about wildlife.

The view that ecosystems consist of interconnected relationships is a model for the structure and organization of this book. Lacking a single overall relationship and recognizing that those that existed were contingent on time, place, circumstance, and need, the history of the human–wild animal interface in prairie Canada is one with many layers and contradictions, all of which were influenced by shared environments and patterns of population growth, economic activity, and technological change. Diverse human wants and circumstances contended with the needs of wildlife, making, for example, holistic and enduring conservation programs seem impossible. To reveal the layers and changing interconnections of this human-wild animal encounter, the book is structured in three parts, with the majority of its chapters addressing the points of connection where interaction with wild animals was acted out. Part 1, consisting of chapter 1, outlines changes that occurred in the region's fauna after 1870 and discusses how the agricultural mindset shaped this change. Part 2 is made up of two chapters that frame the specific relationships that are discussed later in the book. By means of several case studies of Aboriginal and Euro-Canadian thinking, chapter 2 looks at views about animal character, sentience, and intelligence. Chapter 3 details the friends and foes duality that was dominant until at least the 1960s in judging the merit and value of particular wild animals. Part 3 consists of ten chapters, each of which studies particular points of connection. Chapter 4 examines state management of wildlife, which created a formative framework for the public's relationship with wild animals. Chapters 5, 6, and 7 examine, respectively, the use of wild animals for food, the impact of sport hunting, and the introduction of new species (or acclimatization). Chapter 8 studies the farm as a site of connection between wild animals and people, and chapter 9 looks at how the Euro-Canadian view of wild animals as a public resource clashed with treaty promises made to First Nations about hunting and fishing rights. The harvesting of fur-bearing animals as a case study of the commercial use of wildlife is the topic of chapter 10. Chapters 11 and 12 deal with the place of wild animals in national parks and as a source of intellectual pleasure and personal fascination through birdwatching, the work of natural history societies, and the establishment of bird sanctuaries. The final chapter, chapter 13, focuses on the relationships people have had with wild animals at museums and zoos.

This structure helps focus the history of people's relationships with wildlife within its cultural, intellectual, political, and economic context and frames it within the sites of engagement and the activities where these relationships were realized and acted out. In offering such a holistic interpretation of the relationships between people and wild animals in prairie Canada, this book provides a further avenue for the broader study of the history of environment and people. By their nature, the varied dimensions of this human–animal encounter have compelled me to draw on a variety of often scattered and fragmentary sources, and consequently, a broad research strategy in archival and published sources has been essential. Literary work offers another route, one that has been selectively mined here. A wide range of features, reports, and stories about wildlife in newspapers and regional magazines such as the *Farm and Ranch Review*, the *Country Guide*, and the *Western Producer* also reveal public interest and discourse. Although the study of the interaction of people and wild animals has been relatively limited in Canadian historical writing, fish and fur-bearing animals have been studied as aspects of the economic history of the country and as part of Aboriginal-Euro-Canadian relations. There has also been extensive specific study of game conservation programs. While the book's notes and bibliography detail my debt to this scholarship, several works were particularly useful. George Colpitts's 2002 book *Game in the Garden* analyzes themes in the use and regulation of game animals as food in the Prairie provinces and British Columbia during and after the fur-trade era, and it also explores the ways that game animals were used for promoting regional economic development. Colpitts also surveys some aspects of early fish and game conservation. The history of game conservation at the national level has been analyzed by Janet Foster in her 1978 book *Working for Wildlife*, which deals with the federal government's conservation initiatives up to 1922. A second edition of Foster's book was issued in 1998. Game conservation and management at the national level have more recently been the focus of Tina Loo's 2006 book *States of Nature*, which more broadly contextualizes the conservation movement, surveys a broader time period, and provides a sharper focus for understanding its broad social dynamics.

This book, like the scholarship of Foster, Loo, and Colpitts, confirms the observation of historian Harriet Ritvo that historical research on animals has recently been "thriving within the discipline of history" because of growing popular interest in the topic and the broader development of environmental history.[14] The recognition that the natural and humans world cannot be separated offers a richer and fuller understanding of the history of culture and society. It also suggests new approaches and contextualizes current dilemmas and concerns about the ways that humans have shaped and continue to shape the earth. And as George Colpitts has argued, it can help "move the human mind beyond the dated conception of wildlife as a resource to be 'managed,' 'husbanded,' or 'preserved.'"[15] More broadly, the history of people's relationships with wild animals on the Canadian prairies can help us understand that while these relationships have often been sorry ones, more sensitive and respectful models and attitudes have been present all along and can be drawn upon to inform our ongoing interaction with the natural world.

Figure 1.1 While they were most widespread in wooded areas further north, even in the open southern plains areas beavers shaped the landscape and their dams created firebreaks and habitat for other wild animals. This beaver dam was photographed by the North American Boundary Commission on the Souris River in southern Manitoba between 1872 and 1875.

PART 1

THE ANIMALS *and the* PLACE

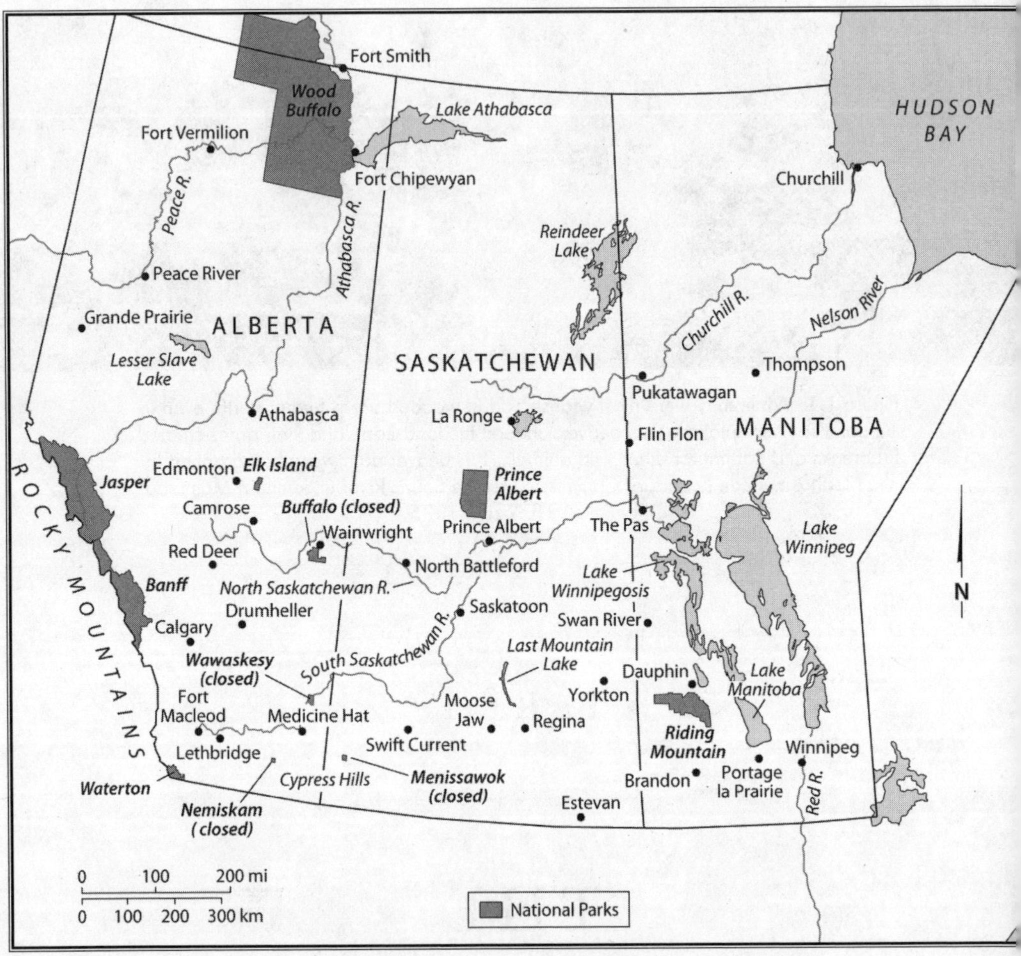

Figure 1.2 Map of prairie region, including present and former national parks. Drawn by Bill Nelson.

1

The State of Things: Wildlife, Land, and People after 1870

> There is not a sound in the air or on the earth; on every side lie spread the relics of a great fight waged by man against the brute creation; all is silent and deserted – the Indian and buffalo gone the settler not yet come.
>
> Major W.F. Butler, 1871

In 1690 explorer Henry Kelsey left York Factory on the shore of Hudson Bay and travelled inland through the boreal forest. The next year in the district north of present-day Yorkton in southeast Saskatchewan, where the parklands begin to give way to the plains, he wrote that he had reached the "Inland Country of Good Report" with its big game and beavers. And there were grizzlies, "an outgrown bear w[hi]ch is good meat … He is man's food and he makes food of man."[1] In the next two centuries there were dramatic changes in the fauna Kelsey had seen, and by the 1880s the bison and grizzlies were gone. But the land remained much the same. In 1870 it featured several belts of vegetation trending like fault lines to the northwest. From south to north, an arc of mixed grass plains north of the United States border formed the northern extension of the great North American grasslands that stretched south to Mexico. The grasslands raggedly gave way on their northern edge to the parklands with their open lands, brush, and aspen. The parklands angled northwest and on their northern edge gradually gave way to a fringe of more heavily forested lands with a mix of different trees, including conifers, which in turn transitioned to the coniferous boreal forest that stretched to the northern boundaries of the region. In the northeast corner of what is now Saskatchewan and across northern Manitoba, a small area of tundra hinted at the beginning of the great interior Arctic lands lying beyond. On the western side of Alberta, foothills gave way to subalpine and alpine environments in the Rocky Mountains. The climate was dry overall but was even drier in southwest Saskatchewan and southern Alberta. Everywhere there were long, cold winters and hot summers.

The region has changed in some particulars in the century after 1870 when the Hudson's Bay Company (HBC) charter lands were transferred to Canada. New boundaries have appeared on the land as the region became defined as the three Prairie provinces of Manitoba, Saskatchewan, and Alberta.[2] The southern prairies we see today are now more uniform. The grasslands are almost all gone, replaced with fields of grain or rangeland. Many of the sloughs that pocketed the land and held water for at least part of the year have been drained. While the great dry area known as Palliser's Triangle in southwest Saskatchewan and southeast Alberta had fewer sloughs and probably even fewer still by the end of the First World War as settlement expanded, land abandonment and water conservation projects after the environmental disaster of the interwar years restored and perhaps even increased water surfaces in the area. When viewed as a single ecozone, the southern grasslands are now arguably the most altered ecozone in Canada, with 95 per cent of the land converted to farmland. The parkland country now probably extends further north than it did in 1870 because of land clearing, although much of the parklands likely became more heavily forested for a time after the extermination of the bison whose grazing had helped keep it open. There remains a fringe of mixed-wood forest separating the parklands from the boreal forest, but these areas too have been increasingly remade by human activity, especially since the Second World War, with hydro, mining and other resource development.[3]

These changes in the land have not left wild animals untouched. Not surprisingly, the southern grasslands now contain a disproportionate number of Canada's threatened and endangered species.[4] In 1870 bison were disappearing and would be gone fifteen years later. And as only a few examples of ongoing change, by 1910 pronghorns were under threat and by the end of the 1920s waterfowl were in precipitous decline. It was said at the time that this was inevitable in the logic of farm settlement, but the decline and near extinguishment of some species was extraordinary and forms a critical aspect of the region's history. Indeed, for some, the environmental history of the Canadian prairies is largely the story of a bloody warfare waged on species after species.[5] But the story entails more than a fall from grace and was inevitable only in the sense that it was a product of the culture of newcomers. And while the transformation of the land had a dramatic negative effect on many wild animals, it also created niches for new ones and for the adaptation of some existing species, a development that added yet further complexity to human relationships with the land and its wildlife.

WILDLIFE ON THE SOUTHERN PLAINS AFTER 1870

In 1871 Major W.F. Butler passed through the Touchwood Hills where the parklands transition to the plains. This was the same area that Henry Kelsey had visited 180 years before, but Kelsey would no longer have found it a country of good report. Butler was awed by what he saw. As he wrote, "there is not a sound in the air or on the earth; on every side lie spread the relics of a great fight waged by man against the brute creation; all is silent and deserted – the Indian and buffalo gone the settler not yet come."[6] This pathos of extinction also haunted travellers in other parts of the prairies. Ten years later, on the trail between St Mary's River Crossing and Fort Walsh in what is now southeast Alberta and southwest Saskatchewan, the land surveyor Lachlan Kennedy and his crew congratulated themselves on seeing a few bison because they had "not been known in the country for three or four years." Kennedy and his crew were keenly aware that bison now mostly existed through the relics of their lives, and he noted that "we counted daily along the trail between 150 to 200 buffalo skulls, and infer that they abounded in equal numbers all over the plains."[7]

While the collapse of bison populations seemed to happen quickly, it in fact took more than half a century. Recognition of what was happening was clouded by natural fluctuations in bison populations and variations in their migration and distribution. In the United States the completion of the Union Pacific Railway in 1869 divided the Great Plains

Figure 1.3 Many travellers in the late 1800s saw bison bones as the remains of a vanished world. When this photograph was taken in 1874 near what is now Lloydminster on the Saskatchewan–Alberta border, bison had disappeared in most places and within ten years almost none remained.

bison into northern and southern herds, which obscured declining numbers but at the same time made hunting more focused. In Canada, bison had mostly disappeared near Winnipeg by about 1820. This forced the Red River Métis to push ever further west to hunt, and the organization of these hunts became part of the identity of the Métis as a people. By about 1860 bison no longer existed anywhere in what is now the province of Manitoba.[8] While they were still encountered further west, they were in decline there too. In 1874 Captain Denny of the North-West Mounted Police (NWMP) reported no bison and little game except pronghorns and ducks between Fort Dufferin on the US border in southern Manitoba and Roche Percee, near present-day Estevan in southeast Saskatchewan, a distance of about 800 kilometres. He found a few bison near Old Wives Lake, west of present-day Regina, a few more further southwest near the Cypress Hills, and greater numbers still further west in what is now southern Alberta. But within five years, there were not enough bison anywhere on the southern plains to feed the people. The last significant hunt took place in 1882–83, and while a few isolated animals remained for a time,

Amédée Forget, clerk of the North-West Council, asserted in 1886 that bison were extinct on the plains.[9]

The decline of the North American bison was evidence, argues historian John Hanner, that major extirpations of wildlife "generally result from either appropriation of habitat or large-scale hunting." In the United States, the opening of commercial markets because of new technology for tanning bison hide that made it stronger and industrially useful, habitat change caused by railway expansion and Euro-American settlement, and the export of hides by rail all contributed to the spiral of destruction. While the killing of bison as part of the US government's strategy to defeat Aboriginal peoples through the destruction of their food supply has become part of American western lore, it probably had little effect on the ultimate destruction of the herds because the policy was applied only after the bison were already in irreversible decline.[10] Canadians tended to argue that the calamity was almost entirely the fault of American expansionism and hide hunters, but the contribution of western Canadian Aboriginals and Euro-Canadians cannot be discounted, although hunting in Canada was likely less important in the ultimate decline of the northern herds than it was in the United States. This was not because Canadians were more virtuous than Americans, but because an industrial-scale hide trade, the greatest stimulus for overhunting in the United States, remained relatively limited in Canada before 1880 because of poor transportation.[11] Even so, while it was known that bison were on the edge of extinction, Canadians showed little hesitation in shooting the few that survived. One of the few left in what is now central Alberta was shot for sport by the Marquis of Lorne in 1881 during his trip west as governor general of Canada, and Amédée Forget testified that he thought it was "the last buffalo killed in that district."[12] In 1888, L.W. Herchmer, commissioner of the NWMP, reported that only six bison were known to exist in the North-West Territories (NWT) and that "three bulls have been killed this summer, their heads selling for $75 each." This was a significant sum at the time, and the collecting and marketing of such relics is one indication of how some members of the public responded to the bison's extinction.[13]

A similar scenario occurred with respect to passenger pigeons, another commonly cited victim of European contact in North America. The birds were common in Manitoba in the 1860s during their migrations to and from nesting areas further north, but by 1870 they were already in decline. Nonetheless, people continued to shoot them when they had the chance, and one resident of Neepawa, northwest of Winnipeg, recalled that around mid-August in 1879 "the wild pigeons came in thousands and fed on the wheat." He remembered that they "were thicker than ever I have seen blackbirds," and he had been able to shoot "12 at one shot, at another turn 11." Since then, he had seen "only odd ones, never any great numbers." By the mid-1880s they were almost never seen in Manitoba, although A.S. Barton of Boissevain on the Manitoba–North Dakota border told the writer Ernest Thompson Seton that in the fall of 1884 he had "shot 3 passenger pigeons out of a flush of 4. These were the last he ever saw in Manitoba."[14]

Whatever lessons its extermination held for later generations, the passenger pigeon was not a keystone species on the prairies like the bison whose disappearance transformed plains Aboriginal life, internal regional economic connections, and prairie ecosystems. What the geographer J.G. Nelson calls the buffalo landscape had provided food for Aboriginal peoples for thousands of years. The bison's disappearance also severed links among plains, parkland, and boreal forest districts by ending almost a century's use of plains bison meat to provision the fur trade. Their disappearance had an impact on the land as well. The grasslands had been continually modified by the grazing, trampling, and wallowing of the herds, which had also carved trails into the land. Bison had ranged north of the open plains into some parts of the parklands where their grazing had likely helped keep down trees and brush. And when the food they provided for other animals dried up, wolves, coyotes, bears, ravens, magpies, and crows vanished from the plains, remaining only as small remnants on the margins of the grasslands, in gullies and valleys, or in unique habitats such as the Cypress Hills. By 1886 there were no grizzlies left on the plains, mainly because the bison had disappeared, although J.G. Nelson argues that the availability of better guns also led to their decline.[15]

Pronghorns too had disappeared from what is now southwest Manitoba by 1886 but remained plentiful further west. In a description of the fauna around Regina at this time, Nicholas Flood Davin, a member of Parliament (MP) and the editor of the Regina newspaper, reported that while there were no moose or elk and only rarely deer, pronghorns were "very numerous" and in some years formed large herds. To his list could also have been added small mammals such as badgers, rabbits, foxes, and various ground squirrels. J.G. Nelson speculates that by this time, in areas with suitable habitat such as the Cypress Hills, beavers had likely been trapped out, although some continued to be found in areas with creeks and aspen until the late 1800s, when they vanished because of trapping and land drainage.[16] But the immense variety of prairie waterfowl remained. Davin reported plentiful waterfowl around Regina in 1886, especially in spring and fall, although many birds stayed throughout the summer too. In 1880 near present-day Brandon in southwest Manitoba, the botanist John Macoun sighted many birds familiar to him from Ontario, such as robins, meadow larks, cowbirds, bitterns, and finches. On the nearby prairie lands that stretched beyond the valley of the Assiniboine River, he also saw various types of plovers and curlews, sandhill and whooping cranes, phalaropes, and other marsh birds. On the open plains to the west he encountered relatively fewer birds but saw several species of blackbirds, buntings, hawks, and gulls, and in favourable habitats such as Last Mountain Lake northwest of Regina, there were "multitudes of feathered game including ducks, geese, teals, pelicans, snipe and plover," which bred "on islands in the lake or along its borders." From mid-August to freeze-up (which came in late October or early November), these birds were joined by migrating waterfowl that stopped to rest and feed for about six weeks on their way south or to the Pacific coast. They filled thousands of sloughs and crowded every pond, and it was "no figure of speech," said Macoun, that everywhere "the ponds and lakelets were alive with them."[17]

As the railway was being built across the southern plains in the early 1880s, many of these waterfowl, as well as mammals like pronghorns, were shot for sport or food by construction workers. When the line reached the eastern slopes of the Rocky Mountains in 1883, crews shot animals they encountered and tossed dynamite into creeks and lakes to kill fish.[18] In 1889 Dr George DeVere wrote that despite "the glowing descriptions to the contrary one occasionally sees in print from the pen of some enthusiasts," the plains, foothills, and mountains of southern Alberta close to his home in Fort Macleod, "which a few years ago were a hunter's paradise, are now barren of all large game." Bison and elk were gone and "the antelope, mountain sheep and goat, black and white tail deer, are fast approaching the same state and are only to be found semi-occasionally in some remote and favoured locality." What mostly remained were waterfowl such as ducks and brant geese and upland game birds such as grouse, "though in sadly lessened numbers." And these declines continued. By 1902 it was reported that mountain sheep, moose, and deer had been seriously depleted by hunting in the mountain areas near Calgary.[19] By 1905 pronghorns were becoming scarce in their traditional territory west of Last Mountain Lake, although mule (black-tailed) deer were found in favourable areas on the plains. But small mammals remained common in many areas. Writing in 1911 about trips he had made, the ornithologist William Rowan noted that he saw little wildlife in southern Alberta but had encountered badgers, coyotes, and swift foxes and plenty of ground squirrels in southern Manitoba.[20] Most types of birds were still numerous in Saskatchewan in 1905, and one commentator argued that some birds were also moving north to escape encroaching settlement in the United States. This was noted in Manitoba as well; scarlet tanagers, rarely seen before, were reported in 1916, lark sparrows were moving north from the Dakotas, yellow-throated vireos and yellow-winged warblers, mourning doves, pinnated grouse, ospreys, and California kites were all new birds reported in the province, and goshawks and golden and bald eagles were "becoming much more numerous than in former years." And in 1907 Andrew Halkett of the federal Department of Marine and Fisheries reported during his survey of fish in Alberta and Saskatchewan that "the valley of the Qu'Appelle [north of Regina] is a regular paradise of birds," leading him to compile a substantial list of the songbirds, waterfowl, gulls, and other birds he encountered.[21] Yet such sightings may have merely been the result of better observation, and judgment about plenitude or loss always depended on one's familiarity with the land and where one had come from. But for those who had not seen earlier times, the plenitude was amazing. In October 1909, as his train broke out onto the flat prairies west of Winnipeg, the Anglican minister Martin Holdom, on his way from England to his posting in central Alberta, marvelled, just as John Macoun had thirty years earlier, that the sloughs "are simply alive with ducks, thousands and thousands. I have never seen anything like it." And when he reached his posting in Castor, in the dry lands east of Red Deer,

Figure 1.4 Newcomers to the region often remarked on the variety and plenitude of bird life. Although bird numbers declined overall, this photograph (probably from Alberta in the 1940s) of Bonaparte's and Franklin's gulls shows that skies filled with birds remained a part of prairie life.

he was similarly entranced. As he wrote to his father, "there are lots of prairie wolves [coyotes] round here; you can hear them at night. Porcupines, muskrat, badgers, lynx, antelope, deer and a few large grey wolves (Dangerous). Also not far off: a herd of wild horses."[22]

Within a decade of Holdom's arrival at Castor, the district was facing the start of what proved to be a long period of environmental change. A number of dry years in the 1920s followed by prolonged drought in the

1930s brought crisis for both wildlife and humans in the southern prairies. Waterfowl were in serious decline in the 1920s because of habitat change and hunting, and this decline intensified in the 1930s when sloughs and ponds dried up and the water level in lakes fell. Land erosion, stimulated by unsuitable cultivation practices, further intensified the impact of the drought on wildlife. From a population said to be 180 million in 1900, the duck population in prairie Canada dropped to an estimated 50 million in 1938.[23] In Saskatchewan in 1937 there were almost no surviving forage or cereal crops in about three-quarters of the province's settled areas. In places, the surface of the soil was bare and game birds that had found food and shelter in stubble fields and straw stacks were "threatened with extinction" and had to be fed by farmers, hunting associations, and the government.[24] Wetland conservation programs to save waterfowl began after 1937 with the establishment of the hunting organization Ducks Unlimited (Canada) (DUC), and wider efforts to turn the environmental disaster around included converting eroded farmland into pastures and building dugouts, dams, and irrigation systems. In 1935 the federal government created the Prairie Farm Rehabilitation Administration (PFRA), and by 1943 PFRA had developed 1.3 million acres of pasture as well as 20,000 small water projects, over half of them in Saskatchewan. Targeted at agricultural needs, these water projects nonetheless created habitat for waterfowl and shore birds, while pasture projects gave shelter to upland game birds and burrowing mammals as well as various small birds. Although conservation of wildlife was incidental to economic objectives, all PFRA community pastures were also made into game preserves where hunting was prohibited, and commentators such as Dewey Soper of the Canadian Wildlife Service (CWS) believed that the PFRA projects were important for the rehabilitation of bird life on the prairies after the drought years.[25] Yet, broad landscape and economic change continued to affect wildlife on the southern plains, even in rehabilitated areas. A study of the ferruginous hawk by biologists C. Stuart Houston and Marc J. Bechard, using historical and contemporary field data, for example, demonstrated these changes. While these hawks had been common on the grasslands of the Prairie provinces before 1900, they were disappearing by the 1940s. By the early 1980s Houston and Bechard found much reduced populations in more restricted territory than a century before because of habitat change. Ploughing the grasslands had reduced the number of ground squirrels, the hawks' primary food. As well, nesting sites disappeared as grasslands were converted in stages to cropland and competing species of hawks gained new advantage. Extensive hunting by farmers at least until the 1930s also intensified depletion.[26] Swift fox populations were affected by the same forces. Once very common on the southern part of the open plains, few were left in Alberta by 1900. Habitat change caused by land breaking and mortality from poisoned bait set out to kill gophers and wolves were significant in the swift fox's decline. As well, they were trapped or shot by ranchers who saw them as pests because they chewed on leather horse gear. Only a few remained in southeast Alberta by 1938, and by 1960 they had been extirpated on the Canadian prairies. A reintroduction program began in 1983, and while it has been somewhat successful, swift foxes remain endangered in Canada.[27] Both the case of the ferruginous hawk and that of the swift fox showed how habitat change played a major role in the decline of certain wild animals and, when combined with other factors such as hunting and the use of chemical poisons, how it had created adverse conditions that were almost impossible to reverse.

WILDLIFE IN PARKLAND AND FOREST DISTRICTS AFTER 1870

While the southern plains were becoming a wasteland for large wild mammals by the 1880s, the parklands largely escaped the dramatic rupture that the southern plains experienced with the disappearance of the bison. Yet there were many changes. The writer Ernest Thompson Seton, who lived between 1882 and 1887 near Carberry in the parkland district east of Brandon, paid a return visit there in 1892. In his journal he recorded some familiar things. He was told that "the wolves are quite common in the winter yet," and many of the bird species he remembered were still around, although in fewer numbers. But the town had grown, fields were larger, more land was broken, fences had been built, many native trees had been cut down, and those planted by settlers had grown. Land had been drained, and he was able to tramp through "a formerly dangerous muskeg and was never wet more than over my ankles, if I had picked my steps I might have passed nearly dry – in five more years at this rate it will be one great flat pasture." Such land drainage was making it difficult for herons, which needed marshy land, and thrashers and towhees – both birds of the grasslands – were decreasing as the land was broken. Badgers too seemed to be rarer, and they, Seton noted, "like most truly prairie things," were disappearing.[28]

Districts further north in all three Prairie provinces remained more closely linked to earlier times. In the forested areas around Norway House at the north end of Lake Winnipeg in 1884, for example, there were still some fur-bearers, a few bears and deer and wolverines. In the parkland area at Fort à la Corne in north-central Saskatchewan, wild animals included beavers, muskrats, otters, lynx, fishers, mink, and marten, as well as black bears and a few grizzlies, deer, and moose. Further, there was a wide range of fish, including sturgeon, whitefish, suckers, pike, and perch.[29] Nearly twenty years later, it was reported that there were moose throughout all northern areas, although a few were found in present-day southeast Saskatchewan as well. Only a small number of elk remained in a few

Figure 1.5 Badgers were part of the bison landscape, but as Ernest Thompson Seton remarked in 1892, "like most truly prairie things" they were disappearing because of habitat change and hunting. This decline continued. By the time this photograph was taken in 1930, they were endangered in many areas.

isolated parkland locations, but deer were found everywhere in wooded areas. Bears were most plentiful in the northwest of what would become Alberta, and there were woodland caribou in parts of northern Manitoba and what would become northern Saskatchewan.[30]

Change that would have significant effect on many of these animals was, however, underway. Andrew Holmes, the assistant provincial game guardian in Prince Albert, noted in 1921 that "settlers are advancing further and further each year into the Hinterland." While this was "unavoidable as civilization must not be retarded," he anticipated that settlement would have a limited impact on wildlife given the large amount of land in northern Saskatchewan that was unfit for farming and would never be settled. The case of woodland caribou, however, showed that such predictions did not take into account the complexities of ecological change. Even when settlement was discontinuous, it broke the remaining habitat into unsustainable units. In Saskatchewan in the early 1900s it was said that there were thousands of woodland caribou in the woody marshy areas between Lac La Ronge, Prince Albert, and the Pasquia Hills, forming populations stretching in a long curve along the forest fringe in the central northeast of the province. Their northern limit seemed to be the Churchill River, but their movements were not understood, and during the interwar years there was puzzlement when the caribou began to push further south in Saskatchewan and Manitoba while seeming to be moving northward in Alberta.[31] Whatever was occurring, the encroachment of settlement and greater accessibility for hunters brought woodland caribou in Manitoba under threat by 1936. In Saskatchewan they had by 1950 been reduced to small isolated herds that no longer migrated, leading to forecasts of their imminent extinction.[32] Further north, barren land caribou occupied a small strip along the Saskatchewan–Northwest Territories border and a somewhat larger area further east along northern Manitoba's border. By the mid-1950s these populations were declining as well. Wolves were blamed, but the reasons were not clearly understood and the matter was under joint study by the CWS and the provinces.[33]

A similar process was occurring with fur-bearing animals, and some were under stress by the 1890s. Beavers, for example, were disappearing so quickly throughout the prairie region because of trapping that it was feared they would soon be extinct. After the First World War, marten and fishers were also depleted in many parts of the Prairie provinces. In Alberta, almost 6,000 marten were trapped in 1919–20, but twenty years later only a handful were taken. Ernest Thompson Seton theorized in the 1920s that the decline of marten was the result of disease, but Dewey Soper, in his book *The Mammals of Alberta*, was sceptical, concluding that "the final downfall of the species was, and is, being sealed by man's clutching hand with guns and traps." Marten had been plentiful in early times, and Soper posited that their decline "coincided with the white man's invasion of North America." Indeed, throughout the interwar years, fur-bearer populations in general declined, and as Manitoba's game commissioner observed in 1930, even when natural population cycles were accounted for, "one cannot escape the conviction that fur bearing animals are annually becoming scarcer."[34]

While the fur trade was an important factor, this decline of fur-bearers in the twentieth century was also part of environmental change. Although it is now difficult to pinpoint where and when particularly vital habitats for fur-bearers changed, clearing forest and brush lands and land drainage in the parklands and forest fringe were clearly important. Yet while clearing was ongoing, it was also piecemeal and geographically discontinuous, and many fur-bearers continued to be found in pockets of suitable habitat throughout these areas. In 1931, for example, about half of Saskatchewan's fur production (mostly muskrat, mink, weasel, and rabbit) came from areas south of the latitude of Prince Albert. But habitat change was occurring on a broad scale. G.W. Malaher of the Manitoba game and fisheries branch noted in a 1946 radio talk that forest fires in northern areas brought huge mortality among fur-bearing animals. Fire was almost always caused by human activity, and its destruction of a forest's "biotic community" wiped out food and shelter that regenerated slowly. Nor could many forest fur-bearers move to new places to find food and shelter. Arguing that many fur-bearing species found in settled districts could also live in forested lands, Malaher contended that species such as marten, lynx, fishers, and wolverine could not "survive for long outside their natural forest habitat." Moreover, when they did move into settled districts, they came into conflict with humans because most fur-bearing animals (except beavers and muskrats) were "predatory and to some extent undesirable around settled communities."[35]

Such broad change continued in the 1950s and early 1960s when clearing and breaking expanded in areas of the parklands that still had some original cover and even further north in adjacent parts of the forest fringe. In areas with mixed-wood forests, such as the Pierceland district near Prince Albert in Saskatchewan and parts of Alberta's Peace River district, land clearing became more systematic. Some of these lands had featured small trees and shrubs with patches of open grasslands and sloughs, but as land was cleared and fire was suppressed, aspen trees filled uncultivated or low-lying areas and underbrush disappeared. Paradoxically, Euro-Canadian settlement was leading in some districts to heavier tree cover. While the margins provided browse for deer, the overall less-open character of the land eliminated habitat for a range of birds, such as sharp-tailed grouse, that needed a mix of shelter and open land.[36] Equally problematic were large industrial projects in the postwar years. In the early 1950s, 130,000 acres of marshy waterfowl and muskrat habitat were lost because of the Pasquia agricultural project near The Pas in northern Manitoba, which was promoted by the province to expand its agricultural production and to employ postwar workers. In 1962 construction was also completed on the Grand Rapids Dam at the point where the Saskatchewan River empties into Lake Winnipeg, flooding one million acres of similar habitat in the Saskatchewan River delta, one of the greatest wetland deltas in the world.[37] Parallel development took place elsewhere as well. The incomparable Peace-Athabasca Delta wetlands in northeast Alberta, a habitat for waterfowl, muskrats, and other marsh animals, were largely destroyed after 1968 by falling water levels because of the completion of the W.A.C. Bennett Dam hundreds of kilometres away on the upper Peace River in British Columbia.

ANIMALS FAMILIAR AND NEW

While human impact on wildlife after the late nineteenth century was uneven across the prairie region, it steadily increased because of population growth and economic, cultural, and technological change. It also intensified because Euro-Canadians did not need wild animals as much or in the same way as earlier peoples had. As geographer J.G. Nelson phrases it, humans had "occupied a much less important place in the [plains] ecosystem" before Europeans arrived; there had been fewer people, and they were "subordinate to other animals in a number of ways." Indeed, animals such as bison had a greater impact on vegetation, soils, and minor land forms than people did. Monoculture, machine technology, and new forms of resource extraction made Euro-Canadians much less reliant on wildlife. As part of their social, economic, and aesthetic preferences and needs, these newcomers, either by accident or design, also brought new animals, plants, and micro-organisms with them. The most common of the animals introduced were domestic cattle, pigs, sheep, and poultry. Historian John Hanner notes that by 1890 the number of cattle on the high plains in the United States was probably greater than the number of bison there two decades before, and the same was likely the case in Canada, demonstrating how bison extermination offered economic gain, not loss, to incoming settlers.[38] These domestic animals had far-reaching consequences. They required new plants for food and had to be protected from wild animals as well as the diseases that such animals sometimes carried. This brought farmers into conflict with elk and deer and with predatory animals such as coyotes, wolves, foxes, and raptors.

In contrast to the case of domestic animals, only a few species of wild animals were successfully transplanted to the prairies. As will be discussed in chapter 7, Hungarian partridge and common pheasants were introduced before the First World War in southern Alberta, and attempts were made to introduce chukar partridges and wild turkeys later. These projects attempted to compensate for declining native game birds such as sharp-tailed grouse by introducing new birds that were exciting to hunt. Various types of game fish were similarly imported and planted in lakes and rivers to improve sport fishing. Such introductions raised concerns in some quarters about their impact on native species and about upsetting the "balance of nature." But it was hard to get rid of new species once they had acclimatized, and they had powerful defenders in the hunting fraternity and provincial game departments, which saw them as offering more economic and social value than ecological threat.

Other new wild animals appeared on the prairies by accident as part of a general diffusion across the continent. As ecologist John Kricher notes, "ecosystems are anything but closed to immigrant species," and not surprisingly, birds spread most easily.[39] Pigeons (rock doves), house sparrows, starlings, and other birds were introduced in eastern North America and spread throughout the continent. This was also the case with invertebrates. Earthworms appeared on the prairies around the beginning of the twentieth century, although it is not clear whether they were deliberately introduced or came accidentally in soil from other places. A variety of insects also arrived in imported soil, plants, and seed, and some became a serious threat to field crops.[40] Vertebrates such as Norway (brown) rats and house mice arrived as early as the late seventeenth century as stowaways on HBC ships and established themselves on the shores of Hudson Bay, where it is claimed that their biological descendants still live.[41] Other animals, birds, and fish introduced elsewhere in North America also spread to the prairies. For example, while the government of Manitoba had tried unsuccessfully to introduce carp in the 1880s to supplement farm diets, natural dispersal succeeded where the government had failed. Carp had been introduced into the Mississippi River system in the 1870s, and by 1938 they had found their way into Manitoba rivers and lakes, likely moving north in complicated stages, perhaps spread across watersheds by sport fishers using carp minnows as bait.

Of greater significance than such new wild species were the various native species that found new advantages in the changing prairie environment. Many native invertebrate species gained new opportunities from habitat change brought by Euro-Canadian settlement: wheat stem sawflies, grasshoppers, pale western cutworms, and, among others, armyworms and wireworms all benefited from monoculture and environmental change. Wheat stem sawflies, for example, profited from the replacement of native grasslands with thousands of hectares of wheat, with its hollow stems.[42] While monoculture was important for insects, wild mammals, birds, and fish benefited as well. The range of some mammals and birds increased when land was cleared and broken in the parkland and forested parts of the region and trees were planted as windbreaks on the plains. As well, showing how ecosystems are open to migrants, animals moved within the region and came from similar environments in neighbouring parts of the United States.

The case of white-tailed (or Virginia) deer was one example of this adaptation and migration. Accounts uncovered by Ernest Thompson

Seton indicated that white-tails (also revealingly called the "Down East Deer") only appeared in Manitoba in about 1881. By the late 1890s, they were becoming more common. It was believed they were moving north from the United States because aspen, saskatoon, dogwood, and other deciduous browse were increasing along the margins of fields in tandem with tree clearing. White-tails at first increased slowly – by the end of the First World War they were still "relatively scarce" – but by 1933 they had become abundant and it was clear that they had "actually increased with encroaching civilization."[43] By the 1940s they had become the dominant big-game animal in southern Manitoba and were beginning to appear in areas further north. Mule deer, which had been found in a few districts in the late 1880s, were by then declining rapidly, and by 1960 they were rarely seen in the province.[44] A similar situation developed in Saskatchewan. In 1928 mule deer were apparently common in parts of the parklands, but deer of any sort were still unusual on the plains. One traveller on the train through the Qu'Appelle Valley near Regina reported that when some passengers spied two deer, the engineer stopped the train so people could watch them, since it was not every day that one "saw a sight like that." Within twenty years, however, white-tails were common in most parts of the province south of La Ronge, with dense populations in the parklands where there was plenty of browse. Their spread was also encouraged by the lack of native predators, especially wolves, which had been exterminated in most of these areas.[45] Similarly, white-tails increased in Alberta, but their history there seems to have been more complex. There were reports of them at Edmonton and along the North Saskatchewan River in the late 1700s, but after about 1870 numbers fell, perhaps because of a series of hard winters. In contrast, mule deer had become plentiful along rivers and streams and in bluffs in central and some other parts of Alberta in the 1890s, but then they began to decline while white tails increased. The provincial game branch banned mule deer hunting in most of the area south of the North Saskatchewan River in 1929, and populations quickly rebounded. Various arguments were advanced to explain these events. Dewey Soper contended that the white-tail range had expanded because white-tails, in the familiar pattern, benefited from "settlement and cultivation," since land clearing increased browse. Jack Ondrack, an authority on big game hunting in Alberta, agreed but added that white-tails had a "higher reproductive capacity" than mule deer; he also credited their proliferation to their greater wariness, which made them more difficult to hunt than mule deer.[46]

Jack rabbits (prairie hare) appeared in Manitoba in the late 1880s, seeming to have moved from the south. Larger than the native snowshoe rabbit (varying hare), jack rabbits are extremely fast runners (on the prairies only pronghorns are faster), and they prefer open country, unlike snowshoe rabbits, which favour parkland and forest areas. Ernest Thompson Seton reported an interview with a Mr John Cadham, who claimed that jack rabbits were unknown when he arrived in the west in 1870. He first saw some near Winnipeg in about 1896 and again saw some near Stony Mountain, north of the city, in 1898. Other observers confirmed that jack rabbits had appeared recently but were now everywhere east of the Red River and steadily spreading. "They are very fond of the ploughed fields," Seton reported, suggesting that farming was creating a favourable environment for them. The same pattern was seen in Alberta: by the late 1800s only a few jack rabbits were observed north of Red Deer, but by 1960 they had spread throughout the treeless and parkland districts.[47] Cottontail rabbits moved into Manitoba from the United States in the early 1930s, and by the mid-1940s grey foxes had come north in response to this new food supply. The first grey foxes were sighted at St Adolphe in the Red River Valley just south of Winnipeg in 1957, and Richard Sutton of the Manitoba Museum argued in 1958 that they were being pushed north into Manitoba because of human pressure on their habitats in adjacent parts of the United States.[48]

Wolf populations also reflected changing conditions. With the decimation of the bison, the wolves that had lived extravagantly on the bison herds declined and their much reduced populations retreated to the foothills and other sheltered areas home to large game.[49] The introduction of cattle grazing on open range land restored a food supply for a time, but ranchers, assisted by government, had exterminated wolves on the plains by about the First World War, and in 1915 wolves were reported to be "very rare" on the eastern slopes of the Rocky Mountains south of the Athabasca River and "not abundant north of it."[50] The great forested areas that stretched north of the farming areas gave wolves a refuge from which they sometimes moved into nearby settled farming areas. In Alberta, for example, the only substantial wolf population by the mid-1920s was in Wood Buffalo National Park, and this population spread southward in the 1930s and 1940s. Low prices for wolf pelts and a decrease in hunting during the Second World War allowed populations to increase, but most were then killed during the anti-rabies campaign of the Alberta government in the early 1950s and populations only increased in the 1960s. By 1969 there were an estimated 3,550 wolves in the province.[51] It was evident, however, that the wolf's cousin, the coyote, was faring better. While the elimination of wolves removed one challenger, increased settlement created more favourable habitat for coyotes, which had ranged in open and parkland areas and occasionally in more heavily timbered areas. As the biologist Norman Criddle argued in his 1925 study of wolves and coyotes, "the destruction of such timbered areas by fire and the gradual clearing of land" for farming "made other territory available to the coyote."[52] And Kerry Wood, an Alberta naturalist, speculated in 1945 that increased coyote populations accompanied the spread of grain farming. Grain was put up in sheaves and left in the fields until threshing, a practice that was superseded by swathing, in which grain was cut into windrows and picked up later with a combine harvester. Both systems provided ideal habitat for mice, to the benefit of coyotes. Another observer claimed that this technological change also brought an increase in foxes because "in former times farmers threshed their grain with threshing machines which left straw in large piles. Now the combine spreads the straw on the fields and the resultant cover presents an ideal place for field mice to winter. Foxes practically live on field mice after freeze-up."[53]

Crows and magpies were also a part of the bison landscape, and they too all but vanished along with the bison. In 1892 Frank Farley, a birdwatcher from Camrose in central Alberta, did not record any crows, but they began to appear there by 1900 and were increasing rapidly within a decade.[54] Crows live in many environments, but they are most common in open wooded areas and therefore benefited from the more open landscape created by land clearing. Unlike crows, magpies are non-migratory, but like crows, they had almost totally disappeared from the prairie region except for a few remnant populations in places such as the Cypress Hills. Amateur ornithologist Lawrence Potter contended that at Eastend in southwest Saskatchewan, there were no magpies in the early 1900s because they had been poisoned by eating bait set out for wolves. By 1911 they were rarely seen, but by 1927 they had returned.[55] By the 1880s there were no magpies left in central Alberta, and they were unknown when the first settlers came to the Red Deer district south of Edmonton. Benefiting from new habitat created by farm settlement and by food such as livestock droppings and feed, a few appeared around the First World War in central Alberta, and by 1922 they were seen near Edmonton, where there had been none forty years before.[56] By 1939 they had spread as far north

Figure 1.6 Some animals, such as crows and magpies, profited from habitat change. This undated painting of a magpie (probably from the 1940s) is by the Saskatchewan painter and nature writer Robert David Symons (1898–1973).

as Grande Prairie in the Peace River country. In Saskatchewan they became widespread in the 1920s but were apparently slower to appear in Manitoba, where it was said in 1954 that they were "comparative newcomers to the province" but were increasing rapidly.[57]

For some observers, these changes seemed to have upset a balanced order in nature. In 1952, reflecting anxiety about this, the Alberta writer John Patrick Gillese discussed the changes in the Mayerthorpe district northwest of Edmonton where he had grown up. He thought they were likely typical of many parts of the prairie region, and "now, by and large," he wrote,

> the wildlife population consists more of hunters than hunted. Coyotes abound in incredible numbers, weasels are thick, man has been everywhere. The deer which used to abound in the Paddle Valley are gone, because the school sections and farm pastures, where once they mated and raised their young in peace, are gone. Ruthless exploitation of the forests has caused the river to shrink to a bare trickle. Where once of an evening I counted more than a hundred muskrats on a bend of river, I saw not a single sign. Only the beavers, increasing in Alberta, take any interest in maintaining the water level, and they work against increasing odds.[58]

The changes Gillese lamented were in fact not so sudden; rather they were the accretions of the expansion of the human niche in the region since the end of the nineteenth century. His lament about a world out of order was a yearning for the animals of his past that had made the world seem plenteous, familiar, and comforting. Yet some of the animals whose absence he regretted and which he thought of as a familiar and perhaps timeless presence had only become prevalent in his own time and place. Ecosystems are not closed or unyielding, and some of the animals Gillese loved would equally have been thought of as strange and perhaps unwelcome in earlier times.

WILDLIFE, HABITAT, AND THE CULTURE AND ECONOMY OF FARMING

The changing distribution and population of many animals and birds and the appearance of new ones were the result of national expansion and the new farm economy on the prairies after 1870. Farming had long been an instrument of national expansion, and as the economist Vernon Fowke argues, Canadian agriculture had been the "basis for economic and political empire" for 300 years.[59] The place of farming in building empire was rarely questioned by Euro-Canadians, and while many on the prairies may have disagreed about how it fit into the national economy, most agreed with the rightness of its imperial purpose. As the NWT Department of Agriculture contended in 1901, there could be no argument that the region was "pre-eminently an agricultural country" and that "all other branches of industry here must necessarily fade into insignificance before those of farming and ranching."[60] Indeed, the region remained dominated by agriculture until after the Second World War. This was critical for wildlife. Of all Euro-Canadian activity, farming brought the most widespread and profound change to the prairies. It so powerfully defined "desirable use" that land unsuitable for farming was popularly and officially called "wild" or "waste" land, as outside culture and economy. This particularly applied to land in northern districts. While having value for trapping and sport hunting and holding promise of an industrial future through lumbering, mining, and hydro development, the categorization of these districts as marginal continued to justify a view that their wildlife, Aboriginal peoples, and environment could be sacrificed for the economic good of agrarian and urban centres. More broadly, as the American historian Donald Worster argues, the role of farming in transforming the earth became increasingly powerful in nineteenth-century North America and in other settler societies. The way in which a society feeds itself involves "the restructuring of the natural world" and the domestication of ecosystems, and the emergence of industrialized farm practices and technology in the nineteenth century was particularly influential in affecting biological diversity and altering landforms and native flora and fauna. He further contends that in North America in the past 150 years, "the land itself evolved into a set of specialised instruments of production" of commodities and farmers became "specialists in production."[61]

Aided by the completion of the railway from Winnipeg to Calgary in 1883 (the transcontinental link above the Great Lakes was completed two years later), prairie lands that a decade before had supported bison were being broken by the late 1880s for crops or turned over to cattle ranching. Ranching was at first largely a substitution of cattle for bison, with cattle grazing on native grasslands in the foothills and plains country of what is now southern Alberta and southwest Saskatchewan. But, as historian Warren Elofson shows, land degradation from overgrazing quickly became a problem in southern Alberta.[62] Elsewhere, tilling the soil brought even more significant landscape change, with large wheat farms on the open plains and mixed farms with livestock, poultry, cereals, oil seeds, and forage crops in parkland districts. The speed at which this took place across the North American west was historically unprecedented, made possible because machine power and industrial methods of organization efficiently moved people, commodities, and capital. Land was surveyed in regular blocks that abutted each other, and on the southern plains, fields of grain followed one after another with unvarying regularity. By the early part of the twentieth century, it was common to describe these areas as seas of wheat stretching to the horizon, and even in the parkland districts with their predominantly mixed farms, the trend was towards uniformity.

This farm settlement boom began in the late 1890s, and over a million migrants arrived on the prairies between 1901 and 1911. Searching for new homes and new futures, these land seekers settled along the ever-expanding rail lines, although their movements were not always a neat progression, as they also followed existing trails or tried to guess the future routes of rail lines.[63] However they reached their land, settlers immediately turned to farming. A farmer using a horse and a walking plough could break twenty acres of grasslands a year. Such change was realized more slowly in the parklands and the forest fringe where nine to eighteen acres of lightly treed land and four acres of heavily treed land could be cleared and broken in a year with human and animal power. The use of steam power followed by tractor power in the early 1900s affected not only breaking and clearing but the management of water through irrigation and drainage. Draining Waterhen Lake near Kinistino, southeast of Prince Albert, Saskatchewan, for example, began in 1920, and by 1922 ditches, dams, and reservoirs had been constructed to convert about 12,000 acres to farmland.[64] While large stretches of land across the region were legally occupied, only 5.6 per cent of such land in 1901 was defined as "improved," that is, broken, cleared, or drained. Rapid development brought this to almost 45 per cent in 1921. Change then slowed. By 1941, 54 per cent was improved, and this rose to almost 58 per cent by 1951.[65] Farm development in parts of the northern parklands was slow because substantial areas with grey-wooded soils were difficult to farm until suitable field husbandry techniques were developed after the Second World War. Farmers everywhere then began to use chemical fertilizers, and the postwar years were particularly notable for the appearance of new chemical pesticides and herbicides.

Breaking the land for cultivation was literally the breaking of one natural world and its replacement with another. It is thus not accidental that the plough has been used by many prairie communities to symbolize their history. In 1912 the Manitoba poet Robert Stead wrote an ode addressed to the plough, describing the prairie "before you came" as dominated by "the red man" and wild animals, and "all lay silent, useless, and unused / And useless 'twas because it was unused."[66] Such a mindset had a predictable impact on wildlife, which declined as habitat disappeared when trees were cut down, sloughs drained, and grasslands broken. But this decline also occurred because of the culture of prairie farming. Wild animals and birds that were not pushed aside by habitat change were gradually killed or driven out if they interfered with farm production or, less commonly, if they posed a threat to human safety. Regardless of farm sector, wildlife was generally seen as another barrier to overcome, no less so than climate, distance from markets, and vulnerability to outside economic forces. Confrontations with wildlife have been waged since the beginnings of agriculture, but farming in an industrial age exacerbated such confrontations through an emphasis on total land use that would see every acre given over to production. This confirmed the general view that any amount of farm production lost to wild animals was too much. Leeway was only given when an animal enhanced production, such as to a particular bird that ate insects and weed seeds. But this was just another dimension of the general view that farmers had the right to consume or send to market every kernel of grain, every head of livestock, whether strong or weak, and every chicken and egg. Yet, to be sure, uneven and low incomes made every dollar count and did not encourage coexistence with wildlife that took even marginal amounts of farm production. Most farmers therefore saw only financial disincentives in maintaining or protecting wildlife or its habitat, and only rarely was it argued that some resources on farmland had to be given to wildlife simply as another operational cost. Instead, economic conditions and demands meshed with an adversarial relationship to confer on the farmer an inherent right to kill wild predators rather than undertake the bother and cost of building fences, shepherding sheep, or following practices that made livestock less vulnerable to predation. During the legislative debate on Alberta's game legislation in 1907, some members argued for "the right of every man to kill" wild animals and birds. On one level this was little more than bloody-mindedness, but it also marked a changed world where wildlife offered no comparative benefit to farming and industry.[67]

While the view that newcomers would have an inherent right to the land, would practise total production, and would domesticate the environment characterized many settler societies, it was manifested within particular cultural contexts. The Canadian west was not idealized, as the American one sometimes was, as a new social experiment.[68] The Canadian ambition was to transplant the institutions and customs of older parts of the country; like Athena, the west would be born fully formed. Historian Andy den Otter shows in his study of David Anderson, the first Anglican bishop of Rupert's Land (1849–65), that this ambition was evident at an early date. Bishop Anderson generally admired the natural beauty of the prairies but deplored its seeming emptiness, "devoid of people, farms, towns and cities." He saw his mandate as transforming a place he compared to a Biblical wilderness into an "idealized form of the British society he had left behind" by converting First Nations people and implanting schooling and farming.[69] While such views were perhaps predictable in a missionary, Anderson was not unique, and in historian Lewis G. Thomas's phrase, Euro-Canadian settlers in the late 1800s and early 1900s saw themselves as colonizers not colonials. And the journalist James Minifie, whose family migrated from England to Saskatchewan in 1909, recalled that "we thought the environment should adapt to us."[70] It is thus not surprising, for example, that many accounts of settlers on the plains around Regina in the 1880s contain little mention of local fauna. Sometimes it was noted that there was good shooting of grouse and ducks and sometimes geese, but wildlife was generally ignored in favour of talk about the fertility of the soil, the remarkable growth of plants in the long days of the prairie summer, the glory of the immense open skies, and the land's great promise for farming. Some of this no doubt was bravado in the face of criticism by outsiders of the stark loneliness of the plains, its fierce winters, the lack of potable water, and a mounting folklore of homesteader failure, and it likely redoubled a determination to confirm the promise of the future for people who had so drastically changed their lives by moving there.[71] Everything would be changed, even the climate, and as Ernest Thompson Seton phrased it in 1907, "in all wheat countries summer frost had fled before the plough." Once the land was broken, the climate would moderate, and just as "summer frost was the curse of the Ontario Peninsula at one time, and of the Bruce Peninsula, and of southern Manitoba," it had vanished in those places except in odd years.[72]

Despite their central role in prairie society and in the reshaping of the land, farmers were not the only ones who saw the future as guided by the conquest of land and nature. Historian Douglas Francis has argued that prairie development featured overlapping phases of fear, euphoria, disillusionment, self-discovery, and reconciliation about the land and the future as people grappled with climate and economic realities.[73] But remaining throughout was a conviction that the resources of the land were the reward for uprooting one's life and coming to a developing society. As with their American neighbours, many migrants to the prairies were convinced that the west would make them rich; they had come to "make good." While common throughout Canada, the quest for wealth from farming, urban and rural land speculation, exploitation of timber and minerals, and trade and commerce found intense expression on the prairies at certain times, perhaps because this quest was continually reinvigorated by the boom and bust cycles of the regional economy that sparked periods of despair followed by euphoria. In 1883 the editor of the *Regina Leader* commented that "we sometimes hear men say 'We have not come up here for the good of our health' and it is said in a connection which would seem to indicate that they mean to make money, honestly if they can, but make it anyhow." And twenty years later Ontario's minister of agriculture was struck during his tour of the prairies by "the all absorbing desire on the part of everybody to make money. Wherever I go … I hear of propositions that are under way and of how much money certain individuals have made."[74] Powerful North American market traditions meshed with these attitudes to ensure that wild animals had little chance of being given an intrinsic right to exist, a view reaffirmed by the drive by farmers to domesticate and subdue the land.

Yet in spite of these dominant discourses of threat, economic benefit, and utility, more commonly than is often credited, some people found different values in the wild creatures of the region. There is a long-standing perception of prairie farmers as particularly close to nature, having an enviable chance while working the fields, as the journalist Miriam Green Ellis phrased it, to think about "the heavens and the earth, and all that is in them."[75] And despite their ambition to domesticate the land, some farmers did indeed look around them and draw pleasure and knowledge from what they saw. The Alberta naturalist Kerry Wood contended in 1946 that farmers were changing and were showing greater respect for wildlife: "They mark the site of the Prairie Chicken's dancing ground and watch the proud drumming of a cock partridge" and marvel at "the bright plumage of the waterfowl." Others thought themselves lucky to see a ruffled grouse "with her brood and see the clever dramatics of the mother as she feigns a broken wing and dragging leg to decoy the human away from her precious chicks." Farmers enjoy "all these intimate glimpses of game birds," and Wood contended that it was thus natural for many of them to protect "the beautiful wild creatures and appoint themselves the birds' guardians." While his conclusions may have been coloured by wishful thinking, there were more than a few farmers and farm families who were notable naturalists or who just took pleasure in watching and thinking about wildlife.[76] And the notion that farm settlement could better accommodate wildlife was not only a postwar phenomenon but had been voiced, albeit by a minority, almost from the beginning of settlement in the region. In a poem written in 1905 by an individual identified as M.E.D from Crossfield, a village in the dry rolling foothills north of Calgary, the settlement transformation promised Arcadia. While native vegetation would be replaced by "juicy grass" on which cattle would graze, wild animals and people would live together in harmony. Alberta was a farming and ranching paradise where "the settler comes and makes his home," but so too was it a place where "The fishes sport in the silvery streams / With no fear of the angler's snare / And devoid of fright from the hunter's gun / The fowls fly everywhere."[77] Such relationships were most often not realized in practice and certainly did little to slow development. Nonetheless, the hope for a more inclusive relationship with wildlife would find voice in references to the aesthetic and emotional benefits such relationships offered to both individuals and society. While they were often grounded in a language of utility that served social and economic needs, efforts were made to promote the survival or re-establishment of native species that were threatened by development. As will be discussed in the following chapters, while each case arose from varied motivations and had different goals – the re-establishment of bison, the saving of beavers, pronghorns, and elk from likely extirpation if not extinction, and the protection of many birds – each one demonstrated the complexity of people's relationships with prairie wildlife. Similarly, the efforts that grew from the intellectual and personal interests and dedication of natural historians, museum curators, and some wildlife administrators and sport hunters – as well as some members of the general public throughout the region – expressed an important divergence from the dominant views that were grounded in a fear of wildlife and the belief that the land's only function was to serve personal gain.

Figure 2.1 Ideas about wild animals ranged across a broad spectrum. Many people saw them simply as threats and barriers to settlement or as creatures that could be used to serve social or economic needs. But there was also a curiosity about their mental capacities, and some people formed intellectual, aesthetic, and emotive relationships with them. Such motivations found expression, for example, in taming animals such as moose, deer, or small mammals and birds. This 1911 photograph shows Billy Day's harnessed team of moose in Wainwright, Alberta.

PART 2

THINKING *about* ANIMALS

2

Some Perspectives on Animal Behaviour

We know now, what was unknown to all the preceding caravan of generations: that men are only fellow voyagers with other creatures in the odyssey of evolution.

Aldo Leopold, 1949[1]

There is a long history of diverse opinion in the European world about animal character and behaviour.[2] St Francis and some other late medieval thinkers variously believed that animals, as part of God's creation, possessed emotions and even moral culpability. All living things operated as part of a great chain of being that formed a cosmic unity.[3] Contrary views gained force during the scientific revolution of the seventeenth century, holding that animals lacked reason and emotion, imagination, spiritual capacity or memory, and a sense of time. These too were old ideas; they drew in part on classical thinking, and some advocates claimed they were consistent with Biblical accounts that put animals at the service of humans.[4] Historian Carolyn Merchant argues that a mechanistic way of thinking about the world gained strength in the seventeenth century and overpowered notions of the earth as an organic entity. Not incidentally, she argues, this view also devalued the role of women in the history of the world by dismissing older associations of the feminine and the organic character of the cosmos. The assertion that everything could be understood and known because the world was governed by mechanical principles that could be explained through science became intellectually and politically dominant, sweeping away, Merchant contends, any lingering restraints on how animals, natural resources, and land could be used for human advantage.[5]

But thinking about animals nonetheless remained divided. Historian Keith Thomas, for example, argues that in seventeenth-century England people who daily worked and lived with animals were especially prone to credit them with reason, memory, emotion, and other attributes. And despite the prevailing belief in a mechanical universe, such views endured, at times becoming popular with the cultural vanguard. Many early nineteenth-century Romantics idealized the world as an organic entity, and some argued that all elements of nature had intrinsic worth.[6] Opinion continued to be variable and changing. Historian Harriet Ritvo notes that some observers in England by the last half of the nineteenth century were contending that animals did not possess intelligence, while others appealed to intellectual and emotional attributes when describing

them. Even so, the Victorians generally tended to reserve the term "intelligence," by which they meant the use of reason and the making of tools, as an exclusively human attribute. Animals were instead popularly said to be "sentient" (that is, they could experience pain and pleasure) and had "sagacity," a diffuse mental power that included emotions as well as cognitive abilities like memory. But Darwin's argument that all living things were part of an interconnected fabric of life shifted the way in which the boundaries between what was human and what was animal could be drawn. As he contended, cognitive similarities (such as memory and reason) could be expected to exist in all animal species given their common origin, and as Ritvo notes, this theoretically "eliminated the unbridgeable gulf that divided reasoning human beings from irrational brutes."[7]

Such discourse about animals was part of the cultural inheritance of Euro-Canadian settlers on the prairies. Most seem not to have accepted Darwin's theories and assumed that humans had unique mental traits and physical abilities that were granted by God and that distinguished and separated them from animals.[8] But, as in European discourse, prairie thinking demonstrated diverse views about animal minds and behaviour. These views are explored here through four topics: Aboriginal ways of knowing animals; representative Euro-Canadian public and literary ideas about instinct, reason, and emotion in animals; the use of behaviourist approaches typical in twentieth-century scientific research about wild animals; and views in popular culture about the limitations of such approaches.

Aboriginal ways of knowing animals are illustrated here by Rock Cree people in northern Manitoba; their thinking about animal character and behaviour is particularly well documented, and their views about animals were echoed in varying degrees in other northern Cree communities in Canada, including Quebec.[9] While it is apparent that there were different emphases and approaches in thinking about animals among Aboriginal groups, it is also evident that Aboriginal communities in the Prairie provinces broadly saw their relationships with animals in a holistic way in that people and animals were connected in spiritual and material ways across space and time. Thus, the Rock Cree provide a case study that is indicative of broader ways of thinking among Aboriginal people in the region.

In contrast to Aboriginal people, Euro-Canadians generally drew much sharper lines to separate past and present and to define differences between species. It was widely assumed that while humans could reason and experience a complex range of emotions, animals operated mostly by instinct. While this was a dominant view, there was varying dissent. At one level, a vague but widespread doubt in the popular culture suggested that animal behaviour did not result merely from instinct. Others were more direct, arguing that animal minds showed all the attributes of human ones, even if at a reduced degree. These Euro-Canadian viewpoints are examined here through the arguments of Robert Bell of the Geological Survey of Canada and the work of writer Ernest Thompson Seton, whose experiences in Manitoba were formative in his career as a writer-scientist. Seton portrayed animals as having various social and emotive attributes and as exercising reason in their daily lives, and his writings enjoyed widespread popularity before the First World War.

Although Seton's work continued to be popular and influential, academic and mainstream scientific thinking about animal behaviour was solidifying around behaviourist explanations. This approach asserted that valid knowledge about animal behaviour could only be had through the controlled observation of factors that could be measured and replicated in experiments. This dismissed the research and conclusions of individuals such as Seton as being little more than anecdotal and anthropocentric fancies that had not been objectively measured and tested. Such a behaviourist approach was dominant in the study of animals in universities and academic forums by the time of the Second World War and is exemplified later in this chapter by the work of the avian biologist William Rowan, who applied it in his investigations of phenomena such as bird migration and life cycles.

This behaviourist approach explicitly and implicitly made consideration of such factors as emotion irrelevant in animal behaviour because they were held to be unmeasurable. While the answers provided by such research approaches to animal behaviour and animal minds were broadly accepted as authoritative, certitude in popular thinking about them nonetheless remained tentative during the interwar and postwar years. The review of mid-century scientific research on animals by the Winnipeg writer Dyson Carter illustrates this scepticism and reflects that of others who also continued to wonder – just as they had before the First World War – if animals might indeed have a mental and social existence that humans could only imagine. Were they part of God's plan? And if they experienced emotions, were they the same ones that humans felt, such as fear, pleasure, contentment, and love for their offspring? Did animals play? Was it possible that they had a moral capacity like humans? Could they communicate with each other? How could one assess their intelligence? Such questions were implied and directly voiced by numerous observers until the early 1960s and spoke to a persistent ambiguity in popular culture about animal lives and capabilities.

AN ABORIGINAL PERSPECTIVE: THE CASE OF THE ROCK CREE

Many North American Aboriginal cultures asserted that animals had souls, intelligence, and emotions that in some respects paralleled those of humans. Animals could communicate among themselves and in some ways with humans, and humans and animals had active and ongoing relationships that were essential for human survival. Of the few systematic studies of these relationships on the prairies, the most focused and comprehensive is Robert Brightman's *Grateful Prey*, a historical-anthropological study first published in 1973 about Rock Cree people at Granville Lake and Pukatawagan, two communities about 220 kilometres north of The Pas in northern Manitoba.

Had they been told about Darwin, the Rock Cree would not have found much that was surprising in his conclusion that all living things were interconnected and related. Indeed, they probably would have only found it surprising that this idea was controversial, because they, like many other North American Aboriginal people, assumed that animals had existed long before humans and that humans shared many characteristics with them. While the matter was expressed in varied ways, it was Brightman's impression that most Rock Cree credited animals with greater "self-awareness, rationality and intelligence" than Euro-Canadians did. Human attributes and skills, such as speech, use of fire and clothing, and food preservation, were said originally to be "the possession of animals, long before human beings existed in the world." Modern animals were not believed to have had an exact moment of creation but were of "broader cosmic provenience." The history of different animals was varied, with some having once been more powerful than humans, and there were various interpretations about their origins. Some Rock Cree believed that contemporary animals had been made by the Supreme Being; others held that they were the descendants of "timeless proto-animals" that had once possessed both human and animal social and linguistic traits. In yet other cases modern animals were thought to be descended from humanoid-like creatures that had been transformed into animals. When explaining differences between animals and humans, some Rock Cree contended that animals, unlike humans, did not need fire or clothing because they had fur, body hair, or feathers, while others held that animals had stronger souls or were cleaner, stronger, or faster than humans. Yet others argued that animals in some respects were less intelligent than humans but smarter in other ways, and that animals could communicate with each other, just not in the manner of human speech. In contrast to the rigid linear sequences of past, present, and future that Euro-Canadians adhered to, the Rock Cree believed that historical animals continued to exist, though not in corporeal form. Hidden from human perception and existing outside linear time, they could be approached through ritual and religious practices that created an association "between the present and this remote past, which continues to exist separately from but concurrently with the present." The explanation that animals had lost their human-like traits was thus relevant only at "one level of human experience, that of perception of the [contemporary] animals' outward form" in waking life. At another level of experience, accessed through dreams, the shaking lodge, and with the guidance of sorcerers, contemporary animals continued "to possess speech, culture and human bodies." When Brightman tried to resolve these various views and distinctions more precisely, his Rock Cree informants simply referred to the "limits of human knowledge."[10]

Brightman argues that Rock Cree relationships with animals did not in practice exist as discrete activities associated with production (for example, hunting methods and equipment) or religion (such as the symbolic and spiritual relationships between animals and humans). Of the various relationships that the Rock Cree had with animals, the case of hunting provides a window into their complexities. Hunting was central in Rock Cree daily life and, in one sense, was a social act involving strict divisions of labour. The higher social status of males was reflected in relationships with animals: men hunted big game such as moose, caribou, and bears and trapped fur-bearing animals, while women fished, collected berries, and caught rabbits and other small animals and birds. But hunting most importantly showed how spiritual life and material life were inextricably bound together. As Brightman observes, "the moral commitments and antagonisms" that hunters experienced with their prey were as pertinent to the human–animal relationship as was the hunter's "knowledge of animal habitat and biomass." Spiritual beliefs and material practices typically expressed a range of views, and this was perhaps even more prevalent in Rock Cree society, which, as Brightman observes, was "exceptionally tolerant of internal diversity in many spheres of practice and belief."[11] While there were contradictions and discontinuities in Rock Cree thinking about the human–animal nexus, Brightman identifies two dominant approaches, which he calls "benefactive" and "adversarial."

The benefactive model was "the most commonly expressed Rock Cree ideology of the hunter–prey relationship" and presumed mutual benefit for human and animal alike through "an endless cycle of gift exchanges between humans and animals." This relationship was founded on an assumption that the souls of animals, like human souls, survived after death. Unlike human souls, however, those of animals were "renewable." When an animal was killed, it was regenerated, either as a foetus or as an identical animal. This regeneration depended on an interrelated series of events and relationships among humans, animals, and spirits. Some Rock Cree argued that animals were governed by spirits that promoted their interests and regulated their populations and made them available to hunters. In the early 1800s it was apparently widely believed that there was a single "game ruler" or animal "boss," or possibly one ruler for each species, which some Rock Cree conceived as a giant version of that species. By the time of Brightman's fieldwork in the 1970s, many Rock Cree no longer found this concept of the game ruler convincing, but many hunters still accepted that an animal could only be killed with its consent, although some still contended that the consent of a game ruler was also needed. In either case, it remained that killing an animal was not "an accident or a contest" but "the result of a deliberate decision of the animal or another being to permit the killing to occur." Animals participated in their own death and gifted their bodies because they pitied humans for needing their flesh. In return, to guarantee the regeneration of the animal, the hunter was obligated to follow rituals and practices that expressed thanks and showed respect.[12]

Hunting was thus a cycle that began with the hunter and the animal confirming what would take place through a dream or some other medium. Dreams were central in this relationship, but they were understood or interpreted in various literal, symbolic, or prophetic ways that did not necessarily reflect the present-day of the dreamer. As Brightman observes, Rock Cree dream theory has no parallels in European thought, neither in "Cartesian dualism of spirit and matter nor [in] the boundary between an external objective reality and a subjective psychic reality to which dreams are assigned." Indeed, an individual could at times direct the content of dreams, and a successful hunt might not only be foretold, but determined in a dream. Brightman notes that the prefiguring of a successful hunt in dreams was sometimes talked about as a sign that permission to kill had been granted – a permission that culminated "when the animal enters a trap or exhibits its body to the hunter for a killing shot." Thus it was logical for some Rock Cree to claim that "animals are sometimes killed twice, first in a dream and subsequently in waking life."[13] Hunting songs accompanied the first part of this process as "a way of praying or asking animals and other nonhuman agencies for gifts of food" and conveying respect and thanks to the animal for the gift of its life. The hunter might offer invocations and prayers during the hunt and while setting a trap. Some held that a hunter was obligated to kill quickly and cleanly to lessen the animal's pain. Meat from animals that had suffered during their death was sometimes considered inedible, which, Brightman notes, suggests that hunters sometimes killed animals against their will.[14] After an animal was killed, the hunter reciprocated for the gift he had been given by treating the carcass of the animal respectfully and in ritually correct ways to ensure that the animal would be regenerated and not resent its death. These rituals took various forms, although they were apparently carried out mainly for animals that were brought into camp and made part of the human social environment. Animal remains left in the bush were simply left there to rot and "naturally to return to life." The souls of killed animals were represented as "honored guests at feasts where food, speeches, music, tobacco and manufactured goods" were given to them. Such offerings could be made before the animal was eaten. Respect was shown in other ways as well: by not dragging an animal's carcass on the bare ground (though it could be dragged over snow), by wrapping the carcass of some animals for the journey back to camp, by butchering only on clean or specially prepared surfaces, by hanging bones in trees (which quietly but effectively demonstrated a hunter's prowess), or by disposing of the carcass in the bush or in running water. Then there were other practices: menstruating women avoided the carcass; men and women alike were prohibited from eating parts of certain animals; and the proper disposal of bones ensured that dogs could not gnaw on them.[15]

Ultimately, the role of humans and animals was complementary, for each gave life to the other. The animal's flesh sustained human life, and the proper treatment of the animal's remains proved respect and restored the animal to life. Thus, respectful and correctly conducted rituals promised both immediate and future success. Because animals watched humans, they saw when another animal's remains were treated in a way that would prevent its regeneration, and they would then refuse to be killed. Thus, while humans did kill animals, they were equally involved in their regeneration.[16] These complex practices and theories perhaps were logical in social and religious terms in a society that accepted animals as beings with souls, history, and agency. The characterization of the animal as benefactor perhaps resolved the moral dilemma that killing and eating creatures with souls, spiritual lives, and a shared history with humans was in effect a form of "murder or cannibalism." Or perhaps it reflected that while the killing of rational creatures possessed of souls was a necessary act of violence for survival, the human nonetheless still felt regret or even shame. Thus, Brightman wondered if the "respect" paid to the dead animal might in fact be "compensation for a crime rather than reciprocations for a gift."[17]

The idea that animals were regenerated presumed that their populations were infinite. Indeed, Rock Cree in the 1700s told traders "that the more animals they killed, the more they would kill in the future," and fur-trade accounts widely reported similar beliefs among other boreal forest people. Since the regeneration of animal populations was theoretically assured by the ritualization of their bodies, animal populations would decline only when the correct methods and rituals associated with using them were ignored. Indeed, because the animal offered much by giving itself, it was "obligatory to take much." Moreover, if a hunter did not kill an animal when it offered itself, the animal would tell its fellow animals that it was not needed. Consequently, wasting meat or using only selected and delectable parts (such as the tongue) were irrelevant to an animal's future regeneration. The only thing that could be "wasted" was the soul of the animal, which would be lost if the rituals that brought its regeneration were ignored. Further, there was no "conception of regional animal populations that could be managed by selective hunting."[18]

In this context, it was logical that the killing did not stop when fur-bearer populations collapsed in the late 1700s and early 1800s under the weight of commercial hunting. Some historians see this depletion as a consequence of more intensive hunting because of more forest fires, new technologies (such as steel traps), the growing demands of the fur trade, and perhaps the concurrent outbreak of disease among beavers.[19] Brightman does not discount these explanations but notes that the Rock Cree perhaps at first did not see fewer animals as an unusual occurrence because animal populations had always fluctuated. But when it became apparent that this was not the case, the culturally logical response would have been to assume that the animals and game rulers were failing to recreate the animals. As Brightman concludes, "the Crees' reaction to the growing scarcity appears retrospectively as a collective profession of faith in the principles of infinite game and immediate return. It exhibited a profound aversion – as much 'religious' as 'economic' – to long-range planning, scheduled consumption and delayed satisfaction." Thus they did not immediately increase their stores of food to deal with shortages, limit killing, or regulate their trapping of fur-bearers. Instead, "treating the catastrophe like any localized food shortage, they continued to attempt to kill as many animals as possible."[20]

Demonstrating the range of thinking among the Rock Cree about animal–human relationships, overhunting would have had a different significance for those Rock Cree who accepted the precepts of the adversarial model. In this case, rather than the relationship of hunter and animal being a matter of negotiation, gestures of respect to the animal were only intended to persuade it to overcome its natural reluctance to be killed. Indeed, killing and eating an animal demonstrated the human's predatory power in a relationship in which animals and humans were opponents. In this view, an animal did not negotiate its death and gift its body but was instead overcome by force or even by trickery, magic, or "medicines." An unsuccessful hunt was not interpreted as a punishment by the animal of the hunter for some transgression, but as the victory of the animal over the hunter. Yet others held that European steel and gun technology had given humans power and "reversed a previously asymmetric power relationship in which human beings were subordinate to animals. Since the arrival of Europeans and their trade goods was widely understood by subarctic Algonquians as part of the intended design of the creator being or transformer, the new technology may have been experienced retrospectively by some Crees as another divine intervention on their behalf." This expressed hostility towards animals and gave humans a preordained superiority because the spirit agencies were interceding "on behalf of human beings." This suggested "a more anthropocentrically oriented world view than is conventionally ascribed to subarctic Indians."[21]

Brightman found it "extremely difficult" to determine the relative weight of the benefactive and adversarial models because people would sometimes express ideas that postulated both models. As he comments, "Crees conceive any event of hunting as morally ambiguous and thus potentially hazardous in its consequences for later foraging. The disposition of the animals toward hunting is an unresolvable question to which the contradictory models of the human–animal relationship provide distinct answers." While admitting that perhaps he had failed to understand the way these models were integrated, he concluded that "the two ideologies are not reconciled in Cree thought and that they provide disparate solutions to the identical moral and practical questions" of hunting and animal–human relationships. Thus, one model might be more appealing during times of crisis, or perhaps each model was seen as more or less effective at different points in the hunting-killing-eating cycles. And perhaps the two models even complemented each other. Animals could be powerful adversaries – they had the power to escape a hunter, they could attack humans, and, in the case of bears, they could kill and eat them. This suggests that hunters were forced to recognize that humans could not overcome animals by will alone and that other forces (such as those underlying the benefactive model) might be at work. Animals had "impressive powers" that were not fully understood; "they live without clothing or fire, move beneath the water, sleep without food throughout the winters." They were also spiritual and mythological presences from which humans might derive safety and well-being. On the other hand, "animal enemies are, in all these respects, potentially very dangerous to human welfare, whatever control that men may momentarily exert over their bodies." Adversarial and benefactive models offered different solutions to these contradictions and reconciled the limits and failures of human agency. And, as Brightman notes, they continued to "coexist in unstable equilibrium" at the time of his study as they seem to have in the 1700s and 1800s.[22]

Whichever model was adhered to, significant changes appeared among the Rock Cree in the mid- and late 1800s with the appearance of population-management strategies for fur-bearers. This shift was likely a response to lower fur-bearer populations, although broader cultural change may also have been at play. Whatever the case, it came to be accepted that killing only a few animals and using all the meat to avoid waste was "an obligation imposed by the animals" and the spirit beings. This foreshadowed concepts of conservation that would later be seen in Euro-Canadian wildlife policy. These radical changes were accompanied by related cultural transformations: "infinite animal resources came to be figured as finite, indiscriminate killing was replaced by scheduled capture, and rates of human predation" came to be seen as affecting animal populations. Typically, however, these management principles were adopted in an "exceedingly diverse manner." While forced to abandon these conservation strategies for a time during the 1930s because of white competition in the trapping grounds, some contemporary Rock Cree by then viewed excessive hunting and the use of only portions of an animal for food "as lamentable or dangerous." But not everyone thought it necessary to manage animal populations, and even those who did saw it in different ways. As Brightman notes, "rather than entirely displacing the earlier conceptions, these innovations have come to coexist with them. For some Crees, the finite animal has been indigenized and management defined as a spiritual technique. For others, management has no spiritual significance."[23]

While the preconditions, obligations, and practices associated with hunting that were followed by the Rock Cree are among the most thoroughly documented, other Aboriginal people in the region, in varying degrees, held similar views. The Chipewyan believed that if a hunter "handles game in a way that goes against the mind of the animal, it is unlucky for people." And in 1913 it was reported that "before killing a beaver" the Chipewyan on the north shore of Lake Athabasca first apologized for the need to kill it. Anthropologist Hugh Brody notes similar concepts of animal–human reciprocity among the Dunne-za in northwest Canada.[24] Among other Cree groups, the Swampy and Thickwood Cree held some of the same views as the Rock Cree, although different environments and resources brought different practices. The Plains Cree (who originally were Woodland Cree but had moved onto the plains in the early 1700s to hunt bison) showed less concern about propitiating animal spirits. In his study of the Plains Cree, David Mandelbaum observes that killing a bear involved only "a modest ceremony wherein smoke offerings and berries were consecrated to Bear spirit power," and by the early 1900s even these rituals had disappeared. Indeed, Mandelbaum notes that hunting observances in general were "loosely observed and count for little" among the Plains Cree. Nor does it seem that the disposal of animal parts was ritualized as it was among the Rock Cree. As Mandelbaum concludes, "the munificent food supply of the plains" likely led to "a reduction in hunting observances and magic"; when the Plains Cree found themselves in an environment with plentiful food, their anxieties about starvation were allayed, leading to the abandonment of efforts "to control game supernaturally."[25] Yet some plains people were concerned about the need to regulate hunting, and one control mechanism was described to historian Hugh Dempsey in 1963 by Solomon Bluehorn. The Pakahoos (or Pakakoos), the Bony Spectre or Skeleton Spirit, was the skeletal remains of a hunter who had frozen to death in the woods. His spirit was said to fly through the air and would stop any hunter it found killing game unnecessarily by paralyzing him with a touch of its bony finger.[26] Blackfoot convictions about animal intelligence also shaped their hunting practices. As archaeologist Jack Brink explains, Blackfoot (like boreal forest people) practised a kill-all policy at buffalo jumps because of a belief that bison communicated with each other – if one escaped, it would tell others of the traps that hunters had prepared for them. Other plains hunters had similar beliefs. As Thomas MacKay of Prince Albert observed in 1887, presumably in a reference to Plains Cree, when bison were caught in a pound, every animal was killed because it was "very naturally" reasoned that if some escaped, any attempt to impound the same animals in future years would fail because "they would know what was up and lead the others astray, and they thought it better to kill everything that came in so they would not profit by their experience."[27]

The beliefs and practices of Aboriginal people reflected the contradictions and complexities arising from their interactions with the animals with which they shared the land and had such intimate and essential connections. For the Rock Cree this was perhaps also a way of trying to reconcile the apparent differences and similarities between humans and animals and a range of questions about consciousness, volition, the origins of life, and the moral ambiguities of the human–animal relationship.

Euro-Canadians approached these problems in various ways, often by ignoring the possibility of sentience or intelligence in animals and assuming that interaction with animals had no moral implications – that people could do with them as they pleased. Indeed, Brightman provocatively wonders if "the mute, soulless, inferior and nonhuman animals of the Oblate missionaries may have been welcomed by some Crees because they posed fewer enigmas." As he concludes, Rock Cree models of the human–animal relationship posited a number of conundrums because "the animals are endlessly regenerated, and yet are finite. I am more powerful than the animal because I kill and eat it. The animal is more powerful than I because it can elude me and cause me to starve. The animal is my benefactor and friend. The animal is my victim and adversary. The animal is different from me, and yet is like me, as much like me as its ancestors were in the earliest time of the world."[28]

INSTINCT, REASON, AND EMOTION IN ANIMALS

Although many biologists were beginning to abandon "instinct" as an explanatory device by the early 1900s because the concept was seen as so variable and imprecise, it was nonetheless widely used in popular thinking to explain animal behaviour.[29] One early twentieth-century source defined "instinct" as "the power by which, independently of all instruction or experience and without deliberation, animals are directed to do spontaneously whatever is necessary" for self-protection or perpetuation of the species. This source further postulated that there were various theories about instinct's origin. One theory, only cursorily described, held that instinct was a product of natural selection. Another held that its origin lay in God's plan to equip each species for life, while yet another held that its origin lay in "the accumulated results of individual experience, fixed by repetition, and transmitted as an inheritance to succeeding races."[30] Another definition simply described instinct as blind behaviour that could not be deflected or controlled. The novelist and political activist Nellie McClung recalled that, as a child in the 1880s in Manitoba, she was alarmed by the cruelty of the family cat, and her mother explained that cats were no different than wild animals, whose instincts made them cruel and violent. As her mother phrased it, "animals will fight, their instinct drives them ... Cats kill birds, just because it is their nature; they are driven by their instincts and they can't be blamed. Nature is cruel, I often wonder why; red in tooth and claws." In this view, instinct explained animal character as well as the lack of morality and mercy in nature. And while the McClung house cat was driven by instinct, wild animals were thought to have a surfeit of it. When the Winnipeg entrepreneur Donald Smith donated some of his captive bison to the national park at Banff in 1898, the *Nor'West Farmer* observed that the animals would not stay put unless they were fenced in, since "it is only too likely that their natural instincts will, in spite of their half-tame condition, reassert themselves and induce them to wander off in any direction." The understanding that instinct was engrained and unconscious was the subject of "The Stupidity of Instinct," a 1902 article about insects in *Saturday Night*, a widely read Canadian magazine. Citing experiments in the United States, the article suggested that "the wonderful results often accomplished by insects are due, not to intelligence, but to automatic obedience to external or internal stimuli." Indeed, one could just as easily "call a watch intelligent as an insect."[31] This echoed Descartes's argument that since language was necessary for rational thought and making moral choices, animals' lack of it made them machines, or in his analogy, like a watch, wondrously complicated but operating only through mechanical force and instinctual movement and behaviour. There was no mystery about animal life; unlike humans who adapted their surroundings to their needs, animals lived at the mercy of their natural environment.[32]

Not everyone was so certain. Robert Bell expressed a different understanding in a talk, "On the Intelligence of Animals," at the Natural History Society of Montreal in 1870, in which he attempted to demolish the popular appeal to instinct as an explanatory device about animal behaviour. The topic of animal cognition was then in the air because of the writings of Darwin and others, and while Bell was not a Darwinist, he would have been familiar with the outline of the current issues. Born in Toronto, he graduated with a degree in applied science from McGill College in 1861 and then took a permanent position with the Geological Survey of Canada in 1869. He completed a medical degree at Queen's University in Kingston in 1878, by which time he was well launched in his career as a geologist, having conducted field surveys in western Canada. While his primary work was in geology, Bell, in the tradition of the time, was also interested in most aspects of natural history, including biology. Keeping prairie connections that he had formed during his professional life, after his retirement in 1908 he divided his time between Ottawa and his farm near Rathwell in south-central Manitoba, where he died in 1917.[33]

Robert Bell opened his Montreal talk by overconfidently asserting that the reasoning abilities of "the higher animals," especially mammals and birds, was "now pretty generally admitted" and that he intended to show that other animals also had a "high degree of intelligence."[34] By intelligence he meant the ability to reason, and he took pains to dispel the popular belief that animals were driven only by instinct, a theory he attributed to the prejudice "against admitting that any creature whatever, except man, was possessed of any reasoning powers." People were "accustomed to witness daily examples of the intelligence of many animals," Bell told his listeners, but were "so prejudiced with rather vague notions about instinct that we do not think of them in their proper light and hence lose much of the interest and sympathy which we would otherwise have for the creatures around us." Indeed, the example of insects showed that "mere instinct" was inadequate to explain animal behaviour. Entomology was then a topic of considerable interest in natural history – a decade later, Charles Darwin would argue that earthworms possessed cognition – but Bell also wanted to show that if insects had intelligence, the case for the reasoning powers of higher forms of animals was even more evident.

Bell asserted that all animals became more intelligent with age and experience. The ability to reason was progressive, arising in the first instance from instinct, which then became intelligence. This was so with insects just as it was with newborn humans and other mammals who learned from their parents. There were many examples of what was said to be instinctual behaviour in insects, and these did not appear to be "essentially different from those of the large animals and which seem to be the offspring of circumstances." He cited well-known examples of what he called "intelligence" among insects, such as the political economy of beehives, the social organization of ants, the ability of bedbugs to observe their surroundings, and the behaviour of trained fleas. He also noted that some insects (as well as some birds and mammals) feigned death to escape predators, a behaviour that he interpreted as demonstrating the use of reason for self-protection. In particular, he detailed what he called "the faculty of reasoning" in spiders. Bell had observed spiders that watched the direction of the wind when spinning and casting their webs and held up "a leg to feel for wind just as a sailor holds up his hand for the same purpose." When a breeze of the right velocity came along, the spiders "would start up full of excitement and purpose to cast a line in case the wind should prove favourable." Further, Bell argued that insects had the ability to use reason to control their environment in order to solve problems. For Bell, this defined rational behaviour, and as he noted, bees and ants

> perform all the ordinary duties of their lives with great skill and industry – this is commonly attributed to instinct. When an accident happens or an emergency arises, however, they are equally skilful and prompt to meet the difficulty. This may be instinct. But when we introduce conditions so completely exceptional and artificial that it is quite impossible the insect could have encountered them without the intervention of man, and we find that they unhesitatingly set about overcoming these conditions, we are forced to conclude that their actions are the result of reason. It is admitted to be a proof of reasoning that an animal on trying one process and finding it does not succeed adopts another. Examples of this appear to be quite as common among insects as mammals.

Given such examples, Bell could only explain people's failure to recognize the reasoning power of animals as a product of religion. "Much of the opposition to admitting that animals are possessed of reasoning powers," he concluded, "is based on religious grounds. It is supposed by some that if we allow that an animal can reason, we must also presume that it must have a future existence." Bell skirted this thorny problem by simply concluding that "the reasoning power in animals may be admitted" without having to believe in "the existence of a future state for them," but he then slyly noted that, anyway, was "there not plenty of room for all?"

Other people made similar arguments. In 1904 the Canadian ornithologist Percy Taverner, one of Canada's best-known ornithologists and the assistant zoologist at the Victoria Memorial Museum (which became the

National Museum in 1926), argued that bird migration reflected accumulated experience. As he explained, "however instinctive their habit may now be, there must have been a time when migrations were intelligent movements, intended to escape some danger or secure some advantage; and through generations of repetition they become fixed into hereditary habits." In other words, a search for advantage based on reason became instinctual, or habitual.[35] And the year before, the Manitoba ornithologist George Atkinson had argued that the "presumptuous idea so generally accepted until recent years" that only humans were "endowed with a higher intelligence had been thoroughly exploded," although many people still refused to accept "this truth." But no one, he contended, could have an association with birds or mammals "without being convinced that their intellectual conditions differ only in degree of development from our own." He adamantly rejected the contention that what appeared to be "higher intelligence" in birds was really only instinct. "Reason is but an evolution of instinct" as part of natural selection "to keep pace with our ever changing struggle for existence." Birds whose environment required it were more intelligent than those whose surroundings did not make such demands. Broadening his argument, Atkinson insisted that many birds also showed "the same characteristics as did humans." Citing specific examples, he argued that birds had memory and reason, sociability, wit and humour, and demonstrated love, vanity, and maternal sentiments as well as personality traits such as indolence, selfishness, industry, and diversity of character. And of course they had song.[36] Other references to the existence of a range of emotions in wild animals had regularly appeared in the late nineteenth century. The poet Charles Mair, for example, noted in an 1890 essay on plains bison that while a bison herd usually stampeded when pursued by hunters, there were instances "of a cow falling, and of a bull remaining by her side, not through accident, but evidently from feeling. He would stay by her, lick her, try to lift her up with his horns, and fiercely charge anyone who ventured to approach long after the chase was over. It is difficult to explain such a scene upon any other grounds than that of personal affection – a sexual altruism which would not be surprising if it were not in opposition to natural habit." People made similar observations about other animals; a US report in the

Regina Leader in 1886, for example, credited dogs with memory and emotional attachments to people who had helped them or treated them well.[37]

The arguments put forward by figures such as Bell, Taverner, and Atkinson were only part of a broader discourse before the First World War that endorsed the existence of reason and emotional capacity in animals. This theme was particularly sharp in Canadian popular fiction in the late nineteenth century when a number of important wildlife writers, such as Ernest Thompson Seton, Charles G.D. Roberts, and Margaret Marshall Saunders, were enjoying national and international success.[38] These literary works touched a nerve with many people, and there has been much debate about the reasons for their success. Margaret Atwood has contended that Canadians identify with these stories about wild animals as well as with their victimization because of a deep fear and anxiety about a hostile environment and Canadian cultural survival under the threat of the ever-greater reach of American influence. The merits of Atwood's argument have been vigorously debated, and other arguments have been put forward to explain the popularity of these stories.[39] For instance, it has been argued that their popularity was consistent with the back-to-nature movement of the 1890s and early 1900s, the growing popularity of nature study, and concerns about the decimation of North America's wildlife. Historian Lisa Mighetto has also contended that the intelligence, sentience, and emotional life of many of the animals in these stories seemed like an antidote to the apparent darkness of Darwinian thought, in which only the fittest survived in a world gripped by unremitting struggle.[40]

Of these stories, those of Ernest Thompson Seton had significant impact. His *Wild Animals I Have Known*, published in 1898, went through nine printings in the first eighteen months. Saskatchewan nature columnist Marion Nixon wrote in 1941 that Seton's stories were her "first introduction to Canadian nature lore," and in 1956, ten years after Seton's death, a correspondent from Brandon wrote to the *Western Producer*'s Youth Page recommending that everyone should read Seton's stories because he was "one of the finest animal story writers who ever lived and who lived for many years in our own prairie country."[41] Indeed, Seton's prairie connections were formative. Born in England in 1860, he came as a boy to

Ontario with his family. Then named Ernest Evan Thompson, he lived first in Lindsay and then in Toronto, where he attended the Ontario College of Art. He later studied art in England and France, and returning to Canada in 1882, he joined his brother on a homestead near Carberry, a short distance east of Brandon in southwest Manitoba, where he lived on and off until 1887. These years were important in his development as a nature writer and illustrator. During the Carberry years he changed his surname to Seton, which confirmed an emerging identity that separated him from his father, whom he hated, and revealed a streak of vanity, since the name supposedly revealed an aristocratic Scottish ancestry. More importantly, as shown in many of his stories and especially in his journals and diaries, the Carberry years provided him with inspiration and a body of research and observation that he drew upon for his entire literary and artistic career. Later in life he inscribed the title page of the first volume of his journal with the lines "Here begins the journal of my life in Manitoba, the best of all my life and when I ended in 1887 my boyhood was gone – my brightest days ended."[42]

Always viewing himself primarily as a naturalist and scientist, Seton published several important field studies, including *Mammals of Manitoba* (1886) and *Birds of Manitoba* (1891), which drew upon his observations at Carberry. In the 1880s and 1890s he published in magazines such as *Forest and Stream*, but it was his short stories that brought him important social connections, fame, and wealth. After leaving Canada for the United States in 1887, he kept his prairie connections for a time. Following his insistent lobbying, Manitoba named him "Provincial Naturalist" in 1892, an honorific position that had no defined duties but conferred both professional credentials and status. In 1907 he helped lead a Canadian government research party in northern Alberta and the Northwest Territories to study wood bison and muskoxen. His scientific work continued to find respect long after he left Canada, and when the first volume of *Lives of Game Animals* appeared in 1925, R.M. Anderson, chief of the biology division at the National Museum of Canada, congratulated him on the book, claiming that "with the virtual extermination of much of our large native mammals which is now being carried on, the opportunity for similar work will never occur again and your work will stand as a monument in its field."[43]

Seton wrote with an easy charm, and his fiction reflected his beliefs about animal character, which he first set out in his earliest and most famous book, *Wild Animals I Have Known*. This book includes stories set in the prairies as well as Ontario and the United States.[44] Poet and author David Arnason argues that, with the "possible exception of Kipling's *Jungle Tales*," it is "the world's best-loved collection of stories about animals in nature." Seton endowed animals and birds with personalities and recognizable lives, in some cases even birth and death dates, and they lived through dramas of loss, fear, family, regret, love, devotion, power, and violence. Always concerned to convey moral lessons, the stories support literary scholar Janice Fiamengo's argument that spiritual and moral themes are crucial in understanding Seton's stories.[45] Seton claimed that his animal stories were true and that while he had "left the strict line of historical truth in many places, the animals in this book were all real characters." He gave them life by treating them as individuals with emotions, character, and the ability to learn, to respond to changing circumstances, to remember, and often to reason. Moreover, he argued, "the fact that these stories are true is the reason why all are tragic. The life of a wild animal *always has a tragic end*." Violence was part of nature and no animal died of old age; their lives consisted of eating others or being eaten, but humans were also a great threat, and the little foxes in one story had an "instinctive hate and fear" of humans, who were the most dangerous enemies because of their technology, their relentlessness, their lack of feeling, and their view that animals did not matter. In the preface to the book, Seton insisted that the moral of these stories was "as old as Scripture – we and the beasts are kin. Man has nothing that the animals have not at least a vestige of, the animals have nothing that man does not in some degree share."[46] As he remarked in a public lecture on "The Minds of Animals," animals differed from humans by having lower reasoning powers and much finer senses such as smell and, in some cases, sight. He also contended that animals could be trained to count (which revealed an ability to think abstractly) and that birds had an aesthetic sense (citing the example of crane dances) as well as traditions (such as migratory routes). And, not insignificantly, animals educated their young.[47] At the same time, Seton had misanthropic tendencies, and in contrast to the empathy he showed for animals, he often portrayed humans as bloody-minded scoundrels and weaklings who were ruled by their emotions and lusts and thought they were above, not part of, nature.

Fiction needs engaging plot and character, and this requirement cannot be discounted as a motive for Seton's depiction of animal minds and lives. But any fiction that purports to depict actual conditions needs to be believable in order to create an emotional and intellectual connection with its plot and characters. In this sense, the popularity of Seton's stories reflected public acceptance of a view of animals as rational and emotionally complex creatures. All the stories in *Wild Animals I Have Known* were built on Seton's belief in the often unrecognized and misconstrued emotional capacity and intelligence of animals. Vixen in "The Springfield Fox" mourned for her dead offspring, while the birds in "Redruff: The Story of the Don Valley Partridge" could name, by reference to climate and the availability of food, the various months and seasons of the year. An equally popular theme was the rational behaviour of animals. The wolf Lobo (in a story set in the southwest United States) used reason to avoid capture; the crow in "Silverspot: The Story of a Crow" solved the problem of obtaining food by watching what was going on around him; the rabbit in "Raggylug: The Story of a Cottontail Rabbit" planned escapes and avoided danger by assessing the demands of particular circumstances; and Scarface the fox in "The Springfield Fox" stood perfectly still while watching a hunter who had lost his trail and "the wonderful part of this is, not that he resembled the round stones and grass, but that he knew he did, and was ready to profit by it."[48] Moreover, animals communicated. For example, in one story Seton maintained that crows had a language and social system "that is wonderfully human in many of its chief points, and in some is better carried out than our own," and he illustrated this story with musical scores to demonstrate how the language of crows worked and what the sounds meant. And he further contended that rabbits did not have "speech as we understand it, but they have ways of conveying ideas by a system of sounds, signs, scents, whisker-touches, movement, and example that answers the purpose of speech." Animals also had fun playing and had emotions identical to human ones, such as maternal and family love. This sometimes tipped Seton into pure melodrama: Mustang, the wild horse in "The Pacing Mustang," committed suicide to avoid capture, while Raggylug, in "Raggylug: The Story of a

Cottontail Rabbit," led another rabbit to its death because it had abused his mother.[49]

The animals in Seton's stories also inhabited a world of social structure and organization. The rabbits in "Raggylug" held claim to the swamp where they lived by virtue of "long successful occupancy," which "was exactly the same as that by which most nations hold their land, and it would be hard to find a better right." Seton was also concerned with leadership and power, and he espoused a conservative political economy. Great leaders protected their community. The intelligence, physical strength, cunning, and wisdom of Lobo in "Lobo: The King of Currumpaw" protected the wolf pack from danger, while the leadership of Silverspot the crow was justified by his strength, age, wisdom, and intelligence. Since crows "are always dependent on each other for life and safety," the discipline and order Silverspot imposed on the community guaranteed its safety, and when he was killed by an owl, the community disintegrated. After his death, the crows still came to their familiar haunts, but "without their famous leader their numbers are dwindling and soon they will be seen no more about the old pine-grove in which they and their forefathers had lived and learned for ages." Equally conservative was Seton's portrayal of female and male roles. Janice Fiamengo has noted the recurring theme of matriarchy in Seton's stories where females trained and protected their young out of maternal love. But if females were nurturing, they could also compromise male strength and leadership. Lobo was undone because of his love for Blanca, a love that clouded his self-discipline and his absolute control of the wolf pack. When the hunters pursuing him finally realized that they could defeat him by controlling Blanca, Lobo's fate was sealed. Once they killed Blanca, a heart-broken Lobo lost his strength and purpose. Less sentimental, Mustang, the wild horse, was brought to disaster because he could not resist the mare set out to decoy him.[50]

The functioning of instinct, emotion, intelligence, and environment is central in Seton's stories, and their interplay is demonstrated in *Wild Animals I Have Known* in the training of newly born animals. Vixen the fox taught her offspring how to hunt by first teaching them to catch animals that were easy to kill and then by hunting more difficult prey as their skills developed. Her teaching method was "example, aided by a deep-set instinct." Crows also rigorously trained their young, and by the fall "the rabble of silly little crows have begun to learn sense" and knew their drill and how to keep sentry duty. They had learned about food and threats as well. "They know that a fat old farmer's wife is much less dangerous, though so much larger, than her fifteen year old son, and they can tell the boy from his sister. They know that an umbrella is not a gun, and they can count up to six." So, too, Molly Cottontail in "Raggylug" taught her offspring what to eat and drink and what was dangerous. "Day by day she worked to train him, putting into his mind hundreds of ideas that her own life or early training had stored in hers, and so equipped him with the knowledge that made life possible to their kind." And among partridges, "their start in life was a good mother, good legs, a few reliable instincts, and a germ of reason." Instinct was "inherited habit" that led the young partridges to hide and follow their mother, but as they grew, "reason entered more and more into their expanding lives."[51]

Among Seton's most closely argued explorations of the relationship of instinct and intelligence was an essay, "The Intelligence of the Wolf," written in about 1900, in which he argued (drawing directly from Darwin) that animal intelligence developed because of natural selection. An animal that obtained food by stratagem, he argued, "will develop greater intelligence than one whose food is taken without art, as is the case with herbivores." As well, an animal that had to save its "own skin by its wits will become more intelligent than one whose strength, speed or inaccessible haunts render it immune." Third, an animal that knew "the value of organization shows a higher development of [the] brain than one which relies on its individual efforts for everything." Animals possessing all three characteristics had the greatest intelligence, and while some had one or two of these attributes, the wolf had all three. Crucial in this argument was the interaction of instinct and intelligence. Seton argued that "all instincts had their origin in reason" and "persisted because subsequently endorsed by reason." Moreover, among wolves, reason was communicated by voice. Wolves trained their young by example, but as the young grew, adult wolves used various calls to teach them. Citing the different wolf howls in hunting and social situations, Seton demonstrated how voice became even more important when wolves formed packs in winter.[52]

Animal communication was a controversial topic, and Seton's views on it, along with some of his contentions about animal intelligence and emotion, led to the criticism that he lacked scientific rigour and an appropriate methodology. Seton was not alone in this – Darwin and his contemporaries had also been criticized for using stories and unverified reports by travellers and writers as evidence of animal behaviour.[53] Although Seton used field research and meticulous observation, he also relied on anecdotal evidence and drew his own conclusions about what his observations meant. Issues about methodology were part of the so-called Nature Faker's controversy in the United States between 1903 and 1907. This pitted the naturalist John Burroughs against Seton, Charles G.D. Roberts, and William John Long, an American writer of the same school. Burroughs was an eminent American nature essayist who built a career writing descriptive accounts about wild animals and nature. In 1903 he published an article criticizing all three authors for taking liberties with the truth about animals. Burroughs charged that their stories were unscientific, sentimental, and fantastic yarns that gave the public a false understanding of nature and animal life. He believed that animals operated almost entirely by instinct and that there was limited evidence of their ability to reason. The controversy dragged on, largely because neither Burroughs nor Long would give it up. It finally dropped out of sight after President Roosevelt (a big game hunter and a firm believer that animals existed to serve human needs) publicly sided with Burroughs in a 1907 article in which he coined the term "Nature Faker" to describe those who presented a view of wild animals as creatures with reason and emotion. While the Nature Fakers controversy may largely have been a literary quarrel that earned public attention because of the great egos pitted against each other, it also reflected mainstream and academic attitudes about acceptable scientific methods.[54] Demonstrating such attitudes, C. Hart Merriam of the Smithsonian Museum wrote to Seton in 1910 after reading Seton's new book, *Life Histories of Northern Mammals*, a study of sixty quadrupeds found in Manitoba and elsewhere in North America. Based on thirty years of observation and study, the book focused on the animals' physiology and life cycles, but it also dealt with "the intimate side" of their lives. It was a great book, Merriam wrote, but while he had learned much from it, he believed that Seton's portrayal of animals as having human-like emotions went too far. As Merriam noted, "then you have a whole lot to say, both in the introduction and at intervals throughout the book about monogamy, polygamy, the Ten Commandments, vice, crime, suicide, and so on. It seems a pity to blemish so good a book with such twaddle."[55] Such criticism pushed Seton towards a more "scientific" stance in his writing and reflected professional biologists' rejection of conclusions about animal behaviour that referenced the "personality" and individuality of animals or their aesthetic qualities.[56]

The animal story was, in any case, beginning to wane as a literary form by the First World War, and the bitterness and destructiveness of the war further turned public attention elsewhere. While wildlife stories continued to appear, by the 1920s nature literature had become dominated by recollections and personal encounters such as those of the naturalist, wannabe Indian, and beaver trainer Grey Owl, who wrote accounts of his life with beavers and as a trapper. Fictional accounts of animals reappeared with growing frequency after the Second World War, reflecting another era of rising concern about and interest in the natural world.[57] In 1954, for example, Fred Bodsworth published *Last of the Curlews*, a novel cast as a tragedy about the extinction of the Eskimo curlew, last sighted ten years before. This is one of the finest pieces of Canadian writing about wild animals – Bodsworth was poetic in his restraint and sound in his science. His portrayal was far from Seton's emotion-filled world, but similar to Seton's generally held views on the treatment of animals by humans, Bodsworth was clear on the point that what the curlews had been sacrificed for did not make up for the loss of such extraordinary creatures.

BEHAVIOURISM AND THINKING ABOUT ANIMALS

Critics of those who believed that the animal world was animated by decision-making, emotionally complex creatures often drew upon a belief that the study of animal behaviour was properly a rigorous and unsentimental activity based on experiments that could be replicated. Such voices strengthened in the first part of the 1900s, and as historians Gregg Mitman and Richard Burkhardt note, the divide in biology between laboratory and field studies had sharpened by the interwar years and was "often drawn along professional versus amateur lines." Thus, despite diverse specializations, laboratory experiments were "the focal point" in the biological sciences, and although fieldwork remained an aspect in some disciplines (such as avian biology), it was largely left to amateur natural historians until the 1940s, when newly emerging methodologies and protocols broadly incorporated it into professional biology.[58] Such professional research approaches in biology thus rejected anecdote and explanations based on subjective observation (such as those Seton employed), leaving the testing of measurable factors as the focus of research. Many of these "behaviourists" also came to be obsessed with anthropomorphism and developed research methodologies that supposedly eliminated the possibility of tainting evidence by subjectivity or the attribution of human qualities to animals.[59]

During most of the period studied in this book, zoologists in prairie universities had limited overall concern with wildlife research. This in part reflected the contemporary character of zoology, which in the early twentieth century was still focused on "the laboratory study of lower animals such as invertebrates, fish and amphibians, rather than birds and mammals." In the case of birds, study had traditionally been left to natural historians and museum curators, who typically had no formal training in biology. For example, Percy Taverner, of the national museum, had no formal science training, having been educated as an architect. Although ornithologists informally created a national network through the individual exchange of specimens and records of observations, there was no national ornithological association in Canada and the field was focused around local natural history societies or American ornithological associations.[60]

It was within this context that William Rowan was hired in 1920 by the University of Alberta in Edmonton to teach zoology. Rowan had taught briefly at the University of Manitoba before taking up the position in Alberta. As an avian biologist, he was not a particularly good fit for the job; many in the academy did not see avian biology as a legitimate concern of zoologists, even though the field by this time had emerged internationally as part of the expansion of the biological sciences and the growth of new subdisciplines. Moreover, Rowan's work and interests did not endear him to Henry Marshall Tory, a mathematician and the university's president, who favoured applied science and saw fieldwork as unscientific and wasteful. While collecting, classification, and field observation of birds remained important, avian biology was nonetheless increasingly concerned with laboratory research and experimentation that tested observable behaviour in order to develop and test hypotheses. Tory did not recognize that avian biology was thus more than the classification of specimens and concluded that studying birds was unsuitable for the university's zoology professor.[61]

Despite this lack of support from the university's president, Rowan carried out research in the 1920s on the diet of birds by studying the contents of their crops or stomachs to measure what they were eating. He also studied their anatomy and embryology. While these were not unfamiliar practices in the study of birds, Rowan moved beyond these traditions by studying such things as the effect of water levels and water salinity on food supply for waterfowl and aquatic mammals. Most significantly, he began to move in a different direction when he began to use controlled groups in experiments on bird migration, a pioneering approach because "most ornithologists did not use manipulative procedures or conduct rigorous biological experiments" at this time. He conducted experiments on the migration of juncos (a member of the sparrow family), and while he could not prove why they migrated, he showed that daylight affected the size of the birds' testes and ovaries, which seemed to influence their impulse to migrate. When their reproductive organs had expanded to a certain stage in the spring, the birds set off on their migration to northern breeding grounds. Conversely, in the fall, when their organs decreased in size, they began their migration southward. Thus, Rowan proposed that bird migration was a phase of reproductive behaviour tied to the amount of daylight. This challenged contemporary scientific theory that migration was triggered by changes in temperature or barometric pressure.[62] It also simply ignored conclusions, such as Taverner's, that migration was a habitual or instinctual activity that had its origin in reason. In 1929 Rowan extended his migration research to a study of eighty-three crows that he had trapped and lodged in pens near the High Level Bridge. The project fascinated many Edmonton residents, some of whom brought food for the birds, a welcome development given Rowan's limited funding. One group of crows was subjected to artificial light that was gradually increased to mimic summer

Figure 2.2 University of Alberta avian biologist William Rowan used crows to test his hypothesis that migration was a response to hormonal changes stimulated by seasonal light conditions. In the 1929 experiments, the crows were confined in these outdoor cages in Edmonton.

conditions, while a control group experienced the natural seasonal shortening of daylight. While suggestive, the results were inconclusive, and it was not until 1931 that Rowan was able to carry out further work on the subject. This time, he managed to obtain over 400 crows. He divided them into eight groups, and each group was tested differently: some were castrated, some were injected with hormones and other extracts, and others were left in their natural state. Each group was subjected to varying levels and durations of light, and when the time came to release them in November 1931, their tails were dyed bright yellow so that they could be identified. A "radio and press blitz" asked the public to help spot them and rewards were offered for every yellow-tailed crow that was shot and delivered to Edmonton. While the project earned great public interest, it ultimately did not enable Rowan to draw a clear-cut conclusion, although it again suggested that migration had a physiological dimension that was affected by light.[63]

In addition to studying bird migration and other topics, Rowan also conducted research on animal and bird population cycles. Fur traders had recorded population cycles in fur-bearing animals on the prairies for two centuries, and those of rabbits and their corresponding effect on predators such as foxes and lynx were also well known.[64] The rabbit cycle was dramatic. As Rowan remarked, during the low point of the cycle, even their skeletons had disintegrated and one began to think that the rabbits could "never come back or restock these denuded districts." Yet, "seven or eight years later the miracle has happened: the country is alive again with rabbits."[65] Extensive research on animal cycles carried out in Britain and the United States in the 1930s and 1940s indicated that while most animal populations were not static in the short term, there was likely an approximate range over longer periods. While similar work in Canada suffered from a lack of funding, research such as Rowan's formed part of a broader international interest in the subject. In about 1925 he began to study cycles among sharp-tailed grouse and snowshoe rabbits, gathering historical and contemporary data and conducting field observations and interviews with trappers, hunters, and game officials. Rowan also exchanged data and corresponded with Charles Elton, whose 1924 book *Animal Ecology* and research in the Spitsbergen Islands in the 1920s on vole populations had made him a pioneering figure in the field.[66]

Rowan's work on population cycles was often intermittent, but he recognized that these events of "such stupendous scale" offered "a challenge to science that had few peers in the realms of wildlife." By 1948 he had concluded that cycles affected a far wider range of birds and mammals than just rabbits and fur-bearers. All members of the grouse family cycled, and it seemed that magpies, evening grosbeaks, blue jays, mice, and squirrels, among others, did so as well. It was apparent that migratory birds, bears, beavers, deer, and wolves were the major exception to population cycling, although at one point Rowan speculated that some migratory waterfowl might be subject to very long population cycles, perhaps in the range of thirty-four years. For Rowan, the discovery that Hungarian partridges, which had been introduced to the prairies as a game bird, were taking on a cyclical character like native birds was of "unusual theoretical interest" and one that seemed "to suggest a good pointer towards a possible solution of the whole problem" of population cycles.[67] Perhaps this helped lead him to conclude by the late 1940s that cycles were climatic and geographical. As he noted, they occurred only in certain northern latitudes and he conjectured that they resulted from weather, possibly nutrition and some climatic factors such as ultraviolet radiation as well. To test this theory, he carried out controlled tests on rabbits by employing ultraviolet radiation and nutritional variants. These tests were inconclusive, but this was not surprising; despite extensive research, no one had yet discovered the reasons for these dramatic events. While there was a well known predator link in the cycles of lynx and rabbits, the exact causes of wildlife population cycles were largely unknown. Nor were their cycles as geographically uniform as it had previously been assumed. Dewey Soper (one of Rowan's former students who worked with the CWS) noted in 1964 that the causes of population cycles were unknown, although various theories suggested that predator–prey relationships, climatic forces, food depletion, and the rapid spread of disease because of high population might all play a role in their occurrence.[68]

Rowan's use of controlled groups, testing, and laboratory techniques to study migration and population cycles demonstrated an important trend in the study of wild animals, one that offered a different approach than the observational ones favoured by most natural historians and writers, such as Ernest Thompson Seton. Such behaviourist methods allowed no room for speculation about reason, emotion, and instinct as explanations of animal behaviour. In her biography of William Rowan, Marianne Gostonyi Ainley does not indicate whether Rowan speculated about the minds of his subjects, but she does note that Ernest Thompson Seton was one of Rowan's childhood heroes. Evidently, by the time he arrived in Edmonton, Rowan had left behind the charms of Seton and an emotive and personality-filled nature for the certitudes of behaviourism. He had a deep love of the outdoors and hunting, which brought him comfort and emotional inspiration, but the rigour of his research did not allow him to go beyond experimentation and testing. Moreover, speculation about spiritual qualities was irrelevant to him; he did not believe that humans had souls, which made the issue an irrelevant consideration in the case of animals as well. In this sense, his practice was a hard and objective one, following the models then accepted as normative in the academic study of animal behaviour.

Figure 2.3 William Rowan feeding a raven in Edmonton in 1953.

QUESTIONING BEHAVIOURIST EXPLANATIONS OF ANIMAL LIFE

By the end of the Second World War behaviourist approaches had penetrated throughout every level of scientific activity. In 1951, for example, students at one Manitoba public school carried out experiments to test the rather obvious theory that nutrition affected health. The school obtained four white rats, two of which were put "on a good diet of brown bread, milk, vegetables, fruit and meat, and two on a poor diet of white bread with jam, cookies, soft drinks and coffee." The rats were weighed weekly. The experiment had its ups and downs. Two of the rats died because they "contracted a disease" from a pupil. And there was shock when another gave birth. Nonetheless, after six weeks the well-fed rats were "contented and friendly" and their fur looked healthy. Not surprisingly, the ones that had in effect been starved were "restless and unfriendly" and had patchy fur and poor appearance. The use of this scientific method led the students to conclude that food played an important part in the development and behaviour of rats and "the result in these rats would be identical in a human being."[69]

While their research protocols lacked rigour, this student experiment illustrated the dominance of behaviourist models in scientific research on animal behaviour. By the late 1930s, numerous theoretically complex and tightly controlled scientific studies of observable animal behaviour had appeared in Europe and North America.[70] In 1942 Dyson Carter undertook a review of some of this research in a two-part article in *Saturday Night* magazine. Carter had earned a master's in chemistry at the University of Manitoba in 1933 and secured registration as a "Professional Engineer" in Manitoba in 1936. While he primarily made his living by teaching and consulting on engineering projects, writing was his greatest passion. Carter wrote extensively and enthusiastically on communism's social and cultural benefits as well as on topics that aimed to popularize science and explore its social connections.[71] His *Saturday Night* article was of the latter ilk and was prompted by observations about primate behaviour by Julian Huxley, the zoo director in London, England, as well as by similar observations noted in a report from India. Huxley had reported that a gorilla in the zoo had used its finger to trace the outline of its shadow on a wall, and the report from India detailed similar behaviour by a monkey. The scientific community raised its "collective eyebrows" at these reports, and Carter set out to explore why they had caused such "a stir among scientists." As he observed, "surely the artistic gorilla and monkey weren't displaying exceptional talent. Haven't we all seen animals do far more interesting things than trace a mere outline?"[72] But, as Carter's survey revealed, contemporary scientific thinking overwhelmingly argued that animals could not imitate, reason, or play. Nor did they possess language.

As Carter found, animal research placed a strong emphasis on learning, reflexive behaviours, and sensory matters, all of which could be tested in controlled experiments. Various studies reviewed by Carter concluded that in regard to the five senses (sight, hearing, taste, smell, and touch) animals and humans were about the same, although some animals had greater acuity than humans in one or more senses. Further, these studies reported that the "synthetic sense" in animals, by which was meant the ability "to take observations of our five senses and draw conclusions about size, distance, depth" and such matters, was also highly developed in many animals and, in cases where "reasoning is not required," was equal to that of humans. But because it was concluded that animals could not count – which was interpreted as proof that they lacked the ability to reason abstractly – the "experimentalists" concluded that "everything they [animals] do is a grouping of sensations and actions." And when cognition was looked at in terms of imitation, Carter noted that the "greatest animal authorities conclude that imitation is impossible," even among "the highest apes," because true imitation involved reasoning, and this was different from "impulsive copying," which was often mistaken by nonscientists for imitation. "Frankly, your reporter is sceptical of this," Carter wrote, since he had observed animals imitating human actions. Nor was he convinced that animals could not reason to some degree. As evidence, he cited Ernest Walker, assistant director of the United States National Zoological Park of the Smithsonian Institution in Washington, who had observed reasoning abilities in mice. Only when political implications were made apparent was Carter tempted to see the logic behind the view that animals could not reason. All the scientific studies he had reviewed suggested that a conservative attitude was a basic characteristic of animals – they mistrusted anything new because they could not reason. As Carter gleefully concluded, this was "truly an alluring line of thought for a student of politics! Are conservatives mere victims of primal emotion?"[73]

In the second instalment of his essay, Carter surveyed additional literature that argued that the attribute that distinguished humans from animals was that "*human beings alone can express thoughts in words.*" But Carter had doubts; what about a flock of crows in a tree in the evening, he asked, "if they aren't talking, what *are* they doing?" Indeed, the Smithsonian's Ernest Walker claimed that birds talked with each other and "any animal lover will rush forward with countless examples" of this. "Still we must be cautious," Carter warned, and noted that "after decades of study, scientists have concluded that all the myriad cries, whines, squeaks, chirps, and grunts in the animal kingdom absolutely do not represent even the crudest of languages." All they expressed were "*emotional states*" such as fear or contentment. Play was another trait that only humans supposedly possessed, and Carter noted that scientists soberly studied this behaviour "with gloomy results indeed." Just as no animal could "imitate, reason, overcome inborn conservative attitudes" or talk, "so too none can play," or at least "not the same way that human children play." Everyone knew that young animals frolicked, but this was excluded from the definitions of play used by scientists because they claimed that play was the ability to build or make something with toys and to indulge in make-believe, which they claimed was possible only in creatures that could reason. Given this definition, Carter admitted the point but concluded that science had missed the most important question concerning animal minds: their emotions. Contemporary science studied animals only with "the older methods of psychology: the static action-and-reaction, conditional reflex, memory, imitation, maze tests and so on." These could not address the question of emotions, and Carter wondered what emotions animals had. Were their emotions different from human ones, and might they even be superior?[74]

Carter was not alone in finding that behaviourist research about animal minds did not square with everyday observations and experiences of animal life and left questions unanswered. Others continued to speculate about the ability of animals to reason and about their emotional and social life. While many articles about wild animals in prairie magazines and newspapers in the interwar and postwar years were descriptive, an appreciable number made reference to reason and emotion in animals, even though these issues were by then, on the whole, irrelevant in the academic study of animal behaviour. Kerry Wood, the Red Deer, Alberta naturalist, noted in 1950 that "some naturalists refuse to credit birds with intelligence, claiming they are creatures of instinctive behavior only and that reasoning is unknown among them." He thought it was impossible to generalize so broadly, and while he thought that bluebirds and juncos operated mainly by what he called instinct, he had observed clear evidence of reasoning ability among crows, magpies, and house sparrows. Dewey Soper, in his book on Alberta mammals, more cautiously drew similar conclusions about the shelters that animals constructed, which he thought "seem to indicate differing degrees of dawning animal 'intelligence,' or at least unequal development of instincts, need and behaviour; simplicity as compared with the more complex; guilelessness versus cunning." At the extreme, the *Country Guide* published a report in 1956 (similar to one of the examples that John Burroughs had ridiculed fifty years earlier during the Nature Fakers controversy) that boldly claimed that wild animals could "act as their own doctors and effect their own cures or [perform] surgery," just like humans.[75]

Such animal reasoning, or at least sagacity, was also commonly cited in reports about hunting encounters with animals. While the retreat of animals to safe places was sometimes described as being instinctual, it was often referred to as purposive action intended to meet changing conditions and survival needs. An article titled "The Instinct of Preservation" in *Good Roads* magazine in 1923, for example, described such behaviour in the case of deer at Buffalo National Park near Wainwright in east-central Alberta. In the fall, it was reported, deer gathered in a coulee near the fence line of the park, and "it is a remarkable fact that directly the shooting season starts, these deer prospect along the fence line seeking a hole to crawl through into the sanctuary. They seem to realise the fact that once inside they are safe from the gun of the sportsman." And when commenting in 1933 on declining hunting opportunities because of habitat change and drought, C.A. Hayden of the Alberta and Canadian Fish and Game Association argued that hunting was becoming more difficult because ducks were deliberately changing their habits to avoid hunters.

"The ducks, more and more," he noted, "are taking to feeding in the fields so late in the evening that shooting is difficult if not impossible." In the early 1900s, he noted, ducks flew from the water to the fields to feed in the morning and mid-afternoon. "Now they feed after dark and sleep and drowse in the sloughs and lakes all day" where hunters could not reach them. "The ducks have learned that to survive, they must change their habits and they have learned their lesson so well" that a sportsman was lucky to get "one shot where he used to get ten."[76] Such anecdotes were a common part of hunting lore. A 1956 pamphlet on white-tailed deer in Saskatchewan described them as highly intelligent animals and claimed that "some hunters insist that these deer know when it's Sunday, or when the hunting season opening-date is. It's practically a certainty that the white-tails know the boundaries of the game preserves in their area," and "tales of the phenomenal cunning and elusiveness of the white-tail are legion among hunters." And in 1964 a report in the *Winnipeg Tribune* claimed that "hunters have long believed that crows, and perhaps some game birds, have a system of surveillance and 'communicate' warning of perils to the flock."[77]

Reports giving evidence of the emotional lives of animals continued as well. In an article called "Nature's Love Stories" in the *Western Producer* in 1934, Mrs L.M. East of Leask in central Saskatchewan wrote that "no one doubts that mother-love exists among the birds, squirrels and in fact all wild life," but most people disagreed with her belief that what she called "mate-love" existed to "any great extent in nature." Nonetheless, she was certain that it did and cited instances that she had witnessed of "love" among crows, squirrels, and woodpeckers on the farm. One account was worthy of Ernest Thompson Seton. A female woodpecker would not leave the nest and she died there. Her mate "called all day; he would not leave the place without her, and later we found him dead too." Even crows, then villains of prairie farm life, showed mate love. While Mrs East tried "to be kind to the birds," when a crow killed one of her chicks, she declared war. Admitting that "there is something of the primitive in each of us, I suppose, that demands 'an eye for an eye,'" she set a leghold trap to catch the crow. As she recounted, "at an early hour the next morning there was a great commotion. I had the crow! Was it the same one that 'stole' my chick? I did not know. He was too badly injured to live even if I had relented at this time, but he had a mate and she cried piteously for him." In all of this, Mrs East made assumptions about these crows on the basis of gendered personality traits: she assumed the aggressor that she had killed was a male, while the one who mourned was female. And when the grieving bird was forced to leave the yard because strong winds came up, Mrs East "cried bitterly, but I never saw her again."[78] In a similar vein, a 1948 description in the *Country Guide* of various animal mating rituals described them with terms such as "love," "affectionate and domestic," and "ardent lover." Such emotional attachments were often ascribed to beavers, whose social lives were popularly seen as having a strong parallel to human ones. When beavers were moved in 1948 from southern to northern areas in Manitoba as part of a restocking and conservation program, one case publicized by the Manitoba government involved transplanting beaver "couples" because "beavers are monogamists and become quickly despondent when separated from their spouses." And as one further example, while most stories and articles written by Kerry Wood were descriptive and made no assertions about emotional states, his 1947 story "Qu'Yuk the Beautiful" told of an Inuit boy who discovers that a swan he has tried to shoot is an old bird that has been with its mate for over twenty years. This knowledge forces the boy to realize that, like people, the "devoted pair" have "a heart-feeling for each other."[79]

The belief that animals could communicate as well as play endured in popular discourse in the interwar and post–Second World War years. Claims suggesting that animals communicated often differed little from the arguments of Ernest Thompson Seton and his contemporaries a half century earlier. In 1928 Alexander Krenakerich of Hamton, near Yorkton in southeast Saskatchewan, wrote in the *Western Producer*'s Nature Lovers column that animals had "their own language which is not understood by men." He believed animals were "made by nature" not to speak the human language and vice versa, which did not mean that they were not intelligent. About the same time, V.W. Jackson of the Manitoba Agricultural College wrote that "there is individuality in birds as in human beings" and some birds could reason and communicate with humans. As he explained, while he was reading one morning, a little wren landed on his thumb and "sputtered long and pinched hard before I recalled that I had plugged the door of the wren house to keep the sparrows out during the winter.

And the following spring the wren got back before the plug was removed, and sputtered before the lady of the house until things were made right." As he added, "every bird observer knows similar cases of bird bravery or feathered intelligence."[80] Such arguments in connection with play also persisted. A 1951 correspondent for the *Farm and Ranch Review* reported observing hawks playing, which showed that "birds and animals have their bit of fun." And the next year the Alberta writer John Patrick Gillese wrote about watching a coyote chasing a rabbit, not to kill it, but "just enjoying the sport." Later that evening the same coyote was seen playing with the horses, and Gillese concluded that "wildlife has as many personalities as the human world, and undoubtedly this coyote enjoys playing as much as another one will enjoy raiding one particular barnyard year after year."[81] And in 1940 Ralph Hedlin, who would later become a journalist and prairie nature writer, reported in the *Western Producer*'s Nature Lovers column that recent studies in the United States described crows playing games and displaying "almost human intelligence" while doing so. This, he thought, was sufficient evidence to show that it was better to study and appreciate crows than to shoot them.[82]

These examples reveal the persistence in popular culture on the prairies of a view that wild animals were able to reason, at least to some extent, and to exercise a degree of control over their lives. The writings of individuals such as Ernest Thompson Seton and the arguments of a range of writers and observers since the end of the nineteenth century both reflected and confirmed this attitude. To be sure, while most Euro-Canadians did not accept, or likely even hear about, the presumptions about animal agency and its intersection with humans that were familiar to Aboriginal people, many credited animals with abilities that academic discourse denied or, at the least, considered irrelevant. Behaviourist research about animals – William Rowan's work on bird migration, for example – was not concerned with such considerations. But while such behaviourist foci remained dominant, their parameters were being questioned by the 1960s by the emerging view that behaviourism "had gone too far, that the methods were too stringent, and that the drive for repeatable and controlled experiments could not be used to uncover all there is to know about the function of animal minds." It was increasingly accepted that researchers should study animals in their natural environment (just as Seton had done fifty years before), although with better protocols to document and transmit the findings. More fundamentally, the moral philosopher Mary Midgley contends that by the 1970s "the difficulty of discussing observation without an observer – the absurdity of enquirers trying to leave themselves out of their own enquiries – increasingly bothered scientists."[83] Yet the disconnect between the assumptions that underlay academic science and those often expressed in popular culture should not be overdrawn. Most people accepted the authority of scientific explanations and ways of knowing in almost all aspects of their lives. Moreover, people's views about animal behaviour continued to be profoundly shaped by the presumption that humans' capacity for speech and their ability to alter their environments and manipulate their food supply legitimized their dominating power. In some cases, such views have been sustained by the creation of ever more finely tuned definitions of what qualifies as an exclusively human behavioural trait.[84] This is not a new phenomenon. In his survey of scientific research, Dyson Carter found that animals were said not to be capable of play because researchers defined play in a way that could not be tested with animals. But conclusions buttressed with ever more esoteric definitions seem increasingly irrelevant in the early twenty-first century when there is growing uncertainty about the boundaries between species. The views of previous generations – that attributes such as the ability to reason, play, and make tools are the exclusive traits of humans that distinguish them from other species – have been eroded or simply found to be false. New research, disciplines, and methodologies that do not shy away from investigating animal character in terms of emotions (and even of social life and organization) are popularizing the view that animals possess emotions such as grief, joy, and love, that they have memory and can learn in complicated ways, and that they use tools, practise teamwork, and cooperate to meet goals and carry out tasks. Still other research has posited social hierarchies among animals as well as filial and family continuities, loyalties, and support that reveal complex patterns of behaviour. While these characteristics do not make animals human, this is a distinction that increasingly seems less important than it once was.[85]

3

Friends and Foes

To fall asleep at night to the love making song of the Screech Owl, to be awakened in the morning by the pleasant gurgle of Purple Martins and pass the day to the songs of wrens and voices of Tree Swallows makes a modern commonplace garden approach in some degree at least that from which we are supposed to have fallen.

Percy Taverner, 1919[1]

The relationship of people with wild animals was often cast as one of friendship or enmity. Certain wild animals were defined by tradition or experience as foes that brought economic loss, spread diseases such as rabies, or otherwise threatened human safety. Friends were most often defined as those animals that enhanced economic activity. These categories were not rigid, and enmity could turn to friendship and friendship to discord – an instability that had its roots in changing values, needs, and perceptions. Such views emphasized people's material needs in their relationship with the natural world, brought certitude about the place of humans in the world, and quantified the distance between people and nature by assessing the value of animal life in terms of human needs. While these utilitarian distinctions between people and animals remained strong, by the 1960s a growing popular sense that all wild animals had value as part of ecosystems undermined this older certitude that their value could be judged so simply. However, this shift did not emerge fully formed but drew upon long-standing views that human relationships with wild animals offered more than mere economic utility. While such economic utility remained a powerful measure, by the late nineteenth century popular literature, commentary, and reports revealed a growing belief that wild animals could be looked upon as offering moral lessons and standards for behaviour. Interacting with wild animals, whether watching them, hunting them, or keeping them as pets, could enrich and improve social life and human character. So, too, the sound of birds, the colour of plumage or fur, the stature of elk and deer, and the form of birds in flight added to the beauty of the world. Social and moral betterment would be enhanced with an appreciation of the intelligence of wild animals that revealed the complexity, interconnections, and fullness of the world. Even so, these attributes provided a tenuous foundation for friendship and conferred no intrinsic rights or value to wild animals. Nor were these attributes of such unquestioned value that they could override material needs. Indeed, perhaps because they succumbed less easily to human domination, wild animals that were said to be the most intelligent, such as wolves, coyotes, crows, and magpies, were often seen as the most implacable foes. And beauty was a cultural artifact that offered only a contingent basis for friendship. Their iridescent feathers and stalwart bodies were not enough to prevent crows from being transformed in some cases from intelligent friends to cunning foes because of changing priorities. Nor was the fine proportion and colouration of magpies sufficient to make them friends. Nor could beauty temper the apparent violence and amorality of nature. As the Manitoba ornithologist George Atkinson admitted, the blue jay was "a gay dashing fellow who carries with him a charm we cannot fail to admire," but even his beauty could not compensate for his cold brutality when he devoured fledglings before their mother "without a tremor save of delight at her suffering and satisfaction with his meal."[2]

Historian Harriet Ritvo argues that wild animals provided fewer social or personal lessons in Victorian England than did domestic animals. While some of them could be eaten or had economic value or provided good sport hunting, wild animals overall were not "even important enough to merit a moral judgement unless they somehow reflected human experience."[3] While prairie people by the early twentieth century showed the same tendency, they nonetheless seemed more willing to find a broader range of value in wild animals. Perhaps because most people lived on farms and in small towns, some wild animals were a larger part of their lives and played a different role in everyday life. Although such increased contact did not entail a corresponding notion that wild animals had intrinsic rights, it did bring to prairie Canada a more intense dynamic to human–wildlife relationships.

THE MORAL AND SOCIAL BENEFITS OF WILD ANIMALS

By the end of the First World War it had become common to see wildlife as contributing to national character. Gordon Hewitt typified these views in his 1921 book *The Conservation of the Wild Life of Canada*, in which he argued that hunting and outdoor activities such as camping and nature appreciation enhanced national life by building character and self-reliance. Nothing, he wrote, "calls for resourcefulness so much as the quest of wild life, when the beaten tracks of a more civilized life, where everything is provided for one, are left and one has to return to the primal competitive habits." Experiencing nature returned people to the wellsprings of national greatness, for "resourcefulness is a characteristic of all those called upon to conquer new lands," as had been shown by the bravery of Canadian troops during the First World War. And in 1920 Saskatchewan's game commissioner Fred Bradshaw quoted Ruskin's claim that "the beauty of nature is the blessedest and most necessary of lessons for men; and that all other efforts of education are futile until you have taught your people to love fields, flowers and birds."[4] Indeed, Manitoba's game and fisheries branch contended in 1927 that if people lost a "love for birds and animals," they would be deprived of "one of the finest influences in life working toward a finer type of citizenship." And wild animals could also help mitigate the impact of catastrophic events. In the midst of the Depression, the director of the Manitoba game and fisheries branch noted that while an increasing number of unemployed people were forced to hunt for food or income, others hunted to relieve stress. It was impossible, he noted, "to appraise the worth of a week's hunt to those overburdened with worldly care, or to the workless," and "in these distressing times, the wilds beckon to many, and offer a hospitality which is not forthcoming from other sources." Later, a young Dick Beddoes, who would go on to become a journalist and writer in Toronto, found respite from the stress he felt during the Second World War in his personal

Figure 3.1 Robins were welcomed as companionable birds that adapted easily to the new farm landscape of the prairies, and their return in the spring was seen as a harbinger of summer and the end of the long prairie winter. This photograph is by Saskatchewan photographer Hans Dommasch (born 1926), c. 1960.

fascination with the great horned owl. In his poem about an owl's soaring flight, Beddoes wrote that it "Lures me to follow him far / Deep in the great lone wilderness / Far from the cares of war."[5]

Wild animals could also serve as symbols of regional identity and appreciation of one's immediate environment. Homesickness and a feeling of being out of place led some early settlers to denigrate their new home. In 1892 Ernest Thompson Seton recorded a conversation with an English friend who repeated the apparently widespread claim that in Manitoba, unlike England, birds were "very rare and never sing." But Seton saw it differently, and as he described it, at the very moment that he and his friend were chatting, half a dozen meadowlarks "were pouring out their fine strains, 20 shorelarks [horned larks] were within hearing, and the chinks of the music were filled with the notes of innumerable Vesper Sparrows, Longspurs, Savannah Sparrows," while "crows and Kingbirds and ducks were all within hearing."[6] Others also looked attentively at what surrounded them. Mildred Galloway of Melfort, a town southeast of Prince Albert, Saskatchewan, for example, contended that robins symbolized pioneer settlement and the future; writing to the Nature Lovers column in the *Western Producer* in 1928, Galloway endorsed a description from the National Association of Audubon Societies of the robin as "distinctly a companion of man, and wherever his hand cleared the wilderness the robin had followed." Another correspondent wrote in 1933 that birds helped confirm her western identity. Prairie birds, she wrote, "are never out of tune, their songs never grow unpopular. These singers are prettier and cheerier than the prettiest and cleverest Hollywood soloist ... The scenery that Nature provides for us is lovelier than any painted for show. We have our own show at home."[7] And wildlife could even symbolize the struggle for prairie co-operative grain marketing. In 1927 and 1928 the *Western Producer*, the newspaper of the prairie co-operative movement, ran a series of advertisements that used wild animals to demonstrate the value of joining the Wheat Pool. One 1928 ad described wild ducks as "the victims of powers beyond their control. They have a lot of sense but it is no use against superior force. If they could get together and take the guns away from the hunters, they would undoubtedly do it." But while ducks could only flee danger, farmers could co-operate and "join the Wheat Pool."[8]

For others, wildlife helped confirm spiritual feelings and beliefs. In recollections of his work in the northwest in the late 1800s, Methodist missionary James Woodsworth celebrated the region's potential for farming and mining but showed remarkably little interest in its fauna and flora. In one of his few references to wild animals, he described being trapped by a storm on an island on Lake Winnipeg with his companions in 1888. When the party finally took the risk and left their "island refuge," looking back a mile out, they saw "a huge flock of pelicans" land on the place they had just left. Woodsworth wrote that "there must have been many hundreds of birds, and when massed on the shore they appeared like a large bank of snow. While driving before a strong wind and bounding over the swells caused by the recent storm, George [one of the Aboriginal guides] was heard to sing very softly, 'God Moves in a Mysterious Way.'" Equally moved, everyone joined in. While Woodsworth's Methodism would have led him to resist the seductions of Catholic and High Church symbolism, this appeal to the ancient and well-known symbol of the pelican as representing Christ's Passion and Resurrection suggested one way that the region's wild animals could relate to spiritual life. More directly, Hugh Duddrige, a resident of Earl Grey, a town north of Regina, wrote a poem in 1928 about the way that a fleeting relationship with a chickadee helped resolve a personal dilemma. A man experiencing a crisis of faith, fear of death, loneliness, and perhaps unrequited love wandered in the woods in mid-winter, wondering if "Life and Light and Love" were but "blank illusion." A clear answer appeared when he persuaded a chickadee to sit on his arm and eat the crumbs of bread he offered. Its vitality, its "thankful chirp," and its hopefulness brought immediately to his mind the example of Jesus and "the lilies of the field" and the "ties of kinship linking all creation" that nature offered to those who searched for it.[9]

Figure 3.2 The encounter with wild animals took on different meanings over time and within particular circumstances. Pelicans, such as these photographed by Hans Dommasch in the 1960s, were variously described as symbols of Christ's Passion, as threats to commercial fishing, and as natural curiosities.

Figure 3.3 For novelist Gabrielle Roy, the recurring uniform pattern of bird nests, such as this duck nest photographed by avian biologist William Rowan, was a comforting confirmation of the steadfastness of the world. Photo not dated.

Wild animals also provided assurance of the regularity and constancy of life. When recalling the houses of French-speaking immigrants in Manitoba in her memoirs, novelist Gabrielle Roy wrote that "those people were amazing, I must say; they'd leave everything behind and begin all over again to make everything just the way it had been at the other end of the earth. I've always been touched by this. It reminds me of the birds, who always build the same nest wherever they go in the vast expanses open to them." Saskatchewan poet J. Arch. McLeod found the same constancy when observing a sparrow's nest and thought that such unchanging "patterns fixed for ages" were evidence that the universe's "master plan is uniformity." But for him, the nest and its unfailing replication of a particular form marked human superiority because they demonstrated that humans could transcend the constraints of tradition and the master plan by inventing new ways of doing things and deciphering the scientific laws of the universe.[10] The constancy that wild animals modelled was, however, most commonly seen in the associations that people had with the migration of birds. Bird enthusiasts faithfully recorded the departure and arrival dates of migratory birds and others wrote poems celebrating their appearance at the end of the long prairie winter. As Saskatchewan poet Vina Chilton wrote in her 1952 poem "The Robins Came," "Instinctive is the flight, or can birds know, / That we who live upon the plains have need / Of singing and of beauty after snow?" While the fall migration with its warning of the coming of winter was more problematic, for the poet Eric Dowson, who in the 1930s farmed near Tisdale southeast of Prince Albert, Saskatchewan, the sight of geese "fleeing from the storm and cold," braving hunters and other perils in their southern journey, made him "Glad that I have no need to roam / Glad for the friendly warmth of home."[11]

The lives of animals were also commonly said to provide models for gender roles, especially for girls and women. Ernest Thompson Seton often employed this device in *Wild Animals I Have Known*; Molly Cottontail, for example, fulfilled her maternal duty by dying to save her offspring, Raggylug. And on the rare occasion when he portrayed male familial devotion (such as when Redruff the partridge took on the care of his siblings after their mother was killed), Seton carefully noted, perhaps mindful of his own father, that such behaviour was exceptional. In 1903 the ornithologist George Atkinson wrote that while evening grosbeaks showed "a selfishness and general discontent with life that is striking and unique among Canadian birds," the worst were cowbirds that laid their eggs in the nests of other birds to avoid maternal responsibility. Atkinson asserted that "the shame of the cowbird at shirking her duties" was a clear lesson that applied to people, one that was repeated by Fred Bradshaw in a 1930 talk to the Farm Girls Convention in Saskatoon, where he warned his young female audience that, "from sheer laziness," the cowbird "sneaks around and slyly deposits eggs in the nests of other birds." And one columnist in *Saturday Night* in 1926 described the maternal defence of the young as a sign of intelligence among birds. "I have never known a really intelligent bird to leave its young to their fate," the writer argued, explaining the lack of maternal affection among some birds as an evolutionary lag, for they "have not advanced through the ages as fast as those which show some bravery and love." For humans, the most intelligent species, the lesson was clear.[12] More broadly, people could also learn about family life from animals. One correspondent in the *Western Producer* in 1926, for example, described how a "family" of grouse consisting of "the father, mother and six children" lived in the bluff near her farm. The female taught the young and lived according to "Nature's Laws," setting "a valuable example of harmony and well being – if only we would heed it."[13] The same doggerel appeared in a poem by Saskatchewan poet Edna Jacques, simply called "Birds": "All summer long they nest in little pairs, / In trees and hedges, corners of the eaves / The little father bringing string and leaves, / For the small mother and her bridal couch." And the biologist Norman Criddle, not usually given to sentimentality, observed that "one could write pages about wrens, their cheerfulness, the house-cleaning of the male who prepared [the nest] for his mate's coming, and a host of other habits, all of which must endear these little birds to any observant person." Such analogies were consistently made over relatively long periods of time, but their emphasis sometimes changed with changing social and political events. Echoing Second World War rhetoric that was being transformed into Cold War propaganda, it was noted in 1953 that "nature" had generously touched many birds with "the magic wand of courage." In showing fierce defence of their nests and young, "from the majestic Canadian goose to the diminutive Ruddy, shines bright the will to resist aggression."[14]

The technique of using wild animals as models for gender roles and family life was employed in the bedtime stories of Blanche Silver, the pen name of Prudence Gruelle, an American children's writer whose stories – populated by the mammals, birds, and insects encountered by three children – appeared as a syndicated feature in the *Western Producer* in 1928–29. The children often talked with the animals, and parallels to human lives were confirmed by the stories' illustrations, which sometimes showed the animals wearing clothing and hats. Although the animals in Silver's stories spoke, showed familiar human emotions, and sometimes demonstrated the ability to reason, they did not have developed personalities as in the Canadian nature writing tradition of Seton and Roberts. Instead, they served as exemplars for virtues such as self-sacrifice, restraint, kindness to animals, proper marital behaviour, sitting down to regular meals, and tolerance of other opinions. In "How Miss Blue Bird Found Blue Feathers," the origin of bluebirds was credited to the transformation of a small drab bird into a beautiful blue bird because of her act of unselfish kindness. And as was discovered in "Doris Meets Mama and Daddy Red Winged Blackbird," birds avoided social conflict by co-operating and respecting one another's property, although they found psychological release from the constraints of civilization by grumbling about each other from time to time. As well, the gender roles of females as mothers and males as protectors and providers were commonly outlined. In "David Has a Chat with Bunny Rabbit's Cousin," Mama Marsh Rabbit took care of the home and the baby rabbits while Daddy Marsh Rabbit protected the family from harm. Motherhood was even important to frogs, as readers learned in a story about female frogs that looked after their "tadpole babies."[15]

The social and individual characteristics that Silver attributed to wild animals offered models of proper behaviour for children, and it was popularly claimed that training children to show kindness to wild animals prepared them for the rigours of adulthood. This was part of a broader animal welfare movement in which various groups after the mid-nineteenth century had attempted to prevent cruelty to domestic animals. Criminal Code provisions were enacted in 1869 to make cruelty to such animals a criminal offence, and the law was amended in 1895 to add penalties for cruelty to wild animals kept in captivity. This, however, was primarily aimed at preventing the baiting of bears, badgers, or other animals as public entertainment. Animals in the wild were given no protection and earned scant regard from animal welfare groups. As it was noted in 1931, it could still be said that while humane treatment of domestic animals and pets was now an ingrained measure of human decency, wild animals did not share in this protective impulse. While such lack of protection benefited various economic and sporting groups, some people claimed that teaching children to be kind to wild animals would improve character and teach consideration that would produce better adults. As the Montreal Society for the Prevention of Cruelty to Animals had argued in the 1860s, boys should be educated not to kill or harm songbirds because it "would make them less inclined to violence as they grew older."[16] This concern with social and gender standards was also suggested by F.J. Fitzpatrick, a provincial game guardian in Saskatchewan, who argued in 1914 that public schools should start teaching "boys that it is cruel and unmanly for them to catch the young songbirds and rob their nests." The persistence of this thinking was further demonstrated by a writer in the *Farm and Ranch Review* in 1922 who urged people to teach their children about the plants and wildlife around them and to discourage "every form of cruelty and unnecessary destruction." If children were not directed towards proper attitudes, their high energy and constant activity would "tend towards destructiveness. A .22 in the hands of the average country boy means death or injury to every wild thing within range. Wanton destruction!" While a boy needed to learn that some wild creatures were useful for farming, the true goal of kindness was character development and the teaching of social propriety. As the writer argued, education would make the boy "a naturalist instead of a hunter and trapper. The study of the ways of furred and feathered creatures would keep many a farm boy content with his lot, and would imbue him with the consideration for helpless things that distinguishes the person of refinement and good breeding from the clod."[17]

The transformative power of rejecting violence was seen in a story by Saskatchewan writer Edward McCourt. In his 1957 story "Walk through the Valley," a boy by chance meets a famous wildlife hunter. Although the boy yearns to have a gun, he is thought to be too young and therefore can only pretend to hunt. Nonetheless, and despite the conventions of male power, the boy does not hesitate to sabotage the famous hunter's attempt to kill a stag. While he immediately feels remorse for betraying "the hunter in the moment of ecstasy," the stag and its "beautiful immobile body, the arrogant upflung antlered head," has "stirred suddenly in the boy a consciousness of power." But it is the power of a transcendent universe, not the power to kill. With strong parallels to Seton's 1899 *Tale of the Sandhill Stag*, the boy later saw the buck standing on a high ridge, "immense and lonely … a creature not of this earth … but a mysterious visitant from some region of gods and heroes of an age long past. And as the boy stared in rapture the buck slowly turned his magnificent antlered head and looked at him across the gulf of earth and time." The boy was brought to tears by the encounter, but when he looked again, the buck was gone. Only the valley and the hilltop and the blue sky remained, "but they were not the same. He knew dimly they would never be the same again. For he had seen them in the light of something great and strange, a vision glimpsed for a moment only and remembered always." At this instant, he realized that his triumph lay not only in aiding the escape of the buck but in his own coming of age.[18]

In McCourt's story, the transformation to adulthood was realized through independent thinking and respect for life. In prairie culture, however, this transformation to adulthood was most often achieved through the loss of a wild animal's life. Embedded in this view were the assumptions that part of an adult's task was to conquer nature and that adolescents gained self-respect by proving a capacity to shoulder this obligation, thus affirming their admission to adulthood and adult responsibility. Byrde McCarthy's story, which was featured in the *Farm and Ranch Review* in 1927, was one example of such a transformation. McCarthy was a sickly child who was not expected to live long. When the family moved to a homestead in the Alberta foothills, Byrde was not allowed to play sports, hike, fish, or hunt because "girls, particularly weak girls, must not do these things." But she "longed to go out into that big outdoors world, believing that there I would find health." And so, "after much coaxing," her parents let her go with her brothers into the foothills. Despite her brothers' scepticism, Byrde learned to shoot and became so proficient that she was permitted to go further into the mountains. Everyone slept outdoors on these trips, and Byrde cooked for the gang, "but that was a simple and pleasant duty." Her health improved rapidly and brought "a surer outlook and [a] grip on myself," and she soon felt so confident that she determined to fulfill an old dream of killing a mountain sheep. Setting off with one of her brothers, she encountered two rams after a five-day trek. Each hunter shot one. The head and horns of Byrde's trophy were of enviable size, but most importantly, the killing demonstrated her newly invigorated confidence and return to health. The death of the ram marked the conquest of her infirmities and history, but it was tempered within the orbit of family and by the control of the senior male siblings and her function as caregiver for the men in the field.[19]

While Byrde McCarthy's hunting experience was an example of the transformative power that some people associated with the killing of wild animals, this was almost exclusively and consistently a male preserve. In his 1901 novel, *The Man from Glengarry*, Manitoba writer Ralph Connor describes how Ranald, one of the novel's central characters, kills wolves that are attacking travellers in the Ottawa Valley. Ranald's defeat of the wolves marks his coming of age and also reinforces the view that life is a primal contest for control and a "life long conflict with the forest and its fierce beasts" that is only resolved when the animals are killed and their habitat is replaced with farms and fields. This theme of transition into manhood through the killing of wild animals was a tenacious one and continued to be the focus of stories into the 1960s. For example, in 1961 the Alberta writer, raconteur, and guide Andy Russell wrote a short story about a boy who shoots a grizzly bear that had killed a steer on the ranch. The boy is conflicted by his victory. While he is awed by the beauty and power of his victim and finds it unbelievable that he has killed such a powerful animal, he knows that "it couldn't be helped – a man had a job to do, he better stick to it till it was done." As a consequence, he returns home with a bold new sense of identity and confidence. This theme was again the subject of a 1962 short story by Eugenie Louise Myles, an Alberta writer and journalist. In this story, a boy kills an old coyote by beating it to death with a stick. This brutal act is part of his coming of age, and when he has gone to bed, he overhears his father confirming the kill. Myles sympathetically writes that "a ghost of a smile played on his face. He stretched out his body full length and pushed his muscles back against the lumpy tick. He slept."[20]

The lessons of such killing were not, however, always so unequivocal, and they could sometimes reflect a more complex meshing of time, place, and maturity. Ralph Hedlin, who grew up in Saskatchewan but whose later career as a journalist and writer took him elsewhere, wrote an autobiographical story in 1956 about a coyote. While "most coyotes are pests and chicken stealers, to be shot or snared or trapped or poisoned, born unwanted and dying unknown," one coyote, which Hedlin named Aya-oo, was different. Hedlin pursued Aya-oo over many years, and although he could have killed him on several occasions, he could not bring himself to do so. Finally, however, he shot Aya-oo, as much by accident as by design.

By then Aya-oo was old and thin and the "slow attrition of the years had worn him down," and Hedlin doubted that he would have lasted the winter. While Hedlin had no concern in principle with hunting, this particular killing was tinged with the melancholy of memory and passing time:

> When I stood over his dead form, I knew why I had not killed him on my two earlier chances. As I looked across the hills, the range I had shared with Aya-oo, I knew that something had gone from these old hunting slopes and that they'd never be quite the same again. When Aya-oo died, the book closed on my boyhood. Death had claimed the coyote, duties far from these hills were claiming me. That is why, instead of pelting Aya-oo, I took a pickaxe and dug a grave in the tough, frozen soil on the highest hill in the area and buried him. I rarely see those prairies anymore, but Aya-oo still looks down over the great sweep of country where we fought our twelve-year battle a long time ago. I like to think of him there, for I left a part of myself there too.[21]

FRIENDS AND FOES – WHO'S WHO

For hard-nosed prairie folk, wildlife that offered beauty, mental comfort, or models of how to live were best when they also offered economic or other material value. Of all wild animal life, birds were the ones most often seen as friends. Game birds were of obvious benefit for hunters, and insectivorous birds and many birds of prey were welcomed and admired as friends of the farmer. While the beauty of many birds was of value in itself, public appreciation of their beauty could also assist in their conservation. As Saskatchewan's game commissioner Fred Bradshaw recommended in 1913, schools should encourage study of "the beauty of the brilliant colours and delicate markings of the birds, the joy that comes from their songs, their wonderful ways during breeding and migration periods." When their importance for destroying weeds, insects, and rodents was also understood, "there will be no fear of these beautiful creatures being slaughtered by thoughtless men and ignorant boys." And as one young observer in Saskatchewan phrased it in 1933, "birds are such gloom-chasers with their different notes and then – to get down to the practical view – they do kill a lot of worms."[22]

From an early date, governments strove to encourage public understanding of the economic value of birds. In 1901 the NWT Department of Agriculture urged farmers to stop worrying about birds eating some of their crops because the loss was minor in comparison to the grain birds saved by eating insects and rodents. Drawing on the findings of American scientific agencies and scientists, it noted that "the food of the cow bird is said to consist of 20 per cent noxious insects, 16 percent grain, about half of which is waste and more than 50 percent consists of weed seeds." Robins and meadowlarks ("often shot for the table") were equally valuable friends, while crows, which "in this country where little corn is

Figure 3.4 It was assumed that cormorants were not worthy of protection because they threatened human use of fish. These birds on a rocky island on Lake Manitoba were photographed in 1888.

grown," also helped the farmer, as did birds that ate rodents. It was, however, also noted that a few hawks and owls caused trouble for farmers, and birds that ate fish meant losses to commercial and sport fishing.[23] In this scenario, friendship was based solely on economic needs: unworthy birds harmed poultry, field crops, and commercial and sport fish, while worthy ones ate insects, gophers, mice, and other small mammals.

RANKING OF VALUE OF SELECTED BIRDS, 1901

Not Worthy of Protection	*Worthy of Protection*
Eagle (all species)	Owl*
Goshawk ("Chicken Hawk")	Hawk (most species)*
Pigeon hawk [Merlin]	Meadowlark†
Cooper's hawk	Cowbird†
Hawk owl	Robin†
Blackbird	Tree sparrow†
English [House] sparrow	Crow†
Loon‡	Snowbird†
Cormorant‡	
Pelican‡	
Merganser‡	

*Ate gophers and mice
†Insectivorous
‡Piscivorous
Source: Compiled from *Annual Report* NWT *Department of Agriculture for 1901*, 65–6.

Insects were of particular concern to farmers and motivated them to view certain birds as friends, especially before the introduction of new poisons in the 1940s as part of the postwar chemical revolution. Prairie farmers increasingly felt threatened by insects, and it was said in Saskatchewan in 1921 that "for many years we knew nothing of insect pests" but now they were "increasing at an alarming rate, and 'old timer' farmers are asking 'what next' as pest after pest makes its appearance."

While most of these insects were native, they had earlier existed "in numbers so small" and caused "so little damage as not to attract attention." Insectivorous birds were useful because their high body temperature and rapid pulse made them "hearty eaters" that consumed hundreds of insects each hour.[24] In 1922 the dominion entomologist estimated that the annual loss of Canadian agricultural production to insects was $200 million and that this loss would have been vastly greater without birds. Similarly, Professor V.W. Jackson of the Brandon Agricultural College argued that "we have little other hope of keeping the balance of nature, than birds. Our greatest pests are insects. They multiply in millions and escape by wing," and only birds could pursue them.[25] Others saw birds as simply another form of field crop husbandry. Charles Nash, in a study on insectivorous birds published by the Ontario government in 1898, remarked that while "the laws of nature" kept "the balance about evenly adjusted" between "the insect world and the vegetable kingdom," this was inadequate to meet human needs, for "man requires that it should be inclined in favor of the plants he cultivates for his own use." People therefore needed to "carefully protect and encourage all the forces that will work on our side against our insect enemies," and birds were especially "important allies" in this struggle. So, too, the ornithologist Percy Taverner contended that while birds did not control insect plagues, they collectively kept insects in check because while each species of bird filled "its own narrow field," all together they covered "every weak point." Indeed, the Dominion Experimental Farm at Scott in west-central Saskatchewan noted in 1935 that birds were so valuable that people should not be waylaid into only "thinking of them in terms of beauty and song" but should keep in the forefront their more important benefit of consuming insects and weed seeds.[26]

Insectivorous and other useful birds were protected by law in Manitoba when the ponderously titled *An Act for the Protection of Insectivorous and Other Birds Useful to Agriculture* was passed in 1890. All birds were protected except those that were named in the legislation. These named birds, which included pelicans, blackbirds, and eagles, could be killed anywhere and at any time. Similar legislation was adopted in the NWT and by Saskatchewan and Alberta after 1905. These initiatives were confirmed in 1917 by federal legislation passed to protect migratory birds. But as bird enthusiast R. Owen Merriman of Hamilton, Ontario, argued in his 1923 federal government pamphlet on the feeding of birds, bird protection legislation was important but not sufficient: "We must use artificial means of attracting birds to our farms, gardens and parks; and give them active help as well as protection to compensate for the many ways in which we have turned the balance of nature against them." Thus, people were advised in government publications and press releases to protect and encourage birds, and talks and articles by bird enthusiasts furthered the movement. People were urged to control cats, provide nesting materials, and plant trees, shrubs, and perennials to offer protection and food. Birdhouses were a particular concern, and a 1919 federal government pamphlet written by Percy Taverner gave detailed plans for nine types of suitable houses for different types of birds. Using Arts and Crafts principles of honest design (Taverner was trained as an architect), his designs promoted the notion that a birdhouse should look like "what it is – a house for birds not a toy human habitation."[27]

As demonstrated by articles and letters in the Nature Lovers column in the *Western Producer*, these recommendations for encouraging birds found a ready audience in the 1920s. Readers urged one another to obtain the Taverner and Merriman pamphlets to learn more about the value of birds and the ways to encourage them. Echoing arguments about the value of birds, one writer noted in 1928 that humans had invented so many "destructive agencies" that they now had dominance "over the most fierce and powerful mammals and the most deadly reptiles" but remained relatively powerless against insects. Salvation came only from their "friends the birds," which had been provided "by the All-wise Provider to keep down the plague of insects and so preserve the 'balance of nature.'" Other correspondents cited statistics from government publications about the value of birds. Using evidence from scientific studies of bird diets, Mary Mracek wrote from west-central Saskatchewan in 1928 that a single meadowlark ate 100 cutworms, 100 grasshoppers, 200 ground beetles, and 50 caterpillars in one day, proving that meadowlarks were "an active and vigilant ally" of the farmer.[28] Other correspondents warned about the need to control cats to protect small birds, and John Kyba of Norquay, near Yorkton in southeast Saskatchewan, reported that he successfully built the nine birdhouses tailored for different birds outlined in Taverner's pamphlet.[29]

Controlling insects and weeds was only one part of the service that birds provided. Predatory and other birds were important too, but their value was less clear-cut. Commercial and sport fishermen saw pelicans, cormorants, mergansers, and other piscivorous birds as a threat, while sport hunters and many farmers saw hawks and owls as a menace to game birds and farm poultry. Most informed observers, however, noted that many farmers did not recognize their true friends. Given the problems that gophers caused, the 1901 Territorial Department of Agriculture annual report deplored that farmers ignorantly and wantonly killed their "much abused and hard-working friends, the hawk and the owl," which ate enormous numbers of gophers and mice. As George Atkinson reported in his 1899 study *Manitoba Birds of Prey*, a pair of Swainson hawks had been observed killing 210 gophers in three weeks. Other hawks, with the exception of three or four species, were equally important, as were owls. Bird enthusiast H.M. Speechley, too, thought it was "sheer craziness" for farmers to shoot raptors merely to save an occasional chicken, which he thought should gladly be sacrificed to friends who gave so much more in return. True, some hawks were a problem, but as Speechley saw it, "here's where knowledge comes in – you realize that among hawks only three out of the dozen or so of Manitoban hawks are responsible for the bad name given by some to the whole tribe." Thus, it was legitimate to kill Cooper's and sharp-shinned hawks and goshawks, but other hawks were valued friends. And among owls, Speechley noted that only the great horned owl posed a real problem for farmers. Thus, "bird knowledge" was of "great practical value to every farmer."[30]

Confusion also surrounded the question of whether crows were friends or foes. This was the source of ongoing debate among hunting, farming, and natural history communities. At one level, the debate was about the priority of sport hunting over other needs, but it also more broadly concerned management of the changing prairie environment because of farm settlement, climatic forces, and the depletion of waterfowl and ground-nesting birds. Crows were said to be friends by some in the farm community because they ate carrion, mice, cutworms, grasshoppers and other

Figure 3.5 Judging which birds were friends and which ones were foes required the ability to identify them. Thus the Saskatchewan government's Better Farming Train in 1920 featured lectures for farm children on bird identification.

insects. But crows also ate the eggs and fledglings of waterfowl as well as those of game and other birds, which made them foes according to sport hunters. While some sport hunters were farmers, most lived in cities and towns. The naturalist Kerry Wood summed up the matter when he remarked that "crows do devour duck and grouse eggs. Sportsmen would like those eggs to hatch into game birds the men themselves could hunt and kill. So sportsmen continually wage war on crows."[31]

The debate about crows occurred across the region, and Saskatchewan's case was typical. In 1901 the territorial government listed crows as the farmer's friend. While crows were then relatively uncommon, by 1915 their numbers were increasing rapidly and they were appearing in districts where they had rarely been seen before. Their status as friends was hotly debated as hunters and their organizations insisted that crows be controlled to protect game birds. Neil Gilmour, the district game guardian in southwest Saskatchewan, observed that the "farmer sportsman" was "particularly outspoken in his hostility," demanding crow bounties to encourage people to kill them. But Gilmour concluded that "if this cheeky black fellow is being maligned," evidence was needed that crows did more harm than good. Never shy to defend birds or question the practices of sport hunters, the provincial game commissioner Fred Bradshaw joined the debate by observing that "many sportsmen blame the crow for everything that is bad in the way of destruction of beneficial birds." But the crow's "virtues are unrecognized by the men who observe its vices," and quoting American scientific studies, he contended that there was "considerable evidence in its favour," especially its consumption of insects. Admittedly, crows were nest robbers, but "we deny the crows the right to kill a few birds for food although we ourselves take the lives of the same birds for sport." In any case, Bradshaw noted that crows enjoyed no legal protection and hunters could kill them as they wanted, and in what seems to have been a miscalculation of public opinion, he sent a questionnaire about crows to 300 game guardians and other observers in 1924. Two hundred of the questionnaires were returned, a high response rate that showed the level of interest in the topic. While Bradshaw claimed that "the crow is not without his friends and defenders among our Saskatchewan people," it was clear that the great majority of respondents were hostile: 83 per cent thought crows were harmful because they ate the eggs and young of game and songbirds, and 75 per cent favoured incentives to kill them. Even so, Bradshaw retorted that "it is an unchallengeable hypothesis that 'Nature makes no mistakes'" and that crows should be assumed to have a "place in the economy of nature." Noting that people had already upset the balance of nature, Bradshaw argued that incentives to kill crows would only upset it further. Something more than predator control was needed. Unrestrained land clearing, for example, had robbed game and insectivorous birds of shelter where they could reproduce in safety. However, Bradshaw reluctantly concluded that changed conditions perhaps justified efforts to reduce crow populations.[32]

The debate about crows was at times acrimonious and tinged with emotion on both sides. It also seems to have been deeply influenced by American and Ontario precedents and rhetoric. Crows had a bad reputation among farmers in the United States and eastern Canada because they ate corn, and this may have influenced views on the prairies even though little corn was grown there. While hunting groups were fairly well united on the issue, farmers were split into various factions. Some farmers favoured crow control because they were also sport hunters or because they thought that crows threatened insectivorous birds, while others opposed it because they believed the birds provided a net benefit for agriculture by consuming more insects and weed seeds than grain or the eggs of other birds.[33] The problem was that there was no clear answer. Crows ate eggs and fledglings of ground-nesting birds and waterfowl at the same time as they ate vast quantities of insects. The latter was demonstrated in the 1920s by biologist Norman Criddle in experiments with his four pet crows. "In one trial, after five hours, the crows had consumed 552 cutworms and were still eating, but there were no more cutworms to be had. A trial with two crows eating white grubs was similarly abandoned after five hours when the birds had eaten 121 – the entire supply of the laboratory." Yet these experiments only revealed one element of the diet of crows. E.R. Kalmbach's studies on crows, published by the United States Department of Agriculture and widely cited by both sides of the debate in Canada, described them as "practically omnivorous," and Kalmbach concluded in 1920 that "from the evidence at hand the crow's merits and shortcomings appear about equally divided." Percy Taverner held much the same opinion, observing in 1938 that crows were neither as bad as their critics painted them nor as good as their defenders claimed.[34] Alluding to the emotions at play, Saskatchewan bird lover Lawrence Potter noted in 1940 that consensus about crows was no more likely than consensus about politics or religion. But he was confident that people, not crows, were "the chief offender, as always," because what "the good birds do is more easily overlooked than the harm they do," and people had upset the balance of nature with their guns, farm machinery, cats, and pigs, which collectively were a greater menace to ground-nesting birds than crows. There were too many crows because people had killed off predators such as horned owls and had created environments favourable to them. As Potter warned, the crow was "in our midst for a purpose, even if his way of life is not acceptable to us." And, as Ernest Thompson Seton had pointed out, Silverspot the crow made scheduled visits to the nests of small birds to eat the newly laid eggs "as regularly as a doctor visiting his patients. But we must not judge him for that, as it is just what we ourselves do to the hens in the barnyard."[35]

Whatever the merits of the debate, champions of the crow lost ground. Bounties and crow-killing campaigns began around the First World War in some provinces, and by the late 1930s hunting lobbies and organizations like Ducks Unlimited (Canada) had for the most part successfully defined crows as predators that needed to be controlled. The Saskatchewan Department of Agriculture for a time opposed crow-control programs, seeing the birds as beneficial for agriculture, but it met direct resistance from the hunting lobby. Although debate was ongoing, by 1936 it was noted that while some champions of the crow could still be found in Saskatchewan, it was generally accepted that the harm crows caused game and songbirds outweighed any benefit they offered.[36]

Crow-control programs on the prairies remained in place well into the 1950s, but by then the public attack on the crow family had become more focused on magpies. Omnivorous like crows, magpies were highly intelligent predators and were said to have a fondness for flesh. In 1927 it was claimed that magpies were welcomed when they began to spread in Saskatchewan because of their beauty, fine proportions, and graceful flight, but hostility to them soon developed. They were said to have a habit of picking at the sores of recently branded cattle and horses, thus blurring the brand or causing more serious injury. They were also said to threaten farm poultry and were more destructive of song- and game-bird eggs and fledglings than crows. As magpies spread and "the cunning, the audacity and the ruthlessness of this robber" became understood, "his advent in any neighbourhood" was looked upon as a calamity. By the 1940s magpies had become popularly defined as vermin. Naturalist Kerry Wood was a particularly harsh critic and in 1949 wrote a booklet, *The Magpie Menace*, that graphically detailed the bird's faults and provided detailed instructions on how to kill them.[37] Such invective was often a reworking of already well-established notions about the evils of crows, and crow-control programs were easily expanded to include magpies. By 1950 crow and magpie control had become almost entirely about protecting hunting interests. By then, too, public discussion about the value of insectivorous, weed seed–eating and predatory birds had dropped off. While claims that birds were the farmer's friend continued to be made, they lacked the vigour of earlier years. Seemingly, people assumed that the new chemical poisons coming out of the Second World War would solve the problem of insects and weeds and that birds were no longer so essential. Insecticides came into widespread use almost overnight, and the availability of pesticides such as DDT and herbicides like 2-4-D seemed to promise total control of most insects. Yet while they offered assurance that human control of nature was more secure, these new chemical weapons sometimes failed. When calculating the astonishing reproductive capacity shown by a major outbreak of aphids in Manitoba in 1955, the *Winnipeg Tribune* presumed an affirmative answer to its question "Can insects take over the earth?"[38] There was, however, no longer any mention that birds, the farmer's old friend, offered a solution to the problem.

FAMILIAR FOES – WOLVES AND COYOTES

The threat said to be posed by crows and magpies to game birds, by hawks and owls to farm poultry, by pelicans and cormorants to commercial and sport fish, and, among others, by ground squirrels to field crops demonstrated an ongoing conflict with wildlife. While opinions on the threat posed by particular wild animals were often tempered over time, wolves and coyotes were consistently targeted as foes by governments and the public.[39] This attitude was relatively new on the plains, where Aboriginal peoples had often (albeit cautiously) admired them. The Blackfoot, for example, believed that wolves were among the first "Earth Beings" to help humans:

> [O]ne winter, when our people were starving, a young man and his family were camped by themselves as they searched for food. The wolves found the family and appeared to them as young men bringing fresh meat to the lodge. The wolves took this family with them, showing the man how to cooperate with other people when he hunted buffalo and other animals. The wolves introduced the people to the other animals in their world. The human beings learned that animals with hooves and horns were all right to eat, but that animals with paws and claws should be left alone.[40]

While often cautious and wary, Euro-American frontiersmen, explorers, and traders on the Great Plains of Canada and the United States before 1870 often had a relatively peaceful relationship with wolves, sometimes seeing them as bothersome but not usually threatening.[41] Agricultural settlers, however, came with European and eastern North American agrarian prejudices and fears about wolves. The danger to livestock posed by strong and intelligent carnivores that hunted in packs was also clear, and it was thus not accidental that hostility to wolves grew as the number of livestock increased in the region. The District of Assiniboia, for example, began paying a bounty on wolves in 1839, the same time that cattle began to be exported from the colony, suggesting a link between the emergence of commercial-scale livestock farming and the hatred of wolves.[42] This hatred was stoked by misconceptions about what wolves ate. While it is now understood that they have a highly varied diet that includes big game, small mammals, ground-nesting birds, plants, and carrion, their threat to livestock was reinforced by a perception that wolves, as the NWT Department of Agriculture claimed in 1898, needed "freshly killed meat for every meal." Such rapacity not only threatened livestock, but fur-bearers and game animals as well. This view continued to play a role in the public's perception of wolf behaviour. In 1935 the *Prince Albert Herald* claimed that wolf packs killed "an unestimated [*sic*] number of moose and caribou daily" to satisfy their "craving for fresh meat." But their savagery went even further, for it was said that they would "tear trapped fox and mink to shreds and at the same time devour considerable [number] of the rabbits upon which fur-bearers depend largely for winter food." When desperate, they were said to devour "animals lying dead in traps" as well.[43]

The characterization of wolves as rapacious and cruel remained popular on the prairies for a remarkably long time. Remembering the accidental death of their milk cow, Lady, on their Manitoba homestead one winter in the 1880s, novelist Nellie McClung recalled that the carcass was dragged to the edge of the field so that it could be buried there in the spring. But the country and its wild animals did not give Lady a dignified end, and McClung wrote that "the nights that followed were terrible when the wolves fought and cursed and cried over her and we had to hear them, and see them too, for it was clear moonlight and they seemed to come from the four corners of the world, snarling, snapping hungry ghouls, grey lean and terrible." Although the wolves had not killed Lady, her death was a tragedy that somehow reflected on them. She had been a "companion" that held the promise of a future herd, and the wolves that ate her carcass symbolized for the McClungs, cowering in the bush garden of their homestead, a wild and dangerous place with an uncertain future. In these mythic portrayals, wolves threatened civilization with their amorality and cruelty. As a wolf hunter in southwest Saskatchewan recalled, "the wolf is not particular whether a man has a thousand head of cattle; he would just as soon kill the only milk cow belonging to a widow with 10 children." A similar view was expressed decades later in a 1946 *Country Guide* article, "Nazis of the North," in which old stereotypes were adapted to current conditions when C.E. Craddock of Big River, northwest of Prince Albert, Saskatchewan, wrote that "the wolf frequently kills for the pure love of killing, just like his human prototype, the Nazi."[44]

Such views, including a widespread apprehension that wolves attacked people, were deeply embedded in popular culture. A short story written in 1930 by an adolescent from Elfros, a town in central Saskatchewan (where there had probably been no wolves for at least fifty years), drew on folklore and fable to portray wolves as creatures with "dripping fangs," a "blood curdling hunting cry," and "gleaming" eyes that were "cruel twin flames of green." Wolves were not merely natural predators, they were murderers, and adolescent boys were not the only ones thrilled by such stories. Clarence Tillenius, a wildlife artist and story writer who was born in the Interlake district of Manitoba in 1913, wrote about wolves in one instalment of his series "Sketch Pad Out-of-Doors" in the *Country Guide* in 1952. While Tillenius usually wrote gently personal descriptive accounts of animals, his imagination ran wild in this article about wolves. A wolf howl, he wrote, was "one of the most desolate sounds on earth. There is loneliness in it, and sadness, and an unnameable, spine-chilling menace. It is like a disembodied voice bewailing the savage vastness of the forest." Pandering to popular prejudices, he imagined with a shiver that reached centuries into the past "the enormous gape of its jaws and the great fangs" and eyes that "glow with a brilliant emerald green when a wolf is angered." He claimed that a wolf was capable of attacking a "bull bison in his prime" and "with one snap can crush a buck deer's spine." Although he mused that they "could obviously be a most dangerous adversary if man were his quarry," he did note that wolves almost never attacked people and avoided them "at all costs."[45] Others also refuted claims that wolves threatened people. While Norman Criddle contended in 1925 that human coexistence with a wolf was impossible because it was "far too dangerous a neighbour to harbour in our midst for account of its destruction of livestock," his research confirmed that wolves did not attack people. In the few cases where trappers had been "found partly eaten by these animals," he was certain that the men had died from other causes and had been eaten by wolves as carrion. Offering further evidence against claims that wolves attacked people, Criddle noted that "we have only to recall the extreme difficulty" in trapping them or even "getting within range of them, to discount the numerous stories which periodically find their way into the press or the fakes in photography which occasionally accompany them." In 1923 James Harkin, commissioner of the national parks, wrote that there was no conclusive evidence of wolves killing people, and almost three decades later the *Winnipeg Tribune* editorialized that the threat wolves posed was exaggerated, for "the 'killer of the North' has a far worse reputation than he deserves." The stories of wolves attacking humans "are the bunk. We incline to the theory held by most Northerners that one [wolf] occasionally stalks a trapper out of idle curiosity and both are equally afraid of the other." And Leslie Sara, wildlife columnist for the *Calgary Herald*, made this argument in 1936 in an even more general way about all wild animals. As he wrote, when "wounded, or in defence of their young, even the mildest mannered animal becomes a fury. But in their natural state, denizens of the wilds shun rather than attempt to come to grips with human beings." He believed that stories of wolves attacking humans and of eagles carrying off babies were only the yarns of storytellers.[46]

Yet popular depictions of the threatening wolf were slow to disappear, and just as the conquest of wild animals was portrayed as part of a coming of age, confrontations with wolves could empower human protagonists and make them part of regional myth. In 1912 the American western painter Charles Russell painted an encounter that George Lane had with wolves near Calgary in 1886. Lane, the scion of a southern Alberta ranching family, is shown on a rearing horse, shooting at a wolf leaping towards him. At the sides of the canvas, four wolves are in retreat. In 1926 the painting was simply titled *The Wolves*, but it played to myths about wolves and the conquest of the land. In 1944 the painting was described in the University of Alberta's folklore project as being about "blood-thirsty brutes" that, when attacking Lane's horse, were greeted "with a hail of bullets." When the Glenbow Museum purchased the painting in 1974, in keeping with the myth that had emerged around it, the museum titled the work *George Lane Attacked by Wolves*. However, as historian Bill Yeo notes, the story was a much simpler one. As subsequently told by his family, Lane was returning to the Bar U Ranch when he saw wolves among his cattle. He shot at them and one of the wounded wolves fought back. This was the subject of the painting. As Yeo observes, it is clear "that the attacking is being done by George Lane, and in fact four of the five wolves in the picture are fleeing."[47] Similar mythmaking was evident in a 1943 article by Z.M. Hamilton, secretary of the Saskatchewan Historical Society, which recounted a story about a ranch foreman in the NWT in the late nineteenth century. Pursued by wolves that tore at his horse's flanks, the foreman fought them off, but he was running out of ammunition and his horse was becoming weak. "Then it struck him that if the horse was left to them he might get away. He was determined, however, not to let the poor beast, who had borne him so bravely, be tortured to death by these cowardly brutes and he put a bullet through its brain." He took off the saddle, slung it over his shoulder and backed down the trail keeping his eyes on the wolves. "He had not gone more than a few paces before they were engaged in their horrid feast." Concluding that "this is the only instance as far as we are aware of wolves attacking human beings," Hamilton then added that it was possible that "it was the horse that attracted the wolves. They always were fond of horse meat."[48]

Coyotes were also the subject of much mythmaking, perhaps because wolves and coyotes are close relatives and observers after the early 1800s on the North American Great Plains often confused them.[49] This confusion is evident in the various names used on the Canadian plains for coyotes; in the late 1800s and early 1900s the terms "buffalo wolf," "little wolf," "bush wolf," and "prairie wolf" were all commonly used, with the latter term still in use as late as the Second World War. Given their physical similarities, the two animals could be easily confused in the wild, especially by people who were unfamiliar with them, such as settlers coming from Ontario, where there were no coyotes until around the turn of the twentieth century.[50] While wolves were often said to be directly aggressive, intelligent, and relentless, coyotes were viewed as sneaky and cunning. Coyotes were said to be particularly destructive of poultry and sheep, but it was also said that they threatened young livestock. Given their cunning, they were not afraid to enter farmyards to kill poultry and the farm dog. In this, they were unlike wolves, but otherwise some observers saw little difference. A farmer from Camrose, Alberta, contended in 1949 that "the coyote is a natural born killer of sheep. He does not always kill when hungry" but sometimes simply for the joy of it. "I have seen them tear the heart out of young lambs with but a single bite. Is it any wonder I have come to hate the coyote?" While coyotes share some of the traits that have helped wolves survive, they are more omnivorous. Their population cycles are tied to those of rabbits, although they also eat carrion, berries, small rodents, and ground-nesting birds such as grouse. As Norman Criddle noted in his 1925 study of wolves and coyotes, this diet benefited farmers by destroying some threats to crops. But when rabbits were scarce, coyotes turned to livestock and poultry, although poultry was always fair game, likely because it resembled a familiar part of the coyote's diet. Criddle thought that their intelligence gave them a strong advantage, but he also noted that they benefited by their adaptability to changed environments. He characterized coyotes as timid, suspicious animals that avoided humans in daytime but would prowl even onto doorsteps at night. But there were few animals, he asserted, that should be feared less than coyotes, and "the old wives stories of their attacking children have never been substantiated by trustworthy evidence." In fact, his experience told him that coyotes had become much shyer by the 1920s because they had learned to fear humans.[51]

One of the most plausible threats from wild carnivores lay in the spread of disease, especially rabies, and public antagonism to coyotes and wolves was further stoked by fear of this possibility. Historian Bruce Hampton notes that while it is difficult to measure how the fear of contracting rabies exacerbated the animosity that many people in the American west already had for wolves, "it probably played no small role in further darkening popular opinion of the animal."[52] And, as will be discussed in chapter 8, public anxiety in farming communities about coyotes in the

1940s and 1950s had intensified negative views about all wild carnivores. Thus, when a rabies outbreak around Fort Chipewyan among foxes (then at the height of their population cycle) and wolves spread south into settled parts of northern Alberta in 1952, it was not surprising that officials reacted strongly. The worst outbreak was in the Fort Vermilion district, west of Fort Chipewyan on the Peace River, but all of Alberta roughly north of the town of Peace River was quarantined. A case of a rabid wolf attacking a human was reported, but no one was infected. There were also reports of "shy foxes, coyotes and wolves" becoming "fearless marauders," of a wolf attacking the tires of a tractor, of foxes that were unafraid of humans, and of the infection of several dogs. By early 1953 the disease had advanced into the whole of the Peace River district and the provincial government launched a massive campaign against all wild carnivores. It hired trappers to create a buffer zone to stop the disease from spreading into farming areas. These trappers used poisoned bait, and by early 1953 a "double trapline on the edge of the forested areas completely surrounded the settled area of the province." One hundred and seventy trappers were employed to run the lines, which stretched for 8,000 kilometres, in order to kill all "predators and other animals which might spread the disease." While this did not immediately end the outbreak, by 1954 the government claimed it had been halted. It was noted that mostly coyotes, wolves, and fur-bearers of "little value such as lynx and bears" had been killed. But the scale of destruction was massive. Almost 120,000 carnivores were killed in the forest fringe areas – about 54,000 foxes, 45,000 coyotes, 10,000 lynx, 5,000 wolves, and 3,500 bears, as well as some skunks, cougars, wolverines, fishers, and badgers. A further 60,000 to 80,000 coyotes were killed in farming areas. By its end, the provincial government spent around $1 million on the campaign.[53]

While Alberta's anti-rabies campaign responded to an immediate threat to public safety and farm interests, it was framed within old fears and often meshed with other concerns and priorities, such as the needs of big game hunters. In 1953 the Alberta Department of Lands and Forests rather sanguinely reported that killing wild carnivores protected the public from rabies but also had "a very beneficial effect" on big game and upland bird populations in the fringe and forested districts. By 1955 the department reported that wolves were no longer a threat to game or farm animals in northern Alberta and that ongoing control programs to ensure that rabies did not reappear helped keep predator populations low to the benefit of farmers and sport hunters alike.[54]

A DIFFERENT KIND OF FRIENDSHIP: WILD ANIMALS AS PETS

During a trip to eastern Canada in 1949, writer Gill Shark was asked to give a public talk about "some interesting aspect of the west." Shark chose to talk about "the westerner's mania for taming wild things," and when he later described his presentation, he noted that "to this day, I don't believe that crowd believed me. But it's a fact."[55] While the practice of taming wild animals was not unique to the prairies, there were plenty of examples. References to fur traders in the 1700s and 1800s trying to domesticate or tame wild animals were common, and some Aboriginal people also kept wild animals as pets.[56] A 1899 report claimed that, except for wolverines and lynx, which were "outside the pale," most boreal forest animals had the potential to be tamed and to become "friends with men on occasions."[57] And there was much interest in doing so. In the late 1940s and early 1950s alone, various western Canadian newspapers told of squirrels, porcupines, coyotes, weasels, skunks, beavers, badgers, falcons, crows, ducks, and, among others, moose and deer being kept as pets. While children and adolescents most commonly tried to tame wild animals, there were many examples of adults doing so as well.

There were many reasons why people wanted wild animals as pets. Curiosity played a role, as did kindness when wounded or orphaned animals were rescued. The practice also reflected a general desire for the companionship of animals, similar to that of having dogs and cats. Young animals were a particular focus; they were accessible, cute, and seemed trainable. In some cases, the effort to bring wild animals into the daily orbit of life crossed the friends and foes duality. But other cases reflected an impulse to dominate and to change the animal's life and character to suit human needs. In this sense, taming a wild creature became a personal challenge, similar to breaking a horse. And then there could be material advantage. An article in the *Farm and Ranch Review* in 1946 advised readers that keeping racoons as pets was fun because they "can perform stunts, which disproves the view of some people that the racoon is stupid." But aside from their cuteness and intelligence, pet racoons could later be killed for their fur or sold to a fur farm or as meat. Altogether, "these little fellows with the patch on their cheeks will not only make money for you, they'll provide something better – many hours of relaxation and fun." And as a correspondent in the *Western Producer*'s Nature Lovers column reported in 1938, she and her family raised coyote pups as pets but "we pelted them as soon as their pelts were prime."[58]

While taming a wild animal potentially involved a proprietary relationship, it usually did not go beyond teaching the animal to behave properly and to not hurt its owners or run away. Such taming did not usually seek to create permanent biological or behavioural changes in the species as it did with domestic animals that were bred to have transmissible traits such as strength, docility, ferocity or better meat, hides, or wool.[59] In this sense, wild animal pets were rarely "useful." Instead, they were curiosities, in part because of their wildness. Wildness was not usually characterized in any particular way but was most often simply thought of as the opposite of being tame or domesticated. In the late nineteenth century wildness was not always viewed positively. As Nicholas Flood Davin, editor of the *Regina Leader* and a member of Parliament, argued in 1887, nature tended towards an instinctual invasive coarseness that was difficult to control or eradicate. This notion endured in the interwar years. Canadian author Archie McKishnie explored this theme in two stories in 1925, both of which described hybrid wild-domestic animals. One of these was the rather improbable offspring of a domestic cat and a bobcat whose domestic traits were demonstrated by devotion to its master and an eagerness to explore, while its wild side came through in a "perpetual watchfulness" and periodic "fits of moody depression." Its hybridity fitted it for neither world, and it could not resist the draw of the wilderness. Similarly, the second story told of a hybrid wolf-dog that felt a deep pull towards humans, but feared them too, and could not leave the wild.[60]

After the Second World War such qualities as untameablity and independence continued to be cited as part of wildness, but some people now viewed them positively as evidence of the value of freedom. This view was suggested in a 1962 story by Saskatchewan writer Arthur G. Storey. Written from the point of view of the bird, the story told of a boy who found an injured Canada goose one autumn. The boy took it home, nurtured it back to health, and named it Honker. In the spring the healed bird became "visibly impatient" and wanted to return to the wild. The boy was hurt by this rejection but his mother knew "that there was something of the ancient Ulysses in all boys and in all wild creatures," and "speaking for the bird," she quoted Tennyson's *Ulysses*, who, home from the Trojan Wars, lamented that he could not "rest from travel" and had been reduced to ruling over a people who "board and sleep and feed, and know not me." Like Ulysses, Honker's restlessness grew, until one day his cries were echoed back from "a great V of Canadas winging northward." Forgetting his fear of hunters, his dread of pain, and his loyalty to the boy who had rescued him, Honker joined the flock that had circled back to wait for him, flying up to them in "wild exultation."[61]

The inability to overcome wildness and to find a good life in the company of humans was an ongoing theme in many reports about wild animal pets. The *Western Producer*'s Nature Lovers column was a popular place for such accounts. Many were submitted by adolescents, most of whom lived in rural and farm settings, which contradicts the argument that, historically, pet keeping was essentially a bourgeois urban phenomenon.[62] Year after year, reports of pet ducks, crows, gophers, and other wild animals were published. One correspondent in 1931 told of hatching wild duck eggs in a hen's nest. Seven hatched but only one lived, and when the duck matured, its wings were clipped so that it could not fly away. It then vanished, probably killed by an owl. Another correspondent, in 1928, had a pet gopher, but it only lived a couple of weeks. "I know that wild animals like to be in the open," she wrote, and it seemed likely to her that it died of an illness because it never left the porch. Yet another correspondent, in 1933, tried on several occasions to raise pet crows but all died. In 1932 another claimed that his hobby was making pets of wild animals. He had kept a seagull, a partridge, and several rabbits but all died. Given this track record, he concluded that he would give up the hobby. And in 1959, Keith Turnbull of Onoway, near Edmonton, had two pet crows, but one died after it flew into a building and the playfulness of the other made it "a nuisance." Because the crow took clothespins from the line and hid them, Turnbull's mother "got fed up with him. Early in October he flew away and never returned."[63]

Other accounts indicated that the appeal of young wild animals waned as they matured. An individual from the Peace River country, for example, wrote in 1954 to Al Oeming, a wild animal expert then associated with the Edmonton zoo, to ask if the zoo would take two coyote pups that the family had raised since infancy. Because they were "such loveable little pets," the family hated to destroy them and "would like to find a good home for them." Oeming declined the offer, noting that coyotes were "a common article" in Canadian zoos, but urged the family to keep them. Oeming had raised a number of coyotes to adulthood and pointed out that they had not lost "their playful and affectionate attitude. They remain, invariably, as lovely pets as long as you care to keep them." Further, coyotes were "really born tame and I have never seen a pup that couldn't get tame and friendly within a very short time." Oeming concluded, "Please don't kill them. There are quite enough people in Alberta … that are doing an all too thorough job of wiping out the interesting and essential coyote."[64]

Oeming's emotional tie to his pet coyotes was clear, and others had similar relationships. Mina Cole of Red Deer, Alberta, raised a porcupine that she named Unky. They had an affectionate relationship, but as he grew, she became ever more concerned that it would be "dangerous to have him climb into my lap," and so he was sent to the Calgary zoo, where he became a "favourite of the young folk." And biologist Norman Criddle successfully raised coyotes from infancy and claimed that they showed "in their playfulness all the little actions familiar to the breeder of domestic dogs. They also become extremely affectionate when treated kindly." "Chipper," a newly born gopher found on the grounds of the Regina city hall by Kathleen Patton, was another success story. "He put his paw on my foot as if to say, 'I'm lost – would you take care of me, please?'" So he came home with her, and she and her husband Bob fed him milk with an eyedropper and made a nest for him in a cloth-lined box. He was fully grown in about a month, and the Pattons released him in the country near an abandoned gopher hole. "In a flash" he was out of his box and "wouldn't let Bob touch him. He was back in his native element, a wild gopher now!" After Chipper investigated the hole, he came back and looked "wistfully" at his box, so Bob tore the cloth into small pieces so Chipper

Figure 3.6 Coyotes were widely seen as a scourge in mixed-farming districts. However, some people, such as this man at Turner Valley south of Calgary, found that they made affectionate pets. Photo not dated.

could carry remnants of the human world into "his new home to make a cozy nest for himself."[65]

For the most part, the public was entranced by wild animals that were successfully raised in human surroundings. This perhaps suggests that they wanted to believe that wild animals were not so wild after all and that people could form meaningful relationships with them and perhaps transcend the human-wild animal divide. This was part of the public appeal of the pet beavers that conservationist and writer Grey Owl kept, which became national and international sensations. Other wild animal pets were also the subject of widespread curiosity everywhere in the region. A group of orphaned skunks taken in by an Alberta family lived with them for seven years and finally died of old age. A local attraction, they were said to be "the most friendly and interesting pets that anyone could wish to have." Another Alberta family had many mice in their house until a weasel moved in and eradicated them. "As thanks the family put food out in the back kitchen and soon had the satisfaction of seeing the slim little animal" eating their offerings. The weasel stayed for several months and became so tame that it had "the run of the whole house." While they never touched it, they thought of it as their cat. "Visitors were often shocked at the sight of a real live weasel scampering across the kitchen floor. Ladies in particular were horrified by its presence." It never killed any farm poultry and when it moved out as suddenly as it had arrived, the family "felt sorry about losing such an interesting and useful house guest."[66] Of such pets, a beaver named Mickey was among the most famous. In 1940 the Forbes family of Red Deer took in a one-month-old beaver kitten that had been mauled by a dog. Ten-year-old Doris and her mother cleaned and dressed his wounds and put him behind the stove in a box lined with hay. Having decided "that babies were the same the world over, whether they had two legs or four," they spoon fed him a mixture of warm milk, water, and sugar every few hours. As he healed, his diet expanded, and while he refused to eat fish or meat, he accepted bread, fruit, cake, toffee, garden vegetables, poplar tree bark, and greens such as clover, grass, and dandelions. Mickey was never caged and lived in a backyard shed where he slept during the day. Around dinner time he came out, and he and Doris played together in the house or in a small wading pool in the backyard. During dinner he wandered around the house and then returned to the shed, where he worked most of the night piling up logs and two-by-fours to make a mock dam. The Forbes family thought he was a perfect pet – he was trained not to gnaw on the furniture and was respectful and unaggressive, demonstrating that nature was not red in tooth and claw. Although Mr Forbes anticipated that he would "get the itchy foot" and leave to join other beavers in a nearby creek where he went swimming from time to time, Mickey always came home. And Mickey grew so close to Doris that he became despondent if she went away for any length of time. Doris carried Mickey around like a kitten when he was young, and she took him to school on occasions when the class was learning about wild animals. They also went swimming together. But with an adult weight of about twenty-five kilograms, such outings soon became more difficult, although Mickey was displayed at the local agricultural fair several times during the Second World War. Visitors paid ten cents to pet him and watch him perform some stunts, and the money raised went to buy tobacco and other comforts for local men serving overseas. But Mickey mostly stayed at home and received visitors. He was a celebrity and during his nine-year life he received over 10,000 visitors. Local town and farm people visited and celebrated him, as did others from far away, including Lady Baden-Powell (the world head of the Girl Guides) and a Swiss family who travelled across Canada to see him. Since beavers were strictly protected and wild fur-bearers could not be kept in captivity without a permit, Alberta's premier arranged for a special permit in Mickey's case. Mickey died of old age in 1948, but he was a true star in that his fame outlived him. Naturalist Kerry Wood wrote about him in the Canadian school reader *Up and Away*, and in 1960, over a decade after Mickey's death, Wood still received letters from schoolchildren across the country asking about him and where Doris was now living.[67]

While pets like Mickey offered social and psychological benefits, they were tamed, not domesticated. This line, however, sometimes became blurred. Fur traders in the 1700s had hoped to domesticate moose, bison, beavers, and other wild animals, but such efforts were unsuccessful and after 1870 were mainly focused on cross-breeding bison with cattle in an attempt to create a new domestic animal. Yet there were instances where wild animals were trained to function much like domesticated ones. One settler in the St Albert area just north of Edmonton, for example, trained a team of moose to pull a buggy and for many years routinely drove the team into Edmonton to get supplies.[68] And in 1903, Walter Anderson, a farmer near Riding Mountain in central Manitoba, obtained two moose calves. The animals were broken and raised by handler Dougall McKinnon, who trained them to draw a buggy, having designed "a harness to fit their rangy, rawbone frames." They were a common sight on the streets of Brandon, and in 1905 McDougall showed them at the Brandon fair, where they were a sensation. The owner of a travelling circus offered to buy them but insisted that they be able to follow a marching band in the grandstand parade. McKinnon trained them to tolerate the music and the noise of the crowds, and the moose were even entered in a horse race, beating all the horses in the field. Their lives, however, ended badly; one of them was kicked by a horse and had to be put down, and the other died shortly afterwards. Some people said it was because of poor husbandry, while others said it was because of a broken heart.[69]

Those who criticized the practice of making pets of wild animals cited the high mortality that frequently resulted and the often precarious position of those that survived. The case of the moose kept by Gus Lindgren, a trapper and prospector in the Cranberry Portage area near Flin Flon in northern Manitoba, provides an example of the latter. Finding an abandoned two-week-old moose calf, Lindgren took her home and named her Patsy. Having no way to feed her, Lindgren, after "careful thought," put her to suckle on two nursing huskies. The experiment worked and Patsy and the huskies became inseparable. When she was about two years old, Patsy left for a time, came back and then left again, returning this time with a calf. This became an annual event, and Lindgren soon not only had Patsy but four of her offspring as well. By then he was under increasing financial pressure – the moose were expensive to keep because they had become habituated to bannock, bread, sugar, rolled oats, and feed oats. And the herd was maturing, and some of the younger females would soon be reproducing. Lindgren hoped to establish a sanctuary to attract tourists and in that way raise money to feed the moose, which would otherwise die, having "been conditioned to a life that must be perpetuated."[70]

While some observers would have lauded Lindgren's work with his pet moose as a mark of compassion, others saw it as a disservice to wild ani-

Figure 3.7 Taming wild animals presumed that the divide that separated species could sometimes be crossed. These two girls at the Sturgeon Heights School in the DeBolt district in northern Alberta in 1918 were clearly intrigued by this tamed young moose.

mals. In the mid-1950s the Saskatchewan government advised against keeping deer as pets, warning that they "almost inevitably" became a menace or were killed by dogs or shot by hunters: "With many of its wild instincts frustrated, and having no fear of humans, the 'tame' buck is a truly dangerous creature. Sooner or later he is almost sure to begin striking or goring humans, and must be destroyed." Prairie writer Gill Shark also argued in 1949 that such experiments with wild animals, "while carried out with kindly intent, often have tragic endings" because the dependent animal could not survive in the wild. Moreover, he argued that most wild animals did not take to captivity, since most were "unreliable" and would return to the wild or become dangerous. Shark spoke from personal experience. He had had a pet coyote that had killed poultry and a pet deer that had become aggressive but could not cope with life in the wild. Unafraid of humans, it had made an easy kill for a hunter. Broadening his argument, Shark noted that "it took us thousands of years to domesticate those [animals] hanging around the barnyard now" and keeping wild animals as pets, no matter how friendly they were or what motives of kindness or generosity inspired their treatment as pets, placed them in an unnatural habitat and broke "their spirit and their ability to look after themselves." His experience had led him to conclude that it was much better "to observe and conserve them in the freedom they themselves are used to and love – and there is much less regret, I find."[71]

Figure 3.8 Although their habituation to human contact often ended badly for wild animals, attempts to tame deer were popular. This fawn was being fed from a baby's milk bottle on the Minor Ranch in the Medicine Hat district in Alberta. Photo not dated.

Figure 4.1 While people were often brutal and aggressive towards wild animals, many were also intrigued and fascinated by them. These diverse relationships were evident in one display in 1910 at the museum run by Dr George in Red Deer, Alberta. This display enshrined the value of wild animals for sport, food, and tourism, but it equally asserted broader intellectual and emotive interests in them.

PART 3

CONNECTIONS

4

A Pursuit of Utility: The State and Wildlife

> When mention is made of the resources of the province, our lands, waterways, mines and forests are lauded to the skies, but the wild life, which has been producing wealth with unremitting regularity for ages is rarely ever mentioned. If it were possible to secure accurate returns showing the value of the game and fur-bearing animals thus far taken in the province, the figures would be astounding, and the wealth represented would exceed that derived from any other of our natural resources.
>
> Saskatchewan Department of Agriculture, *Report of the Chief Game Guardian*, Annual Report, 1915, 236

From the beginning of Canadian control of the prairies in 1870, wild mammals, birds, and fish belonged to the state and were regulated as public property.[1] This public property regime rejected "common use" access that had been a feature of Aboriginal life before 1870 and asserted that every citizen, subject to uniform rules, had a right of access to wildlife as a public resource. This was another demonstration that the Euro-Canadian order after 1870 was a new era in which game, bounty, and bird-protection legislation set rules about the animals that merited protection and how and when they could legally be used. Such arrangements had a formative influence in shaping the human encounter with wildlife on the prairies. Animals seen as friends (generally big game, fur-bearers, and insectivorous and game birds) deserved state protection and encouragement, those that challenged social and economic needs were targeted, and the rest were ignored. Given the goals of state activity at the time, this approach to wildlife conservation was almost exclusively concerned with support for farming, sport and subsistence hunting, government revenue, and, among other things, economic development through tourism.

Public regulation of wild animals in Canada has a long history. Wolf bounties had been used since the eighteenth century, and Upper Canada passed legislation to protect deer in 1821 and a broad range of game species in 1839. With Confederation, the federal government was responsible for inland commercial fisheries and, after 1916, migratory birds. It also controlled wildlife in northern areas that were outside provincial boundaries and on federal lands such as national parks, military bases, and Indian reserves. Canada's responsibility for the hunting rights set out in treaties with First Nations gave it a further stake in wildlife policy. While this division of powers was not logical given the movement of wildlife and fish across political boundaries, the treatment of wildlife as a form of property, and the overriding concern to regulate wild animals for food and sport and in the interests of ranchers, farmers, and fur traders made wildlife a matter of "local concern" and thus within provincial jurisdiction.[2] The federal government was generally content to respect this provincial jurisdiction over wildlife, and its relatively passive role helped minimize federal–provincial conflicts over the matter.

The earliest legislation on the prairies concerning wild animals was the District of Assiniboia's wolf bounty, passed in 1839. Manitoba passed legislation to protect game and fur-bearers in 1876, and similar legislation appeared in short order in the rest of the region. In his history of Canadian wildlife legislation, John Donihee identifies three overlapping periods. A "game management era," from the late 1800s to the early 1960s, focused on game animals and their management through hunting controls and the establishment of game sanctuaries and refuges. This was followed until the mid-1980s by a "wildlife management era" in which the scope of management widened to include non-game animals and a stronger concern with habitat protection through various management techniques. Then followed what Donihee calls a "sustainable wildlife management era" in which concerns about Aboriginal rights, endangered species, and the implementation of stronger regulatory controls with an ecological focus gained importance.[3] The transition from one period to another was often untidy and slow; in the 1960s, for example, the process was halting in some respects even though it was the product of almost two decades of changing scientific and public thinking about ecology and the value of wild animals.

As a major impetus for legislation as well as other government concerns, conservation has been an important topic in the study of the history of Canadian wildlife management. In *Working for Wildlife* Janet Foster stresses the contributions of a handful of federal civil servants in national conservation initiatives before 1922 to national park development, migratory bird conservation, and the work of the federal government's Commission of Conservation. Covering a broader period, Tina Loo in her book *States of Nature* also deals with conservation and management, seeing it as a top-down process in which governments and urban elites used centralized control to impose new values about the use of wildlife to replace those based on local needs and practice. John Sandlos comes to similar conclusions in his *Hunters on the Margin*, which shows federal officials overriding local Aboriginal hunting traditions and patterns of wildlife use in northern Canada. In part, this conservation agenda aimed to maximize the value of wildlife by ending what were said to be wasteful Aboriginal hunting practices. In his treatment of wildlife conservation in his book *Game in the Garden*, George Colpitts suggests that the diverse forces that contributed to wildlife conservation and game management policy were shaped in a matrix of local and regional priorities and through a more active interplay between centre and margin. Through their varied approaches, these studies reveal that the history of conservation and state management is not only about wildlife but also about science as well as economic, political, and social life, all of which form a pervasive interplay that both shaped and reflected the connections between people and wildlife in prairie Canada.

DECIDING WHAT NOT TO CONSERVE: THE CASE OF BISON

The rapid decline of the plains bison in the 1870s brought increasing demands for protective legislation. During the negotiations of Treaty 6 in 1876, the Plains Cree chief Sweet Grass, as well as other First Nations leaders, demanded that the treaty include provisions for bison protection.

Canadian negotiator Alexander Morris refused to make such a commitment, stating only that the government would seriously consider the matter, "but what the law will be I cannot tell."[4] Morris clearly did not see this as a pressing matter; as he wrote in his annual report to the Department of the Interior that year, "a few simple regulations would preserve the herds for many years" and any problems caused by the disappearance of the bison would be met if people took up farming. Many others, however, saw the issue as a more complicated one. While agreeing with Morris that farming had to be encouraged for the time "when the buffalo will be a thing of the past," the Department of the Interior urged immediate conservation of the remnants of the herds to ensure social peace in the region. Otherwise, as the department's M.G. Dickieson phrased it, starving Aboriginal people would commit "outrages" against traders and bring on "an Indian war." The department also feared that with the bison gone and farming in its infancy, the federal government would be burdened with the heavy cost of feeding First Nations people.[5] Nonetheless, the federal cabinet showed little inclination to act on the matter, and David Mills, the minister of the interior, argued that it was the responsibility of the territorial government. Consequently, the NWT Council passed an ordinance in 1877 that prohibited hide hunting and the killing of young and female bison as well as the killing of any bison for "the sole motive of amusement." Only hunting for food was allowed, and traditional Aboriginal hunting methods such as pounds, pits, and buffalo drives, some of which were by then uncommon, were outlawed to reduce the possible scale of slaughter.[6] The Métis greeted this legislation with hostility, and many people felt it was "manifestly absurd" to restrict the killing of bison in districts abutting the United States border "when there was no restriction of the kind on the other side." In any case, the territorial government was too poorly organized to enforce the law and it was soon abandoned.[7] Nonetheless, figures as diverse as Father Lacombe, the eminent Oblate missionary and champion of Aboriginal rights, and John Christian Schultz, the Conservative MP for the Manitoba riding of Lisgar (later a senator) and hardly a defender of such rights, continued to urge federal action to protect the dwindling herds. Schultz argued that, without the bison, plains First Nations people would need to be fed at public expense, and until they "could be weaned from the chase and taught to depend wholly or in part upon agriculture," their poverty and desperation threatened the prospects for orderly Euro-Canadian settlement.[8]

Despite such concerns and the territorial government's inability to manage the issue, the federal government remained on the sidelines. A number of factors were at play. For one, as the territorial experience showed, enforcement of protective legislation was difficult. Moreover, Canadians could only halt the slaughter south of the border through an international treaty, which the American government had no interest in pursuing. Economic concerns were also at stake; as Joseph Ryan, MP for the Manitoba riding of Marquette, observed, if "the buffalo trade in the west were stopped, Manitoba would be deprived of the largest market for her breadstuffs."[9] But the federal government's reluctance to act on bison conservation seems mainly to have reflected a view that extermination was inevitable and would, in any case, serve national priorities. Echoing sentiments much like those in the United States, where the slaughter of bison was accepted as part of economic development, a writer in the *Montreal Gazette* noted approvingly in 1879 that the days of the bison "were numbered and in fifty years hence the wild cattle of the plains will be almost a thing of the past," but by then the prairie region would "have in their place many a flourishing town, picturesque village and cosy farm house."[10] Extirpation came even more quickly than forecast, and one last bit of value was found in gathering bison bones to make fertilizer and for sale as curios. But the bison's true value was reflected in the fact that the climate and soil that had sustained such huge wild populations could equally support vast numbers of domesticated animals. As Thomas Fawcett of the Dominion Land Survey remarked in 1881, "a region which for generations has supported immense herds, even millions, of buffalo, cannot but prove favourable for stock when once acclimated, provided they are permitted to roam over a sufficient area."[11]

And settlement was the priority. Recalling her childhood in Manitoba in the early 1880s, Nellie McClung remembered that her neighbours had mounted bison skulls on their buildings and used them to edge flower beds around their houses as mementos of the past. She also remembered fantasizing about finding a remnant herd of bison grazing below her as she came over a hill. But sentiment was easy and cost nothing, and as McClung frankly admitted, the end of the bison was wasteful and sad, but

Figure 4.2 The commercial gathering of bison bones to make them into fertilizer was one economic use for the remnants of the great herds. Horns were also sold as curios to tourists, as in this 1889 photograph where passengers were offered horns as souvenirs of their rail trip across the prairies.

they "had to go if the country were to be opened up for settlers." In this she was only repeating what was understood by practical folk. Free-ranging bison did not meet the national objective to transform the prairies for the domestic animals, plants, and farms of Euro-Canadian settlers. As John Gunn, a Manitoba provincial politician and member of a prominent Red River family, told the Canadian Senate in 1887, "God intended" and "every patriotic Canadian" hoped that the prairies would soon be filled "with a hard-working, thrifty population who will utilize the boundless grazing stretches of this country in raising improved herds of cattle, sheep and other domestic animals." It was thus just as well that the bison had been "allowed to come to an end" because "farmers and buffalo were not intended to live together," although Gunn conceded that "if a suitable locality could be found somewhat remote from any farming settlements, their propagation and preservation might be desirable and possibly profitable." Ernest Thompson Seton was no less certain. Who, he asked in 1909, was there among the present generation that did not feel "profound regret" at never having seen the great herds? The same regret that Sir Walter Scott had for medieval days was seen in "every boy and young man of our country now, when he hears of the Buffalo days and the stirring times of the bygone Wildest West." But the bison could not be saved. For Seton, their decimation "was absolutely inevitable" for the land was needed by "the outcrowded human swarms of Europe" and farming was a better use of the land than leaving it to wild bison.[12]

Yet nostalgia for the bison ran deep and their disappearance ironically became a part of the region's sense of identity. Once they were no longer a barrier to settlement, a mythical relationship with bison emerged. While the animals were gone, the marks they left on the land endured – in Manitoba in 1885 the bison trails and wallows still remained, as did the buffalo stones, polished smooth by the thousands of bison that had rubbed against them.[13] And some wealthy individuals, such as Donald Smith (Lord Strathcona) in Winnipeg, collected bison as nostalgic curiosities or to crossbreed them with domestic cattle. The federal government also developed an interest in the animals. In 1894 it passed legislation to protect a recently discovered herd of wood bison in the far north of Alberta (an area that later became Wood Buffalo National Park), and in the early twentieth century Canada purchased the last remaining herd of wild plains bison from a rancher in Montana. Neither project reversed earlier priorities; the northern herd lived in a non-agricultural forested region, while the purchased herd was kept fenced within national parks as a tourist attraction. But these projects nonetheless reflected that the loss of the great herds still sat uncomfortably in the public memory. In 1890, when only about 500 plains bison remained in the world, the poet Charles Mair wrote that no act in the natural history of North America brought "such reproach on civilised man as his reckless and almost total destruction of the bison." Indeed, the bison who "spoke" in his 1888 poem, "Lament for the Bison," prophesized that the immensity of the crime would someday be expiated – the character and morality of the "pale" destroyers, with their insatiability and willingness to do anything for immediate gain, could only lead to ruin. Still others assuaged their regret at the loss of the bison by imagining that they continued to be a part of the landscape. The Anglican minister Martin Holdom wrote that when he saw several thousand head of cattle "dotted about in herds right away into the distance" near Calgary in 1910, "I tried to imagine them as buffalo."[14] And such sentiments continued to be expressed decades later. In 1961, when standing beside a buffalo stone in the Cypress Hills, the painter Clarence Tillenius wrote that "in my fancy" there were vast herds of bison: "Beyond them, lost in shimmering haze rode on their painted buffalo runners savage war parties of the past: Crow, Blackfoot and Sioux." Above were cranes, golden eagles, and red-tailed hawks.[15] In contrast to this stereotypical image of Aboriginal people and nostalgia for a vanished animal, Canadian poet Lorna Crozier more recently drew from a scarred but still shining buffalo stone a memory of the vanished herds and First Nations hopes for the future and their belief in the supernatural character of the bison:

> Jim Cuthand who breaks broncos at the ranch
> said the herds moved underground,
> graze just beneath our feet.
> If you press your ear against the grass
> You can hear them breathing.
>
> If his people keep on dancing
> the earth will part like the sea
> and the buffalo rumble out,
> Swinging their heads free of the dark.[16]

UTILITARIANISM AND EARLY GAME REGULATION

Manitoba's Game Protection Act of 1876 was the first such legislation in the region. It arose from concerns that the depletion of game and fur-bearers in the older Canadian provinces and parts of the United States would be repeated in Manitoba. The collapse of bison populations in Manitoba's historical hunting regions on the western plains and of fur-bearing animals in parts of its northern hinterlands confirmed these worries.[17] The province subsequently passed other legislation concerning wild animals, albeit for different reasons: the 1878 Wolf Bounty Act was intended to protect livestock, and the 1890 Act for the Protection of Insectivorous Birds further served farm needs. The Municipal Institutions Act of 1892 enabled local governments to pass bylaws to control predators and to pay bounties on wolves, foxes, gophers, and blackbirds. Although frequently amended, these laws formed the basis for Manitoba's wildlife legislation for almost a century. They also created influential precedents. The NWT's game ordinance in 1883 and its useful birds ordinance in 1902 were modelled on them, and this territorial legislation was in turn adopted by newly formed Saskatchewan and Alberta in 1905.[18]

Historian Tina Loo contends that Canadian game legislation until the end of the nineteenth century consisted of little more than "a number of fragmentary, localised practices concerned with controlling the kinds and numbers of animals killed." She further argues that a "centralized and bureaucratized set of policies aimed at promoting a particular relationship with wildlife" only emerged in the provinces with the consolidation of provincial game laws and the development of centralized policy making and enforcement in the early 1900s. At the federal level, she notes that the creation of the Commission of Conservation in 1909 and the drafting of the Migratory Birds Convention in 1916 confirmed this trend. While these were indeed important developments, they built upon regulatory and social traditions that had a greater coherence than Loo suggests. Manitoba's 1876 Game Protection Act created a centralized legal framework within which the public's relationship with wildlife was applied and developed. It also established uniform provincial standards as part of emerging modern state activity on the prairies after 1870. Such legislation reflected broader North American trends as well. In the 1870s, "an ideology of [game] protectionism" spread in the United States, and as historian James Tober remarks, these early American game laws, "whatever their ultimate merits as instruments of wildlife protection, were hardly frivolous exercises of legislative authority" but were seriously debated and amended over the years.[19]

Manitoba's 1876 Game Protection Act created a foundation that underpinned its approach to wildlife until well after the Second World War. It asserted that the central goal of wildlife regulation was to conserve wild animals that were useful for food and sport and to protect and enhance those that were important for commercial and farm interests. Hunting was regulated to conserve game and fur-bearers, create uniform standards of public access, and ensure adequate future supply. Thus, closed seasons were set for elk, moose, deer, sharp-tailed (commonly called prairie chicken) and pinnated grouse, ducks, snipe, and woodcocks. All of these animals were important for food and sport. The legislation similarly protected muskrats, beavers, mink, martens, racoons, otters, and fishers, which were then important for the fur trade. Rabbits were protected from spring to fall to conserve winter food for settlers. Animals that were not listed were unprotected and people could kill them in any way and at any time and could harass and exterminate them without penalty. The number of such unprotected animals was large and included bears, wolves, badgers, squirrels, all rodents (other than beavers and muskrats), foxes, wolverines, lynx, and weasels, among others. Birds that were not protected included eagles, hawks, pelicans, owls, and blackbirds. These exclusions reflected economic needs, since these animals potentially threatened crops, farm stock, commercial fish, and game animals. The Game Protection Act was amended in 1883 to specifically allow (and perhaps implicitly encourage) the use of nets, traps, cages, and other devices to kill unprotected birds and destroy their nests, eggs, and young. This was reconfirmed in 1890 in the Act for the Protection of Insectivorous Birds. These legislative provisions suggested a management plan of sorts whereby wild animals considered as friends were legally protected, while the public was given an unrestricted right to kill all others. Such categorization was not static and changed over time. Protection of fur-bearers, for example, gradually expanded, often because of emerging economic value brought by changing fashion or because the extirpation of some fur-bearers gave formerly less desirable ones new market value. Thus, foxes and lynx were given protection in 1915, and almost all wild animals with fur, including black bears and red squirrels, were protected by 1963.[20]

While Manitoba's game, insectivorous bird, and predator legislation created important precedents for wildlife regulation on the prairies, it was apparent by the early 1890s that depletion was still ongoing. Weak game law enforcement was blamed by some. Charles Nicolle, a rancher living in what is now southern Saskatchewan, for example, predicted in 1891 that the lack of proper enforcement of the NWT's game laws meant that deer, pronghorns, and other game animals would soon meet the same fate as the bison. Others were convinced that the law itself was at fault. William Hornaday of the Bronx zoo argued in his 1913 book *Our Vanishing Wildlife* that North America, given current rates of decline, would soon be a "gameless" continent. Depletion was happening "according to the law" because game laws were too lax and bag limits were too high. While Hornaday was not the only one to make such predictions, his views were influential in Canada because they shaped the thinking of staff in the dominion parks branch, especially James Harkin, who in 1911 became dominion parks commissioner. The urgency over halting depletion also arose from a worry that the country was nearing a tipping point from which it would not recover. Saskatchewan game commissioner Fred Bradshaw contended in 1913 that if someone fifty years before had prophesized the extermination of millions of passenger pigeons and bison, "he would have been called a lunatic or a crank." But "the rapid disappearance of any species of game is astounding when it sets in," and Saskatchewan sat upon a dangerous precipice: every year, pronghorns declined further, whooping cranes were "all but gone," and geese were declining.[21]

Almost everyone agreed that settlement would continue to bring about the decline of wildlife, but many also began to hope for a different balance. A growing romanticism about the bison persuaded an increasing number of people to express concern that other animals might meet the same fate. Indeed, no sooner were the bison gone than their extirpation became a leitmotif in a broader lament about the ever-widening human impact on North America's wild animals, forests, and land. While the process was in fact incremental, Ernest Thomson Seton pinpointed 1895 as the year when the cumulative consequences of depletion were popularly realized in North America. It was "the low-ebb" year for many game animals, but it was also the year when what he called "the great awakening" occurred. As he argued, "the lesson of the vanished Buffalo had sunk deep in men's minds," leading many people to resolve that wildlife should be treated better in the future than it had been in the past.[22] This was part of what historian Andrew Isenberg calls the development of an "emotional regard" for wildlife in the 1890s. These changing attitudes found expression in demands for better enforcement and tighter legislation as well as in the rising popularity of nature stories, such as those of Seton and Charles G.D. Roberts. For some, the savage treatment of wild animals in North America also reflected social and moral values. In 1905 Manitoba ornithologist George Atkinson commented that North America, by

wiping out animals that were so abundant "as to evoke the wonder and amazement of the entire world," was the paragon of "human selfishness." Driving home that this was a moral issue, he noted that humans had been made the "head of creation" but had neglected their duty "to rise above the animal nature of making all weaker and less resourceful creatures" subservient to their "dominating will."[23] Others thought the decline of wild animals showed a failure of progress. When editorializing in 1912 that depletion of fish in southern Alberta was inevitable because fishing regulations were not being enforced, the *Calgary Herald* noted that "it is astonishing that the world gets no wiser as it gets older."[24]

For many people, however, depletion of wildlife was a sign of progress, and the view that the advance of Euro-Canadian settlement inherently came at the cost of wildlife was widespread. This reasoning had, after all, partly justified the failure of bison conservation in the 1870s. As late as 1953 the *Winnipeg Tribune* remarked that the decline of sharp-tailed grouse (then thought to be endangered because of ongoing land breaking) could not be reversed. But it was only one in a long line of species now gone or diminished because of "civilization's advance," and it was claimed that even had there been no hide trade, the construction of highways, railways, and fences would have "doomed the buffalo," just as fencing the prairies did the pronghorns. These sentiments were only expressions of the belief that wildlife could not be allowed to interfere with the development of an agrarian and industrial economy. This view was privileged in official policy, and one of the bluntest statements of it was made by W.N. Millar, then with the University of Toronto's forestry faculty and formerly a forestry official based in Alberta. In a study he presented to the Commission of Conservation in 1915 on big game in the Rocky Mountains, Millar contended that "the first duty of a country is to its people, not to its wild game." When wildlife interfered or prevented "the establishment of successful homes, it must be destroyed." This included big game animals in the prairie grain belt and in ranching areas where they competed with livestock for food. While some game animals could be accommodated in parkland or fringe areas, the best place for them was in mountainous and forested areas that were useless for agriculture.[25]

Yet even if wild animals were not to be allowed to interfere with the settlement process, some nonetheless were thought to retain value and to offer ongoing benefit if managed properly. Some people viewed this narrowly. As the Manitoba Game and Fish Association phrased it in 1938, "all game is a crop to be harvested like any other crop." A broader view was articulated in 1950 by Ed Russenholt of the Manitoba game branch, who in a speech to the Canadian Club argued for "the use of all our resources of land and water for the greatest good of the greatest number of our people over the longest time." For this goal to be reached, scientific research, sound management, and public education were "the foundation of wise use."[26] This was by then conventional wisdom. By the early 1900s it was increasingly argued that all natural resources – timber, minerals, water, land, and wild animals – needed to be managed more wisely for the benefit of future generations, national power, and wealth. Historians have argued that the emergence of this view in the United States was the result of a belief that science and modern organizational systems could be applied to manage resources and to control their use for long-term national wealth. It has also been contended that this belief was fuelled by urbanization and a growing need among some urban people to reconnect with nature.[27] There was not, however, consensus on how this could be achieved. The debates between the American forester Gifford Pinchot (chief of the United States Forest Service 1905–10) and John Muir (founder of the Sierra Club in 1892) revealed two streams of thinking: Pinchot stressed the scientific conservation of resources to ensure maximum production and future supply, while Muir argued for the appreciation of emotive values in nature. While "preservationist" views such as

Muir's were not absent, historian Janet Foster argues that the "conservationist" approaches put forward by Pinchot enjoyed particular favour in Canada. But as historian Alan MacEachern observes, attempts to draw firm lines between conservation and preservation overplay the differences in early twentieth-century Canadian thinking, which he sees as best characterized by a range of concerns about economic efficiency, the contribution of natural resources to national power and wealth, and public regulation of resources to meet the economic and social needs of current and future generations.[28]

Such utilitarian goals were promoted at the federal level by the Canadian Commission of Conservation, established in 1909 following a conference in Washington DC attended by delegates from Newfoundland, Canada, the United States, and Mexico. Until the commission was wound down in 1921 as a cost-saving measure by the new Conservative government of Arthur Meighen, the commission's meetings and publications served as a national forum for the study and inventorying of Canada's natural resources as public assets.[29] While its most influential figures – Clifford Sifton, for example, the minister of the interior and for a time the chair of the commission – were mostly concerned with urban and land-use planning and the conservation of timber, minerals, and commercial fish, the commission also promoted the understanding of the economic value of wildlife and its contribution to social life and national regeneration. The commission also encouraged provincial governments to ban the sale of wild meat and, as will be seen shortly, ensured Canada's support for the Migratory Birds Convention, which protected birds that were valuable for sport and agriculture.

A BUSINESS PROPOSITION: WILDLIFE, STATE REVENUE, AND TOURISM

Making money from wild animals has a very long history in Canada. Trade in the fur, feathers, hides, flesh, teeth, and fat of wild creatures played a formative role in the founding of the nation, and while such activities had long been superseded in relative terms by farming, mining, forestry, and industry by the early 1900s, they continued to be economically important. As Gordon Hewitt, dominion entomologist and advisor to the Commission of Conservation, noted in 1917, Canada's wildlife was "unsurpassed in its economic value," and its proper use would improve the nation's "natural prosperity and efficiency."[30] Most of this commercial benefit accrued to the private sector through the fur trade and commercial fishing, sport and subsistence hunting, and the defence of livestock and crops from wildlife. But government increasingly saw wildlife as a source of revenue, a trend that the Commission of Conservation endorsed. In the late nineteenth century, governments earned little revenue from wildlife other than occasional fines and miscellaneous fees. As part of a growing belief in the 1890s that public use of wildlife needed to be better controlled, prairie governments increasingly demanded that hunters be licensed. The territorial government began charging non-residents for hunting licences in 1892; Manitoba began doing so for non-farmer resident hunters in 1904; and Saskatchewan saw its game revenues grow from $600 in 1909 to $9,000 in 1910 when it levied a one-dollar hunting licence fee for town and city hunters.[31] This revenue was further enhanced by charging royalties and licence fees on fur. Alberta imposed royalties on wild fur in 1920, and Manitoba and Saskatchewan, the last provinces west of Quebec to do so, implemented them in 1921.[32] Alberta, for example, took in almost $100,000 from the fur industry in 1934–35. About 80 per cent of this revenue came from royalties that were only paid on wild fur, largely by itinerant and Aboriginal trappers, among the poorest people in the province. Belying claims that this revenue aided conservation, the province made no effort to allocate it to improving trapping or fur-bearer conservation.[33] Indeed, in 1923 Alberta's game commissioner Benjamin Lawton noted that government revenue was "not ordinarily considered as having any relation to the protection of game," but "nothing appeals to the general public at the present time, as a good financial showing." Game protection, "if successfully carried out," he noted, "was nothing more nor less than a business proposition." And Alberta, along with other prairie governments, continued to employ these business models. A half century later, it reported revenue of almost $671,000 for the period from 1960 to 1963 from hunting and fishing licences and royalties on fur and commercial fish, net of administrative costs.[34]

Government revenue and private economic spin-offs also came from using wildlife to attract visitors. City hunters brought economic benefit to small towns, and W.G. Ross of the Regina fish and game club claimed

Figure 4.3 Hunting was an attraction for genteel tourist hunters, such as Mackenzie King, the deputy minister of labour and the future Canadian prime minister, photographed on a hunting expedition in the west in 1905.

in 1935 that supplying "the varied wants of this army of sportsmen" was a major industry that merited support through conservation and other assistance. A 1956 survey of sport hunters in Saskatchewan showed that they spent more than $7.5 million, demonstrating that wildlife was "an important economic resource."[35] Great promise therefore lay in attracting visitors from near and far. Hunting forays to the prairies by wealthy sport hunters after the mid-nineteenth century inspired later hopes for the development of a prairie tourist sector based in part on wildlife.[36] One reason for establishing the country's first national park at Banff was to create tourist traffic on the Canadian Pacific Railway (CPR), and the park shaped its wildlife policies in part around the needs of tourists. While the park at first drew elite visitors, a broader range of people began to visit after 1900 because of an expanding middle class and more efficient transportation. Although the aim was still to attract people at the high end of the market, the Alberta game commissioner pointed out in 1908 that Alberta's game regulations offered a seamless succession of hunting opportunities for all: open season began in late August for ducks and geese, on 1 September (when it was still safe to travel in the mountains) for mountain sheep and goats, and on 1 October for game birds. Pronghorns could also be hunted in October, while deer, moose, and caribou rounded out the hunting season in November. Since tourist hunters spent "liberal sums of money," stretching out the length of their visits would "mean more money left in the province." And as the Department of the Interior observed in 1923, game conservation would provide "a perpetual source of profit," since "Canada is destined soon to enjoy a practical monopoly [of big game] on this continent" because of depletion in the United States.[37]

Realizing the value of these tourist attractions required promotion as well as improved infrastructure. Unlike the national parks branch, prairie provincial governments only began promoting wildlife as a tourist attraction during the interwar years. Although its efforts were modest compared to those of some eastern provinces, Alberta had the most extensive advertising programs in the region. In 1932 its publicity bureau placed advertisements in the United States, Ontario, and Quebec and produced tourist brochures and maps. Watching wildlife was one attraction among others at the mountain parks of Banff, Waterton, and Jasper and along with hunting was often advertised as a reason to visit other parts of the province as well. Alberta's fish and game commissioner W.H. Wallace advised Premier Aberhart in 1939 that the province could double or triple its revenue from non-resident hunting licences with increased advertising in the United States and more rigorous conservation enforcement, so that "we have the game for them to shoot in sufficient numbers when they come here." A 1941 Manitoba report similarly recommended that the province expand advertising of its wildlife resources as tourist attractions and increase funding for game bird and fish hatcheries and improved game patrols to conserve fish and game for visitors.[38] Postwar planning underway in most provinces by about 1943 confirmed this thinking; hopes of American tourists flooding north to fish or to watch or hunt wildlife promised government revenue and jobs for people displaced by disarmament.[39] As was argued in Manitoba in 1945, the province had excellent big game and fish resources and "if we provide roads, canoe routes and proper living accommodation, tourists from the large centres of population to the South of us, who have not these advantages, will flock into

these areas."[40] While few roads were built specifically to create tourist access before the 1960s, the overall expansion of highway networks served this purpose. Completion of sections of the TransCanada Highway in Manitoba in the early 1930s had already opened lakes in the eastern part of the province to tourists, while a highway to The Pas later in the decade gave tourist anglers and hunters access to the Saskatchewan River Delta. There was also some growth in tourist fishing in areas accessible by air in northern Manitoba and Saskatchewan, as well as in guided hunting treks in the foothills and mountains of Alberta and in duck and goose hunting by Americans in the southern prairies. In 1947 a highway to Lac La Ronge

Figure 4.4 Although not as large an industry in this region as in eastern Canada, adventure sport hunting in remote districts was an attraction for wealthy tourists. This hunter and his Aboriginal guide brought down a moose in 1924 at Reindeer Lake in northern Saskatchewan.

in northern Saskatchewan was completed and sport angling was encouraged by closing the lake to summer commercial fishing in order to save the lake trout for tourists. Commercial fishing was still allowed in winter but primarily "to reduce whitefish numbers to protect game fish stocks." Even with poor facilities, 6,000 tourists made fishing trips to the village of La Ronge in 1951.[41]

Despite such developments, the wildlife-based tourist sector remained a marginal one in the prairie economy. Since Manitoba and Saskatchewan lacked Alberta's conventionally appealing mountains and foothills, both provinces carefully advertised that most of their land mass consisted

Figure 4.5 Many tourists were drawn to the mountain parks to watch wildlife. These tourists at Banff in 1966 stopped alongside the highway to feed deer.

of lakes and forests that offered unique promise to visitors, especially hunters and anglers.[42] But neither province offered hunting or fishing opportunities that were much different from those in Ontario, Quebec, and the Maritimes, which were closer to the densely populated northeast United States, the largest source of foreign tourists for Canada. In 1955 Saskatchewan attracted less than 1 per cent of auto traffic entering Canada from the United States, while Alberta and Manitoba each drew just under 2 per cent.[43] Nonetheless, governments persisted in promoting a wildlife-based tourist sector in the hope of finding economic diversification. Residents occasionally complained about outsiders taking game or fish that rightfully belonged to prairie hunters and anglers, but such concerns were easily overcome or ignored.[44] Tourist hunting entrenched and confirmed the priorities and work of fish and game associations, government departments, and agencies such as DUC. All saw tourist hunting as adding economic value to game conservation programs and sport fish and game bird stocking programs. It also validated their view of hunting as sport.[45] Although there was evidence to the contrary, one delegate told the 1941 meeting of the Saskatchewan Fish and Game League that American hunters practised "splendid sportsmanship" and "brought their own dogs and hunted more for the sport than for the number of birds they got."[46]

CONSERVING MIGRATORY BIRDS

The utilitarian goals promoted by game regulation were also important objectives for those working to protect migratory birds, an increasingly critical issue by the early 1900s, as habitat change and hunting drove many species towards extinction or extirpation in many parts of the continent. As in the case of bison, Canadians proportionately contributed to depletion, but the main problem arose in the United States. A wide range of birds, from songbirds to raptors, were killed for recreation and food, while cranes, waterfowl, and other birds were shot to stop them from eating crops. Many wild birds with attractive plumage were also shot for the millinery trade. This was curbed in the United States by the work of the Audubon Society (an organization dominated by women), which succeeded in making it socially unacceptable to wear wild bird feathers and the skins of songbirds. In 1915 Canada banned the importation of the plumage of most wild birds, but this apparently only shifted demand to the plumage of native ones. As the dominion entomologist Gordon Hewitt recounted that year, "only the other day, in an electric car, I happened to look down at the hat of a lady in front of me and imagine my disgust to see the skins of two chickadees, one of our most beautiful, most useful and most widely distributed birds, 70 per cent of whose food consists of injurious insects; those beautiful and useful *native* birds had been sacrificed simply to satisfy the thoughtless and wanton craze of fashion."[47]

Solving the migratory birds crisis demanded broad legislative commitment because the birds had to be protected all along their migration routes as well as in their summer and winter nesting grounds. Showing the triumph of good will, the problem was tackled by Canada and the United States in 1916 with the drafting of the Migratory Birds Convention. The Commission of Conservation lobbied for its adoption, and its success in this matter was one of the commission's most important and lasting contributions to Canadian society. The Migratory Birds Convention created a continental system of hunting seasons according to species and location. It banned the killing of insectivorous birds at any time (thus making bird protection legislation such as in Manitoba and Saskatchewan redundant) and limited the length of open seasons on any game bird. It also prohibited the killing of game birds during breeding and migration periods and required the establishment of bird sanctuaries and refuges to mitigate the impact of habitat loss. While controversial among some groups in the United States, the convention found support from local and national bird clubs, fish and game associations, and gun and ammunition manufacturers that recognized that sport hunting, given current rates of killing, had no future without conservation.[48] The convention was less controversial in Canada. While all provinces were required to align their legislation with it, the federal government gave them "as much freedom as possible" to adjust their game laws, "so long as the letter and spirit of the treaty was not violated." The most important revision to existing provincial legislation prohibited the hunting of cranes, swans, and some shorebirds for ten years. Prairie game officials were enthusiastic about the convention, and Saskatchewan's game commissioner called it "the most advanced step ever taken for the legal protection of wildlife."[49] Many provinces were already restricting spring duck hunting, and the convention also banned spring goose hunting, which many provincial game officials favoured but had been unable to implement because of local resistance. Canadian fish and game associations generally supported the convention because it promised to conserve birds for sport hunting. Further demonstrating the convention's utilitarian bent, farmers were brought on side because it protected insectivorous birds and allowed migratory birds that were eating crops to be shot under permit. Moreover, "harmful" birds, identified as including some raptors and corvids like crows and magpies, were excluded from the convention.[50]

The administration of the convention fell to the Department of the Interior's parks branch, which in 1917 created a wildlife section to administer it. Ornithologist Hoyes Lloyd was hired as its head the next year, and in 1919 it became known as the Wild Life Protection section. Its jurisdiction gradually expanded as it took on the administration of regulations under the Migratory Birds Convention as well as the Northwest

Game Act, which set game regulations in the far north. The Wild Life Protection section also advised the national parks branch on wildlife matters. Supported by an advisory board made up of representatives from various federal departments, the section gradually expanded its scope, hosting a federal-provincial conference on game conservation in 1919 and again in 1922.[51] More staff was taken on, including the appointment in 1934 of Dewey Soper (who had studied biology under William Rowan at the University of Alberta) as the chief migratory bird officer for the prairie region. The section's work revealed the gradual emergence of a more scientifically based approach to managing wildlife on federal lands. At first it devoted its energies to coordinating the implementation of the Migratory Birds Convention with the provinces and dealing with ongoing changes in its regulations. It also issued permits for the collection or killing of protected birds, appointed paid and voluntary game guardians, and helped establish bird sanctuaries. Reasoning that aesthetic appreciation and the cultivation of public interest in birds would aid in their protection, the section mounted a number of educational programs. In 1923 almost 90,000 posters and pamphlets on bird protection were distributed, as well as 180,000 copies of pamphlets on birdhouse construction. Further, about 80,000 copies of the pamphlet *Bird Lessons* were reprinted, new titles were under development, and a copy of Gordon Hewitt's *The Conservation of the Wild Life of Canada* was given to every Boy Scout who passed the scout's naturalist test.[52] The Wild Life Protection section also created a new administrative focus on coordinating bird-banding programs and the collection of data about bird populations, movement, and migration patterns, all essential aspects of bird-population management.[53] In 1947 the section was abolished and replaced by an arm's-length agency, the Dominion Wildlife Service (renamed the Canadian Wildlife Service [CWS] in 1950), which consolidated a number of personnel from various agencies into a single unit. The service became responsible for advising the parks branch on wildlife management and also carried out research, surveys, and policy studies on wildlife. This marked an important shift in national wildlife policy and indicated, as historian Graham MacDonald concludes, that "wildlife was starting to be seen in landscape contexts other than those of just parks and wildlife sanctuaries."[54]

Despite its achievements, the Migratory Birds Convention had its critics. While bag limits were added in 1920, Jack Miner, who ran a famous bird sanctuary at Kingsville in southwest Ontario, found the convention too generous to hunters, wherever they lived, and others complained that it merely gave "still further opportunity to the machine-gun artists, our neighbours to the south." While the permitted levels of hunting were indeed high, this was offset by the fact that the convention provided an opportunity to transcend political borders to conserve migratory birds. And there were other benefits as well. Jack Miner recognized that the convention had helped save whistling swans from extinction, and others recognized that the convention was also helpful to frame efforts to save other endangered birds, such as whooping cranes, one of the two crane species of North America.[55]

Efforts to conserve whooping cranes revealed that while bird conservation legislation was largely a utilitarian undertaking, it could also be used to achieve other objectives. While preserving whooping cranes offered no economic benefit, efforts to save them demonstrated an intellectual and emotional commitment to the region's natural heritage in general and to a better future for a magnificent bird in particular. Although some protection was already in place by the time of the convention (Saskatchewan had already banned killing them), the Migratory Birds Convention set a closed season on whooping cranes for ten years. While they apparently were never as plentiful as was later popularly assumed, the birds were careening towards extinction by the mid-1920s. Fred Bradshaw, first as Saskatchewan's game commissioner and then as director of the Saskatchewan Museum of Natural History, played an important role in their preservation. As he lamented in 1924, "one of North America's great birds, great not only in size, but in majesty of appearance, and in beauty and plumage" was being sacrificed "upon the altar of man's innate selfishness and insatiable mania to kill any or all forms of wild life, no matter how harmless, no matter how beautiful, and no matter how little value to the killer." Bradshaw hoped the cranes could be saved by popular effort, and in 1929, on the urging of Percy Taverner of the national museum in Ottawa, he kick-started a campaign to save them, gathering data on the birds by sending 1,500 questionnaires about them to people in North and South America.[56]

The rescue of whooping cranes became a popular cause. By 1949 only thirty-four birds remained worldwide and alarm increased as their number declined to twenty-one in 1952. Adding to their mystique, the location of their summer nesting grounds in Canada was unknown, and it was noted in the summer of 1945 that "somewhere in western Canada these rare birds are now nesting. If the nests can be found and every precaution taken to preserve them, this fine bird may, with good luck, be preserved." By the mid-1950s government and other agencies, as well as volunteers from the public, were working through international networks to protect whooping cranes, and in 1966 a co-operative program between the United States Fish and Wildlife Service and the CWS formalized these conservation efforts. The Aransas Wildlife Refuge in Texas was the focus of American efforts to preserve the cranes in their winter grounds, and their summer grounds, which were discovered in Wood Buffalo National Park in Alberta in 1954, were secured with little difficulty, partly because the CWS kept the exact location a secret. By the 1940s whooping crane rescue had become dominated by scientists and professional wildlife managers, who sometimes acrimoniously debated whether captive (or at least controlled) breeding or broader habitat and protection strategies were the best methods of protection.[57]

To the public, whooping cranes were media stars. Their migration and health were widely reported, and their immense 4,000-kilometre six-week flights in fall and spring across the continent were tracked in newspapers and by official agencies and volunteers in both countries.[58] Hunters were urged not to shoot any large white bird so that they did not accidentally kill a whooping crane. While many people, from the Gulf of Mexico to northern Canada, were committed to the cause, Saskatchewan was a hotbed of fascination. The cranes' flyway passed over the length of the province, and the *Regina Leader-Post* editorialized in 1955 that the number of Saskatchewan people concerned with the birds was "astonishing." It was not, however, only proximity that created interest. Fred Bradshaw's work on crane conservation focused the attention of people in Saskatchewan on the issue, and Fred Bard, Bradshaw's successor as director of the Saskatchewan museum, energetically expanded the effort. For both men, saving the whooping crane was a personal mission and a way of asserting positive social values. Bard confessed that he believed that the birds would likely become extinct despite all the effort, but he felt that attempting to save them was a duty of citizenship. Each generation was obliged to engage the unique problems that it faced, and he thought that the plight of the whooping crane reflected the "changing conditions and complications of our age" that threatened not only wildlife "but our civilization as well." In 1958 such sentiments prompted the prairie writer Dorothy M. Powell to write a story about a boy who prevents the shooting of a whooping crane. In this moral tale, the saving of the bird marks the boy's emerging maturity and ethical sensibility. The story doubtless resonated with the many readers who had also concluded that saving the whooping crane marked one of the ethical tests of their time.[59] And the test was passed. Although still extremely fragile, whooping crane populations have increased and today there are 600 birds in captivity and in the wild. They have come to symbolize that the history of the human encounter with North America's wildlife can be more than a record of despair.

MANAGING WILDLIFE

Game management regulation on the prairies has over time encompassed a sometimes bewildering succession of open and closed seasons, hunting bans, rules about the age and sex of animals that could be hunted, and ever more finely drawn stipulations about how animals could legally be lured and killed. These provisions, however, were not mere technicalities but addressed a variety of concerns. As historian Tina Loo shows, game regulation referenced economic requirements and attitudes about gender, race, and social authority. It also drew, although often imprecisely, on scientific assumptions about the forces that governed animal populations and natural environments.[60] Wildlife management was loosely framed within the view that the animate world operated as a balanced system or as what was sometimes called "the economy of nature." This was an ancient concept. Aristotle, for one, had proposed a version of it. Its mechanics, however, were untested and vague, but even so, as ecologist Frank Egerton notes, it remained a powerful "background assumption" that informed scientific discourse well into the twentieth century.[61]

The idea that nature operated as a balanced system was also sufficiently loose that it could reflect various interpretations. Robert Bell of the Geological Survey, for example, contended in 1901 that "nature maintains a balance," but he also argued that "no species goes on increasing indefinitely," since the geological record indicated that vanished species greatly outnumbered living ones. While he did not speculate about earlier times, Bell concluded that human activity had now probably become the most important cause of extinction. He resolved the apparent contradiction between the evidence of the disappearance of so many ancient animals and the existence of a balance of nature by explaining it in terms of natural theology – as part of God's plan. As he remarked in 1907, because "all the processes of nature and all the properties of matter [are] so exquisitely balanced, that they support each other exactly, and are interdependent on each other," this could only be the result of God's design. Similarly, Saskatchewan's provincial taxidermist Horace Mitchell contended in 1913 that nature was a "regulated and balanced" hierarchy of relationships, with "each species kept within its bounds, in its place intended for it by a wonderful Creator." Humans were "often guilty of seriously interfering with the laws of nature," which made wildlife depletion yet another entry in the list of human moral failings.[62]

There were, of course, other ways to explain the phenomenon. Charles Darwin more or less accepted that nature operated as a balanced system, even though, as ecologist John Kricher observes, this seemed to be "hard to reconcile with his view of natural selection acting opportunistically, adapting populations to changes in their environments." But clearly, new species appeared and others disappeared, and Darwin explained this through what Kricher calls one "of his most famous metaphors, the wedge." Since nature was a balanced system, Darwin presumed that "all of the available space in the economy of nature must be filled." But if this was so, then how did new animals enter already fully occupied habitats? Darwin concluded that they "wedged" their way by force into interstices in the "face of nature," gradually occupying ever larger space. This process also helped define the balance of nature as the "set point" to which an ecosystem stabilized after being disturbed. As Professor V.W. Jackson of the Manitoba Agricultural College explained in 1924, "every living thing bears a biotic relation and dependence upon some other living thing. This stable relationship is called the balance of nature." The balance re-established itself quickly when upset, although for some animals, such as rabbits and their predators, the reset could take several years after the low point in their cycles. A related concept held that forests moved through a series of developmental stages, reaching a climax stage that was stable for long periods.[63] This made the balance of nature a zero sum game, and the arrival of house sparrows and starlings, for example, was commonly condemned as an introduced species that forced out native (and thus natural and rightful) birds and upset the natural balance.[64]

While belief in a balance of nature offered a general explanation of how natural systems worked, it provided imprecise guidance for wildlife management, especially given the limited knowledge about animal biology and ecosystems. Nonetheless, it could be concluded that if nature was an inherently balanced system, disturbance and depletion could largely be explained by forces outside of "nature." The answer was then not hard to find: humans were responsible for upsetting this natural balance. This was confirmed by the well-known examples of the extinction of the passenger pigeon and the near extermination of bison, elk, pronghorns, some fur-bearers, and many species of birds. The task of management thus became the prevention of future depletion by managing human behaviour while respecting the need for economic development and the benefits that "useful" wildlife such as game or fur-bearers provided. Such thinking was evident in Gordon Hewitt's 1921 book on Canadian wildlife conservation. Showing a strong commitment to the accommodation of conservation and economic goals, Hewitt explained that depletion was the result of avoidable and unavoidable causes. Natural forces such as fire, disease, and climate all unavoidably caused depletion, but Hewitt thought that these were of minor importance, which was logical in light of the belief that natural systems were inherently stable. Rather, Hewitt blamed most depletion on human agency. Showing a commitment to utilitarian objectives, he argued that some human activity was unavoidable because an advancing civilization inevitably overwhelmed the flora and fauna that stood in its way. While he thought that its scale "was out of all proportion to its necessity," he believed that the decimation of the bison was unavoidable because "from an economic standpoint the case was against the existence of the buffalo in anything approaching large numbers" given the priority of farm settlement. The decimation of pronghorns was another example of such an unavoidable upset because of farming. In this view, the goal of game conservation was to "maintain the remnant of those animals whose reduction to the verge of extermination" had been caused by the necessary "economic development of the country." Hewitt also believed that the control of predators was another unavoidable consequence of human activity because the resulting upset of the balance of nature brought not just depletion of favoured animals but a disproportionate increase in their predators. But the greatest loss of animals, especially of birds, arose from ignorance and greed. Particularly serious was commercial – or what was called "market" – hunting. Killing wild animals and selling the meat compounded rapacity with carelessness and human aggressiveness and created a vicious circle of destruction. It therefore had to be banned, and since subsistence and sport hunting could also be "disastrous," they, too, had to be "wisely restricted by law." Hunting regulations needed to be crafted to preserve balanced game populations. In keeping with this principle, the killing of female game animals and shooting of game birds before and during nesting in spring needed to be banned. As Hewitt concluded, "let it be remembered that the extermination of any animal cannot be prevented unless such an animal is sufficiently protected to obviate the danger of its destruction at a greater rate than its natural increase."[65]

While the destruction of game by predators was "slight" when compared to "the rapacity of man," it was, in any case, yet another demonstration of human impact. As Hewitt argued, "when animals become reduced in numbers through man's improvidence, then their natural enemies," which had not similarly been reduced, were able to "take an unnatural and abnormal toll. The usual balance of nature is completely upset, and the remnant is exposed to excessive numbers of their enemies," which then had to be reduced.[66] While Hewitt saw this as a minor issue overall, this view gained popularity in the interwar years. Hunters were strong advocates of predator control so that animals could be saved for sport hunting, but the killing of wolves, coyotes, cougars, raptors, and piscivorous birds was part of an ancient competition between farmers, trappers, and fishermen and the wildlife that destroyed crops, livestock, poultry, and fish. Despite a handful of critics who contended that predators were an essential part of the balance of nature and that killing them only further upset this balance, governments and fish and game associations adopted a range of control programs. Justified on the grounds that they helped correct imbalances in nature created by human activity, these programs reflected the apprehension that the human-caused upset of the balance of nature had enhanced the violence of a nature already red in tooth and claw. Indeed, some supporters of predator control simply jettisoned the idea that a natural stability was even still possible. Echoing arguments used in the United States to justify predator control, the Calgary Zoological Society contended in 1931 that the "balance of nature" had been lost and it was necessary "to foster the wild birds and animals of our choice" by killing the predators of song and game birds and by enforcing the game laws. In other words, wild animals and their environments could be managed like farm animals. There was little regret about depletion in this argument; only utilitarian needs mattered and what was left of "nature" could be redesigned to serve humans.[67]

The focus on controlling hunting and predators to prevent depletion left other factors (such as habitat destruction) outside the concerns of wildlife management. There was no willingness by the state to control land development in any significant way to conserve habitat for wild animals. As W.N. Millar frankly noted in his 1915 study of big game in the Rocky Mountains, many factors contributed to the depletion of big game, including climate, disease, and predators as well as loss of habitat and hunting, but only hunting could be controlled "without conflicting with any fundamental rights." In other words, habitat protection would interfere with rights to property and private use of land. The only acceptable solution was to set up sanctuaries where hunting was banned and habitat could be preserved. Gordon Hewitt agreed with this strategy, calling "wildlife preserves, refuges or sanctuaries" essential protective measures. They needed to be large enough "to provide ample summer and winter range" and have clearly defined boundaries that could be patrolled to enforce protection. They also needed to be located in areas where they would not conflict with settlement and industry. Supporting the view that the commercial value of land was greater than the commercial or other value of the wild animals on it, he argued that game preserves should, as a rule, be located in areas "unsuitable for agricultural development" and should not include "mining or other commercial properties that are likely to interfere with their purpose." One proponent argued at the Commission of Conservation in 1913 that the huge scale of environmental change brought by Euro-Canadian settlement and industrial development made it essential to establish sanctuaries, which he described as "places where man is passive and the rest of Nature active. A sanctuary is the same thing to wild life as a spring is to a river" and "its surplus must overflow to stock surrounding areas" to benefit settlers and hunters. Utility thus remained a touchstone, and in keeping with this assumption, it was further reasoned that it was sometimes necessary to kill predatory animals in sanctuaries to ensure "a *maximum* of gain and *minimum* of loss."[68]

Approaches that placed the onus on controlling hunting behaviour, not economic activity, were well established in prairie wildlife management by the end of the First World War. By then, over forty years of legislative activity had created a complex system of hunting restrictions and allowances that aimed to allow sustainable levels of hunting and trapping while maximizing production of animals. This was one goal of Manitoba's 1876 Game Protection Act, for example, and its subsequent amendment reflected the refinement of this principle. Hunting and trapping were prohibited during breeding or nesting periods or when animals were gestating, which in the case of fur-bearers was also when pelts were of the poorest quality and lowest economic value. But each species was regulated as a relatively autonomous unit based on its particular usefulness, which required the ongoing fine-tuning of regulations to maintain an acceptable balance. In the case of moose, for example, in the forty years after 1876, Manitoba's Game Protection Act was amended eleven times to change open hunting seasons for them. Suggesting that regulation was failing to achieve an ideal balanced state, there was a clear trend towards shorter and shorter seasons: a five-month open season in 1876 was reduced to four months in 1879 and to three months by 1883. In what seemed to reflect a growing desperation to make the management system work, in 1896 and 1897 hunting moose (as well as all big game) was banned. When it resumed, open seasons were steadily shortened, dropping to two weeks in 1914. Hunting was then prohibited in 1915 and allowed for only one month in 1916. Killing male animals less than one year of age was banned in 1916. Other restrictions had been imposed in 1900 when a bag limit of two males was set and the killing of fawns and females of any age was banned. Similar restrictions on age, sex, and length of hunting seasons were also set for other big game animals. In the case of deer, the legislation was called the "buck law" because it only permitted the hunting of adult males so that females were left to reproduce. So, too, bag limits for some grouse were imposed in 1898 and were extended to other game birds in 1900.[69] All market (or commercial) hunting was outlawed in Manitoba in 1917, and the prohibition of various "unsporting" practices and devices expanded steadily. These measures applied the ethical standards of sport hunting as a game conservation technique and came to officially establish what was considered to be socially respectable and sustainable game hunting.

In operational terms these wildlife management systems depended, even if loosely, on knowledge about the workings of natural systems and animal biology so that the desired balance of nature could be achieved. Yet despite the remarkable certainty that game officials frequently showed, such knowledge was often limited. In 1909, for example, a study undertaken by the Commission of Conservation on Manitoba's commercial fisheries reported that "considerable haziness exists as to the habits and movements of Manitoba fish." It was not known what whitefish ate, where they spawned, or at what age. This lack of knowledge existed in regard to other fish species as well. Moreover, the Department of Fisheries applied the same regulations across the prairies without any regard to local conditions. Further, in the late 1800s and first half of the 1900s, knowledge of wildlife populations was extremely limited and very little was known about predation. Academic studies of birds and mammals were only emerging, and the cycles of animal populations were largely unknown.[70] In 1931, avian biologist William Rowan noted, for example, that Alberta's waterfowl bag limits were based on errors of fact and interpretation. He contended that waterfowl had been unusually plentiful in the early 1900s because the region was in a wet cycle. Since Alberta's game legislation was drafted during this period of abundance, bag limits had been set higher than was justified by long-range population trends. To his mind, the province's legislation owed more to the accidents of history than scientific evidence.[71]

Preserving a supposed balance of nature to benefit game populations through closed and open hunting seasons was fraught with problems because population counts were difficult and there were no historical population baselines. Open seasons were set by reference to estimates of the size and health of game and fur-bearer populations and levels of hunting, but these were often based on little more than hearsay. Even when full-time game guardians appeared in the field in the early 1900s, the large size of their territories often forced them to base their estimates of game populations on reports from hunters, hardly a neutral source because this information was used to set upcoming bag limits and open seasons. Regulators also used whatever other information they had. The Alberta game branch in the 1920s, for example, used statistics based on the number of game birds that hunters had put in cold storage the year before to set the bag limits for the coming year.[72] A short collective memory owing to the recentness of Euro-Canadian settlement compounded these problems. Research for this book did not uncover any evidence that prairie game managers drew upon the traditional knowledge of Aboriginal people to gain an understanding of animal population dynamics. Geographer Sandie Suchet has argued that the devaluation of indigenous knowledge by wildlife managers in Australia was part of a colonizing discourse that marginalized Aboriginal peoples and imposed a European view of nature. And, certainly, indigenous knowledge was sometimes devalued on the Canadian prairies. Manitoba senator John Sutherland, for example, commented in 1887 that while Euro-Canadians did not know why rabbit cycles happened, Aboriginal people would not know either, and if they did, "it is not at all likely that they would have the correct version of it."[73]

By the interwar years, it was increasingly admitted that the knowledge base for game management was inadequate. In 1927, for example, the Manitoba game and fisheries branch argued that research was "indispensable to effective administration" because knowledge was power and laws "not based on knowledge cannot be effective."[74] Such conclusions were reinforced by the growth in the United States and Britain of scholarship about ecology and zoology. In his book *Nature's Economy* historian Donald Worster details the long, contentious, and often contradictory development of various streams of scientific thinking about ecology in the United States. Yet, throughout these diverse approaches, there was a growing conviction that an understanding of natural dynamics promised more effective ways of maximizing production of wildlife. Interwar research – some of which built on earlier studies – developed new theories about sequences and climax stages in ecosystems, carrying capacity, human activity as an ecological factor, and, among other things, the role of predation in natural systems. Some of these studies purported to draw lessons from the history of deer conservation at the Kaibab National Forest in Arizona where, beginning in 1907, forestry officials had eliminated predators such as wolves and cougars so as to increase deer numbers for hunters. As a result, the deer population exploded and stripped the forest, and in 1925 and 1926 tens of thousands of deer starved to death. It was, as historian Thomas Dunlap terms it, "a conservation horror story."[75]

Changing scientific approaches were also reflected in Charles Elton's *Animal Ecology*, which became a founding text shortly after its publication in 1926. As Dunlap notes, Elton later commented that, at the time of the book's publication, ecology "was just beginning to move from 'easy generalizations about adaptations and the balance of life' to the new world of 'limiting factors,' 'food chains' and 'quantitative measures.'" These concepts "put flesh on the skeleton of the 'balance of nature'" and "provided ways to analyse animal communities and their relation to plants, to order data, and to classify communities."[76] The view that the environment operated as an ecosystem (a term that came into use in the 1930s) confirmed the value of studying the interactions and balances among living things and their environments. The belief that wildlife management was a coherent field of scientific study was further confirmed by the appearance in 1933 of Aldo Leopold's *Game Management*, which became an influential text for wildlife management in the United States and Canada. It proposed that game populations be inventoried and measured and that factors that limited production be assessed. Through such approaches, the lack of knowledge that had hampered the management of game for economic and social production could be overcome, giving new vigour and life to the doctrines of utilitarian conservation.[77]

Following the Second World War, the use of scientific study expanded in prairie wildlife management. Prewar calls for wildlife research that had been stymied by the Depression were now beginning to be heard because of improved government finances and the need to cope with the postwar increase in sport hunting and industrial development. The first aerial survey of deer populations was done in Manitoba in the Turtle Mountain area in 1947. By the late 1940s field and laboratory studies were being carried out by provincial governments and by the CWS on issues such as carrying capacity (the number of animals an area could support), predation, and the influence of ecosystems, weather, and disease on animal populations. Science thus promised to sharpen the ability to increase the number of wild animals. As the Saskatchewan government noted in 1947, most industries had "long accepted the use of science to improve and utilize their resources," and following this lead, the province hired a fish biologist to conduct studies to establish sustainable production levels of fish.[78]

These new ecological and scientific approaches shaped people's views about wild animals in different ways. For some it reasserted and refocused the idea that all wild creatures had value as a part of interconnected natural systems and that they possessed a range of qualities that made them more than mere units of production. At the official and scientific level, however, the new concepts justified the equally long-standing view that the goal of management was the maximization of wildlife production as a natural resource. Efficiency could be had through the ever more finely tuned management of the natural world. In a 1946 radio talk G.W. Malaher, director of the Manitoba game and fisheries branch, noted the promise of such approaches for the more efficient management of fur-bearers. "It is really a regrettable thing and one of which Canadians cannot feel proud," he argued, "that so little is known about many of our fur bearing animals." For 350 years Canadians had exploited them, "yet with many species the scientific knowledge of their life histories, food and shelter requirement and their breeding habits is insufficient for intelligent management to be undertaken. Until this knowledge is obtained, more adequate protection and regulation of the take are the only steps that can be taken towards proper management." Such approaches were quickly adopted in the 1950s. The Saskatchewan Fur Advisory Committee, for example, optimistically argued in 1958 that the management of predators needed to be based "on sound reasoning rather than on opinions." The old style of mass killing of predators had been shown to be unnecessary when control measures were properly calculated in terms of population and environmental conditions. When prey species were in an expanding state, for example, predator control was not needed because predators had little effect on population growth. When a prey species was small or declining, however, predators posed a serious threat that, in the case of barren land caribou, necessitated their control. Further, the committee noted that research on carrying capacity proved the case for adjusting human utilization of wildlife according to the animal's biological needs and an area's carrying capacity in order to maintain populations of "useful species" such as fur-bearers and big game.[79]

Such approaches were also used in an attempt to enhance the commercial value of fish, especially in Manitoba, which had the largest commercial fishery on the prairies. Although the federal and Manitoba governments had undertaken various research projects since the 1930s, these had provided limited practical knowledge for management. As was remarked in 1959, it had been assumed for the first half of the twentieth century that "only the intensity of fishing determined the abundance of marketable fish. Only over-fishing could reduce the stocks." It was now realized, however, that while "fish are looked upon as a crop to be harvested … unfortunately the waters do not lend themselves to the same intensive management as agricultural and range lands" because there were few ways to increase carrying capacity. Fisheries management instead needed "to maintain natural brood stocks to reseed the waters" and to draw upon a wide range of information about size and age of populations, the effects of various environmental factors, and the probable impact of commercial fishing. While regulation remained focused on preventing overfishing, it was now viewed through an ecological lens. Although it was increasingly understood that ocean fish responded to a range of factors, of which commercial fishing was only one element, it was also increasingly clear that the management of freshwater fisheries was even more complex because freshwater environments changed "more violently than those in the ocean." Because freshwater fish were subject to "fluctuations which are of wider amplitude than could be expected from changes in fishing efforts alone," methods of studying fish populations by statistical analysis of catches were inadequate and factors such as water temperature, turbidity, and currents also needed to be assessed. Moreover, as Manitoba's director of fisheries explained, it had been found that "the effect of intensive fishing directed against only one or two of the more valuable commercial species could not be assessed by studying catch statistics and environmental factors alone." Attention also had to be paid to how many fish of each kind were caught, "since changes in abundance of one affect the abundance of other species in their struggle for survival." Age was another relevant factor because more intensive fishing reduced the number of larger, older fish and consequently of desired species. Clearly, further study was needed on the life histories of fish so that appropriate regulations could be developed for commercial fishing and maximization of the harvest.[80]

While lake environments complicated the study of fish, the management of big game in the 1950s drew upon practices that married modern ecology with the history of livestock husbandry. This was revealed in management strategies adopted for white-tailed deer, the most widely hunted big game on the prairies after the Second World War. These strategies endorse Tina Loo's observation that management became more interventionist as it became more professionalized and scientific. As she notes, by the 1950s concepts about carrying capacity had come "to be interpreted in a way that assumed that environments *should* produce a certain amount of biomass. If they did not – if there were too many animals or too few – then something was wrong" and populations had to be culled or increased to match the productive potential of an area.[81] And as historian Thomas Dunlap notes, North American game management now assumed "that deer had to be managed in both directions – to increase numbers when deer were scarce and to cut numbers when deer became too abundant."[82] Drawing an analogy between the management of nature and that of animal husbandry, a pamphlet published by the Saskatchewan government in the mid-1950s argued that white-tailed deer management was "not much different than cattle or sheep management." A rancher let his cattle or sheep herd expand until the range was fully grazed, at which point the animals were moved to a new pasture. The equivalent of pasture overgrazing in deer management was over-browsing, but this was more difficult to judge because it was "a silent, almost imperceptible wave of destruction sweeping through the aspen groves" that could take years to become apparent. Citing the forty-year-old case of the Kaibab in Arizona as "a classic example of unsound attempts at deer conservation" that had caused a population irruption, the pamphlet's authors contended that "the pressing question that must be asked is, 'Can such a catastrophe occur in Saskatchewan?'" The answer was "a definite and resounding 'yes'"; it was "a simple matter of mathematics" because steadily increasing numbers of deer would overwhelm food resources. In this context, the manipulation of open and closed seasons at the local level aimed to manage deer populations so as to maintain sustainable carrying capacity. In the early 1950s Saskatchewan restricted the hunting of deer because populations were low but allowed more hunting when populations had recovered. These methods were now justified by science: "the conservationist, whether professional or non-professional, must be an apostle of the common-sense approach to game management." This sometimes meant control of an entire population through open seasons on all deer, including fawns, bucks, and does. Such "any deer seasons" were said to be the only solution to prevent over-browsing and consequent population collapse. The public needed to "discard the invalid concept of deer management by protection only" in order that sustainable population levels from which everyone would gain could be achieved. Economic spillover in small communities from visiting hunters would grow, the public would continue to enjoy watching deer, and hunters would have long hunting seasons and ample game.[83] There was no assumption in this plan that natural predators like wolves and coyotes (against which Saskatchewan was then waging an extermination campaign) could play a role in deer population control. Similar management policies were in place across the prairies and the rest of North America. Ontario had an open season on does in 1951, and Manitoba used early opening of deer hunting seasons, long open seasons, and "any deer seasons" for most of the 1950s to prevent deer from harming crops and to reduce populations that aerial and ground surveys revealed were too high for the capacity of the land.[84] These policies were aimed to maximize economic and social benefit. Any animal that died because of starvation or was killed by predators was lost to human use. As the Winnipeg Game and Fish Association heard at one of its 1956 meetings, "conservation becomes a farce, if a large portion of the deer crop is allowed to go to waste. It is now fairly well established that a 'both sex' deer season is in reality a conservation project by reason of the fact that an 'only-male' season results in a high percentage of barren does."[85] In practice, this meant that population stability would be achieved through hunting. Natural predators such as wolves and coyotes had been replaced by human predators in a nature constructed by science to serve social and economic needs.

NEITHER A FRIEND NOR A FOE

In the early 1950s William Rowan argued that while hunting and conservation were not at odds, the role of predators was misunderstood by most hunters. Although he agreed that too many predators could "make a sizable dent in the population of prey species," he concluded that irresponsible hunters and poachers did "incomparably more harm" than predators. This presumption that all animals in an ecosystem had a positive role was only slowly adopted by wildlife managers who remained wedded to older and blunter management tools. Nonetheless, a shift in public rhetoric about wildlife management was emerging as part of broader trends in North America. In 1948 the American game management theorist Aldo Leopold published "The Land Ethic," which as historian Donald Worster notes, "brought together a scientific approach to nature, a high level of ecological sophistication and a biocentric, communitarian ethic that challenged the dominant economic attitude toward land use." In part, this essay rejected the centrality of economic value that Leopold had posited in his book *Game Management*, written fifteen years before. This move away from economic value and human domination of nature posited that wildlife had intrinsic values and that control and use of land was not a right but entailed certain obligations. This was a moral issue, and it was also, Leopold now believed, essential for human survival.[86]

After Leopold's death, "The Land Ethic" was included in his posthumous book *The Sand Country Almanac*, published in 1949. The book at first did not resonate with the public, but it gained mass appeal in the early 1960s and became influential in the emerging environmental movement.[87] *Silent Spring*, by the American marine biologist Rachel Carson, had a more immediate and even greater impact when it was published in 1962. Her warnings about the impact of pesticides and herbicides on animals and humans, as well as on water and soil, ignited latent concerns about personal health and environmental conditions. She also further popularized the idea that the elements in any natural system could only be understood in practical and policy terms within an interdisciplinary framework that recognized mutually reinforcing connections. *Silent Spring* had a major influence in Canada and worldwide, and its publication has come to be seen as a marker in the emergence of new sensibilities about environmental issues. Canada, too, contributed an influential manifesto advocating a new relationship with nature. The same year that Carson's book was generating heated debate, Farley Mowat's *Never Cry Wolf* was received with enormous public enthusiasm. Part of a long Canadian literary tradition stretching back to Ernest Thompson Seton and Sir Charles G.D. Roberts, it echoed concerns that had ebbed and flowed in popular discourse for nearly a century. Mowat's criticism of modern wildlife management's quest for efficiency and productivity struck a chord in a society fearful that its natural heritage was being perverted by technocratic and mechanistic dreams of control.[88]

Such writing was part of a broader discourse about managing the encounter with wildlife. New environmental organizations appeared, expressing a huge range of strategies, approaches, and goals. As only one example, the issue of land and habitat was tackled in a new way by the Nature Conservancy of Canada, established in 1962 to conserve wildlife and habitat through acquisition of land or by co-operating with private owners. These organizations pressured governments to pass new environmental legislation, and governments did act on some issues. Manitoba replaced its Game and Fisheries Act with the Wildlife Act in 1963; Alberta followed with a similarly named statute in 1970, and Saskatchewan did the same in 1979. These titles alone indicated a changing attitude, as did the subsequent inclusion of any wild vertebrate, with the exception of fish, in the wildlife legislation of all three provinces. This meant that amphibians and reptiles as well as non-game mammals and birds now merited official concern. So, too, legislation in all three provinces began to include provision for habitat protection and wildlife management areas where land and habitat could be managed in the interests of wildlife.[89] And in 1966 a new national policy for wildlife was tabled in the House of Commons. Concerned with the future of wildlife in an increasingly urbanized and industrialized society, it emphasized research, public education, and federal-provincial co-operation. Taking its cue from ongoing scientific evidence and wilderness legislation and debates about endangered species in the United States, the federal government pledged to encourage research on wildlife and conservation. This seemed to promise changes more in tune with current thinking that stressed the interconnections of the natural world as the basis for wildlife management.[90]

Even so, older priorities and utilitarian attitudes had not disappeared. For example, while the federal-provincial Committee on the Status of Endangered Wildlife was established in 1976, Canada did not pass legislation to protect endangered species until 2002 (unlike the United States, which did so in 1968). And the strength of older priorities was revealed in 1966 when the federal government launched a program to conserve waterfowl habitat by paying prairie farmers to stop draining potholes and burning adjacent vegetation. As Arthur Laing, the minister of northern affairs and national resources, explained, payments to farmers to preserve wetlands were merited because waterfowl were valuable for hunters and drew tourists to Canada.[91] The policy also mitigated conflict with farmers, who would have resisted being compelled to make room for waterfowl on their land. Similar attitudes remained with respect to land and resource development projects. While greater public concern about the effect of environmental change made it increasingly necessary for politicians to articulate a belief that planning for such projects needed to take account of wildlife, this was often of little effect. Prior to the completion of the Grand Rapids hydro station in northern Manitoba in 1962, for example, studies were undertaken about its potential impact on wildlife. These studies had minimal influence on the project; indeed, the Manitoba government argued that because the steadily rising waters behind the dam would flood out moose, a special regulation might be required "to ensure an adequate harvest" before the animals were forced out into surrounding areas that already had large moose populations.[92] While the undertaking of such studies at least marked a new public stance and therefore the possibility of changed approaches in the future, the fact that the environmental studies on the Grand Rapids project were largely ignored and an area of great biological diversity was severely damaged demonstrated the persistence of old attitudes and the lasting influence of utilitarian assumptions about wildlife and its place in the hierarchy of government priorities.

5

Eating Wild Animals

Larding is a necessary matter in cooking game, as wild creatures are so active that they are usually not so fat as barnyard fowls. Larding is simply drawing a small piece of salt pork through the surface of the meat. It is easily done, and improves lean, dry pieces of meat. Ducks are the only wild birds that are never larded. Venison is better roasted in a thick layer of paste made of water and flour … Serve all game very hot.

The New Galt Cookbook (1898), 53[1]

Food is about nutrition and calories. It is also about culture, making eating a complex social act that, historically, took on added dimensions when wild animals were eaten. Why, for example, did people accept or reject certain wild animals as food? How did practical needs for food intersect with these choices? How did depletion, commerce, conservation, and economic and social change shape the use of wild animals for food? In prairie Canada after 1870, these questions were further intercut by the social, legal, and economic differences between Aboriginal and Euro-Canadian communities, each of which showed differences on the basis of need, location, and custom.

While meat and fish were important in the diet of Aboriginal groups, not all wild animals were treated equally as food. The Rock Cree of northern Manitoba, for example, found most fish acceptable, but eating meat involved more complicated judgments. They tended to favour animals that were herbivores or, as they termed it, "clean" animals. Predatory animals and scavengers (except for bears) were "dirty," and the Rock Cree claimed that such animals did not taste good. At the same time, more subtle distinctions were sometimes drawn about an animal's diet in determining its suitability as food. For example, Rock Cree ate lynx even though they were carnivores. But lynx lived on rabbits, which are herbivores, so this made lynx acceptable food. In general, big game herbivores and bears were highly prized food, but fish, muskrats, porcupines, and beavers were also acceptable. Mink, foxes, wolverines, and martens were eaten in the 1700s but were considered inedible by the early 1900s. As Robert Brightman suggests in his study of the Rock Cree, these animals may have always provided marginally acceptable food but the availability of guns improved the Rock Cree's ability to kill preferred animals, thus allowing them to reject those of previously marginal acceptance. Whatever the case, by the 1900s, foxes, wolverines, martens, skunks, fishers, and weasels, all of which were trapped for their fur, were never eaten, even though it took no effort "beyond throwing them into a kettle." Mice, although herbivores, were never seen as acceptable food, and frogs too were avoided, a prohibition that Amédée Forget, clerk of the NWT council, claimed was found throughout Aboriginal communities on the prairies.[2] Other prairie Aboriginal people also showed a preference for eating herbivores. Plains people's reliance on bison was the most obvious example, although pronghorns, elk, and deer were also eaten, along with a range of smaller mammals and plant materials such as berries and Indian turnip.[3]

By tradition and custom, Euro-Canadians favoured eating big game herbivores such as deer, moose, caribou, and pronghorns, and Jimmy Sampson, a hunting guide at Banff, claimed in 1911 that mountain sheep were "bully good eating" and the "finest meat you can get."[4] Like the Rock Cree, Euro-Canadians usually avoided eating animals that ate carrion and/or were predatory, such as cougars, coyotes, and wolves. They did, however, eat bears, even though they are scavengers. In 1958 writer F.A. Twilley observed that early fur traders had eaten lynx and found it a pleasant change from moose, deer, and pemmican, but drawing on cultural associations of his own time, he thought that eating lynx was like eating a pet cat. Mice and rats were never eaten, and other than rabbits, small mammals like weasels, mink, and fur-bearers were usually not seen as acceptable food. Most fish were acceptable, although some people would not eat bony fish or fish with soft flesh or those that were too oily and "fishy." Among the latter were inconnu, a staple Aboriginal food in the

Fort Smith area of northern Alberta. But diet also responded to immediate needs, and pioneer settlers in some parkland areas ate a range of wild animals in the late 1800s, including badgers and porcupines. As naturalist Tony Lascelles wrote about porcupine meat in 1936, although it was "said to be toothsome," eating it was not recommended "unless starvation threatens." He claimed that "it was a custom at one time to protect the animal because it is the only creature that a person lost in the woods without food can kill with a stick. A case of needs must when the devil drives." As Euro-Canadian settlers became established, the range of wild animals they used as food narrowed. For example, they came to see most fur-bearers as not edible, and while it seems that all plains Aboriginal peoples saw nothing wrong with eating gophers, Euro-Canadians saw them as a peculiar but not offensive food. Those who tried it said it tasted like rabbit, and in 1964 one individual recalled that gophers were tasty and that men tramping across the prairies during the Depression ate them. Following their extirpation on the prairies, bison became a novelty meat for Euro-Canadians, who happily ate it when it was available, and Buffalo National Park at Wainwright in central Alberta found ready buyers for bison meat when the herds were culled. While it was only a small stretch for a beef-eating culture to accept bison as a colourful but unchallenging change, dining on other wild animals could convey local colour and provide fun at public gatherings. The dinner at the 1948 annual meeting of the Manitoba Registered Trap Line Officers in The Pas in northern Manitoba, for example, featured a soup made from beaver tails. Ignoring that this was a traditional Aboriginal dish, the *Winnipeg Free Press* reported that many places had unique food: Boston had baked beans, New Orleans had shrimp à la créole, and now Canada had "a steaming brew made from the appendage of the dominion's well-known animal emblem." Prepared by Eddie Hong of the Occidental Club and Cabaret, the soup was a fusion, containing parsnips, onions, carrots, potatoes, and "even tomato soup – a can or two just for good measure." It was amusement with little risk, although the newspaper cast it as an adventure from which the thirty diners suffered "no apparent ill effects."[5]

The general Euro-Canadian tendency to eat herbivores also applied to birds. Predatory birds such as raptors were not acceptable, but most herbivorous birds were welcomed as food. As one report in 1926 explained about mallards, "fish and animal substances form a very small portion of the mallard's diet," which contributed to the "delicate flavor of its flesh and divides it sharply from classification with all fish-eating ducks or mergansers." Grain-fed ducks were best, and in words of advice that would have made grain farmers shudder, it was noted that "when fed from the grain fields of the west, the teal surpasses, in the opinion of many hunters, the flavor of canvasback or mallard." For the same reason, nature columnist Doug Gilroy wrote that ruddy ducks, although small, were apparently good eating because they lived on "delicate grasses" and other vegetation.[6] In the late 1800s and early 1900s a variety of other birds were also seen as acceptable food, and poet Charles Mair enumerated the culinary value of a number of them in his notebook in the early 1900s. Among those earning the simple notation "good to eat" were whooping cranes, curlews, plovers, and grey gulls, while sandhill cranes earned a rating of "fine eating," as did bitterns even though they were "seldom fat." Yellow legs, a shorebird, were poor eating in the summer but were "a lump of fat in October" and were "best boiled in paste like an apple dumpling." The eggs of both black and white gulls were also good eating, but only the flesh of white gulls was good. In the fall, mergansers were "fat and the flesh is good," and when roasted, swans were "excellent," "cygnets very delicate," and a swan's egg was so large "as to make a meal for a man." Mair did not find young pelicans to his taste, although he noted that Aboriginal people ate them and extracted oil from the older birds and put it up "in [a] bladder and mix[ed] [it] with pounded meat but [it] is very rank." In the late 1800s small birds were thought to be edible, and in 1883 the Manitoba Department of Agriculture noted that snow buntings, which migrated from the north during the winter, were "a wondrously good bird when killed and roasted and made into pies." Some Euro-Canadians also ate songbirds in the late 1800s, although it is unclear how widespread this practice was. W. Davies, the MP for the District of Alberta, for example, noted in 1886 that they were good to eat but were too much bother to prepare. By 1920 eating songbirds was no longer acceptable because attitudes towards using birds as food were transformed by the growing commitment to protect insectivorous birds. Eating them – like wearing their feathers – was by then popularly considered not only to be ignorantly wasteful but vulgar as well. And these changing views also encompassed

Figure 5.1 Wild ducks were commonly used for food by homesteaders and workers in mining and lumbering camps. This cook in an unidentified location has prepared the birds for roasting. Photo not dated.

more emotive sentiments. In a letter to his family in the United States, written in 1923 from northern Alberta's Peace River country, Clyde Campbell described hearing a blackbird's "full liquid tones," which were so beautiful that "it gives one a creep up the back like divine music does." Reflecting changing sensibilities, he commented, "and to think that once I thought it rare sport to kill those little joy-bringers."[7]

PLANS FOR REVAMPING THE PLAINS ABORIGINAL DIET

Despite periods of want because of climatic and natural cycles, people on the prairies could still obtain adequate food and income from wild animals and plants when Canada took control of the region in 1870.[8] While almost all Aboriginal people had by this time become dependent on some imported goods and technology, most of those in parkland and wooded districts still largely lived on elk, moose, deer, rabbits, some fur-bearing animals, and birds. Fish (especially whitefish, suckers, and pike) were a staple everywhere in these districts, and a variety of wild plants, including wild rice in parts of Manitoba, supplemented this diet.[9] In contrast, the disappearance of bison had brought a food crisis on the southern plains by 1880 when the Department of Indian Affairs mused that Plains Cree should perhaps be encouraged to move north to areas with adequate food, while Blackfoot and Assiniboine would have to take up farming, as they could not be expected "to leave their own country." Dependent on government rations of flour and bacon as well as some beef (which was introduced to prevent illness that came from a diet of only bacon and flour), plains Aboriginal people also ate whatever plants, fish, wild animals, and birds they could find.[10] Food shortages reinforced the federal government's determination to make plains hunters into farmers and confirmed Euro-Canadian assumptions that an economy reliant on wildlife was unstable and inferior to one based on domesticated animals and crops. Nonetheless, it was anticipated that what remained of plains wildlife, fish, and plants could be managed or redesigned to support Aboriginal people until they became farmers. This would save the government the costs of relief and help smooth the way for Euro-Canadian settlement by stabilizing plains Aboriginal communities and ensuring peace. An investigation of how the region's natural foods could be used to meet these goals was undertaken in 1887 when Senator John Christian Schultz of Manitoba spearheaded the formation of the Select Committee of the Senate on the Existing Natural Food Products of the North-West Territories. The committee held hearings over several months, and government officials, scientists, fur traders, and others testified. Some committee members had already made up their minds before the hearings began. Senator Macdonald foretold that the committee would show how local foods "coupled with agriculture, can be so utilised as to make the Indians self-supporting." So, too, Senator Abbott observed that the committee would show how the NWT's natural resources could support First Nations people who "if they had the provident habits of the whites … would be able to maintain themselves without so large a reference to the government." While the committee was most concerned with plains districts, it did recommend more rigorous fish conservation and the granting of exclusive fishing grounds to First Nations in more northerly districts, but only when this could be done "without any material interference with the vested rights of any considerable number of White men."[11]

This focus on plains areas reflected the committee members' concerns about the cost of relief. Expenditures were highest in the plains and southern parkland areas, the districts of the vanished bison, and lowest in the lakeland and wooded districts of northern and eastern Manitoba and northeast Saskatchewan, with their greater supply of fish, wild meat (especially rabbits), and wild rice. This inspired the committee to adopt a three-part template in its recommendations to replicate the northern wildlife economy on the southern plains. This required extensive manipulation of flora and fauna, but it was confidently expected that such engineering would bring Aboriginal self-sufficiency as well as supplementary food for Euro-Canadian settlers. The plains and southerly parklands would be altered by transplanting to them the "natural food" practices and resources of the more northerly districts. Fish were central in these plans because they reproduced rapidly and were nutritious and easy to catch, regardless of a person's age or sex. Thus the committee recommended stocking lakes with "sturgeon, white-fish, gold-eye, catfish, perch and eels." In lakes with poor water quality, of which there were many on the plains, it recommended the transplanting of jackfish (pike) from other lakes as well as foreign species like muskallongé (a type of pike) and carp. A fish hatchery was needed immediately to breed the species slated for transplanting. Moreover, it was casually recommended that while jackfish were to be transplanted to such poor-quality lakes, they should be eliminated or controlled in good-quality ones through a permanent open fishing season because they were "merely creatures of prey, and destroy large numbers of fish more valuable than themselves." To conserve fish for human use, the extermination of piscivorous birds such as pelicans, which the committee judged were useless for food and "extremely destructive of useful fishes," was also advised.[12]

Adhering to its lakeland template, the committee also recommended that wild rice be sown in suitable places everywhere in the region as human food and to retain water to create suitable environments for waterfowl and fish. Wild rice was a high-yielding nutritious grain that grew readily in relatively shallow water in areas that were of no value for farming. In his testimony to the committee, Robert Bell of the Geological Survey of Canada argued that it was the only native plant suitable for transplanting throughout the region, although he speculated that breeding experiments with wild turnip might also be worthwhile. But overall, he judged that only domesticated Canadian vegetables would prove productive. The committee accepted this advice and recommended that the federal government conduct experiments with wild turnip as well as with crossing native varieties of fruits and berries with domestic ones to produce hardier high-yielding varieties. Although it was outside of its mandate, the committee also recommended that trees and shelter belts be planted to bring "important climatic effects favourable to the increased production of cereals, roots, fruits and grasses in those favored regions."[13]

The committee's last major recommendation concerned rabbits. Like fish, they reproduced rapidly and provided high-quality protein. While committee members closely questioned witnesses about rabbit population cycles, the only thing they learned for certain was that these cycles occurred regularly. While an animal that was only available in appreciable numbers in about three out of every seven years offered limited value as a stable food source, the committee sidestepped this problem by suggesting that rabbit meat could be preserved during times of abundance.

Witnesses told the committee how various wild meats could be preserved as pemmican or by smoking, drying, salting, or freezing, and the committee thus concluded that "in years of plenty Indians should be encouraged to preserve" rabbits for future use. This conclusion failed to take into consideration that there was no tradition of preserving rabbit meat on the prairies and that people could not live well on fresh rabbit meat alone. As the well-known northern saying had it, a person could starve on a steady diet of rabbit meat because it was so lean, fine grained, and hard to digest. These qualities also made it impractical for pemmican. Given these problems, the committee suggested that rabbits of the same species could from time to time be transplanted from districts where they were plentiful to those where they were scarce. While this suggestion evoked the spectre of the Australian continent overrun by millions of rabbits, the committee argued that this would not happen on the prairies, where rabbits were cyclical and their populations limited by the region's cold winters. As for its concern that the rabbits would eat farm and garden crops, the committee accepted the advice of Amédée Forget, who assured the committee that prairie rabbits were not "pests" that destroyed gardens. Even so, the committee recommended that rabbits be transplanted only to "uncultured and wooded districts for he [the rabbit] has a taste for the products of cultivated fields when his own natural food grows scarce." Perhaps of equal importance, an added benefit of a rabbit economy was that it would allow the effective disarming of the Aboriginal population. While rifles were needed to kill big game, only small arms were needed to kill rabbits, and this would therefore assure the military security of the region. As Senator Schultz noted, Aboriginal people armed with rifles had routed Custer and the same weapon had threatened Canadian troops during the North-West Rebellion of 1885.[14]

In addition to its recommendation about rabbits, the committee also considered the potential of domesticating several native mammals, including beavers and bison. Even though some plains First Nations were relatively unfamiliar with beavers, the committee thought that raising them would be appropriate because "the Indians understand the nature" of beavers already, "whereas sheep, pigs and cattle are strangers to them." Beavers were easily domesticated, reproduced rapidly, and provided "better food than either beef or mutton," and their pelts could be sold or bartered for food. With respect to bison, the committee examined the evidence of Samuel Bedson, warden of Stony Mountain Penitentiary near Winnipeg, who was experimenting with crossbreeding bison and domestic cattle to create a new type of livestock better adapted to the prairie climate. Bedson's testimony piqued the interest of the committee members, and while they did not recommend the domestication of beavers in their final report, they recommended that federal experimental farms carry out tests similar to Bedson's bison crossbreeding trials as well as tests with muskoxen and moose. Echoing the concern that bison and farmers did not mix, the committee made it clear that it was not advocating the reintroduction of wild bison. Hybrids that could be fenced and pastured fit into the new west, but wild bison roaming free were part of a past that could not be restored, for, "in the changed condition of the country," they would "disturb the present agricultural training of the Indian, and interfere with the farming and herding efforts" of Euro-Canadian farmers and ranchers.[15]

The committee's report did not break new ground. The Department of Indian Affairs agreed that "the conservation and augmentation of the indigenous animals and plants of the country" would reduce costs and be "conducive to the health of the natives." It thus sent copies of the report to prairie Indian agents and instructed them to implement its recommendations when possible. While it drew particular attention to the recommendations about planting wild rice, it pointedly noted that the department had already been doing this in a number of places, implying that the recommendation was superfluous.[16] The report's suggestions about hybridizing bison were also already underway, and the proposal to introduce foreign fish for food was, as will be seen shortly, not new either. The only original proposals that the committee made were about rabbits, but these were so impractical and speculative that they were of little value. Nonetheless, the report was important in that it revealed, focused, and articulated current thinking about how wild animals could help government to cost effectively support plains First Nations peoples whose old economy had vanished. The report's detailed examination of food storage alternatives also revealed the desire to save public money and to provide plains First Nations people with a more satisfactory alternative to government rations, which they disliked and believed were damaging their health. As well, it articulated the widely held view that food shortages could be ameliorated if Aboriginal people simply adopted Euro-Canadian ways and gave up subsistence hunting to become farmers or labourers. In the interim, the region's flora and fauna could be managed in a way that would sustain them. As Senator Girard observed, the committee's recommendations would promote the "civilization" of First Nations peoples by leading them to self-sufficiency and making them "men." This would stabilize conditions without compromising Euro-Canadian settlement. Indeed, the final report concluded that "nowhere has Nature showered blessings with a more bountiful hand than in the Canadian North-West. About 600,000 square miles of arable and pastoral land seem prepared by the hand of God for the homes of civilized men."[17]

FOOD FROM THE WILD

The Senate committee's recommendations did not change the circumstances of plains Aboriginal peoples, who grappled as best as they could with changed conditions after the bison were gone. In 1888 it was reported that even fish, rabbits, and small game were now scarce in many southern areas, and by 1911 Aboriginal people on the plains and even in some adjacent areas in the parklands had been forced "to confine their hunting to limited areas and fall in with the white man's ways of living and earn a livelihood by cultivating the soil and raising cattle." There was no longer enough wildlife to give them the option of a different lifestyle.[18] While bison-hunting people traditionally ate little fish, it was an important staple for Aboriginal peoples living further north. As anthropologist David Mandelbaum observed, the Plains Cree near Battleford in west-central Saskatchewan in the mid-1930s supplemented their diet of meat and plants with fish. Building weirs, they fished in rivers in the spring, and these occasions also served as important seasonal social gatherings. The fish were dried or smoked and mixed with berries and fat to make pemmican. Catching fish did not have the same status as hunting big game – a ranking that was also evident among the Rock Cree where only women fished with nets. Men primarily used spears for fishing, which bolstered their egos and confirmed their role as big game hunters.[19] Yet fish were vital for survival among all parkland and northern Aboriginal peoples. In the 1880s those near Pigeon Lake, a short distance south of Edmonton, lived chiefly on pike, suckers, and whitefish and ate wild meat when it was available. This was also the case at Lac Ste Anne west of Edmonton, even though overfishing there was already a worry by 1890. Aboriginal peoples in northern Manitoba in the late 1800s also relied on fish. Whitefish was a staple, but it too was beginning to decline in areas close to Winnipeg because of commercial fishing. In northern Alberta, the North-West Mounted Police (NWMP) reported in 1906 that while Aboriginal peoples at Lesser Slave Lake hunted moose and bears and dried the meat for winter, they lived on fish in the summer and winter because other food was "scarce or too expensive."[20] In some areas, changing methods of transportation increased the availability of imported food. By 1902 flour had become a staple of trade, and demand for imported food, except in the furthest northern posts, was growing. Bacon was in demand, as was lard and butter, and sugar and tea were considered essential items. This led some observers to believe that it was becoming less pressing to hunt and fish for food and that more people were instead trapping for income. Others saw it differently. In 1907 the NWMP reported that Aboriginal peoples at Lesser Slave Lake actually "have but little food to buy" because they lived principally on fish, rabbits, and dried moose meat, and "bears and lynx are also eaten when they can be got."[21] Yet the importance of imported food was undoubtedly growing. In describing the Cree at Wabasca in northern Alberta during the interwar years, HBC trader Jack Milne recalled that they lived in the settlement for a good part of the year but still essentially lived off the land. Their staple diet consisted of moose, deer, and sometimes bears, fish, bannock, and tea. Meat and fish were smoked for storage. However, in addition to tea, baking powder, and flour, they also purchased a limited range of other groceries from local traders.[22]

While this way of life lasted for many years across the northern parts of the region, some changes were taking place. By the late 1950s the population of northern Saskatchewan, for example, was becoming focused in a few larger settlements and a number of small bush communities. In the larger settlements, such as Uranium City, most food was brought in from the outside, while people in the smaller bush settlements still got most of their food from the land. Supplemented by waterfowl and ptarmigan, moose and caribou were the staples of many Aboriginal and white trappers and commercial fishers. Given the need for conservation in order to protect local food sources, most of the north was closed to non-resident moose hunting by 1958. Barren land caribou were also in serious decline, and although people were still allowed to hunt them for food, sport hunting was prohibited. The Saskatchewan government also tried to persuade Aboriginal peoples to reserve caribou meat for human consumption and to feed their dogs fish instead. These wildlife food sources had been under constant pressure for some time throughout the region. In 1949 officials in the Department of Indian Affairs argued that "the Indian of Alberta to-day is facing a very serious problem in his efforts to find meat and fish for his family. With the exception of a few residing in the North-East corner of the province who are able to take a fair supply of caribou, the native hunter is finding game becoming very scarce." A recent increase in commercial fishing had, "in most lakes where the Indians usually fish, deprived the native of another source of food," and there were frequent complaints from northern and central Alberta First Nations "that commercial fishing has practically destroyed their normal fishing grounds." These conditions were exacerbated by longer-term depletion of big game and game birds and, more immediately, by a temporary low point in the rabbit cycle.[23]

While many Euro-Canadian farm settlers before the First World War ate wild meat and fish, few relied on it in the same way that northern Aboriginal peoples did. Some settlers in Saskatchewan in the 1890s salted and preserved whatever fish was available (including suckers) as a winter food supply. Some also bought fish in winter from fishers who sold their frozen catch door-to-door. In the late 1880s people in Regina could usually purchase fish year-round that had been caught in Last Mountain Lake, about forty kilometres away. The fish was smoked or dried to get it to market and was intended for immediate consumption.[24] In response to the scarcity of fish in many plains farming areas, the Manitoba government began negotiations with the US government in 1882 to obtain German carp to transplant in small lakes and farm ponds. This had been done in the United States, and Manitoba predicted that carp would "grow fat and savory on decaying vegetation, and its introduction to Manitoba would be a progressive step towards utilizing the thousands of small water tracts" that were difficult to drain and "disfigure[d] so many farms." It was noted that "fresh fish is rather a rarity on farmers' tables" and carp would supplement "ordinary farm products" and sustain the self-sufficiency of the family farm. Although this experiment failed, other attempts to transplant fish into local lakes and farm ponds followed. The federal Department of Fisheries reported in 1911 that it received frequent requests to stock small prairie lakes and ponds with fish to provide food. But many smaller lakes in the prairies dried up in the summer or became very shallow, with water temperatures rising to the point where only "coarse fish" could survive. Aside from extensive commercial fish stocking in large lakes, policies to stock lakes and rivers for sport fishing soon began to take precedence. In this environment, efforts to stock farm ponds fell out of favour until the 1950s, when experiments were undertaken in Alberta to transplant rainbow trout fingerlings to farm dugouts.[25]

Low levels of fish consumption reflected a lack of resources in parts of the region as well as cultural attitudes. Local historian Edith Rowles noted in 1952 in her oral history of pioneer diet in Saskatchewan that one settler told her that he ate very little fish, "as fishing on the river takes time."[26] Limited local consumption perhaps partly explained the dependence of the prairie commercial fishery on exports to the United States. In 1947, for example, only 20 per cent of Alberta's commercial whitefish catch was marketed in the province and the rest was exported to Chicago and New York. In the early 1930s Manitobans ate only about seven kilograms of fish per person per year, equally made up of local commercially produced fish, imported sea fish, and fish caught for sport or home use. In contrast, the Japanese ate almost ninety-one kilograms per person per year and people in Britain ate about nineteen kilograms annually. Some prairie people claimed that eating too much fish was unhealthy, and the Manitoba government tried to counter this notion by noting that the Japanese apparently did "not suffer either physically or mentally" because of their heavy fish diet. In an attempt to increase consumption of regional freshwater fish, the federal and provincial governments jointly sponsored various promotional campaigns, including the lectures and radio talks given by Eveline Spencer, a federal Department of Fisheries cooking specialist. Spencer came west in 1932 to mount demonstrations and deliver radio broadcasts about methods of cooking fish and its nutritional value.[27] These educational programs seemed to help to increase the sale of local fish, and meat rationing during the Second World War further stimulated it. Higher standards and new filleted products that were easier and more convenient to use than whole fish also helped to increase the domestic market for local fish. The Manitoba government had argued in 1932 that local consumption was held back because producers preferred to sell their catch in bulk quantities rather than "cater to the wants of individual families." But quality was also part of the problem. By 1953 standards in the Saskatchewan summer fishery had improved significantly, but those in the winter fishery (which would have contained many small part-time or seasonal producers) were poor and gave the trade in whole fish a bad name. Much of what was offered for sale was "improperly dressed, bleached from exposure to the sun and wind and smeared with blood. In any store where whole fish is displayed, it has an unappetising unappealing appearance." In contrast, local frozen filleted fish, which had been available for a number of years, was growing in popularity. By 1944 most of Lake Winnipeg's walleye (what was locally called pickerel) and an increasing amount of whitefish were filleted, wrapped in cellophane, quick frozen, and then sold in five-pound cartons.[28]

Frozen filleted fish revolutionized the industry because it eliminated seasonal supply and increased consumption by removing the mess and work of preparing whole fish. By the mid-1950s "fish sticks" (a slender pan-ready piece of fish cut from a fillet) were being marketed, although Manitobans, unlike Americans, were slow to accept such prepared foods. But as the Saskatchewan government noted, the popularity of fish sticks in the American market was so great that it was hoped they would do for the fishing industry "what fruit juice concentrates have done for the citrus fruit trade."[29] Such products offered hope for a new future. As the Manitoba government predicted in 1964, they showed the potential of fish such as burbot and suckers, which had formerly been viewed as predatory or invasive fish that should be destroyed or used as mink food. Equally important, mass merchandising of food through supermarkets meant that rigorous quality control was crucial to keep a brand's market share, and with the co-operation of provincial authorities, federal inspection and labelling were tightened to ensure that quality standards

Figure 5.2 Prairie consumption of fish was low and did not form part of most people's customary diet. In this 1949 photograph, Edmonton resident Mary Ferby is carving a fish for Christmas dinner.

were maintained for Canadian fish to enhance its place in urban markets in Canada and the United States.[30]

Whatever efforts were made to increase consumption of local commercially caught fish, most people preferred to eat meat, and wild animals were at times an important source. In *Game in the Garden*, historian George Colpitts argues that the official record ignores or obscures the importance of wild meat in the early settlement history of the prairies. In his study of rural settlement in southwest Manitoba, T.R. Weir also observes that scholars of prairie settlement seldom mention access to wildlife as a factor in how settlers selected a homestead. But, as he argues, "it would appear that the earliest settlers" saw wildlife as "a possible supplement to a meagre diet for the first few years," and its availability was one of the factors that influenced where they settled.[31] In the fall of 1880 John Macoun of the Geological Survey of Canada observed about forty species of game birds at Last Mountain Lake near Regina. All of the ones he shot were "fat and soup made from the various species of snipe and plover" was "a great dainty." Noting that only people who had been on the prairies in autumn could appreciate "the number of birds living or passing through it at that season," Macoun predicted that a future settler would have no problem in laying up enough "fat fowl" as "his winter's meat" in three or four days of hunting. This prediction was borne out.

Frank Farley, an amateur ornithologist from Camrose, for example, observed that two homesteaders in the Battle River district in central Alberta shot between 100 and 200 sharp-tailed grouse in a single day in 1892. After returning home with their wagonload of birds, they froze them for winter food.[32] Mennonite settlers in the Rosthern area east of Saskatoon in 1892 also took full advantage of wildlife, shooting deer, ducks, grouse, and other wildlife and preparing them for winter use. The same practices were followed by many other ethnic groups, including Finnish settlers in the province. Certainly, game legislation encouraged this use of game (as well as fish) by settlers. Manitoba's 1876 Game Protection Act provided that a settler or traveller could kill any protected animal and use the eggs of protected birds to meet an immediate need for food. It also protected rabbits from spring to fall because, as it was said in 1883, they were rarely "hunted for sport, but principally for food," in winter. The NWT game ordinance also encouraged hunting for food through generous bag limits and long open seasons and by allowing the sale of wild meat. Saskatchewan and Alberta adopted similar provisions in their game laws after 1905, although provisions privileging settler use of wildlife for food were soon amended to apply mostly to northern areas.[33]

In later years, nostalgic recollections would portray wild meat as one of the mainstays of homesteading. It was fondly recalled in 1946 that settlers "could tell you that Ruffled Grouse meat is white and delicious while

Sharp Tailed Grouse flesh is dark and rich and they knew that a grain fattened Jack Rabbit was a tasty morsel." As well, threshing crews were so often fed game that many were "heartily sick of Prairie Chicken meals before the threshing season was finished." But such use of wild meat was only a temporary measure for most settlers. As early as 1862 John Palliser had remarked in his report on the prairies that local fish and wildlife would be valuable "during that transition state that a country must endure between the periods when its inhabitants live on wild animals alone, and that period when bread becomes the staff of life and animal food is produced by the care and forethought of civilized man." This was borne out as domestic meat became more readily available. As early as 1883 in Winnipeg, meat from domesticated animals was lower or similar in price to most wild meat. Fresh pork was cheapest, while mutton, veal, ham, and bacon cost slightly more. All could be had for around fifteen to twenty cents per pound. Venison and bison were about the same price, while elk and bear were substantially higher. Rabbits were an exception; when plentiful, such as in 1886, twenty-five rabbits sold for one dollar. Wild birds were usually sold by the pair, making it difficult to compare them with domestic poultry that was sold by weight. However, when plentiful, partridge and grouse were inexpensive, selling for between twenty and twenty-five cents a pair in the Winnipeg area in 1886. Wild ducks and geese were equally inexpensive.[34] In the Peace River country in northern Alberta, Dominion Land Surveyor C.C. Fairchild reported in 1901 that due to the lack of farm settlement, wildlife remained plentiful and the few advance settlers mostly ate moose, bears, and caribou. But hogs were already being raised, and as game became scarcer because of settlement, he foresaw that pork would be produced in sufficient quantities to replace wild meat for home use. In 1918 Saskatchewan's minister of agriculture W.R Motherwell contended that "outside of the Indian population there is very little hunting of big game or even feathered game in Saskatchewan for the exclusive purpose of supplementing our meat supply, the sport being a primary consideration and the meat thus secured being quite secondary." And Saskatchewan historian Edith Rowles's settler informants told her that they "raised their own meat as soon as possible after getting established."[35] Certainly, the domestic production of meat had increased so much that prices for beef and pork were low, which made hunting a comparatively expensive way to obtain meat. In Saskatchewan in 1923, for example, low cattle prices "removed the incentive to stock the larder with cheap [wild] meat." Technological change also played a part. As Kerry Wood wrote in 1946, the use of wild meat had largely disappeared in central Alberta because of fashion and technological change. Cars and improved roads and the popularity of cold storage facilities in the towns made wild meat only a memory, for one could now "slip into town and visit the butcher shop or take out a roast from the cold storage locker."[36] Similarly, reports and promotional literature about the Peace River country in northern Alberta after the First World War frequently and proudly asserted that people could produce their own food on their farms. Indeed, when wildlife was described as part of the district's attractions, it was for sport and not subsistence.[37] F.H. Kitto, in his 1930 promotional booklet about the region, remarked that "nearly every mixed farmer has a varied poultry yard and a thriving kitchen garden," and in 1927 Sheriden Lawrence's home near the town of Peace River was described as similar to a "baronial estate" where, with the exception of some staples, "everything consumed on the place is grown there." And a report in the *Edmonton Journal* in 1932 did not mention hunting as a possible solution for some of the troubles caused by the Depression in the Peace River country; instead, hard times reinforced the dream of self-sufficiency, and low prices for farm production only created "an increasing reliance upon the home farm as a source of food for the farm family."[38]

Despite such increasing self-sufficiency, some settlers continued to hunt and fish for food. Extra money was sometimes the goal; as farmers were advised in 1924, if they used more wild game as food, they could sell more of their poultry, pork, and beef for cash.[39] More typically, however, farmers, especially in fringe northern areas of recent settlement, continued to use wild meat out of necessity. Homesteading with her husband near Grande Prairie in northern Alberta in 1919, Myrle Campbell wrote to her family that they had purchased beef in the early winter but "soon the moose and deer will be coming back from the mountains and I do not think we will want for meat all winter long. There are rabbits here also and they are just like tender chicken. The Wapiti River is only six miles from here and the fish caught through the ice are excellent." Campbell also noted that they traded vegetables for wild meat or sometimes purchased it from local Aboriginal hunters, and the only food supplies they regularly purchased included flour, sugar, tea, dried fruit, and the like. While some of this rhetoric romanticized pioneering, it also reassured distant relatives that the family, despite its poverty, was not in need. In 1928 many settlers in the Debolt area west of Grande Prairie were said to still be fulfilling their winter meat needs with game, and it was again noted in 1932 that in northern Alberta "many people in poor circumstances, especially new settlers," relied on fishing and game for food. So, too, in 1943 residents in the Big River area in north-central Saskatchewan obtained "an appreciable amount of food" by hunting big game, mainly deer, as well as ducks, and as late as 1952, avian biologist William Rowan argued that fringe-area settlers in Alberta still saw game as a source of meat and he contended that the amount they killed exceeded that taken by licensed sport hunters.[40]

Figure 5.3 Game animals such as deer provided food in many settlement fringe areas and in northern Aboriginal communities. In this 1913–14 photograph, hunters in northern Saskatchewan dress a deer killed in the hunt.

Hunting for meat also periodically increased everywhere in the region because of hard times. Dominion entomologist Gordon Hewitt noted that economic dislocation and shortages of consumer goods during the First World War turned people to hunting for "the most primitive and impelling motive – to secure food." The Depression equally forced a return to hunting wild animals for the family table. The Alberta government nonetheless showed little sympathy to pleas that poor settlers be allowed to hunt for food without a licence. Its only concession was to allow destitute families to hunt big game if the municipalities where they lived purchased licences and handed them out as part of their relief allotment. Saskatchewan showed greater compassion and in effect stopped enforcing its game laws in the early 1930s. The province's deputy minister of natural resources contended in 1933 that demands for free game and fish licences, petitions for leniency towards those convicted of illegal hunting, and protests about "the over-zealousness" of game enforcement all revealed a "strong current of public opinion" that could not be ignored. Economic distress had shifted public opinion against "attempts to rigidly enforce Game and Fish protection laws," and 90 per cent of people in the province were "definitely and decidedly opposed to the rigid enforcement of these laws" as well as to other conservation measures that they believed stood in their way during this time of crisis.[41]

SPORT, NOT COMMERCE

In the late 1800s and early 1900s wild meat was readily available for sale in many centres and was sold in butcher shops, on the streets, and at railway stations. Generally, meat hunted during open seasons could be sold legally, and while bag limits frequently changed, they were often high. In 1904 in Manitoba, for example, a hunter could legally kill fifty ducks each day during open season.[42] And where game legislation was poorly enforced, hunters often ignored bag limits or hunted during closed seasons and sold the meat illegally. As a result, criticism of the commercial market in wild meat grew steadily. In 1883 it was claimed that one party in Winnipeg had sold 11,203 ducks, 1,444 grouse, and 1,565 hares between

September and December. I.G. Field Johnson, a resident of Selkirk just north of Winnipeg, informed Premier Greenway in 1894 that a huge number of sharp-tailed grouse had been slaughtered near the town by market hunters during the closed season and that he estimated "that last year fully 1,000 birds were shot during the month of August. And this spring prairie chickens were slaughtered right and left, even when laying." As he reported, "quite a large trade is done at Selkirk in chicken and partridges – I saw a storekeeper buy a sack of them one day last August." As evidence of the ongoing scale of the trade across the region, it was reported in 1911 that a hunter near Moose Jaw in south-central Saskatchewan employed two assistants to help him fill his contracts with Moose Jaw meat dealers for 1,000 birds a week.[43]

Commercial hunting, legal or not, was clearly an ongoing threat to wildlife, just as it had been for bison and passenger pigeons. As a result, temporary bans on market hunting were put in place in the 1880s and 1890s but were lifted when animal populations were thought to have recovered. Manitoba, for example, banned the sale of grouse in 1883 to conserve those that remained for homesteaders. Ten years later the province banned the sale of venison as a conservation measure, and in 1898 the sale of some game birds was prohibited. By the early 1900s, however, demands for a complete ban were increasing. Citing how human destructiveness and the pursuit of commercial gain had upset the balance of nature, the Commission of Conservation argued in 1913 that the passenger pigeon was nearly extinct because of commercial hunting and the same fate appeared to be in store for sharp-tailed grouse and other game birds if market hunting did not stop.[44] Repeating this view more forcefully in 1918, the commission charged that "the market hunter is the recognized guerrilla of destruction with regard to wild life, and the traffic in dead game is responsible for at least three-quarters of the slaughter that has reduced the game birds of North America to a mere remnant of their former abundance." Fish and game associations similarly demonized market hunting, and revealing that it was a moral issue for some, the language used by such opponents was often harsh. Saskatchewan's game commissioner Fred Bradshaw, for example, described market hunting as a "villainous trade" that would "make deep inroads on our game supply that will be hard to retrieve. When the fangs of commercialism take hold of any species of wild life, that species is doomed." Greed was despoiling the public trust, and Bradshaw believed that the "monster" had to be stopped "before the last remnant is devoured." As an example of such greed, Bradshaw reported that one family in southern Saskatchewan had admitted that hunting ducks for sale had provided them with enough money "in two years to purchase and pay for an automobile." Particularly harsh criticism was levelled at Aboriginal hunters, popularly said to be among the most proficient and ruthless hunters, who conspired with unscrupulous urban dealers, construction crews, lumber camps, and mines to clean out a district's game for profit.[45]

Manitoba prohibited the purchase of wild meat from Aboriginal peoples in 1915 and banned all market hunting in 1917. Saskatchewan followed suit the same year, and Alberta did so in 1922.[46] Saskatchewan also amended its game legislation to prohibit construction and lumber camps from having any big game or game bird meat on their premises in order to stop them from employing local hunters to supply the camps with game. There was no justification, the government argued, "for any commercial enterprise to increase its profits at the expense of the game." The strength of this commitment to end market hunting was demonstrated when the Manitoba government upheld its ban despite recommendations that commercially hunted wild meat should be used to supplement wartime food supplies.[47] Even so, the ban was difficult to enforce and market hunting continued. It was claimed, for example, that deer shot near Winnipeg were sold in the city in 1930 and that market hunters, including Métis at Buffalo Lake, supplied hundreds of ducks each week to butchers in Moose Jaw in 1934. And in 1951 it was reported that caribou, then under strict protection from hunting, were killed in northern Manitoba without licence and were sold to mine and construction crews.[48]

The banning of market hunting did not presume that it was wrong to eat wild meat, only that it was wrong to sell it. Fred Bradshaw contended that market hunters provided meat for people who did not need it because if they could afford to purchase wild meat, they could equally afford domestically produced meat. Sport hunters, farmers, and those who used game for their own table were the only people who were now viewed as legitimate hunters. And part of what was said to define respectable and ethical sport hunting was that the hunter and his family or friends ate the animals that were shot, an argument that soon took on class dimensions. For example, Leslie Sara, wildlife reporter for the *Calgary Herald*, did not hesitate to say that sport hunting for urban people was often a display of wealth. As he wrote in 1937, when one added up the costs of hunting equipment, travel, and permits, it was an expensive way to get food. And no matter how well it was cooked, Sara said, it could be tough and "a beefsteak or a chicken which would melt in your mouth could have been bought for a mere pittance" in comparison: "But eat it gratefully and give thanks. You are dining on the food of plutocrats."[49] Even so, hunters often accumulated substantial amounts of meat. As was remarked in 1957, "a good day's shooting yields a mighty fine lot of fine flavoured meat" because Hungarian partridge averaged around 500 grams, mallards around 2.25 kilograms, pheasants 1.4 kilograms, and Canada geese from 3.5 to 6 kilograms. Some of this meat was eaten immediately, but much of it was put into cold storage, where, according to Alberta's game commissioner in 1919, some of it was left for so long that it became unfit to eat. Asserting that cold storage only encouraged excessive hunting, governments began to set limits on how long game could be stored, and by 1949 Manitoba was allowing only fourteen birds to be kept in cold storage at any one time, a limit that the Manitoba Federation of Fish and Game Associations thought was unduly restrictive.[50]

Restricting the amount of game that could be placed in cold storage attempted to ensure that hunters ate the game they killed. But some people increasingly did not want to prepare it for the table. To be sure, some hunters cleaned their own game, but many took it home to have it cleaned by family members. This could be hard and messy work because it was accepted, for example, that ducks should be plucked dry instead of first being scalded. One consequence, as was remarked in 1924, was that while city folk considered game "quite a delicacy," farm women often scorned and turned up their noses "at the all too common game brought in by the men folk of the family" because the unpleasant task of cleaning it fell to them. "This is not at all the way it should be," since men "have had all the sport of hunting so why should they not do a little of the work?"

Women did not, however, find freedom from the task. As was reported in 1957, "all game trails end at the table of hunters, or, rather at the tables of hunters' wives and friends and sweethearts" who had neither the skill nor the inclination to deal with "a mess of birds or a hunk of venison." It was thus not surprising, as Ducks Unlimited (Canada) noted in 1942, that much valuable meat was simply thrown away because no one wanted to clean and prepare ducks brought home by hunters.[51] Clearly, hunting was increasingly becoming more about the chase and the rituals of the hunt than about acquiring food.

6

Hunting as Sport: Ideals and Practice

> No one worthy of the name of a sportsman hunts for the mere joy of killing. It is rather the pitting of skilled marksmanship against a quarry that at least has a chance of taking care of itself that appeals to most men.
>
> *Calgary Herald*, c. 1936.

Hunting was a common point of connection between people and wildlife in prairie Canada. In the late nineteenth century it was increasingly idealized as a sporting rather than a subsistence activity, thus beginning a shift in which it became part of seasonal social activities rather than of the daily routines of life. This emphasis on hunting as sport reflected changing public ideals in the late 1800s. Hunting now stressed competition and the treatment of prey with fairness and decency, and although they were often ignored in practice, these values were reflected in regional game legislation and in the work of fish and game associations. These associations served hunters' needs in a number of ways. As well as lobbying for specific goals, such as predator control, they more generally promoted sport hunting by asserting that some wild mammals and birds were most valuable as game. While this view underpinned provincial game legislation, ongoing lobbying by these associations helped to ensure that hunters' interests were not compromised by over-regulation. A specific concern in this regard arose in the interwar years, a period of unprecedented crisis for prairie waterfowl. While calls to ban duck hunting were unsuccessful, their persistence revealed that sport hunting did not have unquestioned public support and that the values it embodied were not shared by everyone.

Sport hunting in prairie Canada was part of the region's larger history, and like any sporting contest, it presumed both formal rules and social precepts about appropriate behaviour. While it expressed human aggression (and, in particular, male aggression), successful hunting for sport was also predicated on the possession of knowledge about climate, landscape, and animal behaviour that was often difficult to acquire.[1] It was a sport dominated by men, and the pursuit and killing of wild animals shaped and reflected gender relations.[2] While pioneer farm women commonly shot or trapped grouse and small animals like rabbits for the table, this was not sport, and the few women who hunted for sport were sufficiently idiosyncratic to be worthy of remark. Sport hunting was thus largely a male preserve. This was the view of one eighteen-year-old correspondent to the *Western Producer* in 1950 who noted that its youth page rarely featured articles about hunting and instead focused on birds, which he thought made it a "girls" page. This identification of sport hunting with masculinity was said to contribute to the social good through military preparedness, by building self-reliance, personal responsibility, and character, and by teaching competitive skills important in the capitalist marketplace. And echoing a long-held justification, it was remarked in 1954 that hunting made better families because the "thrill of hunting together" strengthened father–son relationships. Male fortitude also gained concrete and lasting recognition in the many photographs, mounted animal heads, and other trophies kept to commemorate the triumph of the hunt. Yet such hunting could be practical as well, and as writer F.A. Twilley recalled, while many settlers in the Swan River district in west-central Manitoba had "moose heads hanging up to show their prowess," the animals had also provided much needed food. And for some, such hunting was simply a male escape. In 1920 M.D. Barker, a provincial game guardian in northeast Saskatchewan, recounted a recent hunting trip of "jolly good fellows" where "formalities and useless etiquette" were left behind. It was like school days again when "just 'we' let loose for a month."[3] Still others saw hunting as part of the unique advantage of prairie life and a relatively open society. Such appraisals of regional advantage emerged at an early date. A Manitoba homesteader, William Lothian, wrote in 1882 to his brother in Britain that game in his district was abundant and legal access was easy. And "to think of you being brought before a magistrate for killing a paltry rabbit, alas, poor Scotland." This remained a leitmotif in descriptions of sport hunting. As was remarked in 1943, "in many countries of the world, only the wealthy can enjoy the pleasure[s] of fishing and hunting, but in Saskatchewan, all can have them for a very small expenditure of money and effort." Sport hunting was also seen as an outdoor activity that strengthened body and mind, a view that sometimes easily lent itself to rhetoric about the Euro-Canadian conquest of the land.[4] Yet for some, it also expressed an emotive connection with the land, wild animals, and place. In 1937 D.G. Stuart, who frequently hunted north of Regina, wrote, "I love that country north of Day Star; there is a wild gamey flavour about it and a charm in its very loneliness; the mallard and the mule deer are its denizens, but you feel that eyes, black, yellow and green, are watching you all the time from the dim recesses of the dense bush. It is reticent rather than unfriendly, but I always give an unconscious sigh of relief when I see the lights of the first dwelling on my way out; and no sooner am I out than I want to go back."[5]

While hunting was popular, it is difficult to quantify how many people hunted and how many animals they killed. In 1913 Saskatchewan's game commissioner Fred Bradshaw asserted that "sportsmen … represent only about five per cent of our population," and in 1927 Jack Miner claimed that only 10 per cent of Canadians hunted. If this was so, it was a significant group, but not even close to a majority of the adult male population.[6] While governments kept records of the hunting licences they issued, these do not accurately reflect the extent of hunting for the years before the Second World War because illegal hunting was common and many people (such as farmers) were exempted from licensing requirements. Clearly, however, the geography of sport hunting expanded with better transportation. Railways were one factor, but motor vehicles were most significant. As early as 1909 people from Manitoba towns and cities drove to the country to shoot grouse on Sundays (even though it was illegal), and similar outings were still popular in 1930. Arthur E. Smith, who farmed at Saskatchewan Landing north of Swift Current in southwest Saskatchewan, observed in 1935 that twenty years before most hunters did not travel more than fifteen or thirty kilometres, but railroads and automobiles now made places with game similar to their "back yard in regard to distances." As a consequence, he wondered, "what chance" did the game have? And during the 1920s, automobile magazines like *Good Roads*, published in Edmonton, often highlighted hunting and fishing as recreational opportunities now made possible for urban men by automobiles. Indeed, a 1929 report noted that on one Sunday more than 200 cars, with an average of three hunters in each car, left Regina for the nearby countryside to hunt.[7]

While big game sport hunting declined during the Depression when incomes fell, this likely did not reduce the number of animals killed because of greater subsistence hunting in fringe areas where settlement had by then expanded. In 1935, for example, it was claimed that areas in Alberta close to settlements and accessible by road were overhunted.[8] Although gas and ammunition rationing during the Second World War also led to a decline in sport hunting and fishing, both increased rapidly after the war. Governments by then seem to have achieved somewhat more effective compliance with licensing requirements than in earlier years, and official statistics in the postwar years indicate trends in sport hunting and fishing. In Saskatchewan, for example, the number of angling licences rose tenfold, from 8,000 in 1946 to 80,000 (or one out of every nine residents in the province) in 1956. There was a similar increase in hunting, and by 1959 Saskatchewan coped with the high demand for licences by issuing them by lottery for certain species and districts. This increase was seen across the region. In Alberta 11,000 hunting licences were issued in 1941 and approximately 60,000 in 1953 (at a time when the province's population was about 1.1 million). The number of animals killed was also significant: in 1956 Alberta reported that hunters had shot 12,805 big game animals and just over 1 million waterfowl and 407,000 upland game birds.[9]

TENSIONS BETWEEN IDEALS AND PRACTICE

There has been a long tradition of treating wild animals on the prairies with casual brutality. They were killed as a matter of course, a practice that the bird lover Lawrence Potter noted in 1935 when he decided to rescue a lone pelican near his farm at Eastend in southwest Saskatchewan and to send it to a sanctuary because "otherwise the poor critter is bound to die when the cold weather comes if some idiot doesn't shoot it first." The prairies were not unique in this respect, and the Commission of Conservation noted in 1919 that Canadians easily tolerated a "pure wanton lust of slaughter." Market hunters were often blamed for this, Aboriginal people were always high on the list of villains, and "white men" were said not to be "free from blame" either. And this love of killing wild creatures did not diminish. In 1963 nature columnist Doug Gilroy observed that bobcats and lynx had begun showing up in parts of the prairies, and even though they were still rare, "whenever one is seen for the first time by anybody the first thing they do is to race for a gun." Gilroy did not understand why "we humans are this way," but he was convinced that the animals posed no threat and in fact were useful for agriculture as predators of mice.[10]

It was widely hoped that transforming hunting into a sport would ennoble the motives for hunting and thereby end the bloody-minded slaughter of wildlife. Of course, hunting as a sport had ancient roots; medieval aristocratic hunting, for example, featured highly ritualized behaviour. And while British hunting forays in Asia and Africa had revelled in the slaughter of wild animals in the 1830s, the pressure of depletion and changing sentiments about killing wild animals (as well as a desire to demonstrate class exclusivity and imperial mission) led to the view that hunting should be founded on sporting ethics.[11] Drawing heavily on this British model, sport hunting in the United States gained strength after the mid-nineteenth century. In Canada this American example likely meshed with the country's British connections and Canadians came to idealize hunting as a sport governed by a sportsman's code of conduct that stressed fair play, self-control, and restraint. The amateur tradition of sport was drawn upon in the contention that sport hunting was never for monetary gain and that the superior culture and character of the amateur sportsman were demonstrated by his participation in sport only for pleasure and the glory of victory. Thus, such practices as unfairly luring big game animals or shooting them when they were in the water or mired in snow or mud were all prohibited. Upland game birds were only to be shot on the wing, and the hunter, as part of a communion with nature and natural history, needed to know the species of bird he was pursuing. A clean killing shot was prized because sporting ethics demanded that a wounded animal be tracked, possibly for hours if necessary, to put it out of its misery. Nor should hunters kill an excessive number of animals. On the prairies, this sportsman's code of conduct also dictated that a sport hunter should eat what he killed or give it to someone who would do so.[12]

Ernest Thompson Seton provides one of the clearest articulations of these idealized sporting relationships between hunter and prey. In *Wild Animals I Have Known* he sometimes portrayed the hunt as an intelligent contest between hunter and prey. In the story "Lobo, the King of Currumpaw," a wolf of great strength and intelligence found gratification in defeating his pursuers' efforts to capture him and seemingly understood the dynamic nature of this relationship with them. Thus, when he was caught and his jaws were tied with rope, "he made no further resistance, and uttered no sound, but looked calmly at us and seemed to say, 'Well, you have got me at last, do as you please with me.' And from that time he

took no more notice of us." Although Seton became more critical of sport hunting as he grew older, he nonetheless believed that it was a legitimate and even necessary social activity. But hunting for pleasure was only acceptable when it followed the strict rules of sport and reciprocity. In a rough parallel to the thinking of some Rock Cree, the hunter and the animal formed an emotional relationship during the hunt that gave dignity to both and validated the death of the animal. Sport hunting also served social priorities. Seton argued that strenuous outdoor sport was the best way to dissipate energy and aggression and to build health, and he saw hunting and fishing as among the most important examples of such sport. As he noted in an article in about 1920, "the hunting of an animal may be set forth in ten chapters, of which the first nine, the pursuit – are wholly delightful." Outdoor exercise, "life among the beauties of nature, the good fellowship, the evident harvest of health," and "the joys of rivalry with a wary foe" all made for "manhood and health of body and mind." But then came the "last unpleasant chapter – the destruction of a harmless and beautiful animal – that part it must be admitted no sportsman enjoys." But death could not be avoided, for "the chase without the capture would scarcely have the same appeal." In contrast was the type of hunting in which "the nine noble chapters" of the chase were neglected and the killing became the whole story. This was "unquestionably degrading and devoid of any redeeming elements." Bloody-minded slaughter brought debasement, while sport hunting enriched social mores. To advance these values, wildlife needed to be conserved through well-crafted game laws, low bag limits, and "the absolute prohibition" of any practices that reduced the idealized balance between hunter and prey. Killing deer in the water and luring them with salt licks were "abominations," as "the chase is the glorious thing and by eliminating that, we are defeating the object for which the sport is permitted. The more chase and less destruction the nobler the sport." Indeed, Seton thought that "for some kinds of game I would go farther by tabooing firearms" and permitting only bow hunting. "In this way the amount of sport would be increased and the destruction greatly reduced."[13]

While Seton's ethical standards were more rigorous than most, and while the moral and psychological interplay he envisioned between animal and human was perhaps also more complex, his idealization of the chase had been part of the sportsman's code of conduct in North America since the middle of the 1800s. This view made certain animals "game" and others not worthy of the contest. On the prairies, white-tailed deer were particularly valued. Their horns made fine trophies and their wariness and speed made them an ideal prey that conferred the greatest social benefit. As Seton wrote in 1909, "the hunter makes the highest type of soldier, and the Whitetail makes the highest type of hunter that is widely possible today." Advertisements echoed this argument when they claimed that only the best ammunition would deliver the "well placed shot necessary" to kill wary white-tails.[14] The intelligence of other creatures equally made them suitable prey. It was claimed in 1926, for example, that the mallard drake made for fine sport because "by nature he is a wary bird and capable of holding his own against the wiliest hunter, once his instincts for self-preservation are thoroughly aroused." Manitoba writer Dick Fairfax also asserted in 1956 that "I think that all game know when they are being hunted and when they are not. I have had geese fly less than 60 feet over me when I was on the road driving a team, or out plowing. But just try to get that close when you are loaded and ready for them." And in 1962 another observer contended that mallards were "no more cunning than other ducks in avoiding the hunter," since all were "pretty smart in this respect, otherwise duck hunting would not be a sport."[15]

The fair treatment of such prey was said to equalize the relationship between hunter and hunted and to transform sport hunting into an ethical practice. In the mid-1930s the *Calgary Herald* editorialized that "no one worthy of the name of a sportsman hunts for the mere joy of killing. It is rather the pitting of skilled marksmanship against a quarry that at least has a chance of taking care of itself that appeals to most men."[16] Such values were incorporated in game laws and served the dual purpose of setting standards of respectable behaviour and promoting game conservation. As one Saskatchewan game guardian remarked in 1921, hunters increasingly realized that hunting was not about the number of birds shot but about playing "the game [and] taking no unfair advantage of the birds, and getting his pleasure in securing a few of them under difficult conditions." Fish and game associations also espoused this code of ethical conduct, and it found its way into popular fiction. A 1949 story by the outdoors writer Frederic M. Baker, for example, tells of a hunter who weighs the ethical demands of "sport" so that he can shoot a bear with "a clear conscience." The story ends with the hunter's pleasure at having "got my grizzly" and his even greater satisfaction that "I hadn't smashed my ethics to get him."[17]

The pairing of sporting ethics and conservation was ongoing, and a sampling of Manitoba's and Saskatchewan's legislation indicates this process. In 1879 Manitoba banned the use of snares, ropes, cages, and traps in regard to all game except rabbits. In 1883 swivel guns and night lights (which blinded an animal long enough for the hunter to shoot it) were outlawed, and in 1898 shooting at night was prohibited. In 1904 the use of weapons grouped in batteries and sunken punts (in which hunters hid in deeper water close to birds) was prohibited. In 1911 hunting waterfowl from power boats was made illegal, and the use of blinds and decoys was subsequently regulated, as was the gauge of guns, shooting from horseback, and the transporting of loaded guns in motor vehicles. In keeping with changing technology, Manitoba later banned the shooting of waterfowl from planes, powerboats, and snowmobiles, while Saskatchewan hunters were reminded in 1945 that shooting from a moving vehicle was unfair and that they "should at least give the bird a chance."[18] Similarly, by 1957 electronic duck and goose lures had become highly effective and were banned in Canada and the United States under the

Migratory Birds Convention. As Winston Mair of the CWS remarked, "seasons and bag limits in Canada are set in the assumption that there are certain handicaps in hunting" and prohibiting electronic luring devices would help to keep duck hunting in Canada "as near sporting as possible." And to encourage what Seton had called more chase and less destruction, Manitoba allowed bow hunting in 1957, as Saskatchewan had in 1952. Saskatchewan, however, took no chances in trusting the sportsmanship of bow hunters and prohibited the use of cross bows and any bow with less than a forty-pound draw as well as the use of poisoned arrows and arrows with explosive heads. Although the success rates of bow hunters improved, there was indeed more chase than killing: in 1952 Saskatchewan's 19 licensed bow hunters reported no kills and in 1955 there were eight kills by 152 licence holders.[19] Another element of game legislation that linked the sport hunting creed with conservation was the banning of market hunting because amateurs participated for the sake of the sport and never for material gain. So, too, the prohibition against shooting female big game animals was justified as part of the ethics of sport hunting. As Saskatchewan's minister of agriculture W.R. Motherwell stated in 1917, it was "inhumane and unsportsmanlike" to kill any "female deer of any kind."[20] And H.L. Felt of Findlater, northwest of Regina, confirmed this view when he argued that "no sportsman will kill a female, and I can see no more sport in such a thing than there would be in going out into some farmer's field and shooting one of his old cows. The harder the big game animal is to obtain the more it will be appreciated by the successful hunter."[21]

This sportsman's code of conduct, however, was often more honoured in the breach than in the observance. Hunting was deeply embedded in prairie social and economic life, and it was difficult to regulate because it took place across a large land mass. In 1932, for example, the Manitoba government frankly admitted that game protection depended on moral suasion and public support because the laws could not in practice be enforced and therefore "the conservation of our game supply lies very much more in the hands of the sportsmen than in the hands of the officers responsible for game protection."[22] Yet such a mindful sport-hunting culture was far from reality. A 1904 report noted that the NWT's game ordinance was ignored by a wide range of residents as well as by commercial travellers from Winnipeg who hunted without a licence while travelling between towns. In 1909, as part of a broader campaign to have the duck hunting season moved later in the autumn, newspaper reports and fish and game associations in Alberta complained that the province's game law allowed hunters to kill ducks that were too young to be used for food. As they pointed out, some people herded flappers (ducks too young to fly) and, for the sheer pleasure of killing, beat the young birds with sticks and poles, leaving hundreds of them dead in the water. It was, they claimed, behaviour of which no "man calling himself a sportsman might be justly proud." Yet many communities not only tolerated such practices but celebrated them. As the provincial game department noted in 1908, many towns in Alberta declared "a civic holiday on the twenty-third of August of each year, in order that the so-called sportsman may get out and destroy the innocent flapper" that was too young to get away. By 1919, even though the season had been pushed slightly later, the slaughter of young ducks still took place and the *Drumheller Mail* thought it was "a shame to permit it."[23]

Such brutality occurred everywhere and was directed against a great variety of wildlife over a long period of time. For example, after the running of big game with dogs was banned in Saskatchewan, hunters in Moose Mountain in the southeast corner of the province devised an alternative method. It was reported in 1907 that a group of hunters worked together, clapping hand bells to drive elk and moose into locations where they could be easily shot. The area was "said to have been swarmed by hunters," and it was predicted that elk and moose would soon be extirpated. As an indication that such behaviour was socially acceptable, it was often recounted as hunting lore. In 1926 Francis Dickie, a journalist and writer born in Carberry in southwest Manitoba, published an account in the *Grain Growers' Guide* about his encounter with four bears. A photograph of Dickie kneeling over the dead bears illustrated the article, in which he told how he and his hunting partner had spotted the bears on a mountainside in British Columbia. After a long climb, they found them lying asleep "in a cuddling heap." But "our hearts were made hard by the age old hunting instinct of mankind," and while the bears "lay like a litter of careless puppies," he and his partner shot two of them. The other two bears ran, but Dickie and his partner shot them as well. The one Dickie killed was "magnificent in his courage and ferocity," and Dickie paid "inward homage to the gameness of my adversary" as he aimed his gun at it. Revealing a gendered interpretation of the contest, Dickie observed that his shot created a state of hysteria in the bear – "with a queer convulsive motion the beast raised his forepaws to its face, the action like a hysterical girl giving way to a sudden burst of tears. Then he shuddered, crumpled, lay still." Meanwhile, his companion had killed the fourth bear. When he reached Dickie, he pointed to his watch and said, "'It's took us thirty seconds, and four of them – and they're grizzlies!' In his voice was that ecstasy of a man when he speaks of sacred things."[24]

This was precisely the sort of behaviour that the sportsman's code of conduct was supposed to prevent, and it was far from a unique case. In 1939 one central Saskatchewan correspondent told readers of the Nature Lovers page in the *Western Producer* that a lot of illegal hunting was going on and wild animals were being killed just for fun in his district. People trapped badgers even though they were becoming scarce, and they shot porcupines because the slow-moving animals presented "such a splendid target to a person with a .22." In 1947 the newspaper in Claresholm, a town south of Calgary, reported that some hunters bragged around town that they were out shooting before the open season and were ignoring bag limits. And, as only one more example among many, writer F.A. Twilley recalled in 1966 that while some people in his district near Swan River in west-central Manitoba had hunted because they needed food, others had shot "for the love of it. One man boasted of shooting 47 moose during one winter. Of course it served some purpose. He sold most of it but much was wasted." In another case, a man "shot four moose for little reason. They were standing at the side of the trail and all of them were cows, and proved to be in calf. As a result no doubt, we have had no moose for many years."[25]

A further problem lay in the fact that such hunting continued in the face of depletion because hunters turned to other species once choice ones were scarce. In Alberta in 1908 it was feared that if waterfowl were exterminated, "the sportsman will then turn his attention to song and insectivorous birds." Such behaviour was evident across the region. In his hunting memoir, Manitoba writer Dick Fairfax recalled that he had turned to hunting sandhill cranes, since "all game, including migratory birds,

Figure 6.1 Whatever strictures were mandated by the sportsman's code of conduct, this undated hunting party near Gladstone, Manitoba, proudly advertised that they had shot any wild animals they encountered.

were getting scarcer" because of "more settlers, more guns, more barbed wire." And indicating that killing was sometimes the only goal, one hunter was quoted as saying in 1926 that "there are just two kinds of ducks – sitting ducks and flying ducks" and if ducks were not around, American coots could be shot "in excuse for something better."[26] Indeed, when discussing the ban on hunting certain species of ducks under the Migratory Birds Convention Act, avian biologist William Rowan noted in 1931 that "our sportsmen are merely interested in quantity, not quality, and many of them do not even know a mallard for certain when they see it." This mindset was further demonstrated in the 1950s and 1960s as hunting escalated, and some observers increasingly thought that the whole culture of hunting was becoming ever more debased. There were ongoing reports of hunting with night lights in Manitoba and Saskatchewan, and although the Manitoba government promised to stamp out this "despicable practice," sustained complaints showed that it continued to be a problem. Having a loaded gun in a moving vehicle so that animals encountered along the road could be shot was another familiar practice, and by 1959 this was the most common infraction of Manitoba's game law.[27] One commentary published in the Saskatchewan Fish and Game League's magazine in 1958 further noted that camaraderie among hunters was breaking down and that hunters were behaving badly. The birds were "pursued from daylight to dark with little, if any, chance to feed and rest." Moreover, some hunters would "'rent' a likely piece of land so that hunting may be assured without interferences," and "at once the situation worsens – lead flies, mostly at targets out of range and the cripple loss rises – hunter success drops. Goose hunting has given way to goose shooting. This is the pattern and it has showed up in Saskatchewan as it has in other places." This neglect of "general sportsmanship" was also made evident in reports that "cars and tractors have been known to be used to herd the birds towards waiting guns." Concluding that hunters seemed to be "bent on fouling their own nest so far as the future of this fine sport is concerned," the correspondent argued that better regulations (such as allowing shooting for only half a day or prohibiting shooting over water) were needed to encourage more ethical practices and create a sporting relationship between the hunters and the birds. But such changes were hard to achieve. In Manitoba in 1962, for example, when it was proposed that the hunting of waterfowl be restricted to half a day, the proposal was rejected because it conflicted with the needs of city hunters.[28]

HELPING SET PRIORITIES: FISH AND GAME ASSOCIATIONS

Hunting enjoyed political support and social status because of the importance of game as food in the settlement period and deeply engrained cultural and recreational traditions and customs. Lobbying by fish and game associations also helped to maintain sport hunting as one of the most important uses of wildlife on the prairies. Similar organizations existed in many parts of North America from the 1850s onward and their influence has led some historians to argue that the wellspring of conservation in the United States lay in the work of sport hunters who popularized sport hunting and tried to halt the depletion of game upon which it relied. While game protection was indeed an important aspect of conservation thinking, this interpretation ignores the reality that game associations in the United States had a very narrow understanding of conservation and had few concerns other than the defence of hunting or the conservation of game animals.[29] This was also the case on the Canadian prairies, where fish and game associations were typically concerned only with game animals and paid little attention to fur-bearers and other non-game wildlife like songbirds and small mammals unless they enhanced sport hunting and angling.

Prairie fish and game associations (sometimes called rod and gun clubs) were typically dominated by city- and townsfolk from the English-speaking dominant culture. They appeared in many communities soon after settlement, and their meetings, dinners, shooting matches, and other gatherings were important community social events. But they were more than social clubs. They often had an overlapping membership and close relationship with rifle clubs, but while members of both groups shared an interest in guns and marksmanship, fish and game associations had broader interests. The Manitoba Game Protective League, a Winnipeg organization, for example, was formed in 1882 by hunters and anglers to promote the conservation of game and sport fish in the province. Similarly, the Macleod Fish and Game Association was established about the same time to try to halt depletion of game in southern Alberta. By the 1890s a similar club was operating in Edmonton.[30] All had similar goals. As C.A. Hayden of the Alberta Fish and Game Association observed in 1933, hunters backed conservation programs, including predator control and the introduction of foreign game birds into the province, to ensure a continued supply of game.[31] The Saskatchewan Fish and Game League, too, argued in 1943 that its members shared the goal of conserving and restoring game so that current and future generations could enjoy hunting and fishing. Equally important was the defence of public access to game on Crown land. This effort at times served Euro-Canadian interests because it entailed lobbying for restriction of First Nations treaty hunting rights, which were said to violate the principle of equal access. Such equal access was seen as paramount; as the Saskatchewan league argued in the early 1940s, "all game and fishery laws and regulations" were "tools of wildlife management employed to try and give all sportsmen an equal chance to take a fair share of the harvestable portion of these resources." Or as Ben Rosnau, president of the Alberta Fish and Game Association, put it more bluntly in 1963, every hunter had a "God given right to hunt on Crown land."[32] Because of the difficulty of enforcing game laws, fish and game associations often claimed that their presence in a community allowed them to act as proxies for the state in this regard.[33] Concrete evidence of this was rarely provided, however, and some observers were sceptical of such claims; as Saskatchewan's game commissioner Fred Bradshaw argued in 1913, hunters were untrustworthy guardians because they would almost never chose preservation at the cost of a successful hunt.[34]

Fish and game associations also tried to mediate conflict between hunters and farmers. Many farmers saw hunters as serial trespassers who caused damage to fences and crops and, most seriously, shot cattle and horses by mistake. Some of this animosity masked urban-rural tensions. As early as 1905 it was said that "near some of our towns, this over running of the farmers' fields is becoming an annual nuisance" and "a great deal too many young town chaps, in whose hands firearms are not any too safe, are allowed the run of the country during the fall months." Some farmers posted no trespassing notices or signs demanding that hunters get permission to hunt on their land, but the problem did not disappear and the Saskatchewan Fish and Game League lamented in the early 1940s that the breach between farmers and hunters appeared to be widening.[35] Prairie governments, with their large rural constituencies and their influential hunting lobbies, were caught in the middle, and tensions grew as sport hunting increased in the 1950s. Governments urged dialogue to solve conflicts, and some fish and game associations took steps of their own. The Saskatchewan league, for example, took out a blanket insurance policy in 1957 to cover lawsuits against members who admitted that they had damaged farm property when hunting. But not all hunters were association members, and associations often did not have the resources to deal with the issue. This forced the state to assume the burden. Manitoba began paying compensation in 1955 for livestock killed by sport hunters, and Saskatchewan initiated a livestock compensation program in 1959. In the first year, it received fifty claims of which thirty-three were paid.[36]

The influence of fish and game associations expanded with the federation of locals into province-wide organizations. The Saskatchewan Fish and Game Protective Association (it would later be called the Saskatchewan Fish and Game League) was formed in 1925 to give local fish and game organizations a unified voice. By 1943 it had about 1,800 members across the province. The association gained further influence by claiming that it spoke for all legal sport hunters and anglers and that it was the guardian of respectable sport-hunting behaviour. The same pattern was seen elsewhere. The Manitoba Game Protective League was operating province-wide in the early 1920s, although local groups were only effectively brought together in the Manitoba Federation of Game and Fish Associations in 1946. By 1930 Alberta was "the best organized province in the Dominion as regards fish and game associations" with forty-four local fish and game associations represented in the recently formed Alberta Fish and Game Association.[37]

The creation of provincial federations of local fish and game associations revealed an already existing co-operative relationship between government and local associations that gave the latter a voice on provincial game legislation and policy. In 1919 the Commission of Conservation had urged the provinces to encourage the formation of new fish and game associations and to promote existing ones so that they could assist with public education, law enforcement, and the revision of game laws.[38] This had been the goal in 1906 when the Saskatchewan game branch had organized a meeting in Regina to form the Saskatchewan Game Protective Association to help in the "saving of our game from extermination." Provincial game officials appear to have drafted the association's terms of reference and to have sought to attract members by offering a silver trophy in a trap shooting competition. While the organization seems to have been short-lived, the Saskatchewan government continued to co-operate with fish and game association locals, and as the province noted in 1921, these organizations collectively helped with "co-ordinating and representing local opinion and reporting it to the provincial level." Although there were bumps along the way, this co-operative relationship continued once provincial federations were formed. As the Saskatchewan Fish and Game League noted in 1943, while relations with the previous government had been difficult, good relations had been re-established and the government had, "for the most part, carried out our requests and acted favourably on our recommendations." Indeed, two years later the Saskatchewan Department of Natural Resources confirmed that its game protection work relied on close co-operation with the league in stocking game birds, maintaining game preserves and public shooting grounds, encouraging "clean sportsmanship," and smoothing hunter–farmer relations.[39]

The organization of the Alberta Fish and Game Association had similar goals and alliances. In 1928 Alberta premier John Brownlee urged the Calgary and Edmonton fish and game associations to spearhead the formation of a federation of Alberta's local associations. This became the Alberta Fish and Game Association, and as Brownlee had anticipated, provincial authorities could now deal with a single organization on game administration instead of a myriad of local and often fractious ones.[40] Such co-operative relationships were entrenched when the provinces began to set up fish and game advisory committees. In Alberta in 1944, for example, of the nine members on its advisory board, four were from the Alberta Fish and Game Association, three from the fish and game branch, and two from the University of Alberta. Indicating a concern for broader involvement, five members had been added by 1949, one each representing farmers, the Department of Indian Affairs, fur dealers, trappers, and outfitters. The fish and game group nonetheless remained the largest block on the board, giving it direct and substantial input into provincial game policy.[41]

Such political influence revealed the broader framework in which game policy was formed. In his survey of big game conditions in the Rocky Mountains in 1915, W.N. Millar, of the University of Toronto's forestry faculty and a former forestry official based in Alberta, commented that the politics of Canadian game preservation was shaped by three groups. First in importance was the small but influential minority made up of hunters and commercial interests connected with hunting. This group had "practically exclusive control of the making and amending of game laws." A second and smaller group was not interested in hunting but tried "for various reasons to shape the game laws so as to retain the game undiminished." In Millar's view, these naturalists and nature lovers were "not very effective." The third group was the great majority of the population that did not hunt and had no interest in it. Many probably also had little interest in wildlife, but Millar noted that they supported "restrictive game legislation" or at least did not oppose sport hunting overall. As Millar further observed, while the hunting group was a minority of the population, it was able to exercise power because it was highly organized, it claimed to serve a range of social (often gendered) values, and it was, among other things, of service to pioneer settlement and economic activity.[42] But its influence depended on the tacit agreement of the non-hunting majority, and when the majority began to withdraw its agreement, hunting suffered a loss of political power and social acceptability. This likely helps explain the decline of hunting after the 1960s, when more people in this non-hunting population shifted towards a more critical view of hunting because of changing concerns about environmental issues.

"THE 'PREDATOR' COMES FIRST, LAST AND ALWAYS": KEEPING WILDLIFE FOR HUNTERS

Late nineteenth-century game legislation informally promoted predator control by not protecting certain wild animals. This allowed people to kill such animals at any time and at will, and the introduction of inducements to kill certain animals, such as bounties on wolves, foxes, gophers, and blackbirds, formalized this as state policy. The policy served various sectors, including sport hunting, and by the First World War many fish and game associations were lobbying for control of predators that competed with hunters for game. These predator policies became an important part of the identity and goals of the associations and found ready and willing support from government agencies that independently and co-operatively supported similar goals.

Predator control was in theory warranted by the belief that the balance of nature had to be readjusted. It was also justified by self-interest. At times, sport hunters, mixed and grain farmers, ranchers, fishers, fur traders, and parks authorities all saw the elimination of wild animal competitors as an important tool in the pursuit of their own concerns. But of all the proponents of predator control, sport hunters and farmers showed the greatest vigour, at times becoming obsessed with the elimination of wild competitors. D.G. Colls of the Canadian Wildlife Service noted in 1948 that the members of the Saskatchewan Fish and Game League in Saskatoon were a "nice bunch of fellows" but had a narrow view of wildlife conservation. Colls believed wildlife conservation should encompass a wide range of measures, but such a view was "extremely controversial" with this group. Expressing his exasperation, Colls confessed,

> [M]y only personal objection is the way in which the word "conservation" is continually tied in with the word "predator." Conservation apparently means "kill all things that prey on birds or animals that we want to kill." I suppose they are quite conscious of the other, more important, concepts of conservation, i.e. land and water use, cover and food, etc. but they rarely seem to mention them. The "predator" comes first, last and always. It is my hope that that idea can be overcome someday – not that I don't believe that "predators" may require some control in certain areas – it is the overall slaughter that rubs me.[43]

The Saskatchewan league's fixation on predators was shared by hunters and their associations elsewhere on the prairies. Some concerns were localized, such as the threat to game animals of cougars in Alberta. While cougars were found in coniferous forests in many parts of the prairies in the 1800s, by the interwar years sizable numbers lived only in the Rocky Mountains, although they were occasionally seen in southern Alberta as far east as the Cypress Hills. Due to an apparent increase in numbers in the 1930s, they were killed in the Rocky Mountain parks, and Alberta offered a bounty on them until 1963.[44] More broadly, wolves, raptors, crows, and magpies were of general concern across the region. Campaigns against wolves were ongoing, and after their elimination on the southern plains by the end of the First World War, efforts against them shifted north to fringe and forested areas. Although protection of farm stock remained a justification for killing wolves in these areas, the conservation of big game was now the most important goal. As F.J. Fitzpatrick of the Saskatchewan game branch argued in 1916, while the government tried to protect big game by regulating hunting, "the real depleter is the wolf, not the man. Our north country would supply game for the whole Dominion for many years to come if no game was killed but that taken by man. Man kills game at a time when he thinks it is best but the wolf is after it all the time." He concluded, "Mr. Wolf is on the job three hundred and sixty-five days and nights every year and as long as he is allowed to work his will on our game we might as well save our time as waste it prohibiting the hunter by law from taking the small proportion he takes annually." Consistent with such views, the provincial government paid bounties on wolves, increasingly in northern and fringe settlement districts. These bounty programs were ongoing. Saskatchewan paid bounty on 1,200 wolves for the period between 1907 and 1917 and on 3,000 wolves, at a cost of $48,000, for the years between 1943 and 1948. Other provinces did the same, with Alberta, for example, paying bounties in this latter period on 6,148 wolves and 294 cougars.[45]

When bounties were officially abandoned in Ontario and western Canada in 1954, more finely targeted systems of wolf control were already being employed. For example, Saskatchewan allowed the use of strychnine to kill wolves in northern areas in 1945 because they were "responsible for the slaughter of our big game animals" and were threatening to move south to prey on domestic livestock. Concerns about the collapse of barren land caribou in northern Saskatchewan, Manitoba, and the Northwest Territories in the late 1940s also led to expanded wolf-control programs and the use of poison in northern districts. And demonstrating the more ecologically based wildlife management approaches that were being adopted after the Second World War, the Manitoba government by the late 1950s was tailoring its wolf-control programs around the specific conditions and needs of different districts. Wolf control in fringe areas was designed to protect livestock in adjacent agricultural districts; in more remote areas its purpose was to protect big game for hunters and fur-bearers for trappers, and in the far north it served to protect barren land caribou as a local food source and in the longer term for sport hunters. Since the use of poison endangered domestic stock in farm districts, professional hunters were employed to kill game there, but to protect caribou, poison was freely used in the far north to kill wolves.[46]

Although wolves were important high-level predators, some people believed that wolves were scapegoats and that over-hunting was the real culprit. The *Winnipeg Tribune* editorialized in 1950 that moose populations had declined "thanks to the automobile, high powered rifles and great increase in the number of hunters," and while "we like to fool ourselves that the wolf is responsible," the facts "may as well be faced – senseless slaughter by man is chiefly to blame." Clearly, wolves did not "fit in with man's scheme of things." Confessing a "sneaking regard" for the wolves, the *Tribune*'s editor further discounted "most of the stories" of their ferociousness. If big game animals were to survive, more was needed than to "thin out the wolves" and the hunter needed "to check the urge to kill in the name of sport every creature of the wilds that he sees." A month later, a follow-up editorial quoting a letter to the editor in the Toronto *Globe and Mail* made the same point when it argued that hunters "would like to substitute the rifle for the wolf. This can never be satisfactory because collectively the hunters' appetite is insatiable and a rifle doesn't distinguish between a sick animal and a strong one." The only way to protect big game was to control hunting, and the editorial concluded that "too many species of wildlife have been exterminated" in Canada, "and certainly the wolf hasn't been responsible."[47]

A similar situation existed in regard to raptors, which were blamed for killing game birds. Although only a few species were recognized as predators in most provincial game legislation, hunters (like farmers) saw all of them as a threat. And even had they wanted to distinguish the "good" from the "bad" species, few people could do so. As Canadian ornithologist Percy Taverner noted in 1926, "eagle bounties have been paid on Red-tailed Hawks and Horned Owl bounties on specimens of Long-eared Owls." And in 1937 Selby Walker of the Inglewood Bird Sanctuary in Calgary wrote to Dewey Soper that "through the efforts of uninformed sportsmen and others, local 'Shoot the Hawk Campaigns' have almost cleaned out the Rough-leg and Swainsons from this district while the small Falcons listed for destruction in the Game Act are becoming more not less common." Soper agreed, confessing that "it has always seemed such a crime that our beautiful, beneficial raptors have been so wantonly and idiotically slaughtered off. Despite all that is done to educate the public along these lines still abysmal ignorance and prejudice prevails."[48] Indeed, in a speech to the Alberta Fish and Game Association annual meeting in Red Deer in 1955, avian biologist William Rowan deplored the widespread shooting of owls and hawks and especially singled out the popular custom of men shooting any raptors that they spied sitting on fence posts or telephone poles while they drove along the country roads. Although hunters claimed they were protecting game birds, Rowan observed that hawks ate gophers and the only benefit of killing them was entertainment and higher profits for ammunition manufacturers. As Rowan concluded, the anti-hawk "fanatics" of the fish and game association did not care about the facts and indeed wanted to legalize their war on these birds and were lobbying (unsuccessfully it turned out) the Alberta government to delist protected hawks.[49]

While efforts to eliminate wolves and raptors continued, the most widespread and popular predator campaigns mounted to benefit hunters were against crows. The number and range of crows expanded along with farm settlement, and hunters saw them as a threat because they ate the eggs and young of upland game birds and waterfowl. As previously discussed, farmers and hunters sometimes disagreed about their value to humans. Many farmers insisted that crows served their interests by eating worms and insects, and the Saskatchewan government argued in 1909 that, "from the standpoint of the game protector" the crow might seem harmful, "but it is a friend of the farmer." But hunters and their organizations successfully turned public opinion against crows. They lobbied government for control programs and by the interwar years were organizing and mounting "crow shoots," promoting government incentives to kill them, and issuing a steady stream of anti-crow articles and announcements. Private companies also encouraged the campaigns. By 1929 the Dominion Ammunition Company, for example, was awarding a prize to the fish and game association local that killed the most crows.[50]

One of the earliest references to crow control as a field for state policy was made in 1915 when the Manitoba game branch blamed the shortage of sharp-tailed grouse on crows and hawks and proposed a bounty on them. This was implemented in 1919, by which time the province was claiming that crows had become "a powerful enemy and menace" of insectivorous as well as game birds. In other words, crow control would serve farmers as well as hunters. A bounty of ten cents for each pair of crow legs and five cents for each egg was offered. In response, 42,100 crows were killed and 327,227 eggs were destroyed, which cost the province over $20,000, a considerable sum at the time. The program was intended to be a one-time effort to keep "this bird of prey down to some sort of even balance." When this failed to happen, an ongoing program was established in 1925. In the next five years the provincial treasurer paid bounty on 664,444 eggs and 378,035 pairs of legs.[51] By this point, Alberta had a similar program. It included magpies as well because they had expanded their range in the interwar years and were viewed as an enemy of game birds. Wary of the open-ended expenditures that bounties could bring, Alberta offered only fixed cash prizes in several categories. This proved to be a sufficient incentive, and 229,455 crows and a lesser number of magpies were killed and 527,386 eggs were destroyed under the program from 1924 to 1929.[52]

These policies became more exclusively connected with hunting because of the decline of waterfowl populations in the 1920s and their subsequent collapse in the 1930s. Hunting organizations often claimed that crows bore a good deal of blame for the disaster. This stance was part of a wider discussion about what was happening with waterfowl populations, and as will be seen shortly, debate broadly revolved around hunting and habitat change as probable causes. Those who blamed habitat change also tended to blame crows (and increasingly magpies as well) for this upset. Amateur ornithologist Frank Farley believed that there was a direct link between fewer ducks and more crows. In 1932, he published evidence of the large number of duck eggs and goslings eaten by crows. Arguing that the sooner people admitted that crows were responsible for "the present deplorable plight of our wild ducks" and implemented measures to control them, "the better will be our chance of saving the ducks from total extinction." And showing an urban bias in the sport-hunting fraternity and in anti-crow sentiment, the Red Deer Board of Trade sent a circular letter to newspapers in Alberta in 1935 declaring that crow control was essential if game birds, especially waterfowl, were to be "saved for the pleasure and profit of ourselves and posterity." While the board admitted that many factors influenced waterfowl populations, "a vast amount of evidence" showed that crows were "one of the most detrimental factors." Since hunting should not – and could not – be restricted, crows had to be controlled. The board quoted various authorities on the "cruel, wicked, murderous ways of these black, old nest-robbers" and called opponents of crow eradication "a few long-haired anti-everything fellows." Citing statements by Ontario conservationist Jack Miner, a rabid crow hater and admirer of waterfowl, the board urged the public to help "lessen the vandalism of the thuggish crow."[53]

After its establishment in 1937, Ducks Unlimited (Canada) added its resources to the anti-crow campaigns. While it ran its own control programs, including ones against northern pike, mink, weasels, and other predators of ducklings and duck eggs, crow and magpie control was its main concern. As it noted in a 1940 newsletter, "the war on crows is in full swing over the Farmlands. Again DU sparks activities, in coordinated campaigns with governments and organized sportsmen."[54] It also provided grants for this purpose. For example, DUC's chief naturalist B.W. Cartwright advised the Manitoba Game and Fish Association that "I can assure you that we are prepared to make the customary grant to the Crow Control Committee for their 1944 campaign. We would, in fact, be pleased to see a further extension of the campaign, and would be prepared to increase the funds for that purpose." This support was intended solely for the defence of hunters and their interests; as the organization argued in 1949, "more ducks are destroyed by crows and magpies in western Canadian provinces alone, than are shot by hunters throughout North America each year."[55] And as the Saskatchewan Fish and Game League phrased it in 1943, crows and magpies killed at least 60 per cent of the province's duck hatch each year, suggesting that all government needed to do was to "give us the funds each year, eliminate the crow and let the sportsmen reap the harvest."[56] Such rhetoric once again revealed that crow campaigns were now deliberately and solely intended to serve the interests of hunters. As a result, governments were increasingly willing to let hunting associations assume more campaign administrative and financial responsibilities. By 1936 the Saskatchewan Department of Natural Resources and the Saskatchewan Fish and Game League were formally co-operating in running campaigns, and by the 1940s they were also sharing costs, although the bulk of the funds still came from government. In 1947, for example, Saskatchewan paid $10,500, or two-thirds of the bounty, on about 79,000 magpies and about 70,000 crows. The province used a portion of the yearly game licence fee to fund this campaign, which suggested that hunting licences were not a tax but a conservation levy that benefited hunters.[57] Similarly, by 1945 crow and magpie campaigns in Alberta were entirely run by the Alberta Fish and Game Association; the province simply provided a grant for half the costs. Such campaigns continued, and even war did not stop them. In 1944 the Ration Administration (a wartime federal agency) made shells available to the Saskatchewan Fish and Game League for its predatory bird-control campaign.[58]

The success of these campaigns was largely due to their integration into community social life and their broad public acceptance. The Drumheller Valley Fish and Game Association in 1928, for example, organized monthly outings in the spring where teams throughout the district competed for prizes for the greatest number of legs and eggs collected. As an indication of the public's engagement with the venture, the local team's progress was charted in the Drumheller newspaper, and by the end of June it appeared that the Drumheller team was well on the way to victory, having destroyed 10,000 eggs and 2,000 birds. As a sign of the camaraderie of the competition, the losers bought dinner for the winners at a local restaurant.[59] The social respectability and credibility of such events were further reflected by their sponsorship by government, fish and game associations, and DUC. While cash and other incentives helped entice public support, the campaigns were also successful because they appealed to people beyond the hunting fraternity. For example, in 1931 Ethel Webster of Seven Persons, near Medicine Hat in southeast Alberta, wrote (under the letterhead "Floral Friends Club") to Percy Taverner at the national museum in Ottawa to express her satisfaction with the Medicine Hat Fish and Game Association's crow and magpie bounty. "These pests are greatly reduced in this neck of the woods," she wrote, "and we have seen more young flocks of birds reach maturity than ever for some time." Songbirds were more plentiful, as were grouse, because of the campaign, and farmers too had gained by the reduction of crows, as "people who have been tormented with them eating the poultry report practically no molestation this year."[60]

Such views drew easily on moral judgments about good and bad birds, and as was so often the case in predator control, ethical behaviour and sportsmanship were sidelined. Hunting ethics apparently applied only to game animals and no quarter needed to be given to other creatures. Magpies were difficult to shoot, and as one correspondent to the Nature

Lovers page of the *Western Producer* reported in 1954, she and her family therefore went out at night and shot upwards through their nests to kill the sleeping magpies. In 1945 another writer revealed that blinding roosting birds at night with a bright flashlight was an efficient way to kill them.[61] And Red Deer naturalist and writer Kerry Wood, who usually had a balanced view of things, promoted a device known as a "crow bomb." Used in the United States, this crow bomb was a "fiendish contraption invented for the laudable purpose of destroying crows." Gravel, chilled shot, and a stick of dynamite were encased in wet cement. When the cement was set, the contraption was hung in a tree, usually in the autumn when large numbers of crows were roosting in preparation for their winter migration. The operator hid close by behind a screen, and once the crows were settled in for the night, the bomb was detonated, with the blast of shot and gravel doing "deadly havoc to the roosting crows." Wood also wrote a pamphlet about trapping and shooting magpies, again developing elaborate and complex systems to trap "these evil birds." He also suggested that they could be killed with poison, although he piously qualified this advice by noting that the required permits should be obtained.[62]

The success of the crow and magpie campaigns owed much to their appeal to schoolchildren. In special competitions, prizes were awarded to the children who killed the most crows and magpies or collected the most eggs. There were also competitions among schools in which prizes were given to the school whose students collectively gathered the most eggs and killed the most birds. Some schools organized specific outings for this purpose, and in 1928 one Saskatchewan schoolgirl reported that "we do so want to get rid of that black pest of Saskatchewan." Her school was collecting eggs and legs in hopes of winning a prize, and she noted that "next Saturday we are going on a hiking trip through the woods to gather eggs." The students would also have a picnic, and she anticipated that they would "have lots of fun, for what is more delightful than a day in the woods?"[63] Fun it may have been, but it provided no lessons in human kindness. In Alberta in 1929 schoolchildren received about 66 per cent of the total prize money paid out for killing crows and magpies and just over 76 per cent of the money awarded for egg collection. Such programs were carried on for decades; the Saskatchewan government, for example, launched a new competition in rural schools in 1949. While it no longer gave cash prizes, the ten students who killed the greatest number of crows and magpies received an all-expenses-paid trip to northern Saskatchewan, which the province was then promoting as a tourist destination.[64]

While the campaigns' focus on schoolchildren may have been an effective tactic for stoking public antagonism towards the birds, not everyone approved of them. In 1928 the editor of the Nature Lovers page in the *Western Producer* argued that such behaviour should not be encouraged, claiming that it was a perversion of "a child's impulses to be kind to all living creatures to give them prizes for tormenting any kind of bird, whether useful or otherwise." There was a political point to be made as well. Since the *Western Producer* was the newspaper of the Saskatchewan Wheat Pool, the editor further argued that "kindness and co-operation are inseparable – one cannot be cruel or unkind and yet practice the true spirit of co-operation." If crows were to be killed in the interests of farming, then "the wisest and most human [*sic*] way" would be to hire professional hunters to kill them as quickly as possible. Others objected for more personal reasons. In 1929 one Saskatchewan correspondent wrote to the Nature Lovers page that she hated seeing posters offering prizes for crow legs and eggs; she had pet crows and they were intelligent and played pranks and she wanted to have more of them.[65] Still others thought it was unnecessary to kill crows. V.W. Jackson of the Manitoba Agricultural College suggested in 1924 that "we don't need to exterminate criminals but we should frighten them into good behaviour. A gunshot early in the morning will frighten a lot of guilty crows into better ways of living." And rather than shooting them, one could deter them, perhaps by hanging twine around the poultry yard, as they would not fly under anything that they might get entangled in. There were others as well who had simply revised their view on the subject. Kerry Wood, for one, changed his mind about crows (but not magpies) in the early 1950s when crow populations were declining because of predator-control programs in Canada and the United States. Wood also grew increasingly disturbed by the ethics of encouraging children to kill crows and magpies. As he wrote in 1953, he was opposed to "having children go out and destroy crow eggs and tear the feet off young fledgling birds. Pest control is at its best a cruel business – and I for one am loathe to encourage youngsters to practise such cruelty."[66]

Although crow- and magpie-control programs had continued unabated for decades, it was evident by the 1950s that these birds were still a force to be reckoned with. In 1953 the government of Alberta reported that despite the expenditure of much money to eliminate them, crows and especially magpies continued to be a problem and efforts against them needed to be stepped up.[67] By the end of the decade, professional management instead of amateur hunters and egg gatherers increasingly came into favour, and in 1959 the province began employing professionals to lay strychnine baits to kill magpies. Some people found this strategy offensive, and as J.D. Herbert of Calgary wrote, the new policy was "received with shock by most conservation-minded people in this province. The use of any poison for the destruction of a species of wildlife against which the indictment can never be final should not be condoned by thinking people." He further argued that the use of strychnine was "morally criminal" because it indiscriminately killed so many other animals. In reply, Minister of Agriculture L.C. Halmrast assured Herbert that scientific tests were clear about strychnine's benefits and that it was an efficient tool because a "large number of magpies can be taken with a minimum of effort, an economy of cost and with maximum safety of domestic animals and man."[68] Apparently, modern organizational techniques and industrial processes would achieve what older approaches had not, all without having to change a fundamental part of the human–animal encounter.

HABITAT OR HUNTING? DUCKS AND HUNTERS IN THE INTERWAR YEARS

Ducks and geese were the most plentiful game birds in Canada's prairie region, which was crossed by a number of important waterfowl flyways. They made fine sport and in 1930 it was claimed that "the joy of flight shooting with wild full grown ducks coming down wind with half a gale behind them" was unmatched "by anything else on the continent."[69] Yet concern about depletion grew steadily after about 1900. While goose

Figure 6.2 For duck hunters, flight shooting offered an exciting hunting experience. In this photograph (likely from the early 1950s), hunters wait in a marsh near Edmonton for another shooting opportunity.

hunting in the spring continued everywhere until 1917, spring duck hunting had been banned in Manitoba in 1898 and in Alberta and Saskatchewan in 1907. The opening date and length of the fall hunting seasons were controversial, with a host of observers charging that they opened too late or too early and stayed open too long or not long enough. Debate intensified as waterfowl populations fell, and pessimism was growing in North America by the late 1920s when field studies and reports showed dramatically lower duck populations. As the decline accelerated in the 1930s, some species appeared to be endangered, and demands grew for lower bag limits and shorter open seasons and even for a total ban on hunting ducks and geese.[70]

In this debate, avian biologist William Rowan stood almost alone in suggesting that low duck populations were likely the result, at least in part, of natural cycles. He believed that these cycles were about thirty-three years in length and were tied to the regular rise and fall of lakes, of which the drop in lake levels caused by the drought of the 1930s was only the most recent instance.[71] Very little was known about the matter, but Rowan's theory seemed to be borne out when duck populations on the prairies increased dramatically after 1939 as water conditions improved. However, even Rowan was unsure of this, and as he wrote in 1943, perhaps the ducks were "merely back here in their former numbers at the expense of other migratory routes."[72] In other explanations for the population collapse, hunting or habitat destruction was blamed, and the habitat-versus-hunting divide came to characterize debate about duck depletion and solutions for it.

In the early 1930s Luther Goldman of the United States Biological Survey estimated that duck populations had fallen by half in the previous thirty years, and he argued that while habitat change may have driven ducks further north to nest, the decline was also the result of hunting.[73] Ontario conservationist Jack Miner had no doubt that it was caused by hunting. As he wrote in early 1927, "every place I lectured, persons were asking, 'What has become of our wild geese that used to come here by the tens of thousands? Where are they? Have they changed their migrating route or what has happened to them?' I answered them by saying, 'Automatic guns, and pump guns and systematic shooting is where they are migrating to.'" And much of this hunting was legal; bag limits were too high and gave "a lawful privilege for any individual to shoot twenty-five hundred ducks and eight hundred geese in one season." Indeed, Miner argued that from the bird's vantage, the open season was very long: it opened in Canada in September and did not close in the southern United States until six months later.[74]

There was little agreement about reducing bag limits or the length of open seasons. Facing a worsening situation, the US government reduced the open season on ducks to one month in 1931. Using its authority under the Migratory Birds Convention Act, the government of Canada urged (but did not mandate) the provinces to do the same.[75] By this point, demands for the temporary banning of duck hunting were growing. In 1932 the Manitoba game league recommended a totally closed season on ducks, and in 1935 the *Calgary Herald* editorialized that duck hunting should be banned until populations recovered. As it noted, this approach had helped save pronghorns and it might save ducks as well. But Fred Bradshaw, then the director of Saskatchewan's provincial museum, pointed out in 1934 that such action was unlikely. While some Saskatchewan sportsmen now appreciated "the calamity that has befallen the ducks" and were advocating the temporary banning of duck hunting,

Figure 6.3 Even during the unprecedented collapse of waterfowl populations in the interwar years, bag limits remained high. These two hunters from Cardston in southern Alberta in 1928 likely had not exceeded their bag limit by shooting these eighty-six birds.

Bradshaw thought it was "somewhat doubtful" that "the large majority of sportsmen are willing to go to such lengths, however desirable."[76] Indeed, despite ongoing discussion, the three Prairie provinces were unable to agree on coordinating lower bag limits or open seasons, and individual provinces were loath to act alone, possibly giving advantage to their neighbours.[77] Other factors were in play as well. As Saskatchewan Game Commissioner E.A. Etter wrote Hoyes Lloyd, the supervisor of wild life protection in Canada in 1933, if he was to describe duck conservation in Saskatchewan, he would have to say "that the only influence at work here is that for longer open seasons and larger bag limits. The Department of Natural Resources are apparently determined to lease all available bird sanctuaries and public shooting grounds for grazing, and are doing so … Game protection and conservation has now become a d—d nuisance. Revenue is the one and only thing."[78]

Habitat loss during the drought in the 1930s added to the crisis and for some observers explained the collapse in waterfowl populations. Jack Miner had little patience with this argument, calling it "another trigger finger excuse advanced by the thoughtless shooter." Who, he asked, "hasn't been duck hunting the last ten years and seen thousands of acres of green marsh with more guns than ducks?" But habitat change on the prairies became a genuine part of the crisis when the dry years of the 1920s, followed by prolonged drought in the 1930s, destroyed bird sanctuaries that had been a primary shelter for many birds. In 1940 Dewey Soper of the CWS recalled that before 1930 the region had scores of lakes and "many famous waterfowl and wader areas" that were now ruined and dust dry.[79] The view that habitat change was the most important cause of the waterfowl crisis took some of the pressure off hunters to change their ways and suggested that they could continue, perhaps at a reduced level, to hunt. This thinking received a huge boost with the establishment in 1937 of DUC, which argued that drought and habitat change on the prairies, not hunting, was the cause of the decline of North American duck populations. DUC's mission was to encourage the killing of waterfowl predators, to restore former breeding grounds, to protect existing ones from destruction, and to manage such areas "with a view to obtaining maximum waterfowl production." One of its favourite slogans was "More Water – More Ducks!" The organization restored wetlands by damming, dyking, diverting waterways, and dredging sloughs and lakes, and projects were fenced, fireguards were built, and predators were controlled. Wetlands were also cleaned to rid them of duck botulism, and beavers were transplanted to help with water retention.[80] DUC for a time advocated the setting up of huge incubators in each province to hatch wild ducks for release in the wild. This idea was not a new one. Alberta's game commissioner Benjamin Lawton had argued in 1928 that it might be a good idea to establish an area where wild ducks could be "reared both naturally and artificially" for release into the wild. Ongoing interest in this idea built on research at Delta (located at the southern end of Lake Manitoba), where experimentation had begun in 1931 under the patronage of James Bell, a wealthy Minneapolis duck hunter who was alarmed by declining duck populations. Waterfowl biology was very poorly understood, and the Delta station (which was formally incorporated as the Delta Research Station in 1938) conducted various experiments and studies. While it did research on botulism and other diseases, its main area of experimentation was the artificial hatching of wild ducks to increase North American duck populations.[81]

Prairie wetlands were said to be the home of about 80 per cent of the migratory waterfowl shot by American hunters, and DUC primarily served the needs of these American hunters in its attempt to increase continental duck populations.[82] DUC and its parent organization also served these hunters' interests by opposing restrictions on duck hunting. In 1937 John Huntington, president of Ducks Unlimited Inc., the American parent organization, commented to E. DuPont (head of the large American ammunition and chemical company and a Ducks Unlimited Inc. supporter) that there had for a number of years been a "well organized movement to stop duck hunting in the United States. Without questioning the sincerity of the motives of such organizations as the National Association of Audubon Societies, the Emergency Conservation Committee, the Wildlife Protection Fund, the American Nature Association and the American Forestry Association," Huntington contended that a closed season would only speed up extermination. While he did not explain the logic behind this conclusion, Huntington was clearly of the opinion that "the movement for a closed season in 1937 is stronger and better organized than at any previous time and those of us who believe that closing the season would hurt rather than help the ducks will have to band together to prevent this happening."[83] DUC echoed this opposition to tighter regulation. In a radio broadcast in the late 1930s, W.G. Ross, MLA from Moose Jaw and president of DUC, told listeners that "here was the land, where once the buffalo and antelope roamed; where wildfowl flourished in untold millions; where native upland game thrived in plentitude [*sic*]. Nature took its own care of the proper balance." While governments had responded to the upset of the balance of nature with game laws, Ross believed that these laws had failed and that government had contributed further to the problem by allowing land drainage, overgrazing, and forest and prairie fires. In his view, conservation required "constructive, practical leadership," and by applying its "special knowledge of wildlife requirements," DUC would manage to do what government had failed to do. In other words, habitat restoration and the technical expertise of DUC rather than further regulation of hunting would save North America's duck populations.[84]

DUC typified utilitarian relationships with wildlife, and its publications often employed a chattiness that echoed boys' adventure stories. They also often reflected the assumption that waterfowl existed to serve hunters in the same way that workers served industrial capitalism. Ducks were frequently referred to as workers (which presumably made sportsmen the bosses), western Canada was described as a factory or a farm that produced ducks, and hunting seasons were called "production seasons." The language of one 1940 DUC newsletter was typical. "The last issue of this report went out November 10, 1939. *Then*, the ducks were hitting south. *Now*, they are coming north … coming back 'with their sleeves rolled up' – eager to get to work turning out another big duck crop. The vast duck factory of the Canadian west is alive with their wings a-hum [and] with their quacking – as they surge northward on the heels of the spring break-up."[85] Such messages enjoyed hearty support from many Canadian hunters and their associations, whose needs, after all, were the same as those of American hunters. DUC benefited from skilful management and promotion and earned a reputation for getting things done. It was also politically influential because of its energetic and committed membership and its cultivation of prominent individuals. Like many fish and game associations, DUC understood the value of political and bureaucratic alliances in achieving its goals. Its Canadian board over the years featured provincial premiers, cabinet ministers, and leading professional and business figures. James Richardson, the influential and wealthy Winnipeg grain merchant, O. Leigh Spencer, director of the *Calgary Herald*, and W.G. Ross, MLA (Moose Jaw), were among the first members of its Canadian board in 1937.[86] This pattern continued, and in 1940 Manitoba premier John Bracken accepted a directorship. But whatever influential alliances DUC formed in Canada, the members of the US parent organization, which provided most of the funding for the Canadian organization, were especially important. As John Huntington of the parent organization remarked, "unless the Canadian operating corporation is functioning to the entire satisfaction of the trustees of Ducks Unlimited Inc. it will be an easy matter to straighten out matters by the simple expedient of withholding funds."[87]

DUC was welcomed by many people who worked professionally in wildlife conservation in Canada. Echoing its familiar alliances with other sport-hunting groups, the Alberta government supported DUC's work in the province by providing office space and by collaborating in its work.[88] The federal government worked with DUC as well. In 1937 Dewey Soper at the federal government's wildlife agency told Selby Walker of the Inglewood Bird Sanctuary in Calgary that the $2 or $3 million DUC planned to invest in habitat restoration was "no small matter and our divisions could never hope to get so much money to spend on the restoration of waterfowl." Noting that the drought had been terrible for water birds, Soper wrote that DUC's intention to co-operate with the federal government's land reclamation agency, PFRA, to restore wetlands was a potentially productive approach.[89] Yet such endorsements for DUC ebbed and flowed, and overly slick manoeuvring by the American organization sometimes raised hackles in Canada. Writing somewhat defensively to Hoyes Lloyd in 1943, Soper noted that while it was true "that a good many people do not entertain a very high opinion of Ducks Unlimited," it had achieved habitat restoration that no one else had been able to. Yet he was by this time concerned that DUC was making unjustified claims about its success. This concern grew, and Soper observed in 1946 that his suspicions had been confirmed by the findings of two US Wildlife Service officers who had inspected ninety-four DUC projects in the prairies. "I have the whole story. It is a sorry one, with terrific misrepresentation."[90]

Duck populations began to increase in North America in 1937, the year DUC was established. This suggests that the organization benefited from good timing: its Canadian projects began when duck populations had at least stabilized and were beginning to increase slowly.[91] But concerns persisted even though population numbers looked better, and in 1940 demands for a temporary ban on duck hunting were again heard in Canada and the United States. Most hunters continued to oppose this, arguing instead for limited habitat protection and crow and other predator controls. But as the bird lover Lawrence Potter remarked, he had "far more sympathy with the crow plundering a duck's nest" than with sport hunters who refused to stop killing ducks when populations were collapsing.[92] In *The Ducks, Geese and Swans of North America* (1943), one of the most influential books on waterfowl at the time, one contributor, Ira Gabrielson, argued that waterfowl populations had fallen because of the drainage of 100 million acres of land in North America in the previous fifty to seventy-five years. Admitting that it was "absolutely true" that strong populations depended on having suitable habitat, Gabrielson held that it was "equally true that all the marsh restoration in the world will be useless" if hunting was not restricted.[93]

Prairie naturalist Tony Lascelles took this argument one step further. In 1940 he argued that the decline of duck populations was caused by over-hunting, not drought. Lascelles noted that fieldwork by American and Canadian ornithologists in northern Canada, where 70 per cent of North American waterfowl bred, showed that there was more suitable habitat than was occupied by waterfowl. In other words, "the breeding stock has been so reduced that there are too few birds left to occupy fully the available breeding grounds, even where conditions are still favourable." As Lascelles pointed out,

> Sportsmen and others interested in the future of North American waterfowl are thus confronted with two conflicting arguments. "Ducks Unlimited," according to their catechism, aver that the shortage of waterfowl is caused through lack of water in and about the arid regions of western Canada, and all will be well, without recourse to restrictions reducing the kill, when water is restored. The U.S. Biological Survey says there is no lack of water in the

> north, where the majority of waterfowl now breed, but a shortage of birds to occupy available terrain, a condition that can only be rectified by reducing the number of birds killed annually. Both cannot be right.

Lascelles was convinced that the ornithologists, not DUC, were correct. He asked, what was the benefit of habitat restoration on the prairies "when there is ample water in the north awaiting tenancy by the hordes of waterfowl that filled the skyways in happier days." Habitat restoration on the prairies could offer no benefit "unless one believes in defying the simple laws of biological arithmetic. Water cannot breed waterfowl without parent stock, however much such a miracle is desired." The only way to restore any species "suffering from excessive exploitation" was to restrict hunting. As Lascelles concluded, "waterfowl and not water is the crying need of the hour."[94]

"WE ARE DONE WITH HUNTING"

The controversy over sport hunting and waterfowl conservation revealed that among the general public, critics of sport hunting were fewer in number and less influential than its defenders. Even so, there was a considerable record of concern about the practice. Some people had long argued that killing wild animals for pleasure was morally wrong, that hunting was cruel, that it led to depletion, and that it was a selfish act that interfered with the enjoyment of wildlife and the outdoors by non-hunters. But such critics were generally disadvantaged in a society in which the dominant discourse saw sport hunting as culturally normative and as forming a legitimate relationship with wildlife governed by ethical guidelines and bag limits, closed seasons, and other rules that served conservation needs. Those who criticized sport hunting in principle thus not only faced the challenge of deeply rooted cultural practices and norms but also the possibility of being seen as eccentrics as well. Less controversial than opposing sport hunting in principle was the argument that non-lethal pursuit as well as visual appreciation could provide people with a satisfying engagement with wild animals. Recreational fishing provided one example where catch and release programs were increasingly seen as a solution to overfishing and as a mark of good sporting behaviour. As a Manitoba government tourist promotion pamphlet stated in 1941, catch and release of rainbow trout allowed the fisher to appreciate nature's vigour and beauty and "when the effort is rewarded with an iridescent, thrashing beauty, as in the rainbow, the thrill of capture must be such that a desire is felt to release the fighting beauty. All power to such emotions! They emanate from a real sportsmanship!"[95] Similar arguments were put forward in the case of mammals and wild birds. In 1920 M.D. Barker, a game guardian in northeast Saskatchewan, argued that "the pleasure in killing" a deer or the value of its flesh could in no way compare to the pleasure of seeing "these lovely creatures in happy gambol among our bluffs and dells." And Jack Miner warned in 1927 that a hunter should be aware "that the fall of one bird out of the air from his deadly aim can give pleasure to one only, while thousands are deprived of seeing it alive and that less than ten per cent of the people of North America want to shoot with a gun."[96]

While Miner was critical of sport hunting, his criticism was selective – he believed it was unreasonable to oppose moderate levels of hunting because ducks and geese multiplied rapidly and he supported the destruction of hawks and crows to protect ducks. Other people were similarly selective. After camping beside a pair of Canada geese for a month in 1946, for example, Kerry Wood concluded that he would never again point "a weapon at a Canada goose" because of the respect he had gained for the intelligence and social habits of "this beautiful, freedom loving" bird. So, too, while he had little difficulty with hunting for food and had done so himself when young, Wood could no longer shoot a deer because it "was too interesting in its live state." Such sentiments about deer were widespread. After the shooting of does was prohibited in the NWT in the late 1890s, deer increased rapidly in what is now southeast Saskatchewan, and it was said that they sometimes "became almost domesticated." When a temporary open season on does was declared in 1903 to reduce the population, game officials noted that hunters who shot them were confronted with "a feeling of animosity" from some members of the public.[97] This also occurred after the Second World War when game policy moved towards "any deer seasons" and deer hunting increased rapidly.[98] In response, government officials tried to lessen public criticism by arguing that "the often heard cry, 'Protect the deer, cut down hunting,'" would only bring over population and a population crash. Deer hunting, they argued, was in fact an act of kindness that its critics did not recognise, and "the nature lover must be prepared to understand that a swift, lethal bullet is often the only alternative to equally lethal and infinitely more cruel starvation. Only if there is adequate hunter harvest will the nature lover be able to find plenty of deer for his enjoyment."[99] But scepticism endured. In 1952, for example, the provincial government planned an open season on deer in the Whiteshell Game Preserve in eastern Manitoba. W.G. Hamilton of the Whiteshell District Association surveyed residents in the area about this open season and found them uniformly opposed. Revealing a sharp resentment of the proposal, Hamilton wrote that "surely we are entitled to one spot in Manitoba where nature can take its course without the interference of hunters." Indeed, he argued that "the deer in this district are worth a whole lot more on the hoof to the province as a whole as a tourist attraction than they are to the individual hung up as meat."[100]

While Hamilton's criticism was directed at a specific government initiative, other critics went further and rejected sport hunting in principle. Reflecting on moose hunting, writer F.A. Twilley noted that he had never been a hunter. While he admitted that he "couldn't hit the side of a barn" with a rifle, this was not the reason he did not hunt. Rather, he "had no liking for the sport" for "[m]oose are God's own horses and a live one looks better than a dead one." In a 1924 fictionalized account of the 1844 Red River bison hunt, W. J. Healy, the provincial librarian of Manitoba, voiced his opposition to sport hunting by drawing a clear line between killing animals for food and killing them for recreation. Admitting the undoubted excitement and challenge of the bison hunt, he nonetheless wondered why anyone "has a right to do for his pleasure what he feels, or must know, to be cruel? There is much of the brute yet in many of us, but surely we have, since 1844, advanced towards the time to come, let us hope, when it will no longer be considered a sign of superior nature to find pleasure and satisfaction in the slaughter of animals."[101] Demonstrating the same point of view, Charles Frederick, the editor of the *Peace River Record*, wrote a *mea maxima culpa* in 1922 that began, "[W]e are done with hunting. We are done with sport which involves the taking of life from one of God's creatures for the mere sport of killing." While he had hunted for many years, he had replaced his rifle with "a camera or at most a fishing pole." This change of heart had come about because of an early-winter hunting expedition for moose with a couple of friends. While the search for moose was futile, they came upon a spot "where the soft snow had been padded down as if by some animal rolling on it." After following some tracks, they came upon a shallow den where a large black bear raised itself up and, "with blinking curiosity and an odd snorting not unlike a pig," peered at them. But "he did not stare long," for one of the hunters immediately shot him in the face. The bear threw himself against the side of his den, moaning and "pawing with frightened agony at his nose," but a second bullet had "a more merciful effect." The bear was immediately skinned and its body cut up and the fat taken off. "It was sport such as one might read about in the field magazines," Frederick wrote, but it was a "rather gruesome sort of sport." But more than this was the moral affront:

> Every characteristic of the animal, its habitation so carefully selected and prepared, its body, so carefully prepared by nature with enormous layers of fat upon which full life and vitality would be maintained through the long winter months, the very condition of the entrails, devoid of an excess in which disease might lurk, bespoke the watchful care of the Creator of all things, over even beasts of the wild forest, those same beasts for which men in the name of sport underwent hardship for the sake of hunting down and killing. No one wanted the meat; not one of the party needed the fur robe taken from the animal. But it was a great day's sport and in years to come the handsome hide, mounted with head and claws intact, will adorn a floor as a souvenir of the chase, and a tribute to the wonderful game preserves between the Smoky and the Peace rivers.

So, "someone else can have our rifle," Frederick concluded, lamenting that he did not have a photograph of the living bear and its den instead.[102]

The agony of the bear that Frederick describes is remarkably similar to that described by Francis Dickie earlier in this chapter. For both men, the death of the bear was a transcendent experience, but with entirely different conclusions and implications. Frederick moved from the death of the bear to a rejection of sport hunting, while the killing of the bear in Dickie's case was exhilarating and changed nothing about his relationship with wild animals. But everything changed for Frederick, as his plans to use a camera to embrace his new-found relationship with nature and wildlife revealed. This was a personal choice for Frederick, but as a machine, the camera served all users equally and could represent a range of attitudes, needs, and assumptions. Writers about the history of wild animal photography often stress the ways that the camera has been used to construct scenes to celebrate the conquest of nature, reflect masculine power, embody cultural longings, fears, and hopes, or lament the loss of a vanishing wilderness and preserve its memory.[103] As they had since the late 1800s, sport hunters photographed animals after they had shot them to provide a sort of double trophy. Many others simply photographed their successful hunts for posterity, as is attested by the thousands of photographs of prairie hunters proudly exhibiting – in carefully arranged tableaux – the animals they had killed. Such connections between photography and hunting continued to be important; in 1962 an article in the *Farm Ranch Review* advised hunters to take a camera to the field to create a record of their outing and kills.[104] But like Charles Frederick, others saw the camera as a tool that offered an opportunity for a non-lethal sporting relationship with wild animals. Photographing wildlife became increasingly more accessible as technical innovations allowed the short exposure times needed to photograph quickly moving animals and birds. As Dewey Soper found out when photographing a duckling featured on his 1946 Christmas card, it could be fun and rewarding to learn new technical skills and ways to work with tricky and difficult subjects. He explained to a friend how he had kept the duckling quiet long enough to photograph it: "Camera (Graflex) on tripod focussed on a given point. Place duckling on the spot, cup hand over him until he is quiet; quickly remove hand and press the trigger as soon as he looks intelligent. Usually the head goes up first for about one second to get bearings and then he is away. If your reflexes have been inadequate run like ___, secure your subject again and try once more. Success is guaranteed with enough patience. Some are a sinecure, others are not. Individual temperament varies considerably."[105] The sheer chance of getting a good image and the pleasure found in the technical demands of the craft also attracted the Saskatchewan nature writer Doug Gilroy, who found wildlife photography to be a year-round hobby with no closed seasons. As he pointed out, it took "time and patience, but outwitting them [the animals] is where the fun and satisfaction comes in." As well, the skill of the craft appealed to Winnipeg wildlife photographer C. Dettloff, who, when writing an account about photographing migrating waterfowl in 1933, quipped that photos were "a better bag than a good dinner and a lot of feathers."[106]

Dettloff's humour slyly suggested a criticism of sport hunting. Photography could be a substitute for hunting and a way to preserve and appreciate wildlife without harming it. Dominion Commissioner of Fisheries J.W. Robertson made this point at a meeting of the Commission of Conservation in 1915 when he took on the argument that sport hunting built

male character. Arguing that hunting bison, pronghorns, and birds should be banned, he contended that "if we encourage hunting with a camera and discourage hunting with a gun, we shall not be any less courageous or resourceful. Canada ought to be a bird sanctuary from ocean to ocean, so far as guns are concerned. I know no feathered creature that any man is entitled to shoot with a gun in these days." Others tried to redefine the social and psychological value of hunting by transferring its supposed virtues to the camera. As the *Calgary Herald* editorialized in 1936, photographing wildlife was difficult and "far outrivals shooting done with a rifle or shotgun. Vastly more skill is required to get a close-up of a deer drinking at a waterhole, or leaping a fallen log with exquisite grace than merely bringing it down at long range with a high powered rifle." As well as needing all "the woodcraft knowledge and wiles of the successful game hunter," wildlife photographers had to resolve many technical difficulties to get a good shot, and while "their shots may fail more often," they would "at least have the satisfaction of knowing that no wounded animal will creep away to die miserably."[107]

Growing arguments for a more humane relationship with wildlife after the Second World War were typified by the work of nature writer Dan McGowan. As he wrote in 1948, the camera offered "a pleasant way to hunt." No licence was required, no regulations had to be obeyed, and it could be done all year. "Best of all one does not have to slay an animal or bird in

order to secure a worthwhile trophy." Nor did the camera require travel to remote places; cities had plenty of wildlife suitable for photographing. Regina, for example, had 145 species of birds in and around the city. Indeed, the number and type of wild animals brought into a relationship with humans extended beyond the conventional definition of "game." While moose, elk, deer, bears, and other charismatic wildlife made great subjects, seagulls, newly hatched birds, chipmunks, and other animals in their natural setting were possibilities too. But even McGowan found that there were limits; he concluded that the porcupine was impossible to photograph because "in form it resembles a haggis and, unless in a moving picture, you cannot tell whether it is coming or going."[108] This trend towards wildlife photography was further reinforced as movie cameras became more accessible. In keeping with the view that cameras were preferable to guns, the 1952 film *Hunting with a Camera*, produced by the Canadian Government Travel Bureau for the National Film Board in co-
operation with Montreal's Redpath Museum, claimed that "the thrills of the chase can be recaptured time and again when hunting is done with a motion-picture camera." And for some, the movie camera led to life-changing decisions. As the Alberta guide and outdoors writer Andy Russell wrote in 1960, filming and watching bighorn sheep opened for him "a shining adventure with cameras instead of guns. I suddenly realized that I could never kill another bighorn."[109]

The view that the camera enabled a new ethical relationship with wild animals (as demonstrated by people such as Charles Frederick in the 1920s, Dan McGowan in the 1940s, and Andy Russell in the 1960s) provides some context for the contemporary decline of sport hunting. While the state has continued to sanction it, sport hunting has declined both in the number of hunters and, perhaps most significantly, in social respectability.[110] While there are still a great many sport hunters and sport hunting continues to have many defenders, there is a growing defensiveness about the sport.[111] Although there has been discomfort with sport hunting for decades, most hunters, confident in the rightness of their conduct, have ignored opposing views. But the fall of an ancient sport, from significant social standing to a defensive stance in only a few decades, represents an extraordinary social change. Why sport hunting has declined and lost respectability remains a matter of debate. Some observers point to urbanization as the cause, others single out the rise of electronic media (including hunting games) that have captured male attention, and still others see government hunting regulation and the banning of lead shot as contributors.[112] While these may well be relevant factors, they are not sufficient to fully explain this change. Urban males were traditionally an important part of the sport-hunting fraternity, the impact of electronic media on male behaviour remains insufficiently understood, and government has regulated hunting for well over a century. A more important factor lies in the cultural shift that has occurred among the majority of people who had never hunted for sport but until then had tolerated it or at least accepted its legitimacy. Historian Thomas Dunlap argues that this change in majority opinion came about in the United States because "what we have lost is the attraction to the hunt, to the ritual of pursuit, the demonstration of woodcraft, and the mastery of nature that hunting implied. We no longer see the virtues of the pioneers, at least in this form, as an essential part of the nation's culture."[113] In Canada the linking of national history and identity with hunting is not so evident, but a similar shift in majority opinion towards a more critical stance about sport hunting does seem to reflect broad changes in thinking about environment and the place of humans in nature. How long this trend will last is unclear, but greater numbers of people in the silent majority who had previously accepted the regulated culture of hunting are now finding themselves increasingly uncomfortable with the killing of wild animals for sport.

7

Reconstructing Nature: Acclimatizing Animals for the Wild

> The novelty of importing foreign species of game birds seems to have a special attraction for most sportsmen but we have so many evidences of native species giving way to the aggressiveness of alien birds that many reputable game conservationists now consider it a dangerous policy to encourage such probable replacements. There are no better game birds anywhere than our native grouse and it is far better to provide for them ample protection than run the risk of their depletion by introduced diseases or other causes as the result of experimenting with imported birds.
>
> Fred Bradshaw, 1926[1]

As people have moved around the world, they have brought new plants, animals, and diseases with them. Historian Alfred Crosby terms this "ecological imperialism" and notes that the diseases, plants, and animals that accompanied people in their movement were as much a part of their baggage as their household goods. In what he calls the neo-Europes – Canada, the United States, Australia, Argentina, and similar settler societies in which incoming Europeans became dominant – domestic animals and plants that suited the newcomers' aesthetic preferences or economic needs and customs became foundational in the economy and society. Other animals and plants were introduced by accident or design, sometimes forming new wild populations. Historian Thomas Dunlap observes that this biological invasion was part of the conquest of new lands by Europeans as well as their "continuing attempt to come to terms with their new lands, to find their place in the country and its place in them."

These transfers were particularly extensive in the Antipodes, where an extensive array of plants and animals were transplanted, such as deer in New Zealand and rabbits in Australia, which were introduced as game animals. Various songbirds were also transplanted in both countries as familiar reminders of home and because they were said to be more beautiful than many indigenous birds. This hope of recreating "home" seems to have been a weaker sentiment in North America, perhaps because its landscapes more closely mirrored European ones.[2] While such traffic was heavily weighted towards the movement of domestic animals and plants from Europe to the rest of the world, many plants and some animals were also transferred to Europe from abroad. This activity was so popular that societies were established in England and France to encourage and organize it.

"Acclimatization" was a broad concept that included a number of activities and theories, including the environmental adjustment of an organism – the common meaning of the term today. But as historian Michael Osborne notes, in the British world of the eighteenth and nineteenth centuries, the term referred to "the transfer of so-called exotic organisms from one location to another with a similar climate." While the most important transfers were of domestic animals, other animals were also introduced with the intention of establishing wild populations. The goal was to supplant as well as to supplement native species. Historian Harriet Ritvo observes that, in comparison to the transfer of domestic animals, transfers of "exotic organisms" in the late nineteenth century were on a smaller scale and resulted from the work of a few individuals and organizations. On the Canadian prairies, Hungarian partridges, common pheasants, and other birds as well as various game fish were introduced to supplant indigenous ones or to supplement native species that were in decline. While some concerns were voiced that such introductions were misguided, acclimatization was sanctioned by the dominant view that wild animals and the environment could be managed to serve social and economic needs. Often influential in fish and game associations, proponents of acclimatization on the prairies were confident that they were improving the world by making it more useful in meeting human needs. Moreover, they often showed a casual confidence that transplanting new species into the wild was a simple matter. As Ritvo concludes, however, the task was not always easy or achievable.[3] With the exception of the transplanting of Hungarian partridges and certain fish, such projects on the prairies often failed or proved to be more difficult than anticipated.

New non-native species also appeared on the prairies as the result of acclimatization efforts elsewhere. Having been introduced by the French at Port Royal in 1606, rock doves, the ubiquitous pigeons of every urban area in North America today, were among the earliest introduced birds that have dispersed across the North American continent.[4] House sparrows (popularly called English sparrows at the time) were introduced in the mid-1800s in the eastern United States for sentimental and commercial reasons – to remind people of Europe and to introduce birds that would eat more insects than indigenous ones did. Later introductions were made in California, and as can be seen in every part of North America every day, the experiments were highly successful. While the sparrows were welcomed at first, they soon came to be seen as a threat to indigenous songbirds and were sometimes called the "British bullies" in the United States.[5] Negative opinions about them guaranteed that their spread was closely tracked. While it was reported in 1887 that they had not yet reached the prairies, they had appeared in several towns along the railway between Calgary and Edmonton by 1903, which suggests that they were hitching rides on boxcars on the steadily expanding regional rail networks.[6] However they spread, they were a familiar part of the prairie landscape by the First World War. House sparrows were hardy, smart, aggressive, and prolific. In 1920 one observer called them "noisy, quarrelsome and dirty" and argued that they drove out native sparrows, wrens, purple martins, and bluebirds, while in 1950 another called them "tough little bird gangsters." Many people objected to their droppings and nests that dirtied barns and other buildings.[7] Yet house sparrows found a few defenders. As one correspondent in the *Western Producer* ironically observed in 1930, hostility to them proved that "a civilized nation" was one in which people tried to kill the sparrows that ate weed seeds and insects and then promptly turned around and spent "millions of dollars to fight insect pests."[8]

Starlings had a similar history on the prairies. After several failed attempts, they were successfully acclimatized in 1890 and 1891 by a New

Yorker who it is said wanted to introduce the birds mentioned in Shakespeare's writings to the city.[9] Although they had not yet been spotted, birdwatchers in areas around Saskatoon were looking out for these birds by 1929. They adapted slowly to the prairies, and as late as 1954 Marion Nixon, the bird columnist for the Regina *Leader-Post*, reported that there was no permanent starling population in her district in southeast Saskatchewan and she hoped "they do not become hardy enough to survive our winters. Until now, a severe spell in winter has always finished the last ones off."[10] By 1952, however, they had established a foothold in the somewhat milder climate of Alberta. Writer John Patrick Gillese described them as intelligent birds that could "think" and imitate a wide range of sounds. He further claimed that they would return to pick up something they had dropped and that such a demonstration of "thrift" was as admirable a trait in birds as in humans. But they were also aggressive, and Gillese admitted that they might become a problem and therefore "man must be careful when he meddles with the laws of nature." Indeed, starlings soon joined house sparrows as unwelcome birds, and by the mid-1960s they were commonly viewed as pests in warehouses and on farms.[11]

HUNGARIAN PARTRIDGES: COMPENSATING FOR THE DECLINE OF SHARP-TAILED GROUSE

Fears about the future availability of native upland game birds stimulated the introduction of Hungarian partridges, which proved to be the most successful acclimatization effort on the prairies. The grouse family includes the most common indigenous upland game birds on the prairies. There are a number of species, some with restricted range. Historically, the most widely distributed of these included ruffled grouse (sometimes simply referred to as partridges, found throughout the region in aspen-wooded areas), sharp-tailed grouse (popularly called prairie chickens, found almost anywhere with open land and patches of low brush), and spruce grouse (sometimes called "fool hen," found in open areas or along the margins in forested districts). Pinnated grouse moved from the United States into southern Manitoba and Saskatchewan in about 1900 and spread rapidly for a time, but they had vanished from Alberta by 1929 and were extirpated in Saskatchewan and endangered in Manitoba by the 1940s.[12] Sharp-tailed grouse were especially common in the late 1800s and their periodic migrations were sometimes vast; Ernest Thompson Seton reported that people in southern Manitoba in the late 1800s occasionally saw "a stream of small flocks and individuals flying past for hours on end, totalling thousands or tens of thousands of birds." Habitat loss because of land breaking and clearing led to a steep decline in sharp-tailed grouse numbers by the early 1900s. It also isolated breeding populations from each other and, equally important, destroyed their dancing grounds (called leks). Crucial for mating rituals, some leks are believed to have been continuously used for hundreds of years. Habitat loss was ongoing, and after the Second World War, hard-edged habitat, rather than the brushy variety that grouse needed, continued to expand in parkland and forested areas as roads were improved and fence lines were cleared to provide a few more acres of agricultural land. And although burning would have created brushy open habitat, the rigorous suppression of fires further limited suitable habitat.[13] The birds' decline was also exacerbated by relentless hunting. They were hunted for food as well as for sport, although some people thought there was little sport in it. As writer Nellie McClung recalled, sharp-tailed grouse in Manitoba in the 1880s "were as thick as you ever saw black birds in the oat fields in Ontario, and so tame, it was a crime to shoot them." Some observers thought high bag limits legalized over-hunting, and the editor of the newspaper in Blairmore (in the Crowsnest Pass west of Fort Macleod in southern Alberta) wrote in 1910 that while bags of 200 sharp-tailed grouse were legal, they were excessive, for he believed in "bird shooting as sport," not "wholesale slaughter."[14]

All indigenous prairie grouse have cyclical populations, sometimes with local variations in timing. The reasons for such cycling continue to be studied, but it is known that populations tend to peak about every ten years, followed about two years later by sudden collapse. Few birds are seen for several years but then they quickly reappear. While everyone recognized that rabbits were cyclical (carcasses in the bush provided the evidence), grouse left "practically no trace of death behind them: they just disappear, their skeletons rarely found in evidence."[15] Without physical

Figure 7.1 The population cycles of sharp-tailed grouse – popularly called prairie chickens – were not understood for many years, and the low points of their cycles triggered recurring apprehensions about their imminent extinction. Such fears in part helped spark the introduction of foreign upland game birds such as Hungarian partridges and pheasants. People sometimes fed game birds in winter, which is likely the case with this flock of sharp-tails in 1928.

evidence of their death and not recognizing that they were cyclical, Euro-Canadians over the years interpreted the sudden drops in grouse numbers in various ways, and this lack of knowledge was significant in the early history of efforts to acclimatize Hungarian partridges and other game birds. One popular explanation blamed human activity, which fit logically with the common theory that wildlife populations were typically stable and that humans were the primary cause of their upset. Thus, depletion was usually blamed on hunting and governments responded with closed seasons. In 1907 Saskatchewan promised to ban hunting to save grouse from extinction. Their subsequent strong population increase was attributed to this policy, and by 1911 farmers were complaining that sharp-tails in particular were a nuisance.[16] But the birds disappeared again in 1914–15, and it was later recalled that this collapse "created a feeling of considerable uneasiness in the prairie provinces. Man's ruthlessness, combined perhaps with inimical natural forces, was again manifesting itself, and to avoid the repetition of the unfortunate history of the Western States, where Pinnated Grouse was almost exterminated by the market hunter, all Prairie Provinces combined in establishing an absolute closed season." The efficacy of banning hunting was said to be demonstrated once again when populations increased rapidly.[17]

At the same time, it was recognized that hunting alone did not always explain falling populations. Saskatchewan's game commissioner Fred Bradshaw contended in 1915 that while human destructiveness was "at the head of the list" for the shortage of birds that year, other factors included bad weather, destruction of habitat, and predation by crows and coyotes. He discounted disease as a factor since carcasses were never found, "which one would expect to be quite common if the disease were of a serious character." It was also said that grouse were killed when they ate strychnine-laced grain that was set out to kill gophers, but this was disproved in 1918 when the Saskatchewan game branch conducted experiments which proved that sharp-tailed grouse had "some power of resisting the poisonous effect of strychnine."[18] The suggestion that grouse migrated periodically was also popular. In 1909, "to the surprise of many," sharp-tailed grouse became fairly common in districts where they had been "scarce or entirely absent." This sudden upswing was explained by "the semi-migratory nature of this bird, which is influenced in its movements by the supply of food and other conditions."[19] In 1915 the Alberta game branch appealed to this view to explain recent shortages, as did Charles Barber, Manitoba's chief game guardian who reported in 1919 that many types of grouse had "been coming from the north in large covies [*sic*], indicating that their extreme scarcity in the past few years may have been due to that migration to the North Country."[20]

The periodic collapse of grouse populations, confusion about their life cycles, and fears that they faced extinction sparked efforts to supplant them with hardier birds. Consequently, Austin de B. Winter and members of the Calgary Fish and Game Association released seventy pairs of Hungarian partridges in 1908. The birds came from Hungary and were brokered by a company in Pennsylvania. Demonstrating a casual attitude about the introduction of new species, the birds were apparently raised in the backyards of local hunting enthusiasts. More birds were released the next year, and about 150 birds in total were released at Calgary and others were later released at Edmonton.[21] Their promise as game birds was quickly apparent, and it was claimed by one hunter in 1933 that Hungarian partridges did not have "a single fault" and now provided "the best shooting on the prairies."[22]

Hungarian partridges spread rapidly throughout Alberta from this initial planting at Calgary, and by 1919 they had reached Cardston in southern Alberta. The local newspaper welcomed their arrival because "it will not be long before what we call the prairie chicken here, but which is in reality the Sharp-Tailed Grouse, will be disappearing, and then no doubt the Hungarian Partridges will take their place." The disappearance of sharp-tailed grouse was accepted in a rather matter-of-fact way, perhaps in part because it was thought that Hungarian partridges were better game birds. Fast fliers, they were more evasive than sharp-tailed grouse and demanded higher sport-hunting skills because they were "not so easy for the pothunters to annihilate in the winter as they will not sit quietly up in the trees for the poor wing shots to murder at close quarters." By 1926 they had reached High Prairie in northern Alberta where one hunter noted that sharp-tailed grouse were scarce in areas where land was broken but remained plentiful on range land. Declining habitat meant that declaring a closed season would "not bring them back," but in any case, Hungarian partridge were plentiful.[23]

While Hungarian partridges naturally spread throughout Alberta, their dispersion was not left entirely to chance. In 1923 the Alberta game branch transferred Hungarian partridges from southern Alberta to the Edmonton district where populations had seriously declined. Then in 1927 a reverse transfer took place. Birds were now plentiful in most areas of the province except around Calgary, and on the recommendation of the Alberta Fish and Game Association, the government transferred Hungarian partridges from other districts to "replenish" the Calgary area.[24] Notwithstanding such management, Hungarian partridges were generally vigorous and spread rapidly eastward beyond Alberta. Although none were released in the province, they reached a few areas in south-central Saskatchewan by 1922, just fourteen years after their first release in Alberta. By 1926 they were nearing the Saskatchewan–Manitoba border and appeared to be moving steadily eastward at about forty kilometres per year. Nonetheless, Manitoba sportsmen were not content to wait for the birds to reach the province through natural dispersal, and in early 1926 they brought birds at their own expense from Czechoslovakia and Alberta. Further plantings were also apparently made in 1927 or 1928. As a result of such plantings, along with natural dispersal, Hungarian partridges were found by the early 1940s throughout the agricultural districts of Saskatchewan, Manitoba and the Dakotas.[25]

The enthusiasm for Hungarian partridges was not shared by everyone. While not always clear-cut or consistent, concerns arose on a number of fronts. One Alberta farmer refused to allow the release of Hungarian partridges on his farm because he believed they would only attract hunters who would damage his property.[26] This was an indication of poor hunter–farmer relations and not a criticism of the birds. Indeed, it was often noted that Hungarian partridges were insectivorous and a friend to farmers; in 1934, representatives of thirteen municipalities in southern Manitoba passed a resolution asking that the hunting of Hungarian partridges and sharp-tailed grouse be prohibited in drought areas because the birds were useful for controlling grasshoppers and other insects. But not all farmers agreed. For example, the Saskatchewan Rural Municipalities Association resolved in 1944 that Hungarian partridges should be removed from the "protected list of useful and game birds" because they damaged crops.[27] There was also concern about the general principle of introducing non-native birds. Revealing trends in thinking about environmental management, the Ecological Society of America resolved at its 1921 meeting in Toronto that the introduction of non-native plants and animals in US national parks should be banned. Similar concerns came to be associated with the introduction of Hungarian partridges. In 1922 Saskatchewan game commissioner Fred Bradshaw noted that while the introduction of foreign birds was "a very dangerous proceeding," he was not concerned about Hungarian partridges (which by then had been reported in southwest Saskatchewan), calling them "Alberta's praiseworthy contribution to the establishment of this bird east of the Rockies." Saskatchewan also encouraged their spread by temporarily declaring a closed season on them.[28] Yet Bradshaw became warier as rumours surfaced that Hungarian partridges behaved aggressively towards native grouse, and while the evidence was contradictory, he noted in 1925 that "we fancy not many people would care to see our hardy and native bird supplanted by a foreigner." Such worries were perhaps enhanced because sharp-tailed grouse were then at a low point in their cycle, but as Saskatchewan game guardian Neil Gilmour noted in 1926, whatever was said about the aggressiveness of the Hungarian partridge, nothing could be done, since he "came into Saskatchewan uninvited" and has come to stay and "there is nothing we can do to improve his manners."[29] Some people, however, tried to control these newcomers. In 1924 Neil Pratt of Ravenscrag in southwest Saskatchewan told the Saskatchewan bird enthusiast Lawrence Potter that Hungarian partridges had become quite numerous in the district. Claiming that they were "the sworn enemy of our prairie chickens," which were "just terrified away" by the Hungarians, Pratt had all the Hungarian partridges trapped on some property he was managing. As a result, Pratt noted, "the prairie chickens immediately returned" and he did not think that the Hungarians "ought to be allowed to increase." Still others were concerned about the cultural attitudes and priorities that lay behind their introduction. Marion Nixon, a birdwatcher and nature columnist who lived at Wauchope in southeast Saskatchewan, saw game bird introductions, tourist hunting, and habitat change as having a cumulatively negative effect on native prairie fauna. As she wrote in 1941, "I sometimes wonder if our native grouse will be crowded out by these imported species. Here there are certainly more partridge this winter."[30]

Although concern about the decline of sharp-tail grouse continued well into the 1950s, the grave concern expressed in the early 1900s about their future weakened in the interwar years.[31] By this point, Hungarian partridges were treated like any other game bird – in 1946 Saskatchewan's Department of Natural Resources even claimed that they were "a native of Saskatchewan." While hardly a native bird, they were becoming indigenized as was shown in 1936 when their populations in the prairies crashed. This was surprising; Hungarian partridge populations were not cyclical in their homeland, but something on the prairies was leading them to behave like native prairie grouse. Ten years later, this happened again: sharp-tailed grouse populations fell drastically in Saskatchewan, as did those of Hungarian partridges, although the latter's decline was less serious.[32] Whatever the cause, it was by then increasingly understood that sharp-tailed grouse populations were cyclical and that their sudden decline was not a prelude to extinction. This led some to argue that hunting had a negligible long-term impact on grouse. Indeed, the existence of natural cycles only further justified hunting for some. As was noted in 1953, native grouse "will decline whether there is an open season on them or not," and "even if they are not shot by hunters, experience shows that they would be gone next year anyway." Indeed, by the 1960s some studies suggested that hunting had never been as detrimental to the birds' survival as people had believed and that habitat loss was more critical in the overall decline of all of the prairie grouse family. Food shortages and climatic variations year-to-year also caused loss of local populations, and while hunting led to temporary declines, it likely contributed little to long-term fluctuations. In this context, it was argued in 1963 that, given natural fluctuations, controlled sport hunting was "a way of harvesting birds which could very soon be lost to use by natural fluctuations in population."[33]

OTHER NEW GAME BIRDS

Even though Hungarian partridges were successfully dispersing across the prairies, other attempts to add to the stock of game birds seemed ever more pressing as depletion of native upland game birds and waterfowl continued during the interwar years. As C.A. Hayden of the Calgary Fish and Game Association noted in 1933, this made the "replacement of native game species by other species which will survive alongside civilization and which will meet the discriminating demands of the sportsman" a pressing task.[34] In addition to Hungarian partridges, other possibilities were therefore pursued. In 1930, for example, the Manitoba game branch investigated the feasibility of importing giant and black grouse from Czechoslovakia but found the cost too high. As well, the Moose Jaw Wild Animal Park distributed wild turkey eggs for hatching in the mid-1930s, but the effort did not lead to the creation of self-sustaining wild turkey populations in Saskatchewan. And although adult wild turkeys were apparently introduced in Manitoba in 1941, the earliest recorded planting of them was in 1958–59 in the Pembina Mountains in the southern part of the province. Others were also introduced in Alberta in the early 1960s.[35]

Other than these attempts to introduce exotic grouse and wild turkeys, most attention to new game birds focused on pheasants and chukar partridges. Common (often called ring-necked) pheasants had been introduced in the United States in the late 1800s, and like Hungarian partridges, they were seen as a promising replacement for vanishing prairie native game birds. In about 1900, sport enthusiasts at Brandon introduced pheasants in the area, and while the populations were not vigorous, they seemed to be self-sustaining, as a few birds were seen in 1913 around the city and further north near Dauphin. By 1926 other attempts to bring these birds to Manitoba were said, with considerable exaggeration, to demonstrate their suitability for the province's climate.[36] In 1934 the Manitoba game branch released 1,000 birds that it obtained from Wisconsin, and more birds were released in 1936 and 1945. The Saskatchewan Fish and Game League also sponsored their introduction. In 1931–32 eggs were purchased and distributed to "old Country Gamekeepers" and to turkey farmers for hatching. Despite lacklustre results, the project was said to have been worthwhile as a public relations exercise because it had promoted the league and saved it "from stagnation and demise" by attracting new members and renewing the commitment of current ones.[37] More directly, the project sparked interest among provincial game officials, who began to collaborate with the league in implementing a co-operative arrangement with North Dakota to strengthen game resources on both sides of the border. In exchange for 5 million walleye eggs and several hundred wild Hungarian partridges, North Dakota sent 1,700 pheasants captured from the wild and 2,000 eggs to Saskatchewan in 1933. A further 1,000 birds were shipped the next year. Four hundred pheasants were kept at the Moose Jaw Wild Animal Park for laying, and the rest were released. Eggs were hatched in the backyard of Dr Merkley, a Moose Jaw Game Association member, and were also sold at cost to Saskatchewan residents for hatching. While the Saskatchewan government owned the birds and officially ran the project, the Saskatchewan Fish and Game League was the driving force behind the project and coordinated the hatching and release of the birds. The project received extensive positive press coverage, but some people saw it as elitist. In the fall of 1933 some individuals calling themselves the "Square Shooters" stole almost 1,400 chicks from Dr Merkley's backyard over a series of ten trips. In a note they left behind, the thieves asserted "to hell with your gun clubs getting all the birds. We think we have as much right to them. We are placing the 1,380 chicks we took in twelve places with from 75 to 125 in each place. We smothered a lot in the first trip but the others came through OK."[38]

Saskatchewan's attempt to introduce pheasants was unsuccessful in the long term, and the only real success on the prairies occurred in Alberta. First introduced in southern Alberta in 1909, pheasants were repeatedly released there over the next twenty-five years. Demonstrating the international trade in game birds, the Calgary Fish and Game Association acquired 600 pheasants from an Oregon game farm in 1927 and another 1,800 a short time later. In 1929 more were set out by the Medicine Hat Fish and Game Association in the Cypress Hills, about eighty kilometres away, which revealed the extent of the town's sport hunting hinterland. Among other local efforts in Alberta at the time, the Drumheller Valley Fish and Game Association raised funds in 1928 to buy pheasants to restock "this portion of the province with a good game bird." The birds were raised by a local farmer, but the *Drumheller Mail* noted that they had "none of the wild characteristics" of good game birds because they followed people around and were "always ready to perch on hands that hold food for them." As a result, the farmer's family "now look upon them as pets."[39] Even so, self-sustaining populations were established by the 1940s in the Eastern Irrigation District near Brooks, east of Calgary, where the birds found suitable habitat and a somewhat milder climate. Nonetheless, regular stocking was undertaken in the 1940s and 1950s. To create opportunities for "good shooting for this very coveted bird," the Alberta Fish and Game Association persuaded the provincial government in 1945 to share the costs of establishing a pheasant-rearing farm at the provincial horticultural farm at Brooks. Sixteen buildings were constructed to hatch and brood the chicks, and a 6,000-egg-capacity incubator was purchased. In districts with established pheasant populations, schoolchildren were paid two cents for each egg they gathered. Almost 1,500 eggs were gathered in

1945, and the surplus were distributed to various fish and game associations for hatching. Some association members used broody hens for the job, and the associations then released all of the young birds in suitable areas.[40] While this effort encountered some local opposition because the gathering of eggs was thought to reduce already small local pheasant populations, the government dismissed such fears, claiming that a hen would lay more eggs when the first batch was lost. Entrepreneurial instincts caused some difficulties when payments had to be doubled in 1950 because children demanded a better price for gathering eggs. With some disruptions, the Brooks facility released just over 51,000 pheasants between 1953 and 1962.[41] Saskatchewan also carried on with its pheasant-rearing projects during these years, as did Manitoba, but neither was effective. In 1949 the Manitoba game branch and the University of Manitoba co-operated in a short-lived project to breed a hardier strain of pheasant that could tolerate Manitoba's winters. In 1951 the province admitted only limited success with pheasant stocking, and while the birds survived mild winters, there was high mortality most years. Even so, 10,700 birds (half purchased in the United States and the balance raised from eggs secured in Saskatchewan and Alberta) were released in 1955, but a hard winter again brought disaster.[42]

Notwithstanding some success in parts of southern Alberta, pheasant populations generally could not be maintained without ongoing stocking. Attempts to introduce chukar partridges, which like Hungarian partridges were prized as game birds because of their wiliness and fast flight, proved even less successful. The favoured game birds of Mogul emperors, the first chukars in North America were brought to Illinois from Karachi in 1893, and about 795,000 and 10,600 birds were released in the United States and Canada respectively between 1931 and 1970.[43] The first attempt to introduce them in Canada occurred when a number of birds were released near Calgary in 1937 and 1938. Subsequently, eggs were sourced on the prairies from far and wide, purchased, for example, from Kansas, Minnesota, Wisconsin, and Pennsylvania. While a number of prairie hatcheries busily produced chicks, chukar acclimatization efforts were most highly developed in Saskatchewan. In 1937 chukar eggs were hatched at the Moose Jaw Wild Animal Park, and the Saskatchewan government and the Saskatoon branch of the Saskatchewan Fish and Game League set up a breeding station at Beaver Creek near Saskatoon the next year. In 1939 and 1940 the province exchanged – with Kansas and Pennsylvania – Hungarian partridges and sharp-tailed grouse for chukar partridges.[44] The birds were kept for laying, and the plan was to hatch the eggs at the station and distribute surplus eggs to the public. The Saskatoon fish and game group supervised the station, and the government spent about $10,000 on the project in the first eighteen months. By March 1941 there were 408 birds at the station, and their eggs were hatched in incubators. Professor Rae of the Department of Poultry Husbandry at the University of Saskatchewan carried out experiments on crossing chukars with Hungarian partridges.[45]

Chukars raised at the station were released at various locations in the province in 1941 and 1942. The project looked promising – some of the birds laid eggs in the wild – and the station began distributing eggs to interested parties for hatching. But significant problems followed. Unlike Hungarian partridges, chukars tend to remain in one area, sometimes forming large populations. While domestic cats were reported as a threat, a more serious problem arose because the birds had difficulty adapting to prairie conditions, and even when numbers increased, they only did so slowly. By this time, the Saskatoon branch of the Fish and Game League had shifted its interests to raising Hungarian partridges and sharp-tailed grouse, and by the end of 1943 the province decided that the chukar breeding results did not justify the cost or effort. It appears that the last birds were shipped north to Prince Albert for release in 1944. Interest revived later, and chukar plantings were attempted again from 1956 to 1962 but with similarly poor results.[46] Attempts in the 1940s and 1950s to introduce the birds in Alberta and Manitoba also failed, and it was generally admitted by the 1960s that attempts to acclimatize them on the prairies should be abandoned.[47] In the end, the thousands of chukar partridges hatched and released over several decades only demonstrated the perseverance of their promoters and the hope that the birds would at some point constitute a new game bird population, one that would supplement and replace native game birds that could no longer be relied upon for sporting opportunities.

The introduction of chukar partridges, like that of Hungarian partridges, was an attempt to reconcile the environmental change brought by settlement with demands for an adequate supply of game birds. It was a clear choice: because settlement, property rights, and economic development could not be restricted, the region's upland game bird populations had to be revamped to sustain sport hunting. Yet new views on wildlife management were beginning to emerge. In 1944 Alberta's fish and game commissioner E.S. Huestis speculated that habitat restoration, not "artificial reproduction" at hatcheries, was the best method for increasing game bird populations. "As far as possible," he wrote, the natural habitat of trees, shrubs and natural grasses that agriculture had destroyed had to be brought back. Using trees and shrubs for hedges instead of barbed-wire fences would create cover and food for birds and mammals and would beautify and improve rural life as well.[48] And in 1941 the Manitoba government experimented with permanent feed plots to provide food for upland game birds. Small plots of up to an acre along bluffs on Crown land were seeded with millet and buckwheat to provide winter feed, especially for introduced pheasants.[49] It followed this up in 1946 with about twenty-five "feed plots" in southwest Manitoba. Between one and two acres in size, they were planted with fruit- or berry-producing shrubs to provide shelter and supplementary winter feed. Annual seed plants such as sunflowers and grain were also planted to provide food until the shrubs matured. This mix of food sources recognized the needs of both native upland game birds (which could survive the winter on seeds as well as on berries, buds, and rose hips) and new upland game birds such as pheasants (which were more reliant on seeds). A similar but larger project was established in 1948 at Melita, in the southwest corner of Manitoba, as an experimental upland game bird winter feeding area.[50] But such approaches remained extremely limited, and even the simplest provisions for habitat protection were not mandated or broadly promoted anywhere in the region to reverse the decline of native game birds or to enhance the prospects for game birds that had been introduced.

NEW FISH FOR SPORT

As with game birds, prairie fish populations were often manipulated to meet sporting needs. While the acclimatization of birds involved the introduction of foreign species, fish acclimatization showed a more mixed approach. In some cases, native fish species (such as rainbow trout) were transplanted from one part of the region into lakes and rivers in another where they had not existed before. This brought new species into competition with existing ones or introduced fish into areas (most notably in the Rocky Mountains) that historically did not have fish at all. In other cases, native fish of a particular species were transplanted to supplement existing populations of the same species, and in still other cases, non-native fish species were imported from outside the region with the hope that they would form self-sustaining populations or at least offer short-term angling opportunities.[51]

Like the game bird–stocking projects, early fish stocking programs were planned without knowledge of their potential environmental impact and with an equally remarkable casualness about remaking prairie river and lake environments for the benefit of Euro-Canadian settlers. In 1889, for example, the *Regina Leader* recommended that nearby Last Mountain Lake be stocked with new varieties of fish, including salmon, trout, bass, and sturgeon as substitutes for the lake's pike, suckers, whitefish, and walleye. Showing the breadth of the proposal, the newspaper also recommended the introduction of broad mullet, a fish of the South China Sea. The federal Department of Fisheries agreed with the principle underlying such recommendations for a number of reasons. As it will be recalled, Manitoba had tried to transplant carp from the United States to supplement rural food supplies in the early 1880s, and the federal fisheries department supported the view that fish caught locally should be promoted for food and especially for commercial production. Thus, when the federal government established a fish hatchery at Selkirk, north of Winnipeg, in 1893, it planned to stock lakes with "a new and better kind of food fish than jackfish and suckers, which are now so common in these waters." Accordingly, the hatchery emphasized the production of whitefish to meet domestic and commercial fishing needs.[52] The replacement of indigenous fish with familiar game fish was also a popular suggestion from an early date, and as Inspector Gilchrist of the fisheries department remarked in 1887, the NWT did not have "good angling fish, for although the pike and pickerel [walleye] are large and bite eagerly, they afford poor sport." As he further noted, many prairie settlers wanted to replace "pike and less valuable fish" with bass, a familiar and popular sport fish in eastern North America. While he thought largemouth bass might be suitable for lakes with muddy bottoms and weedy shorelines, he favoured smallmouth bass because he thought they were tastier and better "upon the hook." Moreover, largemouth bass were more aggressive and were "voracious the entire year" and should not be put in waters containing whitefish or other species useful for food.[53]

Notwithstanding Gilchrist's cautionary advice about transplanting bass, the public remained fixated on them. They were planted in Shoal Lake in southwest Manitoba in 1888, and the fisheries department reported in 1906 that it continued to receive applications from settlers in Alberta for stocking bass and other game fish in lakes that only had "suckers and pike" or had no fish at all.[54] Bass were apparently introduced at the tourist resort of Sylvan Lake west of Red Deer in 1908, and the department set up a floating fish hatchery there in 1925 to raise bass and walleye for sport fishing. Transplanting trout was also popular. The *Calgary Herald* recommended in 1926 that more cutthroat, brook, and steelhead trout be stocked in the Bow River to encourage tourism, for "our rivers are so adapted to trout fishing, that once they are fully stocked they would vie with the best in the world in luring ardent anglers from many quarters of the globe."[55] The hope of creating a trout fishery was widespread; as one example among many, Frenchman's Creek in the Cypress Hills area of southwest Saskatchewan was planted annually (unsuccessfully it seems) with trout between 1924 and 1930. By 1934 trout had been successfully planted in a number of lakes and rivers in Alberta where they had not existed before, and lakes such as Fish Lake in the east-central part of the province near Drumheller, which could not support trout, were planted with native game fish like northern pike.[56]

Fish transplanting was also undertaken from an early date in the mountain national parks to improve recreational opportunities for tourists. A number of lakes in the Rockies did not have any fish, while native species (such as cutthroat and bull trout) were being depleted in others. The scarcity of sport fish in 1899 at Banff was said to be "the one defect that is complained of in the many attractions of the park." Accordingly, bass were planted in Lake Minnewanka and "Nipigon trout" (most likely brook trout) were introduced into the Bow River.[57] The aura and popularity of fly fishing as a recreational activity among tourists led park authorities to press for the establishment of a fish hatchery to raise trout. As the park's superintendent Howard Douglas argued in 1902, stocking would serve to avoid the need to place any restrictions on fishing by visitors, as "the streams can be kept so full of trout that it will be impossible for the tourists to deplete them."[58] Banff Park officials continued to lobby for a hatchery until one was established there by the federal government in 1913. By the mid-1920s federal fish hatcheries in the region included three in Manitoba, one at Fort Qu'Appelle in southern Saskatchewan, which produced whitefish and other species for the commercial fishery, and the Banff hatchery, which produced game fish – mostly locally indigenous cutthroat, whose population was declining – and rainbow trout, which were not indigenous in the park. By the late 1920s surplus production from the Banff hatchery was offered free of charge to fish and game associations in southern Alberta.[59]

The transfer of natural resources to the provinces in 1930 shifted responsibility for stocking to provincial authorities. While commercial fishing needs remained important and native species such as whitefish and walleye were planted to enhance commercial production, recreational needs continued to be critical as well. In 1932, the Manitoba government pledged to increase the supply of "the better and more popular varieties of game fish in suitable lakes and streams of the Province," since "nothing will give greater satisfaction to a large majority of the citizens of the Province."[60] While the Depression tempered these plans, they revived in the 1940s owing to the growing popularity of recreational fishing and the province's efforts to stimulate tourism. As an example of how foreign and local species were part of the overall strategy, the Manitoba government by 1940 was stocking various lakes for sport use with rainbow trout (obtained from a commercial hatchery in Montana in exchange for walleye eggs), largemouth bass (traded for walleye eggs as part of an exchange program with North Dakota), and walleye and whitefish fry (produced at its own hatcheries using locally gathered eggs). In 1942 the government expanded these programs by setting up a trout hatchery at Whiteshell Provincial Park in the eastern part of the province, the only one of its several hatcheries that produced only game fish. By 1947 the hatchery was producing a range of fish, including lake, brown, and rainbow trout. Brown and rainbow trout did not reproduce in the wild in Manitoba and continual replanting was necessary.[61]

Similar programs operated in the other two Prairie provinces. In 1932, for example, the government of Saskatchewan released walleye and perch fry (grown at its Fort Qu'Appelle hatchery, which also produced fry for commercial fishing) into twenty lakes in the southern and central parts of the province to "restore depleted lakes and introduce more desirable species of fish." Brown trout (obtained from an American hatchery) were also planted in a number of lakes, and it was anticipated that a visitor would soon be able to "pitch camp to find a private paradise where he may enjoy the freshly caught trout in the pan over the open fire." The next year, the government repeated this undertaking in the northern part of the province when it stocked 1.3 million walleye and perch fry in districts with summer resorts or that were accessible by road. Then, in 1934, walleye from the Fort Qu'Appelle hatchery, speckled trout fry from the Banff fish hatchery, and black bass from North Dakota were planted in fifty-one lakes, some of which previously had no fish at all.[62] These projects enjoyed popular support; in 1941, for example, the Saskatchewan Fish and Game League endorsed them as well as the province's construction of such infrastructure as fish ladders around dams. Provincial revenue from angling licences tripled between 1945 and 1949, and the government noted that this boom was the result of easier access to outdoor recreation thanks to better transportation and "increased leisure and the universality of the annual vacation."[63] Promotion of specific tourist destinations was sometimes a motivating factor for these projects. In 1949, for example, attempts were made to stock Lac la Ronge in northern Saskatchewan (then being touted by the province as a tourist destination) with Arctic grayling, "the only native species which is an excellent fly fish." This transplanting was purely speculative; it was admitted that "very little" was known about grayling, and it seems not to have been a surprise that the undertaking failed.[64]

The same trends were seen in Alberta. The province stocked lakes in response to local requests; in 1932, for example, it adjudicated thirty-seven applications and approved fifteen for stocking with trout and nine for stocking with northern pike, perch, or walleye.[65] As in the other two Prairie provinces, Alberta carried out extensive planting. In 1950, for instance, the provincial government planted about 7.3 million pike, perch, and walleye fry or eyed eggs in thirty-seven lakes and about 750,000 trout in various rivers. A decade later it was experimenting with planting "the exotic golden trout" in lakes and rivers in the foothills near Pincher Creek (south of Calgary) and Kokanee (a freshwater salmon) in the Glenmore Reservoir in Calgary to improve game fishing opportunities for Calgarians. As elsewhere on the prairies, the growth of recreational angling brought steadily increasing pressure on fishing areas. In 1957 it was said that many Alberta trout streams had become so crowded on weekends that "confirmed outdoor enthusiasts" were staying away, which reemphasized the need for an expanded trout stocking program.[66]

As with game bird acclimatization projects, it was initially believed that the establishment of new fish populations would be relatively easy. It was assumed that success would come by keeping the eggs and fry healthy until their release into a suitable environment. By the 1920s, however, it was becoming apparent that this was not sufficient; losses of stocked fry (that is, newly hatched fish that were free swimming and could take food) appeared to range from 85 to 100 per cent, while the stocking of fingerlings (which were around ten centimetres in length, were up to eight weeks old, and had begun to develop scales) resulted in a much lower loss. Thus, in 1930 the Calgary Fish and Game Association built rearing ponds where it could grow cutthroat and rainbow trout fry, obtained from the Banff hatchery, to the fingerling stage.[67] By the early 1950s new studies suggested that various factors throughout the whole process – from the gathering of spawn to incubating, hatching, and planting – could also affect the success of stocking efforts.[68]

Attempts to increase the survival chances of transplanted fish were predicated upon ideas about the balance of nature. Since it was believed that an ecosystem would be, by definition, fully occupied, it was imperative to remove competing species so as to leave food for newly planted ones. Predatory fish also had to be removed to create a safe environment for the newcomers or for existing desirable species. This applied to commercial and game fisheries alike. By the early 1950s a bounty was offered for burbot (also called "Mariah"), a high-level predator, to protect declining commercial whitefish stocks in Lake Manitoba. Despite growing scientific evidence to the contrary, one commercial fisher in the area dismissed the suggestion that overfishing was a cause of whitefish depletion, claiming instead that burbot "destroyed more commercial fish than all those caught by fishermen."[69] Similar attitudes existed in regard to game fish populations. It was contended in Saskatchewan in 1959, for example, that rainbow trout could not be stocked in lakes containing native species such as northern pike, perch, and walleye because they merely became "food for the native predator fish."[70] What were called "coarse" or "rough" fish also had to be removed from lakes to create a healthier game fishery. In 1936 about 635,000 kilograms of tullibee (cisco) were removed from Last Mountain Lake in south-central Saskatchewan to make room for sport fish. At about the same time, Alberta wildlife officials removed as many northern pike, suckers, and other fish as they could from waters they wanted to improve for trout.[71] In 1959 just under 500,000 kilograms of fish, mostly tullibee and several species of suckers, were removed under contract by commercial fishers or private contractors from several of the largest lakes in southern Saskatchewan and were sold as mink food. While it was admitted that such efforts were difficult to evaluate in terms of improved sport fishing, they at least made use of "a resource that would be otherwise wasted."[72] Another coarse fish removal project concerned carp, which had appeared in Manitoba in the late 1930s. Within two decades they were moving west into Saskatchewan, and the Saskatchewan government and the province's fish and game associations resolved to stop the invasion. With some hyperbole, the project was deemed "the challenge of the century in Saskatchewan resource management." Various techniques were used: screens and barriers were placed across rivers to trap them, a ban was imposed on the use of live fish as bait (it was feared that carp were being used for this purpose), and, finally, short stretches of the Qu'Appelle River were poisoned in an attempt to kill the invaders. All of these tactics failed.[73]

Saskatchewan's use of poison to eliminate carp was part of a popular fish management policy called "lake rehabilitation" in which a water body was poisoned so that it could then be repopulated with more desirable species. Manitoba was the first Prairie province to use this procedure when it poisoned a lake in 1946. When all the native fish in the lake were dead, it was replanted with brown trout.[74] Rotenone, derived from the root of the derris plant, was used because it was lethal to fish but not mammals. This experiment with rotenone was repeated in 1951 at a small lake in Whiteshell Provincial Park, which was subsequently replanted with rainbow trout, and similar lake rehabilitation was underway in small lakes in some districts in the province for at least the next decade.[75] Saskatchewan adopted this approach in 1959, and Alberta tested it in 1960 when it used rotenone to poison a stretch of the Sheep River south of Calgary and restocked it with rainbow trout. Saskatchewan was enthusiastic about this "important fisheries management technique" for replanting lakes with "more desirable species of game fish such as rainbow trout." While Alberta and Manitoba continued to use rotenone, Saskatchewan soon switched to toxaphene (now banned everywhere in the world). Toxaphene kills every living organism, and a lake treated with it remains toxic for two years, after which it can be restocked.[76] Such poisoning techniques brought success where previous fish acclimatization attempts had failed. One example was Lac Beauvert, a small mountain lake at the Jasper Park Lodge, a luxury resort in Jasper National Park. As part of the development of the lodge, the lake had been stocked with cutthroat trout in 1917 to supplement its whitefish and northern pike. They did not take, and more cutthroat, as well as brook, rainbow, and lake trout, were stocked the next year. This also failed. Lac Beauvert did not have the types of fish desired by sport fishers until 1964 when it was poisoned with rotenone and replanted with species that better met the sporting demands of Jasper Park Lodge guests and other visitors to the park.[77]

Despite such profound manipulation of the natural world, there seems to have been little negative public reaction to lake rehabilitation or fish acclimatization projects in general. Negative comment was rare, although some criticism arose indirectly because of admiration for the sporting qualities of indigenous game fish. For example, there were suggestions in a 1953 report that some Manitoba sport fishers disliked such projects because they were happy with indigenous sport fish.[78] It was also occasionally argued that non-native sport fish were not needed because some indigenous fish already offered excellent sport. In 1949, for example, naturalist Kerry Wood argued that northern pike were "fresh water tigers" that grew up to 4.5 kilograms, making them worthy of pursuit. Other commentators thought walleye provided even better sport than northern pike.[79] But these were simply comments about the characteristics that made a species a good game fish. Unlike many mammals and birds, to which a broad range of characteristics were sometimes attributed, cold-blooded creatures like fish were seen as unknown and mysterious. Without the appeal of fur or feathers, they at best had only what was thought of as a cold beauty. This perception – in conjunction with postwar wildlife management's interventionism and its mechanical calculation of value – no doubt contributed to the popularly held view that entire water environments and the wild creatures they contained should in some cases be destroyed so they could be redesigned to make them more useful. Indeed, a number of the fish species found in prairie lakes and rivers today have been introduced there. In Alberta, for example, brown, brook, and golden trout have all been introduced to the province, while grayling, bull trout, cutthroat trout, and rainbow trout are indigenous in some locations but were introduced in others. In Manitoba, eight native species have been transplanted outside their native range, ten species have been introduced, and two are human-made hybrids.[80]

The consequences of such manipulation have sometimes been unanticipated. In the Rocky Mountains, for example, the long-term impact of fish planting has in some locations endangered indigenous bull and cutthroat trout. It likely has also had an impact, although it is poorly understood, on invertebrate populations. Cumulatively, these changes have been significant. In 2002, of the nineteen species of fish in Banff National Park, only eleven are native to the area. Native cutthroat trout are disappearing on the eastern slopes of the Rockies because of interbreeding with introduced rainbow trout, and in 2009 they occupied only 5 per cent of their former range. While a few populations of pure cutthroats exist in areas where the water is too cold for rainbows, it is very difficult to keep them apart in other locations and it is anticipated that their ongoing hybridization with rainbows will eventually wipe out pure cutthroats in most places. This situation exemplifies the inherent dilemmas involved in such manipulation. While it is judged that rainbows will not be destroyed in locations where they have been introduced because they are too important for sport fishing, Parks Canada has been removing brook trout that were stocked in the 1960s in the remote Devon Lakes in Banff National Park in order to restore the "natural state" of the lakes and assert what was defined as the ecological integrity of the site.[81]

Such official concerns about introduced species are a departure from earlier attitudes. While the introduction of new species into the wild to benefit hunters and sport fishers and the alteration of natural lake and other environments to accommodate them had been questioned by a few observers, the dominant attitude held that nature should be managed to create economic or social value. This was at the heart of official policy, the concerns of wildlife managers, and the thinking of influential public groups. All saw it as a progressive and realistic approach, and the various acclimatization projects reflected this stance and extended its reach. As Manitoba game officials argued in 1926, while land breaking and clearing was leading to the depletion of wildlife, such changes to the land could not be stopped. Wildlife managers could deal with the consequences of this by "negative means" such as by controlling hunting, but a better approach was a "constructive" one, including "replenishing the supply" of game birds with introduced birds such as Hungarian partridges. In keeping with this attitude, the state was slow to bring in controls on the introduction of new species. Manitoba, for example, passed legislation in 1956 requiring permits for the import of exotic birds and animals, but regulations to implement this legislation were only drafted in 1959 so as to prevent the importation of nutria, a "get rich quick" fur farm animal that regulators feared would end up in the wild and compete with commercially valuable muskrats.[82] The concern thus remained with the utilitarian values of wildlife, not with the integrity of wild animals within a complex natural environment.

8

It's a War: Farming and Wildlife

> I beg to suggest there is no more important work that deserves immediate attention … than the destruction of the gopher, wolf and badger.
>
> Ex Farmer, Letter to the Editor, *Regina Leader*, 23 August 1894

Farming created critical connections between people and wildlife in prairie Canada. Most occupied land was given over to it, and farmers typically viewed wildlife in terms of its impact on farm production. Aside from a few wild animals such as insectivorous birds, most farmers felt they faced more foes than friends. This conflict was in some ways inevitable. Wolves killed cattle and sheep; gophers destroyed grain; coyotes, foxes, skunks, and weasels killed poultry; hawks and owls scooped up chickens and turkey chicks and even the family cat; bears destroyed apiaries; deer and elk ate crops in the fall; rabbits killed shrubs and small trees by nibbling on the bark and twigs in winter; and ducks, blackbirds, and other birds ate crops in the fields in the spring and fall.

Confronted by these challenges, farmers dismissed undesirable wildlife as pests, vermin, or a nuisance. Ducks that ate grain in the fields were "feathered thieves," wolves were evil phantoms that slaughtered livestock, and coyotes were marauders, chicken thieves, and lamb killers. And as farmer Oliphant in one of Ernest Thompson Seton's stories claimed, it was his compensatory right to craft a whip from the hide of a woodchuck that had eaten his crops because "that hide was raised on stolen feed." Competition for resources thus led effortlessly to moral judgments about wild animals that justified their extermination. This was consistent with the triumphalist language used to describe changes to the land as historically inevitable. Hugh Brody's exploration of the conflict between agrarian and subsistence cultures in *The Other Side of Eden* offers a portrait of the consequences of such beliefs in northwest Canada where hunter-gathers were denigrated and marginalized as unproductive, nomadic, and a barrier to the proper use of land through cultivation. And from the start, land breaking on the prairies was held up not only as one of the rewards of physical effort but also as an assertion of moral and psychological qualities. Saskatchewan poet Edna Jacques demonstrated this view in her 1939 poem "The Homesteaders," an undimmed celebration of the conquest of the land by those who devoted their lives to "Pushing the frontier further / Back with relentless hands / … / Holding fast to their birthright / Born to the realm of toil / … / Holding the country's future / Safe in their calloused hands."[1] This birthright arose from a sense of European prerogative that was sustained by hard work. Yet Jacques's celebration of cultivation was also a backhanded justification of the whole settlement project even though it had turned out so badly in the dry years of the 1930s for everyone and everything on the southern plains. While the drought could not have been prevented, erosion could have been mitigated with better husbandry, but it was too late for that, and all that could be done was to repair what had been destroyed.

FARM CULTURE AND WILDLIFE

Economic need, family responsibilities, and personal pride demanded that the farmer's wild competitors be eliminated or controlled. This was yet another dimension of the profound impact that farming brought to the prairies, along with the clearing, breaking, and draining of land. The historical record is full of farmers' complaints of wild animals threatening their livelihood and of their ongoing war with them. Yet many prairie farmers teetered on the edge of insolvency, which reinforced their resentment of wildlife that lived at the cost of any farm production. These threats varied in their details by type of agriculture. Cattle ranchers had an antagonistic relationship with wolves, grain farmers unceasingly battled gophers, and the varied production of mixed farmers triggered negative relationships with an even wider range of wild animals. None of these sectoral relationships were watertight, and all were continually reframed within the particular economic and social needs and types of farm practice. Yet all farmers shared a view of wild animals as competitors. This perspective was reinforced by social and cultural attitudes that saw settlement as a conquest of the land and a process in which farmers had the right to use every acre in their possession for production. But prairie farmers were also susceptible to a culture of blame. Pioneers had come with the belief that success would be theirs in return for hard work and sacrifice, and many indeed worked hard and risked and sacrificed much. Failure was thus a troubling personal rupture that was often blamed on others. And there were many others to blame: the CPR's high freight rates and its arrogance, the federal government's tariffs and insularity, and the grain merchants and their greed. And wild animals were blamed too. An individual calling himself "Ex Farmer" wrote in the *Regina Leader* in 1894 that the NWT's farm economy was depressed and people were abandoning their land: "I beg to suggest there is no more important work that deserves immediate attention" to correct this problem "than the destruction of the gopher, wolf and badger."[2]

This picture of conflict between farmers and wild animals was nonetheless more complex in its origins and context than a simple adversarial stance suggests. Farm settlement and land clearing brought varied and dramatic environmental changes, and while they caused the reduction or extirpation of many species, others remained and profited from the new niches created by this tremendous transformation of the land. Farmers thus came to see many of these wild animals – such as wolves, gophers, coyotes, crows, and magpies – as a threat to the very activities that had given them such new options. This attitude was also integrated into community life. Since wildlife showed no respect for property lines, no one wanted to be seen as someone who sheltered animals that preyed on their neighbours' animals or crops. As a consequence, bloody-minded community campaigns against predators were hard to resist, especially given the dominant view that wild animals could be killed with little compunction. These negative views were also reinforced by such state initiatives as predator- and pest-control programs, which were often focused around protecting various farm sectors or specific local circumstances. In 1926, for example, the Alberta government organized community hunts after an eruption of jackrabbits destroyed crops in the Taber area south of Calgary. Because the weather was cold, the rabbits were huddled together in bunches, and men on horseback or on foot easily drove them into specially constructed corrals where they were killed. "In this way as many as a thousand rabbits were killed in one drive," and the corrals were then moved and the process repeated in another district.[3] Foxes that raided farmyards could also be a problem, as could bears. In 1943 George Renouf, the MLA for the Porcupine and Duck Mountain districts in northwest Manitoba, asked the government to declare a special bounty on bears because he was receiving "phone calls and letters almost continually" from farmers "asking for protection against these pests who have no value and serve no purpose."[4] In Alberta, black and brown bears were unprotected and could be shot at any time, and in Saskatchewan, the message that it was acceptable to shoot bears was confirmed in a photo of two hunters with a dead bear in the 1949 annual report of the Department of Natural Resources. The photo's caption triumphantly asserted, "This Bear Robbed His Last Bee Hive Near Prince Albert." To further protect farm interests, in 1950 Saskatchewan allowed the use of night lights in hunting non-game animals like skunks, which killed chickens, and bears, which destroyed apiaries.[5]

The spread of disease from wild to domestic animals also prompted the state to take decisive action to protect farmers' interests. It will be recalled, for example, that a rabies outbreak in northern Alberta led to the widespread destruction of predators in an effort to protect farm animals, while deer were shot in southeast Saskatchewan to control an outbreak of foot-and-mouth disease that threatened livestock markets.[6] And when governments attempted to balance the needs of other constituencies in wildlife policies, farmers sometimes felt disadvantaged, but there was never much doubt that all prairie governments privileged farmers in many respects. This was reflected in various programs to control coyotes, wolves, gophers, and other wild animals that threatened farm and ranch production. It was also shown in the special rights of access to game that were granted to farmers. For example, when hunting licences were first instituted, farmers in Saskatchewan and Manitoba successfully argued that they should be able to hunt without one. This was justified on the grounds that farm settlers needed to hunt for food, and they continued to enjoy this privilege long after this was no longer the case. This privilege was defended on the grounds that farmers' fields provided "feed for the prairie chickens and ducks and geese." The notion that farmers were owed concessions because wildlife ate their crops presumed that farming was the only legitimate use for arable land. In 1956 the *Farm and Ranch Review* editorialized that "when it comes to the claim that the state owns the game birds, it might be suggested that if such is the case the state should pay for the feed the birds take from farm crops. If the farmer feeds the birds surely he has something like a vested right in their ownership." This remark was also a criticism of state support for sport hunting to the supposed disadvantage of farmers. As the editorial pointed out, one farmer near Calgary "lamented that he did not seem to own his farm as yet, as everything that is any good below the surface belongs to the government and it has been argued that everything that flies above it belongs to the hunters."[7] Such resentments continued despite the fact that farmers' demands were accommodated at almost every turn – provisions for hunting for food were made to assist farm settlement; insectivorous birds were protected; farmers were given the right, under the Migratory Birds Convention Act, to kill birds that damaged their crops; wolves and coyotes were controlled at public expense in the interest of protecting livestock; bag limits for deer were in part set in order to protect farmers' crops; and the state did not interfere with changes that farmers made to the land, even if they had a negative impact on wildlife. Yet farming was often a marginal business in which any loss mattered a great deal, and farmers' relationships with wild animals were established within a specific historical context in which the remaking of the land was celebrated as a mark of an advancing civilization. Circumstance, culture, and economic life thus combined to create a mindset of antagonism and resentment that shaped farmers' relationships with the wild creatures around them.

Figure 8.1 Waterfowl were commonly seen as a threat to grain in the fields. This flock of Canada geese was feeding on the stubble near stooked grain in Alberta in 1928.

Crop damage caused by migratory birds was the subject of ongoing complaint by farmers from the start of settlement. In their view, the problem was compounded by the Migratory Birds Convention Act, which restricted hunting seasons, set bag limits, and banned the hunting of some species entirely. Most of the problems with waterfowl occurred in areas near marshes or large bodies of water on migratory routes, and while the convention provided that farmers could obtain permits to shoot migratory birds that were damaging crops, the damage was often done and the birds had moved on by the time they received their permit. In 1919 a game guardian noted that farmers in southeast Saskatchewan "complained bitterly" about the protection given cranes and whistling swans that damaged their crops. However, the guardian confidently reported that he had persuaded the farmers that swans did not eat grain, and as they were such long-lived birds, few were fit to eat because of their age and thus were "much more beautiful alive than dead."[8] While the collapse of waterfowl populations in the southern prairies in the late 1920s meant that the problem was encountered in fewer areas, resentment nonetheless persisted. Popular farm rhetoric presumed that increased duck populations could only be had at the expense of farm profits. In 1932 one Saskatchewan newspaper called them "feathered thieves," and a resolution at the United Farmers of Alberta convention, "protesting against any reduction of the hunting season for wild ducks," hoped to see their numbers reduced through increased hunting. Expressing the same sentiments, the *Peace River Record* editorialized in 1936 that open seasons needed to be set earlier because every year grain was lost to the "flocks of wild ducks and geese that infest farmers' fields" and farmers were "powerless before the onslaught." But many farmers did not stand idly by. As one Saskatchewan farmer wrote in 1935, "we in Saskatchewan who raise the ducks from the egg stage to maturity are the ones on whom the cost of production falls the heaviest." The "farmer thinks and knows" that ducks are "his enemy and does his best to destroy all he can knowing that when fall comes his crop of oats and wheat will be destroyed especially if he lives close to a large lake."[9]

Conflict over ducks increased in the late 1940s when North American duck populations continued to rise from their historic lows. The development of farming in fringe areas near marshy land and lakes created further opportunities for conflict, and near The Pas in northern Manitoba in 1949, for example, ducks damaged crops on about eighty farms (all near marshlands). Resentment about ducks feeding on grain in autumn increased when peace time brought more rigorous enforcement of game laws. In 1946 the annual convention of the United Farmers of Alberta passed a resolution demanding that more duck hunting licences be issued and that the resulting fees be given to farmers to compensate them for having to deal with the "unlimited production of ducks as encouraged by Ducks Unlimited." To deal with the growing number of complaints, Alberta formulated regulations that would allow farmers to bring in hunters to protect crops, but these provisions were so badly abused that they subsequently had to be revised.[10] Governments also fielded more complaints from farmers because the greater use of swathing made grain more accessible to ducks, and a series of wet years in the late 1940s and early 1950s exacerbated the problem by delaying the harvest. Nonetheless, international agreements continued to limit the killing of ducks outside of open seasons or in excess of bag limits. Various alternatives were investigated, including the use of noise-making devices, searchlights, repellents, and scarecrows. Both the CWS and DUC tested various scaring devices, but they were not effective long-term solutions as the ducks simply became accustomed to them. Nor did farmers have the patience or time to install, operate, and maintain them; one critic remarked that scaring devices were impractical to use and that the real solution was to stop the "glorification and undue protection to [*sic*] our so-called web-footed friends" and begin treating them as farm pests.[11] "Lure crops" planted to divert waterfowl away from grain fields were more promising. Tested in both Canada and the United States, they were used along with feeding stations on the edge of lakes and marshes to keep ducks out of the fields.[12]

Given these problems, farmers increasingly argued that government needed to take responsibility for them, and some people, such as John McDiarmid, Manitoba's minister of mines and natural resources agreed that such claims had merit. In 1946 McDiarmid wrote that while the state was not obligated to protect its citizens from natural hazards, it was obligated to attempt to remove or alleviate hazards when they occurred. Manitoba natural resources officials, as well, believed that government bore some responsibility because it had enacted laws that protected waterfowl and derived "considerable revenue from sport shooting." Farmers interpreted such comments to mean that government should compensate them for crop damage caused by waterfowl and other game. In 1952 the Manitoba Farmers Union proposed that this be done, and the *Winnipeg Tribune* editorialized that the only practical alternative to compensation – whose costs would be catastrophic – was to give farmers the right to shoot game that threatened their livelihood.[13] A more practical solution was government-organized insurance for crop damage caused by game. Saskatchewan set up the first such program in 1952. Farmers paid premiums for coverage, and a $1 fee was added to each hunting licence to help subsidize the program. It was not immediately popular, perhaps because the crop damage was not great enough to make the purchase of insurance cost-effective. In 1956 the program received only 122 claims and paid an average of $328 per claim, while 275 claimants in 1959 were compensated an average of $545. Alberta set up a similar premium-funded scheme in 1960, but it also received a lukewarm response.[14] By the early 1960s, an uneasy peace had been brokered between waterfowl and farmers. Manitoba's approach typified the techniques employed across the region. Among other tactics, scarecrows and scaring devices continued to be used and feeding stations were set up near marshes to keep waterfowl away from fields. As well, decoys were placed in stubble fields to lure waterfowl to eat grain left there by combines, and in some cases, grain was spread on harvested fields to lure birds away from unharvested ones. The use of such approaches indicated a weakening of the customary abetting by the state of the adversarial relationship between wildlife and farmers and the adoption of a more nuanced effort to protect farmers' interests.[15]

"AN ORGANIZED WARFARE AGAINST THE WOLVES"

In 1839 the council of Assiniboia ordered strychnine from England to poison wolves and offered a bounty of five shillings for each wolf killed within eight kilometres of the Red River settlement. By 1850 a shortage of strychnine and settlers' complaints of "great loss by the depredations of wolves" persuaded the council to renew the bounty. Over the next twenty years bounties for wolves were periodically offered in Red River, but the system was repealed in 1868 as unproductive.[16] The Red River experience was the opening salvo in what became an ongoing conflict on the prairies between farmers and wolves. Over the next 125 years wolves were harassed, bountied, poisoned, and killed by professional hunters. This long struggle was a testament to the resilience of both parties. Driven by their need to protect livestock, a long-standing culture of animosity to wolves, and their commitment to tame the land, people employed their technology, organizational skills, and inventiveness to kill wolves. As for the wolves, their intelligence, endurance, adaptability, pack behaviour, and ability to travel long distances and live on a varied diet made them formidable foes.

Wolf-control measures were put in place relatively quickly after the prairies were brought into Canada. Manitoba's Wolf Bounty Act of 1878 allowed provincial and local governments to pay bounties on a cost-shared basis on various animals, including wolves and coyotes. By the mid-1880s the development of ranching in what is now southern Alberta and southwest Saskatchewan also brought with it demands for protection from wolves. Ranching relied on an extravagant use of large tracts of public land that were leased at low rates to ranchers, a practice that influenced animal husbandry. In 1886 William Pearce of the Department of the Interior observed that "no one keeps his stock on his own land" and cattle were left on the range, often for several years, until ready for market. This open-grazing system brought wolves back to the plains, where, as Pearce explained, they "subsist on the cattle as they did on the buffalo, by killing lame or sick stock, calves, etc; they also destroy colts." In 1885 the NWT Assembly members from these ranching districts demanded that the government offer bounties for wolves, but the proposal was rejected because of objections by members from other parts of the territories where wolves were not a problem. Left to their own devices, a number of southern Alberta ranchers imported "large, powerful and ferocious-looking wolf and stag hounds" from Europe to kill wolves in their district. This tactic, it was claimed in 1888, helped mitigate the "loss and annoyance" from wolves and at the same time offered "excellent sport." Some ranchers also set out poison and others organized community hunting campaigns or hired professional wolf hunters.[17] While these efforts removed immediate local threats, wolves cleared from one area were soon replaced by those that moved in from adjacent districts. Thus calls were made for what the *Macleod Gazette* in 1891 called "an organized warfare against the wolves."[18]

While the rejection of wolf bounties in the NWT Assembly in 1885 revealed regional differences, it also reflected worries about their efficacy. Although bounties were a venerable tradition in Europe and North America, effective bounty systems were costly because they had to be applied for a long period over an animal's entire range. Bounty programs were also often riddled by fraud because payments were frequently made for coyotes rather than wolves (especially in the case of bounties for wolf pups whose pelts were difficult to distinguish from those of coyotes). Critics thus argued that it was better to hire professional hunters (as was done in many parts of the United States), to use poison in a systematic way, or to raise bounties sufficiently so that hunters could hunt full time and make a living from it.[19] Nonetheless, the territorial government established a bounty system in 1893 in response to a constant stream of complaints "about the damage done to young stock by timber wolves, particularly in the foot-hills."[20] It excluded coyotes, which were of less overall concern for ranchers. Applicants had to offer up the pelt with the head attached as proof of the animal and its sex and age. While the bounties varied over time, in 1898 an applicant received seven dollars for an adult female, five dollars for an adult male, and two dollars for a pup. Although he believed the bounty was worthwhile, William Pearce noted in 1896 that it only applied to about 30 per cent of the stock-grazing districts and therefore needed to be extended. This was achieved when the territorial government began to offer bounties on a cost-shared basis in co-operation with stock growers' associations. The Western Stock Growers' Association, formed in 1896, was made up of ranchers in the Fort Macleod, Calgary, and Maple Creek areas, and the Willow Bunch Stock Growers' Association represented those in what is now south-central Saskatchewan. This government–stock growers' alliance forestalled criticism from other parts of the territories where wolves were not a major problem, and the government argued that those benefiting from wolf control would be happy to share the costs, since any rancher "would consider a payment of $20.00 or $30.00 for the destruction of a wolf a very profitable transaction."[21]

With this public-private partnership, the territorial government and the stock growers' associations collaborated to rid the region of wolves. The territorial government funded its share of the bounty, and the stock growers' associations administered the program. The Western Stock Growers' Association increased the bounty for female wolves from its own funds, and some individual stockmen raised funds to pay additional bounties in their districts while others paid out of their own pockets to hire wolf hunters to kill wolves on their land.[22] This arrangement did not eliminate all complaints; the stock associations generally represented big ranchers and many of them resented that many small ranchers who did not belong to the associations still benefited from wolf control. Moreover, the requirement for cost-sharing reignited sub-regional resentments. As the *Macleod Gazette* editorialized in 1899, it looked as though the NWT government thought farmers were "more deserving" than ranchers because it did not demand matching contributions from farmers in the eastern NWT for noxious weed-control programs as it did for wolf control, even though wolves were as "noxious" as weeds.[23]

In some locations, such as at the Walrond Ranch in the foothills south of Calgary, wolves caused significant losses of calves and colts. Historian Warren Elofson argues that wolves were "one of the stockmen's most formidable enemies" in the 1890s and, although control became more effective, they continued to be seen as a threat in the early 1900s. While wolves were a genuine danger to livestock on the range, they were at times also a convenient scapegoat. Elofson notes a case in 1901 where a ranch manager blamed wolves (among other things) to counter criticism of his mismanagement. And government statistics sometimes belied the extent of the problem. NWT government reports on bounty payments between 1899 and 1904, for example, indicate a relatively modest number of bounty payments. In these six years bounty was paid on a total of 2,248 wolves. While this was a significant number, it applied to an immense territory that included all of what is now southern Alberta and a large part of southern Saskatchewan. Also significant, just over 72 per cent of these bountied animals were pups, a good number of which, without doubt, were coyotes. In 1900 bounty was paid on 394 animals, only 130 of which were adults, even though the territorial government had increased the bounty that year because of reports of "serious losses" from wolves in districts in southern Saskatchewan.[24] Bounties paid on adult wolves reached a low in 1903 when bounty was paid on only 42 adults. Even so, after 1905 the new governments of Alberta and Saskatchewan continued to encourage the killing of wolves through bounties and prizes. While a few were reported in the Alberta foothills and in the Medicine Hat area in 1909, wolves by then had been eliminated in plains areas like the Drumheller Valley. In 1943 a Mr H. Wilson recalled that in 1903 he had killed a den of wolf pups near Swift Current in southwest Saskatchewan and "was proud to think that I was one of the last to kill wolves in this vicinity."[25]

While bounties helped to reduce wolf populations, changes in cattle ranching also contributed to their decline. A harsh winter in 1906–07 (which repeated a similar disaster in 1886–87) led to the bankruptcy of some ranchers when cattle froze to death on the open range and in snow-filled gullies. Confronting the second such disaster in twenty years, ranchers began to rear cattle in fenced areas where they could be fed in winter and, incidentally, were somewhat more protected from predators. By the First World War such changing animal husbandry practices, along with hunting and poisoning, had in effect extirpated wolves on the southern plains. As noted earlier, the war on wolves then shifted to forest fringes and northern districts to protect fur-bearing animals and big game. After 1940, reports of wolves in farm areas periodically surfaced in northern forest fringe districts in Manitoba and in areas north of the town of Peace River in Alberta.[26] There were also complaints that some national parks – Prince Albert National Park, for example – were breeding grounds for wolves, which would move out from the protection of parks to kill livestock. But wolf control now only marginally served agricultural interests. Their effective eradication in ranching districts by the time of the First World War meant the elimination of another key species of the old bison landscape and measured the region's ever greater remoteness from that world.

CUNNING FOES: COYOTES

The virtual disappearance of wolves on the southern plains and their reduction in other settled parts of the prairies brought to an end only one part of the war against carnivores in defence of agriculture. Coyotes quickly replaced wolves as public enemies, drawing hostility from broader geographical areas and farm sectors. Poultry on mixed farms was especially vulnerable to coyote predation, while calves, piglets, and colts were also vulnerable to some degree. Lambs and sheep on sheep ranches, mostly found in the southern arid regions of the prairies, were also said to be at risk from coyotes. As early as 1885 this threat led the *Regina Leader* to editorialize that "we are becoming a sheep-raising country" but "the little wolves [coyotes] have already shown themselves dangerous. Let the North-West Council give a bounty for the slaughter of them. So much a scalp." While it was in fact a marginal business for many reasons aside from coyote predation, Elizabeth Mitchell, in the account of her visit to the prairies just before the First World War, reported that many people believed that coyotes were the main barrier to successful sheep farming in the region.[27]

Cattle ranchers, on the other hand, typically did not worry about coyotes. While some saw them as a potential threat to weak or very young calves, others did not, apparently agreeing with Montana ranchers who in 1901 were said to view coyotes as "almost harmless" to cattle because they were primarily scavengers. Such views help to explain why the territorial government did not pay bounties on them. But Manitoba, with a stronger mixed-farming sector, included coyotes in its 1878 Wolf Bounty Act even though they were not widely reported as a problem. Further west, coyote populations were increasing in the early 1900s, and at the time when the provinces of Alberta and Saskatchewan were established, expanding grain farming on the plains brought more rodents and more food for coyotes, while land clearing in parkland areas created improved habitat there. This raised alarms and by the early 1900s coyotes were widely cited as a threat. Saskatchewan's chief game guardian warned in 1905 that they caused extensive damage to farm operations, and his counterpart in Alberta reported in 1916 that farmers were complaining that coyotes caused "more damage than all the other species of carnivorous animals combined." Not even children were safe, and the Alberta guardian cited two cases of coyote attacks. While he did not think that coyotes were more numerous overall, he believed they were increasing in settled areas because rabbits

Figure 8.2 Coyotes were widely said to be a threat to mixed and sheep farming and were killed with hounds, traps, and poison. This coyote was caught in a trap on Walter Davies's farm in Crawling Valley, east of Calgary, in 1910.

were scarce that year, forcing coyotes to hunt for food in farmyards.[28] The newspaper in Peace River in northern Alberta also editorialized that year that coyotes "were becoming more and more a curse" and one could not travel anywhere "without seeing one to a dozen of the brutes, who boldly patrol the roads and forage on the poultry yards and pig pens of the farms."[29] While losses to coyotes did occur, coyotes were also scapegoats. Investigations of coyote depredation found that many of the kills were committed by farm or stray dogs. Some sheep ranchers claimed that hounds that were kept to kill coyotes were worse than the coyotes, and they went "as far as to declare that they prefer coyotes to hounds on the principle that 'of two evils, choose the least.'" In Saskatchewan, predation by dogs was so widespread that the province passed legislation in 1909 to allow the killing of any dog that was destroying or harassing livestock or poultry.[30]

Concerns about coyotes rose and fell along with cycles in their populations, and these cycles were tied to those of rabbits. Complaints about coyotes often increased for about a three-year period when rabbit populations were falling and coyotes were starving. Complaints in many parts of the prairies, for example, rose in the late 1910s, then fell off from 1921 to 1924 when rabbit populations were high, and then rose again a couple of years later. This provided the context for reports in 1926 that coyotes were "exceptionally numerous" in northern Alberta's Peace River Country and had killed about half of the farm turkeys in the Grande Prairie district. Reflecting this recurring pattern, complaints again rose in the late 1940s, and in 1949 the Alberta government forecast that rabbits would not begin to increase until 1952 and that, until then, desperate coyotes would move closer to farmyards looking for food. Indeed, biologist Norman Criddle argued that farmers should study these patterns and use knowledge about cycles so that they could take steps to secure their livestock and poultry during periods when starving coyotes would be searching for food.[31]

Farmers were not, however, willing to put up with even temporary increases in coyote populations, and government policy therefore swayed back and forth in time with the population cycles of rabbits and coyotes and the rise and fall of public complaints. Although each province had its own approach to coyote control, bounties were typically used before the late 1940s but were inconsistently applied. Governments sometimes stopped paying bounties when budgets were exhausted, when public complaints declined, during winter, or when the price for long-haired fur was high in the hope that trappers would kill coyotes for their fur instead of for the bounty. These inconsistencies made the coyote bounty system inefficient. Indeed, the Saskatchewan government claimed in 1914 that one of the few benefits of the system was that it gave some rural people a bit of money in hard times or when mild weather made hunting easier and more fun.[32] And killing coyotes was a popular sport. For example, extensive photographic evidence of hunting coyotes for sport testifies to its popularity in southern Alberta in the late 1800s, and it continued to be popular there at least until the Second World War.[33] In 1939 one sportsman in southern Alberta observed that even though people did not go fox hunting on the prairies, hunting coyotes with greyhounds was "far more thrilling." In fact, it was becoming ever more exciting because of natural selection, since "incessant coursing throughout the West effectually eliminated the slow and dumb of the coyote fraternity, leaving only the fast and wise to reproduce." The hunt took only five minutes from the slip to the kill, and since greyhounds killed quickly, he claimed that it was not as cruel as when a fox was "harried by foxhounds for hours or [a] bird [was] wounded by gun-fire who escapes to die a lingering death." Not everyone saw it that way; in 1929 a number of letters to the Nature Lovers page of the *Western Producer* called for a ban on "such an indiscriminate and cruel so-called sport."[34] But prairie society generally tolerated and even celebrated any method of killing coyotes in much the same way that it did in regard to magpies and crows. Without a tinge of disapproval, for example, the *Edmonton Journal* reported in 1929 that Martin Anderson, a farmer near Bassano, east of Calgary, used his new automobile to kill thirty-five coyotes on the prairie – "without using guns or dogs." When chased by the car, the coyotes became confused and circled in front of it "until almost exhausted." Then they took off in a straight line, "making it comparatively easy to manoeuvre the car" to run them over. This so-called sport was sufficiently common that the Alberta government amended its game legislation in 1942 to legally sanction it as a method for the "eradication" of coyotes.[35] So, too, hunting wolves and coyotes from airplanes, a practice that was not only illegal but also considered vulgar in regard to game animals, was officially championed and celebrated. The Saskatchewan government encouraged it and in 1948 reported that "a number of Americans and local residents increased their activities along these lines with very favourable results." The next year the Alberta government collaborated with some rural municipalities and livestock associations to encourage the shooting of coyotes from aircraft in winter, "when snow-covered ground makes the coyotes good targets for low-flying marksmen." Almost 2,500 coyotes were killed in this way in a two-month period, and the forays were "so spectacular" that the National Film Board filmed them.[36]

A similar mindset inspired community coyote hunts. These were first held in 1949 in central and southern Alberta and were soon underway throughout the province. One report noted that "there are upwards of six hunts going on every half-holiday and Sunday afternoon" in almost

every district in central Alberta. The Alberta government encouraged the hunts by publishing a pamphlet with tips on how to organize them to help "rid the districts of predators which are taking much too heavy a toll of poultry, livestock and wild game." Military models and language were grafted onto the myth of the pioneer, and the hunts were said to embody "the art of Community Effort, practised in pioneer days when a problem common to all required attention." Hunts required "Leadership, Organization, Discipline," and success depended on advanced planning and organization, including naming a Hunt Master and Hunt Leaders and setting "Zero Hour – the time when the actual hunt moved off from the field positions." Meat was scattered in the area several days in advance to draw coyotes and assure "more complete extermination." Armed with shotguns, fifteen hunters per mile lined up in formation along the four sides of the hunt area. Advancing systematically towards the centre, they shot the coyotes they encountered and drove the rest towards the centre of the square, where they could be shot when the hunters converged there.[37]

This organizational model was widely copied across the region. One community coyote hunt, in 1949 in Disley, northwest of Regina, saw 300 men in twelve trucks roll "out of town to round up and destroy the district's coyotes." The "plan of the campaign was to circle an area 12 square miles [31 square kilometres] and gradually tighten the line until the coy-

Figure 8.3 The war against coyotes ramped up in the late 1940s with the organization of community coyote hunts in rural areas. These men were gathering at the start of one such hunt near Islay in east-central Alberta in 1949.

otes were herded into a small central ravine which had been designated as the 'Killing Grounds.'" This plan did not succeed; the coyotes "simply ignored the rules of the hunt" and "took advantage of the thin, broken line of hunters to sneak their way back through gaps in the circle." But the outing was said to have been fun, and it concluded with a bean supper at the curling rink.[38] Such community hunts were still uncommon at this time in Manitoba, where coyotes were seen as "small fry" in predator control in comparison to wolves. By the mid-1950s, however, they were occurring there as well. One of the first of such hunts was organized in 1954 by the Neepawa Game and Fish Association in co-operation with the provincial game branch and the RCMP. Six hundred hunters showed up for "Operation Predator Roundup," which promised to be a one-day "blitzkrieg" on timber wolves, coyotes, magpies, and bears. The hunt took place over nine sections of scrub land "infested with magpies and coyotes." The results were less dramatic than promised: no bears were encountered, a timber wolf apparently "escaped," and five coyotes were killed as well as "dozens" of magpies.[39]

Community coyote hunts were social outings and a time for camaraderie, but naturalist Kerry Wood, for one, found them thoroughly objectionable. He believed that many protected game birds and animals were killed, but he also thought that there was something horrible about "the trapped helplessness of the wild creatures, fleeing ahead of the advancing hunters, unable to win their way out of the encircling trap." Citing a case of vicious cruelty to a magpie at one hunt, he found "an unsporty element about such hunts that is part and parcel of the old mob violence we should abhor." But the hunts revealed a high level of collective anxiety about coyotes as well as rural–urban tensions, especially when city people criticized the hunt as "slaughters." Alfred Goff of Lipton, northeast of Regina, contended in a letter to the *Farm and Ranch Review* that such criticism merely showed the ignorance of urban people who did not know how damaging coyotes were. He claimed that critics of the hunts had been "led astray" in believing that game birds and everything in sight were killed, and defended the hunts by pointing out that game officials advised the hunters about appropriate behaviour. Tensions about rights over wildlife were also expressed. Edith Bowman, who farmed southwest of Edmonton near Wenham Valley, made the familiar argument that farmers fed wildlife creatures with their crops and thus had a right to kill them. Even if the odd game bird was killed, Bowman thought that it was "the farmers that protect and feed these birds throughout the summer. So what harm is there if a farmer does get an odd bird in the coyote hunt. The city people go around shooting in the fall time so why shouldn't a farmer be able to get a few birds." Moreover, she thought ridding the country of coyotes was a good idea and "the more of them killed the better."[40]

Community hunts and other methods rarely achieved effective coyote control before 1950. Between 1947 and 1949, for example, even though

Saskatchewan paid over $100,000 in coyote bounties, public complaints about coyotes continued to grow. The war against them seemed stalemated, and as the *Regina Leader-Post* noted, "the mournful howl of the coyote – the No. 1 enemy of farmers, ranchers and poultry raisers in Alberta and Saskatchewan – rings with seemingly undiminished defiance over the prairies despite costly and elaborate measures aimed at extinction of the predator."[41] The situation changed dramatically in the early 1950s with the adoption of new control methods. In 1949 Saskatchewan officials learned of an approach used by the US Fish and Wildlife Service that relied on the use of poison, a device called "coyote getters," and government-hired hunters. Saskatchewan officials immediately travelled south of the border to learn more. They were impressed by what they found, and the province tested the poison (sodium fluoroacetate, commonly referred to as "1080," its brand name) as well as coyote getters in two southern districts.[42] Isolated in the early 1940s, 1080 was one of the new generation chemicals that was remaking postwar life. Highly toxic to all mammals, insects, and birds, this water-soluble, odourless, tasteless, and "very deadly poison" was injected into a recently killed animal to make a poisoned bait. The carcass of the animal that ate this bait also became toxic, thus allowing the poison to move through the food chain. The coyote getter was a mechanical device, the mechanism of which was placed in the ground "with the shell portion protruding above the surface. The tip was covered with a rag or piece of fur treated with a scent to attract coyotes and wolves." When an animal bit into the tip, the mechanism discharged sodium cyanide in a .32-calibre shell directly into its mouth. While sodium cyanide was lethal, it did not carry over (as did 1080) to other animals that ate the poisoned carcass.[43]

The success of these tests and their widespread press coverage stoked public support for these new control methods. The Saskatchewan Sheep Breeders Association and the Co-operative Wool Growers' Association were enthusiastic and urged full-scale implementation, and a survey found that 80 per cent of rural municipal officials supported their use, which promised to reduce the costs of coyote control.[44] Thus, in 1950 the province implemented the distribution of 1080 and coyote getters in most settled parts of the province and employed professional hunters in a number of districts. Although Saskatchewan generally did not report kill

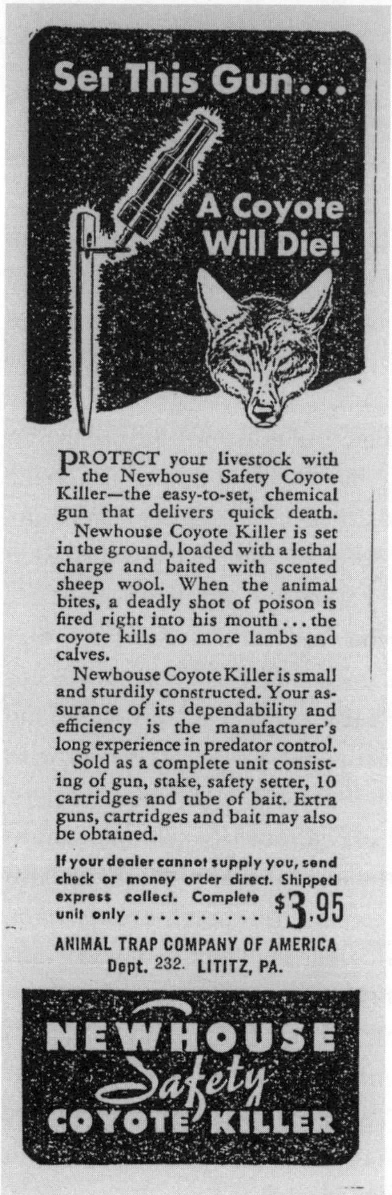

Figure 8.4 New approaches to coyote control in the late 1940s included the use of poison, professional hunters, and a device popularly called a "coyote getter." This advertisement for one such device appeared in the *Farm and Ranch Review* in March 1947.

statistics, it claimed that 2,000 coyotes were killed in the spring of 1950 alone.[45] Mortality mounted as the program matured, and by 1952 the Saskatchewan government claimed that a very substantial number of the 100,000 coyotes it estimated had existed in 1950 had been killed. Coyotes were now "under complete control" in all settled areas of the province. The government was congratulated on its success; as a report from the municipality of King George southwest of Saskatoon stated in 1952, the program had been "very satisfactory, the result of which is that there are no coyotes left." Saskatchewan then gradually scaled back the program and shifted more of the costs onto local governments. Nonetheless, 1080 and coyote getters, as well as hired hunters in some districts, were used throughout the 1950s and 1960s. Strychnine was also added to the arsenal, but the program was by then clearly in a maintenance phase.[46] These techniques were also applied in northern areas to eliminate predators of fur-bearing and big game animals.

No other province used 1080 as lavishly or sought total control of coyotes as relentlessly as Saskatchewan. Manitoba was reluctant to authorize the use of poison and instead used professional hunters more extensively, and in 1959 it only allowed provincial officials limited use of 1080. In Alberta in 1950, James McFall, secretary of the Alberta Federation of Agriculture, urged the province to approve 1080 for use against coyotes. While community coyote hunts had "provided fun for the participants" and the government-sponsored program of shooting them from aircraft had killed many more, he argued that these measures had not provided the "final answer" that 1080 offered.[47] But even the Alberta government, never a stranger to extreme measures in predator control, was leery of the Saskatchewan model. It resisted approving 1080 because of its toxicity and instead declared coyotes an agricultural pest (along with rats and various insects) in 1951. This made them public enemies, and the government urged the public to take active measures to destroy them. Instead of approving the use of 1080, it adopted a policy under which only problem coyotes would be killed, reasoning that only 20 per cent of coyotes turned "outlaw" and killed poultry or livestock. It also encouraged the extensive use of coyote getters and distributed them free of charge for use around farmyards. This approach killed a significant number of animals; in 1961 it was estimated that 410,300 coyotes had been killed in the previous decade.[48]

These new coyote-control programs allowed the Prairie provinces to eliminate bounties on coyotes, and those on wolves were formally abandoned at a Calgary conference of game officials from Ontario and the west in 1954. While Alberta kept its bounty on cougars to protect big game and while bounties remained a policy option as an "emergency" measure in all three provinces, it was clear that the chemical revolution had to all intents and purposes brought the long history of predator bounties to a close.[49] In the history of people's encounters with wildlife in prairie Canada, the coyote-control programs were among the basest. While the programs were officially about "control," their scale and ferocity suggested that extermination was the goal. The use of 1080 was irresponsible because it was known that this toxic substance moved through the food chain. Although there were complaints of the inadvertent poisoning of domestic cats and dogs as well as of birds, weasels, and other wild animals, the official position in Saskatchewan was that the transfer of toxicity was overdrawn because only the mucous membrane and stomach of a poisoned animal were dangerous. In any event, such concerns were largely peripheral given the general hysteria that fuelled the use of this poison. As one farmer wrote in the *Western Producer* in 1951, trapping was the "retail method" of coyote control and poison was the "wholesale" one, and farmers insisted on having it. The use of 1080 and coyote getters generally received high praise, and as the *Farm and Ranch Review* commented in 1958, they could, along with other tactics, "make the poor prairie coyote a rarity except in zoos."[50]

Although the use of these methods was generally popular, there nonetheless was some ongoing concern about coyote-control programs. Some people were concerned that the programs were misguided in attempting to eradicate a key species. Dennis Sizer, a farmer near Landis, west of Saskatoon, observed in a letter to the *Western Producer* in 1951 that one could not pick up a newspaper without seeing an article about the coyote menace. But Sizer had trapped and hunted the animals for years, and he believed they were "more beneficial than destructive." They mostly ate mice, gophers, and rabbits, and he argued that the value of the grain such animals ate "far exceeds the loss suffered from coyotes." Coyotes had been "nearly wiped out" in many districts, and Sizer worried that "the balance of nature cannot be interfered with on this scale without having bad results as all rodents are sure to increase" while other animals would be unintentionally killed by the poison. It was therefore up to everyone with "an ounce of sporting blood in him" to publicly oppose this campaign because the government would otherwise be swayed by those who had lost a few chickens to coyotes. In any case, Sizer thought many of such losses were not the fault of coyotes but of "some dog that had gone bad," and he contended that he had "lost more chickens to a stray dog than I have lost to coyotes in five years."[51]

Sizer's concerns seemed to be confirmed in the Red Deer district the next year. Coyotes had been almost eliminated there in 1950 and 1951, and some people argued that extensive crop damage by mice the following year was the consequence.[52] Whether this was so was unclear; the cycle for mice in Alberta seemed to be a four-year one, with major irruptions every ten or twelve years, and the 1952 plague was a repeat of one seen about ten years before. Moreover, the 1952 high point in the cycle was experienced throughout the province and was so extraordinary that it drew national comment.[53] It was thus unlikely that fewer coyotes was the only cause, although more coyotes clearly would have helped control the situation. Whatever the case, critics blamed the mouse problem on coyote control. One such critic asked, "And what do the farmers intend to do about it? They don't seem to know." And James Gray, a Calgary journalist and historian, remarked that these "tiny vegetarians," unmolested by coyotes, had waxed "exceedingly fat on the stooked, swathed and stacked grain." In Gray's mind, it was Darwin's "cycle of expansion and contraction of animal and insect species," for, as the mice increased, so did the coyotes, which then reduced rodent populations so drastically that the coyotes began to starve and their numbers fell and "the whole process" started again. The interruption of this process by the killing of coyotes fell hardest on the grain farmer, and Gray noted that "as the coyote never did him any harm he is inclined to take a dim view of the coyote shooters," unlike his neighbours who raised livestock and poultry.[54] Naturalist Kerry Wood opposed coyote-control programs on the same grounds. Later admitting that farmers mostly shouted him down when he defended coyotes, he wrote in the *Country Guide* in 1945 that coyotes were omnivorous, and "despite the howl there will be to the contrary, it is safe to say that the farmer's poultry and young stock do not provide coyotes with enough meat to live on." Indeed, American studies of the stomach contents of coyotes cited by "An Observer" in a letter to the *Western Producer* in 1952 showed that mice and other rodents were their main source of food.[55]

By 1960 criticism of the scale of the coyote-control programs was solidifying throughout the region. While it had no objection to predator control in principle, the Manitoba Game and Fish Federation worried that the "extermination" of coyotes would trigger broader negative environmental impacts. This view had been voiced by Manitoba's chief game official in 1959 in what seemed to be a veiled criticism of Saskatchewan's extremism. As he stated, Manitoba had only just authorized the use of 1080 in a very restricted way, and the province had not pushed coyote control "to the point of extermination" because this was "contrary to all concepts of good farm and game management."[56] And in 1960, the Saskatchewan Department of Natural Resources reported that there were now more skunks and red foxes "than ever before," and circumstantial evidence suggested it was because there were fewer coyotes and "Nature has a way of filling a vacuum." The same year, nature columnist Doug Gilroy claimed that "even farmers who pressed for a vigorous coyote-control program are now asking whether their action was altogether wise. Certainly control has been so effective that the coyote population is now very low." As he concluded, "we are coming more and more to recognize the function of predators in maintaining a healthy wildlife community" by eliminating weak and surplus animals.[57] Whatever the case, the extremism and widespread support for the war on coyotes suggest that it was about more than just coyotes. Perhaps farmers' sense of themselves as an economic and social vanguard that was taming the land was challenged by their failure to subdue coyotes or control their spread. This came on the heels of the environmental disaster of the 1930s, and while the war years had not been bad for farmers overall, the late 1940s and 1950s, despite postwar optimism, were poor ones, with bad weather, early winters, and low commodity prices. In this context, the farmers' frenzy about coyotes perhaps demonstrated an attempt to exercise at least some control over an environment and economy that had been a challenge for many years.

GOPHERS: POTENTIALLY AS GREAT A PEST AS RABBITS IN AUSTRALIA

While ranchers and mixed farmers battled wolves and coyotes, grain farmers waged war on gophers. On the prairies the term "gopher" usually referred to Richardson's ground squirrels (which were also known as flickers, flickertails, or prairie gophers), but it also commonly referred to other ground squirrels (such as striped and Franklin ground squirrels) as well as to various types of pocket gophers. Collectively, these animals were

Figure 8.5 There are many species of ground squirrels in various parts of prairie Canada. All, such as the striped ground squirrels commonly found in parkland districts, are important as a food source for raptors, coyotes, and other predators and for conditioning the soil. Photo not dated.

keystone species; they ate vegetation that stimulated new growth, their digging aerated the soil, they were an important food source for badgers and many raptors, and their burrows provided homes for other animals and insects. Their importance was recognized from an early point. Robert Bell of the Geological Survey of Canada argued in 1895 that while there were as yet no earthworms on the prairies to help aerate the soil, bison had churned it with their hooves and when wallowing. More importantly, however, moles, ground squirrels, and other burrowing animals had refined the soil for centuries. "In working their burrows," Bell noted, "they dig out only the fine materials and cast it upon the surface, leaving behind all the layers of stones and as much of the gravel as possible." The fertility of prairie soils was thus partly the product of burrowing animals, and Bell contended that "these hitherto apparently insignificant creatures are really the most important animals in the country, ranking in the services to man far ahead of the lordly buffalo, moose or red deer of the same region, and also ahead of all the fur bearing animals combined."[58]

Even as Bell was asserting the value of burrowing animals, most of them were already defined as farmers' enemies and what would prove to be a long war against them was well underway. From the start of farm settlement, Richardson's ground squirrels in particular were seen as a threat because they destroyed grain in the fields. Other species of ground squirrels were also considered a problem in some locations.[59] The Manitoba Department of Agriculture observed in 1883 that gophers, not insects, were the most "formidable enemy the farmers have to contend with." And in the dry year of 1889, the commissioner of dominion lands reported that crops in some districts on the southern plains had been ravaged by gophers and unless the animals were controlled, they would become "as great an evil as the rabbit pest in Australia." In the same year, Angus McKay of the Indian Head Experimental Farm implied that the apocalypse had already arrived and that gophers were found everywhere in the agricultural districts of the NWT, especially in areas with light soils during dry years. They caused much damage, "in many cases clearing off entire fields."[60] Gophers also damaged home gardens and burrowed through canal embankments, causing washouts in irrigated districts in southern Alberta. One estimate in 1925 claimed that gopher damage totalled $25 million a year in the three Prairie provinces. This included lost grain and the damage caused by badgers, which dug large holes to get gophers out of their burrows. Losses also increased as prairie farming became more mechanized and haying and harvesting equipment was damaged when it hit gopher holes and mounds. While they were a problem in ranching areas, since they destroyed grass and since stock and riders could be injured by stepping into their burrows, gophers posed the greatest threat to grain farming.[61]

With the exception of an extensive campaign after 1941 to control Columbian ground squirrels that were damaging gardens and crops in southern Alberta foothill ranching and farming areas, efforts to control gophers were most generally focused on the southern plains.[62] While there were complaints that they invaded wheat fields from nearby unoccupied lands such as those owned by the CPR (adding yet another complaint about the corporation), gophers favoured cereal and forage crop areas where sprouting crops provided spring food and tillage made their lives easier. Ernest Thompson Seton wrote that Richardson's ground squirrels had increased with tillage, and while there had been plenty before, there were even more after the land was broken.[63] What is now known about their life history suggests that changing habitat was important in their new-found success and that tillage and the cultivation of cereal crops favoured them. It is now recognized that litter sizes among Richardson's ground squirrels are almost half the size (five to six young) on native grasslands as they are on tame forage crop fields (nine to ten young), and that they prefer open fields with relatively low vegetation under 30 centimetres in height.[64]

Figure 8.6 The term "gopher" was a broad one but usually referred to the Richardson's ground squirrel, such as this one photographed in Alberta by Dewey Soper of the Canadian Wildlife Service. Photo not dated.

The spread of gophers was evidence enough for some people to believe that they were under siege by a species of amazing fecundity. In 1916 the newspaper in Cardston in southwest Alberta made the preposterous claim that "where there are two gophers this spring there will be nearly 200 next year."[65] Such fear of being overwhelmed by nature was strong, and this siege mentality led people to employ almost any method to kill gophers – they were shot, snared, trapped, gassed, poisoned, and drowned with water poured into their burrows. Local industry sometimes contributed to the arsenal of weapons. In 1895 a Mr McCormick of Innisfail in central Alberta invented a "gopher exterminator" that used large fans to force air "through a tube holding burning sulphur and charcoal." The nozzle of this tube was inserted in the gopher burrow, and the fumes were driven down "into the homes of Gopher and Company with fatal results." Hope was also held out for non-mechanical techniques. In 1900 one correspondent to the newspaper in Sintaluta, east of Regina, urged the NWT government to offer a $20,000 reward to "the inventor of some process by which disease could be introduced into them by some process of inoculation" to wipe them out. But this tactic had already been found wanting: the territorial government noted in 1901 that European experiments to exterminate ground squirrels by infecting them with chicken cholera and mouse typhus had been reviewed but were not adopted because the epidemics did not spread quickly.[66] And new suggestions continually emerged. In 1953, for example, the Alberta Department of Agriculture sponsored demonstrations on how to use methyl-bromide capsules to kill gophers in gardens and also advised people to try piping car exhaust into gopher burrows.[67]

While shooting and trapping were popular, using poison was the easiest method and appeared to be the most effective. Strychnine was the poison of choice; it was inexpensive, readily available, and highly toxic, and instructions on how to make various concoctions with it were printed and exchanged like kitchen recipes. In 1922 the field crops branch of the Saskatchewan Department of Agriculture tested and distributed a number of recipes for such baits. Generally, strychnine was mixed with molasses, corn syrup, or other sugars and was sometimes flavoured with anise or other substances that supposedly appealed to gophers. The mixture was stirred up in a washtub along with oats until the kernels were infused with it. The efficacy of poisons in liquid, dry, and gaseous forms was also debated and tested, and in 1922 one study (inspired by evidence from the trenches of the First World War) concluded that rodents were particularly susceptible to poison in gaseous forms.[68] An array of commercial products, most containing strychnine, were sold over the years under brand names such as Gophercide (which promised "Seven Dead Gophers for One Cent"), Kill-Em-Quick (whose ads described gophers as thieves that, unlike human ones, could not be scared away, "you have to kill them"), and Cynogas, a poison gas in powder form that pledged that "[g]ophers are dead in five seconds!"[69]

Poison was sometimes used indiscriminately, and as the Saskatchewan Department of Agriculture reported in 1915, some farmers simply scattered poisoned grain in the field as they drove along in a buggy. In addition to the threat this posed to useful birds, this approach was wasteful because only a few poisoned grains needed to be placed at the burrow openings in early spring when there was little other food around. Such timing also killed females before they bred. But, as the report further noted, everyone in a district had to use poison consistently and properly to gain control of a district's gopher population.[70] This was, however, difficult to achieve, and the inadequacy of individual efforts led to demands for help from government. Some people doubted that such help would be forthcoming. Nicholas Flood Davin, the editor of the *Regina Leader* and a Conservative MP, asserted in 1887 that the government would not help because some animal-loving "sentimental philanthropist would get into a state of fury and dithyrambic madness from this, and the hand of government might be stayed."[71] Notwithstanding his flair for the dramatic, Davin was entirely wrong: all levels of government willingly responded to public demands for help in killing gophers. In 1890 the Department of the Interior purchased 8,000 traps and lent them to agricultural societies for distribution to homesteaders in the territories. The territorial government subsequently took over this program and in the early 1890s addressed the uneven quality and high price of strychnine by purchasing strychnine of proven purity for distribution to farmers. Although the program was dropped in 1897 as a result of the lobbying efforts of territorial druggists who claimed it was unfair competition, the government had accomplished its goal by imposing higher standards and lower retail prices.

Good-quality strychnine could now be had cheaply everywhere, including by mail order, and by 1898 a sixty-grain bottle of strychnine – enough to kill 300 gophers – could be purchased for twenty-five cents. This was not a great expense because few people used more than two or three bottles a year. Although it only did so for a short time, by 1907 the Saskatchewan government had entered into the strychnine business, giving the poison free of charge to rural municipalities for distribution.[72]

Other time-honoured solutions were suggested to deal with the menace. Expressing the same sub-regional and sectoral resentments that were associated with the wolf bounty programs, one correspondent wrote to the *Regina Leader* in 1894 that if the government paid bounties on wolves, it should do so for gophers too: "I have never heard of the whole of a rancher's crop of calves being eaten by wolves but the total loss of a year's crop of grain by gophers is no uncommon thing." But the precedent of Montana, where the government had been nearly bankrupted by its gopher bounty program, made NWT government officials cautious and thus the matter was generally left to local governments, even though they could ill afford it.[73] The Municipal District of Indian Head and Wolseley (east of Regina), for example, paid $2,300 in gopher bounties in 1887 but soon found that it could not afford to carry on, and the district's MP suggested to Prime Minister Macdonald that a federal government grant was needed for the district to "effectively destroy" the gophers.[74] Aside from the question of affordability, local government bounties were also ineffective because not every municipality paid them. As one correspondent noted in the *Regina Leader* in 1894, a uniform effort throughout the country was required to control gophers, and in the favoured analogy of a nature out of control, he claimed that gophers were increasing at such a rate that the NWT would "soon be as badly off as the Australian colonies are with the rabbit pest." A four- or five-cent bounty would ensure that "we should then hear little more of the Gopher nuisance."[75]

Concerns about gophers rose and fell, but they ramped up during the First World War along with efforts to increase wartime food production. One Manitoba advertisement showed gophers dressed as invading German soldiers, and after the war, new enemies (inspired perhaps by the turmoil of the Winnipeg General Strike) led to the characterization of gophers as Bolsheviks. By this time, both municipal and provincial governments were participating in various control programs. Typically, a cash reward was paid per gopher tail. Some of these programs were costly. In the 1919 campaign, for example, Manitoba paid a two-and-a-half-cent bounty on 351,394 gophers as part of its five-day-gopher-killing blitz. But as the manager of the program, Professor Jackson of the Manitoba Agricultural College, noted, the 1919 blitz was a great success in making Manitoba "a clean and thrifty province." As with the crow and magpie campaigns, schoolchildren played an essential role, and as the dominion entomologist Gordon Hewitt reported about the campaign in 1917, "essays on gophers were written in the country schools and many of the manufacturers of gopher poison offered prizes for the best essays and for the greatest numbers of gophers killed. One boy claimed to have killed 1,082 gophers with one packet of poison." While poisoning seems to have been the method of choice for adults, most children killed gophers by drowning, trapping, or shooting them. The best result in Manitoba's 1917 gopher-killing campaign "was obtained by Alex Henry of Rapid City, Manitoba; although only fourteen years of age, he shot 386 gophers in one day." Girls showed a similar aptitude: two schoolgirls from Glenboro, southwest of Winnipeg, shot 141 and 132 gophers in one day.[76]

A similar approach was used in Saskatchewan, which declared 1 May "Gopher Day" in the years between 1916 and 1921. Rural municipalities offered bounties and sometimes distributed gopher poison, and the provincial government awarded prizes and medals. Schools whose students turned in the largest number of gophers received awards. Huge numbers of gophers were killed; about 514,000 in 1917, about 1.2 million in 1920, and about 2 million in 1921. In 1917 the first prize was a gold watch donated by the minister of agriculture; runners-up received silver and bronze medals; and schools with the largest number of kills earned bronze shields. By 1919 prizes in the annual events had become extravagant. The Heintzman Company donated a Victrola to the highest-placing school, and individual prizes included a bicycle from T. Eaton and Company, a violin from the Mason Rich Company, a wrist watch from the *Regina Post* newspaper, and, perhaps most suitably, four bushels of registered Marquis wheat from the Angus McKay Seed Company. Six Shetland ponies were awarded in 1920, and one pony was awarded by the Kill-Em-Quick gopher poison company in 1921. Showing that such corporate sponsorship had spinoff benefits, the province also distributed 3,200 cans of Kill-Em-Quick.[77]

Children were also major players in Alberta, and in 1929 schoolchildren in the province killed almost 200,000 gophers compared to the 61,000 killed by adults. The gopher problem was considered to be of such public significance that some people wanted to give it national standing. In 1919 one Alberta farmer proposed that the federal government declare a "Gofer" day around 10 May to establish a holiday when people could kill gophers. And one prairie MP suggested in 1920 that .22 rifle shells should be exempted from federal tax because "very often a boy is given a rifle and is told to shoot gophers, and it would aid production if these creatures were killed for a gopher will eat one bushel of wheat in a year." While the government agreed that killing gophers was a good policy, it concluded that the administrative costs of such an exemption would use up any savings it brought through higher production.[78]

While gopher campaigns continued to be socially normative events (and remain so in a limited fashion today), the frenzy against these animals died down after the mid-1920s.[79] Although rural municipalities continued to pay bounties when they could afford it or to meet public demand, historian Thomas Isern speculates that "as the country settled up, people still considered gophers a nuisance, but a routine and less critical one." In the 1930s many municipalities in all three provinces offered bounties on gophers, although, as Isern notes, this might also have been an aspect of rural Depression relief. By the 1940s, as Isern further notes, gophers were seen as "merely one of a number of pests deserving control," though a "persistent, somewhat costly, but not momentous" one.[80] Yet killing gophers as a public service remained popular. The Homemakers Club in Bracken in southwest Saskatchewan, for example, paid $108.53 in bounty for 21,530 tails (or a half cent per tail) to seventy-five children in 1931. But this was just a part of its routine social service activities, which included such projects as helping to sponsor a music festival and funding the skating rink. And the rural municipality of Mountain View southwest of Saskatoon offered children a gopher bounty in 1941 in the form of war stamps to "enlist gophers to beat Hitler."[81]

Killing gophers was part of prairie culture, most notably during childhood and especially for boys. Gopher snaring was a common pastime at school recess in many districts. A typical schoolyard scene was described in the *Country Guide*. A piece of twine or rope about two metres long was

Figure 8.7 Killing gophers was a popular boyhood sport and money-maker that was encouraged by governments and various organizations. These boys were hunting gophers near Rosedale in the Drumheller Valley in southeastern Alberta in the 1920s.

tied at one end into a slip knot to form a noose. This noose was placed over the gopher hole and the hunter lay on the ground a short distance away holding the other end of the twine or rope. When a gopher poked its head out the hole and through the noose, the hunter pulled hard on the string to snare the animal. The hunter then swung the string like a lariat up into the air and brought it down sharply, killing the animal by dashing it on the ground.[82] It was not a merciful business, and another incident showed how it sometimes slipped into torture when the gopher was given "a lingering death" while the boys "enjoy[ed] the process." It began with the "merry outdoor boy" rejoicing as he caught the gopher. But innocence soon shifted to cruelty as he "teases the trembling animal, torments it, makes hilarious sport, delights in the wild veil of fear across the bright eyes, enjoys the shriek of panic and thinks the best part of the game is seeing how horrible a frenzy he can work his furry captive into."[83]

This second report implied that encouraging children to kill gophers (as well as magpies and crows) was a questionable practice because it was not only inhumane but also brutalized the child. In addition, another argument suggested that nature would look after gophers if people just let it do so. This argument did not dispute the value of gopher control, only that there were better natural methods. Paralleling criticism of the killing of raptors to protect game birds, the Manitoba ornithologist A.E. Atkinson argued in 1899 that the "undisturbed increase" of mice and gophers that were damaging crops was the result of the ruthless slaughter of raptors and other natural enemies. But the average farmer saw all hawks and owls as a threat to farmyard poultry. This prejudice was "so firmly rooted that it is impossible to eradicate it," and when given the opportunity, the farmer would "never fail to vent his spleen upon the unfortunate hawk, and all simply because of the loss of an occasional fowl." In 1919 the Saskatchewan game branch purchased 5,000 copies of Percy Taverner's pamphlet *Hawks of the Prairie Provinces in Their Relation to Agriculture* to be used in schools to help promote understanding of the value of some raptors for farming.[84] This seems to have had little effect. In 1932 the Alberta ornithologist Frank Farley noted that there had been a "ruthless slaughter" of Swainson's hawks, which were often shot as "they watch for gophers from telephone poles." Farley claimed that as the number of hawks declined, the number of gophers increased, and so did the bounties that municipal governments paid. He cited the example of the small municipal district of Lloyd George in central Alberta, which in 1931 alone had paid $519.90 for gopher bounties on 51,990 gophers. Calculating that a pair of large hawks would kill at least 350 gophers in a summer, only 148 pairs could kill, at no charge, the same number of gophers as the district had paid bounty on. For Farley, this calculation proved the "folly" of paying bounties "in vain attempts to control or exterminate the pests" while at the same time allowing the "indiscriminate killing of hawks that are and always have been the gopher's natural enemies!"[85] Young Walter Schowalter of Hayter in central Alberta employed a similar cost-benefit analysis in a letter to the Nature Lovers column in the *Western Producer* in 1934. In a scene that he claimed was constantly "enacted everywhere on our prairies," a hawk slowly circled a farmyard, effortlessly lifting on the wind. "Mrs. Farmer comes to the window. Does she see anything graceful or beautiful? Hardly! She lets out a terrible scream. 'A Hen Hawk! A Hen Hawk! Oh, my precious chickens.' Mr. Farmer drops his newspaper, plumps a 12 gauge into his old shotgun and rushes out. Boom! and Mr. Hawk drops to earth full of holes." Schowalter then set out his calculation

Figure 8.8 Despite their value as predators of mice and ground squirrels, hawks were indiscriminately killed by many farmers and sport hunters. Northern harriers (then known as marsh hawks), such as this one photographed by Saskatchewan photographer Hans Dommasch in the 1960s, were common on the prairies.

of what this single shot had actually cost Mr Farmer. After accounting for all factors, he concluded that even after "feeding the hawk a scanty gopher ration and charging against him a generous bill for chicken damage," Mr Farmer would still be $80.70 ahead if he had held his fire because of the grain that would otherwise have been eaten by the gophers that Mr Hawk had killed. "Kill the hawk and you are losing money."[86] While Schowalter's calculations may have been sound, they did not take into account that short-term goals were sometimes a priority on prairie farms. Even if grain could produce more income on a yearly basis, chickens provided ongoing food and a small but necessary trickle of income throughout the year, which was essential on many cash-short farms.

Such arguments about the economic value of hawks had little impact. Repeating what been contended for over half a century, naturalist Kerry Wood wrote an article in the *Country Guide* in 1944 deploring the common rural practice of shooting hawks at every chance. The article drew a number of letters in reply, and the magazine's editor wryly noted that "if space permitted in this great family journal we could have a whale of a controversy over this hawk question." Similar debates centred on weasels and other natural predators.[87] As early as 1915 the biologist Norman Criddle noted that weasels were "generally regarded only in the light of destroyers of poultry, whereas in reality they are only casual enemies in this respect and very great friends in every other" because of the mice and gophers they killed. Weasel fur was then in fashion and weasels were trapped widely, which Criddle saw as self-defeating. "At the present time hundreds of men and boys are busily engaged in trapping those animals for the price of their fur" but were "causing a loss far in excess of the value paid for the skins" because of the increased numbers of rodents."[88] Others continued to make this point. Kerry Wood described weasels as predators of ground squirrels and mice in 1948, and in 1951 the Alberta government made them a protected species in the buffer zone along the Saskatchewan border to help keep rats out of the province.[89] But old habits died hard, and as Wood phrased it in 1961, people still had "a senseless hatred of the swift little creatures," which remained one of "the most maligned mammals in Canada." Skunks were valuable for the same reason, as were badgers, which were often blamed for killing chickens and for digging holes that horses might stumble in and break a leg.[90]

While arguments like these had a long tradition, they only gained broader popularity in the 1960s. They were, however, in the vanguard of what would prove to be the rehabilitation of skunks, weasels, coyotes, and other predators that came with the halting public acceptance of an ecological view that placed wild animals within a set of relationships with the land and one another. While bloody-mindedness endured and many farmers continued to see wild animals as competitors in their struggle to make a living, attitudes were being tempered by changing cultural standards. In 1960, for example, the Alberta Department of Lands and Forests responded to complaints about deer that were eating crops by advising farmers to use scaring devices to deal with the problem. And the same year, complaints about extensive damage to sunflower crops in southern Alberta by blackbirds were met with the same recommendation. The blackbird problem persisted into the mid-1960s, and lure crops, scaring devices, and the old favourite – poison bait – were tested.[91] While these approaches reflected in part a growing respect for new scientific knowledge about wildlife feeding habits, they were also a small but significant indication of changing standards because, only twenty years before, guns would have been offered up as the first line of defence. Although the rights and needs of farmers remained a powerful influence on wildlife policy, and while the long-standing belief that animals that interfered with farming should be killed retained currency in some quarters, the relationship that farmers had with wildlife was changing and old practices were being challenged in new ways.

9

First Nations, the State, and the Economy of Wildlife

From the point of view of the Indians the most important – and, as it turned out, the most misleading promise – was, "you will have the right of hunting and fishing just as you have now until the land is actually taken up."

Irene Spry[1]

Since the signing of the first treaties in the 1870s, the right of First Nations peoples to use wildlife has been a source of controversy and conflict. While statements made during treaty negotiations seemed to promise First Nations unique rights to wildlife that respected their needs and histories, various administrative and political decisions about treaty hunting rights and the management of fur-bearers rejected these pledges. This forms one part of a complex legacy of differing views about treaty hunting rights, Aboriginal economic and social needs, interpretations of the terms of the treaties, and the ownership and regulation of wildlife.

At the outset, the public property regimes that governed the ownership of wildlife after 1870 presumed a relationship with wildlife founded on state regulation to benefit private and individual use. This consciously overrode competing "common use" (or what can also be called group stewardship) practices in many Aboriginal communities. While common use approaches took different forms, they involved the allocation of use to particular groups or individuals or the regulation of use by means of various community conventions, customary patterns, and understandings. The great bison hunts on the western plains by the Red River Métis after about 1821 drew on these traditions by regulating and policing the hunt through rules and conventions created and enforced by the community. Euro-Canadian settlers after 1870 rejected such local and community approaches because of their different assumptions about property rights and the ownership and management of natural resources. The typical Euro-Canadian solution for conflict over natural resources such as land was to establish private property rights to them. This did not, of course, work with wildlife that did not respect artificially created boundaries, which further confirmed the need for the state to govern its use as public property.[2] Under public property regimes, individuals could gain ownership of wild animals by legally capturing or killing them, but until that moment, they remained public property. The state set rules about access to and the killing of wildlife on the basis of law, custom, public opinion, and the relative influence of different groups and individuals. The state also granted individuals and corporations the right to use wildlife under specified conditions, such as in licensed trapping areas.

A foundational concept in the public property regime was that everyone had access on identical terms. Often phrased as "the equality of all citizens before the law," this equality principle came into conflict with promises made during the treaty negotiations that First Nations would have broad access to wildlife on public land. The Department of Indian Affairs, as the federal department responsible for First Nations, took varying positions on the matter, but the provinces on the whole argued that treaty rights ran afoul of equality of access and public ownership of wildlife as well as of conservation programs. First Nations were thus expected to follow the same rules as everyone else. Métis people were in a different position because they had no treaty rights and their story thus is only incidentally touched upon in this chapter.

FIRST NATIONS HUNTING RIGHTS

Treaties 1 to 8 (between 1871 and 1899) and Treaty 10 (signed in 1906) freed the land in the Prairie provinces for farming, commercial fishing, mining, lumbering, and other activities. While the treaties dealt with a number of issues, those regarding wildlife were among the most important. Verbal promises made in treaties 1 and 2 negotiations and written and verbal commitments made in respect to treaties 3 to 8 promised that First Nations peoples and their descendants could, as it was written in treaties 3 to 8, "pursue their usual vocations of hunting, trapping and fishing," subject to regulations that the government might make "from time to time." These rights applied only on Crown land and did not include lands that "may be required or taken up from time to time for settlement, mining, lumbering, trading or other purposes." Ensuring these rights was a very important concern for First Nations peoples during the treaty negotiations; economic security was the most pressing matter for them in this time of change. Historian Jean Friesen argues that without the verbal assurances that people would have the right to trap, fish, and hunt as before on Crown land, Manitoba First Nations would not have agreed to treaties 1 and 2. The same conclusion can be drawn for most of the other prairie treaties that followed.[3] These priorities also reflected

the assumption by First Nations that the treaties were not mere surrenders of land but agreements among equals that would enable First Nations to control their future relationships with Euro-Canadians and help them to adapt to coming economic and social change. In this sense, the wildlife provisions in the treaties promised the protection of hunting, trapping, and fishing as part of traditional vocations and ways of life.[4]

These promises quickly came under attack. By the early 1880s the government of Manitoba was challenging treaty hunting rights, arguing that the province's recent game legislation applied equally to everyone in the province and that the treaties conferred no special rights to hunt on Crown land or for exemption from closed seasons and other conservation provisions. The First Nations strongly disagreed, as did the Department of Indian Affairs, which countered that the province's position violated the terms of the treaties. In 1884 Lawrence Vankoughnet, the deputy superintendent general of Indian Affairs, observed that in other provinces "Indians are allowed to kill game at any season of the year and anywhere for sustenance, but are not permitted to kill in close season for sale." But Manitoba refused to yield, arguing that the purpose of the game laws was to conserve game and "if 'exceptions' were to be made for Indians, it would be impossible to secure observance of the law by settlers." Indian Affairs in effect bowed to the province's intransigence. In 1886 the Indian Act was amended to allow the minister of Indian Affairs to issue an order making a particular band subject to provincial game legislation. This surrender to provincial demands was further illustrated in 1890 when Vankoughnet "gently suggested that his [Indian] agents use their influence to 'prevent wanton killing by Indians' and also to keep them from hunting in the closed season 'whenever possible.'"[5]

The government of Manitoba was not the only one to challenge First Nations treaty rights. Disputes also arose between the federal Department of Fisheries and the Department of Indian Affairs. In terms of day-to-day life at the time, fishing was probably a more important concern than hunting for many First Nations in Manitoba. While fish provided critical food for people as well as their dogs, the Department of Fisheries was eager to encourage commercial fishing, then primarily on Lakes Winnipeg and Manitoba, and thus, beginning in about 1882, it tried to restrict First Nations fishing to particular locations. First Nations resisted because, as they pointed out, fish changed locations and the treaty had in any case given them the right to fish wherever they liked. This position was challenged by Euro-Canadian settlers, commercial fishing interests, and the Department of Fisheries, all of whom charged that First Nations fishing practices (not commercial operations) were causing depletion. Noting that First Nations complained that dam construction and commercial fishing by Euro-Canadians in Manitoba prevented them from fishing "as they had previously" done, the Department of Indian Affairs argued in 1884 that they should "be confirmed in the use and enjoyment of the fisheries which they have been in the habit of using from time immemorial, as the Government, in allowing them this privilege, will accomplish a large annual saving." It further noted that while it was desirable to end conflict between Euro-Canadian settlers and First Nations over fishing, the most important consideration was to prevent "dissatisfaction among our Indians, and the existence of a sentiment that the Crown has not kept faith with them in a matter, to them, of such vital importance as their fishing rights."[6]

But the Crown did not keep this faith. Disputes occurred throughout the 1880s and 1890s when First Nations opposed Department of Fisheries attempts to enforce such things as closed seasons, licence requirements, and restrictions on feeding fish to dogs. Ultimately, First Nations lost out, although the Department of Indian Affairs claimed something of a victory in 1891 when new fisheries regulations restricted commercial fishing to deep water, while fishing closer to the shore and at the mouths of rivers was reserved "to Indians and white residents of the country." The rights of First Nations peoples were also partially recognized (though a domestic licence was required) by the granting of "the privilege to Indians of catching fish for their own consumption, but not for barter or sale, during the close season." Nonetheless, Department of Fisheries officials continued to show an undimmed eagerness to limit Aboriginal fishing in the coming years. In 1913 its inspector of fisheries for Saskatchewan and Alberta argued that First Nations traditional fishing practices had to be curbed because fish stocks could withstand such practices only if there was no commercial fishing. The inspector argued that because Euro-Canadian commercial fishing could not "reasonably be restricted," "wasteful" traditional methods (such as fishing during spawning season) must "give way to the more orthodox." Commercial fishing, which served the "best advantage and general welfare" of Canada as a whole, could not be restricted by the "special claims" of First Nations peoples even if "necessities have to be provided in other ways." Presumably this meant that government welfare, wage labour, and activities such as farming would compensate for the loss of treaty fishing rights and pave the way for the expansion of commercial fishing. The displacement of First Nations subsistence use of fish by Euro-Canadian commercial use was thus a deliberate strategy that justified the use of public funds to mitigate any resulting social costs.[7]

By the close of the 1890s, treaty hunting rights had been curbed through Manitoba's assertion of its jurisdiction over game, while the Department of Fisheries at the federal level had effectively achieved the same result with regard to fishing. As Jean Friesen concludes, "in the end, the Department of Indian Affairs found it easier and preferable to restrain Indians than it did to disallow provincial legislation or challenge federal departments." The federal government did defend treaty hunting rights in the NWT in 1890 when it disallowed amendments to the territorial game ordinance that would have put First Nations "on the same footing as white men" by permitting them to hunt outside of open seasons only when they were starving.[8] But the territorial government was a weak adversary with its limited resources and restricted constitutional power, and after 1905 the Department of Indian Affairs faced increasing conflict over treaty hunting rights and provincial game legislation with Saskatchewan and Alberta as it had with Manitoba. In a typical statement of provincial policy, Alberta's chief game guardian, Benjamin Lawton, asserted in 1908 that its game legislation applied to all "persons" and even if the treaties permitted unrestricted hunting on Crown land, First Nations peoples were still subject to the provincial laws because they were "persons."[9] In regard to hunting for food, First Nations rights were most often an issue in certain southern parts of the newly created provinces of Saskatchewan and Alberta because both provinces allowed hunting for food without a licence (or issued it free of charge) in unorganized northern areas for most of the years leading up to the Second World War. This approach was also adopted by Manitoba for about the same time period, and it met the needs of Euro-Canadian settlers in fringe areas as well as those of prospectors, lumbering interests, and others involved in frontier natural resource development. While it also met the subsistence needs of First Nations people in these areas, it did not constitute a recognition of treaty hunting rights because it applied to everyone equally.[10] When licences were required, the only concession that the provinces would make was to issue them free of charge to First Nations people, but this implicitly identified treaty hunting provisions as privileges and not as rights.[11]

While the Department of Indian Affairs had, by the close of the nineteenth century, in essence conceded that provincial jurisdiction over game could override treaty hunting rights, it continued to argue that the provinces should make special hunting concessions to First Nations, if not as a matter of right then at least to help them avoid dependency. This held little sway with the provinces. In 1908 Alberta's deputy attorney general told the deputy superintendent general of Indian Affairs that "the protection of game is, and has always been a matter regarded as one of the matters falling within the exclusive jurisdiction of the Provinces" and that Alberta's Game Act was "binding upon all persons, irrespective of 'race, creed or previous condition of servitude' within the Province." In reply, the Department of Indian Affairs could only threaten that it would "intervene to the extent of its ability" to guarantee food for "its wards" to prevent starvation. The hollowness of this threat was apparent to everyone, forcing the department to urge respect for the needs of First

Nations people on compassionate grounds. In 1912 J.D. McLean of Indian Affairs wrote Benjamin Lawton that he regretted that Alberta was "not acting very generously towards the Indians who were the original owners of their country, and who in the past made their living by hunting and fishing, and in the North do so still to a large extent." In this matter, Indian Affairs was a supplicant. In 1914 Indian Affair's deputy superintendent Duncan Campbell Scott wrote Benjamin Lawton that some First Nations were having great difficulty making ends meet because of low fur prices. As a result, they would "be compelled to follow the pursuit of game for food with greater thoroughness than previously," and the department believed that "the game laws of the Province should not be rigidly enforced" against First Nations to give them "the chance to maintain themselves."[12] Although Indian Affairs officials regularly assessed the situation and sought legal advice on the constitutional issues that it posed, the department shied away from encouraging First Nations to demand that their treaty rights be respected. In a carefully worded reply to Chief James Sennum of Whitefish Lake in northeast Alberta, who had asked why the hunting terms of the treaty were not honoured and why the Alberta government had "the power to pass laws which would annul this treaty right," Indian Affairs noted that the provinces had "the sole right of legislating with regard to their game, and as the courts seem to uphold them in this contention, the Department can do nothing, even if it thought it advisable to save Indians from the consequences of disregarding the provisions of the Provincial Game Ordinances." Indian Affairs continued with this diplomatic approach, reporting in 1919 that it was attempting to persuade the provinces "that the game laws should be to some extent a non-irritant in their application to Indians in view of the fact that hunting is their natural means of livelihood" and that First Nations should be granted "special privileges ... with regard to hunting and trapping in so far as consistent with adequate protection for the preservation of game."[13] What this meant in practice was not clear, nor was it apparent how the provinces could be convinced in light of their uncompromising insistence on the universal application of their game laws. This standoff satisfied no one, especially First Nations people. Consequently, many of them flouted provincial game legislation as well as federal fisheries regulations – a response that was ongoing and widespread. In 1894, when the Department of Fisheries demanded that the RCMP enforce fisheries regulations at Lac La Biche in northern Alberta, Superintendent Griesbach replied that a detachment of at least twelve men would be needed "as these people are determined to fish, otherwise they will starve, and this I do not think they will do." And in 1909 a Manitoba game guardian complained that it was difficult to enforce game legislation near reserves because First Nations claimed the right to hunt anywhere throughout the year. And the same year, Superintendent General of Indian Affairs Frank Pedley frankly summarized the general situation when he remarked that game laws did "not excite much sympathy in the breasts of the Indians, who regard them as an infringement of their hereditary and treaty rights."[14]

The transfer of natural resources to the Prairie provinces in 1930 seemed to reaffirm treaty hunting rights, at least in the way that the federal government perceived them. The transfer agreements signed by each province and the federal government included identical clauses that stated that provincial game laws would apply to First Nations "provided, however, that the said Indians shall have the right, which the Province hereby assures to them, of hunting, trapping and fishing game and fish for food at all seasons of the year on all unoccupied Crown lands and on any other lands to which the said Indians may have a right of access."[15] The minister of the Department of the Interior observed that this signalled that the provinces had now accepted what he called "existing treaty rights." In keeping with its understanding that the provinces were now committed to upholding First Nations hunting rights for food on unoccupied Crown lands, Indian Affairs instructed its Indian agents in 1931 to make it clear to First Nations peoples that hunting on Crown land was allowed only for food for their own use and that "they must not, in any case, take game for commercial purposes of any kind in any way contrary to law, and that wanton slaughter will not be tolerated."[16] In subsequent years, the federal government continued to view the Natural Resources Transfer Agreement (NRTA) as having enshrined treaty rights. As the Department of Indian Affairs explained in 1950, when the NRTA was signed in 1930, "it was recognized that the Dominion could not, because of their statutory obligation to the Indians, transfer the 'property' without let or hindrance. In the Indian treaties there was a definite lien on the game resources," and by signing the NRTA, the provinces had "agreed that the lien existed and assured the Indians of their continued right to hunt, fish and trap for food at all seasons of the year on vacant Crown lands or lands to which they had a right of access."[17]

The provinces, however, took the well-worn path of opposing First Nations treaty rights, and Alberta's position provides one example of this opposition. Although it subsequently dropped this demand in the face of opposition from the Department of Indian Affairs, Alberta immediately proposed that First Nations people who hunted for food outside of open seasons would need a permit. It also sought to enforce other provisions of its game legislation in the case of First Nations hunters. Alberta suffered a setback to this stance in 1932 when the Supreme Court of Alberta, Appellate Division, overturned the conviction of William Wesley, a Nakoda (Stoney) who had been charged with hunting for food on Crown land contrary to Alberta's game legislation. The Department of Indian Affairs supported Wesley's appeal and engaged counsel to represent him. In his ruling on the appeal, J.A. McGillivray stated that he believed that First Nations peoples, "like the white man," should be subject to the game laws with respect to commercial or sport hunting, but hunting for food was a different matter because "the white man ... generally speaking, does not hunt for food." He ruled that the NRTA clearly guaranteed the right of First Nations to hunt for food and that they had an "unrestricted right to hunt for food in those unsettled places where game may be found."[18] This ruling outraged the Alberta Fish and Game Association, which passed a resolution "utterly" condemning any laws or any rulings giving First Nations special hunting rights. It also began a long-lasting lobbying effort against the decision. While the province was chagrined by the decision, it still had a powerful weapon at its disposal because it retained the power to define what was meant by "unoccupied Crown land" to which the hunting-for-food provisions applied.[19] By the 1950s, land defined by Alberta as "unoccupied" had decreased substantially, and it decreased further in the following years. Between 1963 and 1973 alone, access by First Nations to Crown land in Alberta fell between 25 and 33 per cent because land used for registered trapping lines (some of which were held by First Nations and Métis), as well as for various resource projects and forest reserves, was no longer defined as unoccupied.[20] These developments were part of a complex legal and political history that is beyond the scope of this discussion, but it bears noting that conflicts over access to wildlife continue with varying intensity to this day. For example, Métis, too, claimed hunting rights, and while they shared a history and often had interchangeable economic roles and lifestyles with some First Nations peoples, governments continued to reject demands for recognition of their rights. In 2003 the Supreme Court of Canada ruled in their favour, although the negotiations implementing the judgment have proven to be a drawn-out process.[21]

ACCESS TO WILDLIFE AND THE LOGIC OF MARGINALIZATION

Conflict over access to Crown land and how wildlife could legally be hunted had concrete ramifications for First Nations' economic and social life. By providing rights of access to wildlife, the treaties theoretically offered flexibility and economic security by confirming traditional subsistence strategies for food and income. This was important for sustaining a mixed economy in some communities, especially in northern areas. Historical geographer Frank Tough points out that in northern Manitoba many people lived by hunting and trapping, sometimes supplemented by wage labour and other activities. He describes these varied ways of making a living as a web of intricate responses to the economic and social integration of Aboriginal communities into the Canadian economy and to the effects of metropolitan control and social and economic inequality. There was, of course, no single pattern in all parts of the region; reliance on wildlife as part of a mixed economic strategy depended on location and availability of resources. Descriptions of economic life in various First Nations communities in 1919 revealed that farming and ranching and occasional off-reserve labour were typical of many First Nations people on the southern plains, while those in more northerly areas made a living hunting, trapping, and fishing. In some of these northern areas, wage labour, guiding, lumbering and selling hay and wood provided additional income. The interwar years brought critical changes to these economic and social patterns. In northern Manitoba, Tough argues, while "no

Figure 9.1 *Top* These Chipewyan trappers at Brochet in northern Manitoba in 1924 posed with what would have been a profitable catch of foxes and beavers.

Figure 9.2 *Bottom* While plains Aboriginal peoples had lost their traditional economy by the end of the 1800s, many Aboriginal people in parkland and forest districts were able to continue to live off the land. These women and girls were dressing a moose hide in the Candle Lake district east of Prince Albert, Saskatchewan, in 1910.

sudden event clearly indicates the point in time when the diversified Native economy ceased to provide growth and security," the 1920s and 1930s were significant because of the depletion of wildlife, declining opportunities for wage labour, and increased competition with Euro-Canadians for game, fur, and fish. The growing importance of cash and a greater dependence on new consumer goods further entrenched these changes, as did the historic under-capitalization of Aboriginal economic activity in sectors like commercial fishing.[22]

The case of the Lesser Slave Lake agency reveals similar patterns in northern Alberta. In 1911 it was reported that while the various First Nations in the district "do some gardening and farming in a small way, fishing and hunting are their only real occupations." By 1915 shortages of game and fur-bearers, along with the emergence of new wage-labour opportunities, led more people to farm, work for surveyors, sell hay or wood, or work on the river boats. These options expanded, although unevenly, during the First World War and the early 1920s. As the monthly reports of the Indian agent for the Lesser Slave Lake Agency in 1930–31 show, the Depression hit the area hard, bringing high unemployment and a return to hunting and trapping, which, despite a scarcity of game, now offered the only option for making a living.[23]

In the next decade, even this limited use of wildlife became more difficult as farm settlement and government restrictions on First Nations hunting played out. This was illustrated by the history of First Nations peoples in the Nordegg and Baptiste River areas on the eastern slopes of the Rocky Mountains southeast of Jasper National Park. In 1900 they had access to an extensive hunting and trapping range, but as land was taken up by Euro-Canadian farmers and designated as forest preserves, game sanctuaries, and parks where treaty hunting was not allowed, the territory in which these First Nations people could hunt and trap became ever more restricted. By 1943 they usually moved to other parts of the province in the spring to work as farm labourers and moved back to the Nordegg district after harvesting wrapped up in the fall. Although a few people found work in local lumber camps and mines, most hunted and trapped during the winter. By this time, however, expanding Euro-Canadian farm settlement and a growing number of white trappers encroached even more on traditional hunting lands and reinforced First Nations people's need to work outside the district.[24] In the next two decades these patterns were seen elsewhere as well. By the mid-1950s fewer First Nations people in northern Manitoba trapped. Instead, they worked as labourers in the south or at the nickel mine at Thompson or on northern hydro and highway projects. These trends were confirmed by low fur prices and extensive competition. In northern Saskatchewan, trapping and fishing districts had become so overcrowded by 1962 that it was impossible to make a living. Increasingly, First Nations people got by with government winter works programs or left the area for seasonal work in the sugar beet fields of Alberta or at the port of Churchill.[25]

The declining viability of hunting, trapping, and fishing as an economic option for First Nations people after the Second World War was the result of wildlife depletion caused by habitat change, population growth, and over-competition for resources. While these factors were complex and difficult to control, the decline was also exacerbated by provincial governments' ongoing attacks on treaty hunting rights, by federal deference to provincial jurisdiction, and by the federal government's overall policies towards indigenous peoples. It also reflected the ramifications of the dominant view that the principle of public property was the only acceptable concept for governing the use of wildlife. Euro-Canadians' unbending insistence on equal access to wildlife was a significant demonstration of their refusal to accommodate the needs of First Nations people. The Euro-Canadian principle of equality before the law was paramount in this view, and the provinces consequently argued that they could not and would not show what they called "partiality" to "special interests," by which they meant First Nations rights. But whatever principles of equality such attitudes embodied, they arose from a diffuse antagonism towards the idea of First Nations hunting rights as well as from a refusal to recognize that First Nations and Euro-Canadians had different relationships with and needs for wildlife. For game officials and politicians, it was also a matter of trying to enforce the law and maintain political power in the face of widespread Euro-Canadian resentment of treaty hunting rights. This meshing of ideals and realities was commonly reflected in the ongoing argument that had been employed in Manitoba as early as the 1880s – if First Nations were not forced to obey the game laws, settlers could not be forced to obey them either.[26]

While few Euro-Canadians on the prairies dissented from the anti-treaty hunting rights position taken by provincial governments, there were complicated cross-currents within the debate. For example, while fish and game associations were vocal and often harsh critics of First Nations hunting rights, individual hunters were sometimes more ambivalent and some others voiced more nuanced views as well. Although Saskatchewan's game commissioner Fred Bradshaw argued in 1919 that the hunting practices of First Nations needed to be controlled to conserve game, he admitted that there was, even among some critics, "an unmistakable trace of sympathy for the Indian" that was "in a greater or less degree" found "in every person who recognizes the red man's former privileges, and his present day improvidence."[27] And a Mr Norman Foster of Edmonton observed during the interprovincial wildlife conference in Saskatoon in 1936 that since there seemed to be insufficient game in Alberta for both First Nations and sportsmen, First Nations people's "right to shoot could not be denied," for their existence depended on hunting.[28] Such accommodation, however, was complicated by location and need. Andrew Holmes, the game guardian in the Prince Albert district in northern Saskatchewan, reported in 1922 that game laws were strictly enforced in southern districts where First Nations people lived on reserves and were learning to farm and not be "dependent on the chase." In northern regions, however, people had no option but to hunt, trap, and fish for a living. Suggesting the importance of official discretion at the local level, he noted that the Saskatchewan game branch did not interfere with First Nations' use of wildlife as long as they did not "wantonly destroy game" and made "good use" of what they secured. This was a matter of leniency and pragmatism and left the judgment of what was defined as waste and proper use of wildlife to the game official in the field. It was not a recognition of treaty hunting rights; Holmes was adamant that First Nations peoples had "no more rights [to game] than the white man" except that they did not need a hunting licence, only a treaty card and a certificate of permission from the Indian agent. While there had formerly been many complaints about First Nations big game hunting during mating season, he noted that this had now fallen off, even though hunting after the end of the open season still went on. Holmes blamed the Indian agents for this practice because they gave out letters of permission to trap, which First Nations hunters thought entitled them to hunt big game as well. In any case, Holmes believed these practices were declining because some offenders had been prosecuted and Indian Affairs was trying to control illegal hunting among First Nations.[29]

The informal balancing of needs that Holmes described did little to mitigate the general criticism of treaty hunting rights by governments, fish and game associations, and other members of the public whose religious, political, economic, and social attitudes coalesced in a presumption that Euro-Canadians had the right to force the assimilation of all Aboriginal people. For some observers, the treaties had only interrupted a natural progression towards this end. Frank Pedley of Indian Affairs remarked in 1909 that hunting and fishing, "which may be regarded as the natural aboriginal pursuits, are becoming largely, if gradually, curtailed." The passage of time and Euro-Canadian contact "have among many communities eradicated the natural craving for the excitement of the chase, and substituted other pursuits." But the number of Aboriginal people who "cling to these pursuits" had "been augmented by the treaties made during the last few years," and these people would continue to hunt and trap as long as it gave them "even a precarious means of existence." Nonetheless, assimilation was inevitable, and "matters will adjust themselves through time as they have done elsewhere."[30] Of course, Pedley knew that assimilation was being hurried along through residential schools, farm instruction, and other policies and programs. And controlling First Nations' use of wildlife had long been recognized as part of this process. For example, in 1887 Senator Schultz described enforcement of the game laws as one way to integrate the new western territories into Canada. He asserted that the decline of prairie wildlife was the result of Aboriginal overhunting and that "wise counsel" and the "strong arm of the administration to prevent the continuance of this habit" was needed "to restore the balance of productive power." And Robert Bell of the Geological Survey argued in 1886 that "the Indians are so improvident that they do not see the effect of the destruction of game and fish as we would. They are fond of destroying life when they can make no use of the game," but such habits would be corrected by the enforcement of game laws.[31]

The loudest and most emotional debate about hunting rights arose, however, over setting priorities between sport hunting and First Nations' use of wildlife. The *Macleod Gazette* editorialized in 1887 that "it is too bad that the future game prospects of this country should be sacrificed for the Indians" who refused to obey the game laws. "If they were in need of the game, there might be some excuse for it," wrote the editor, "but they receive ample rations" from the government and would not starve if prevented from hunting.[32] In this view, providing rations in times of need (another treaty promise) gave the state the right to conserve game for sport hunters to the detriment of First Nations people. Yet criticism of this priority was sometimes voiced. The Oblate missionary J.B. Ducharme wrote from La Loche in northern Saskatchewan in 1939 that "notwithstanding the solemn promises of the Government," First Nations people were prevented from shooting ducks "for their living in order to permit sportsmen to shoot at their fancy those birds, so valuable to the inhabitants of the north." But such criticism rarely made it into the public record, and the valorization of hunting for sport and the devaluation of subsistence hunting made a broadly accepted principle out of the Euro-Canadian claim that sport hunting was one of the best uses of wildlife. As Dan Gottesman notes in his history of First Nations hunting rights and migratory birds legislation, there was an "inability or refusal of policy-makers to distinguish between native and white hunters on utilitarian, psychological or cultural grounds. Both natives and whites were *users* of game, but their understandings of 'utility' were worlds apart. The former relied on game for clothing, subsistence or dietary supplement, the latter for emotional gratification or economical gain." As he notes, "few if any Indians hunted for 'sport,'" and while "hunting was one way in which Indian men could demonstrate their 'manliness,'" this was "usually understood in terms of their value as good providers. It was not understood, as it was by whites, in the narrow individualist sense of 'self-reliance' and 'resourcefulness.'" As one example, many Rock Cree people found the concept of sport hunting an alien one, which is not to say that they did not take pleasure from hunting. But the Euro-Canadian division of time into "leisure" and "labour," in which sport was a leisure-time recreational activity, had little relevance for them. The sporting mindset was also missing – killing for sport without using the animal for food or income was disparagingly called "he plays with an animal."[33]

Many Euro-Canadians of course used wildlife in the settlement era as part of a subsistence economy, but the future was idealized as one of self-sufficiency through farm production and hunting for recreation and not need. The view that farming defined a boundary between civilization and barbarism and was part of God's plan for humankind confirmed assumptions about the best relationship of wild animals and First Nations. As historian Sarah Carter has shown, efforts to transform First Nations people into farmers reflected Euro-Canadians' views about the cultural value of agriculture compared to that of subsistence hunting and trapping. The latter were not considered to be serious or "real" work but only acceptable activities for people in areas without arable land or other options.[34] While Euro-Canadians were supposedly culturally and socially invigorated by hunting and by watching and interacting with wild animals, such relationships were said to corrupt and hold back First Nations people. As the Department of Indian Affairs noted in 1888, the Blackfoot were starting to farm, which was a remarkable sign of progress, since only a decade before they had been "veritable Ishmaelites," constantly at war and "depending on the buffalo for subsistence." Howard Sibbald, the Indian agent at Morley, west of Calgary, argued in connection with the Nakoda (Stoney) that "as long as they can hunt you cannot civilize them."[35] And criticism of those who persisted in subsistence hunting was often sharp. The *Macleod Gazette* claimed in 1895 that the Nakoda had exterminated game in the southern Rockies (although they were still being accused of doing so a decade later), and a NWT game guardian wrote in 1902 that "I have great trouble with the Stony [*sic*] Indians who have an idea that they are a privileged class and who are allowed to wander around at their own sweet will. They destroy more game in a week than all the sportsmen will in a season," and they should be confined to their reserve. And in 1908 the Alberta game commissioner reported that although Nakoda land was "fit for nothing but the raising of horses and cattle," they were close to big game country and "are consequently hunting at all seasons of the year instead of endeavouring to earn a living by other means." This issue was "a sore one with many persons in the south-western part of the province," and the game commissioner's office faced a steady stream of complaints. This criticism did not acknowledge that the Nakoda were responding to changed circumstances. As the Department of Indian Affairs had noted in 1886, the federal government was providing them with rations because the arrival of the railway a few years before had driven wild animals "to much more distant parts," forcing the Nakoda to travel long distances to hunt for food and income.[36]

The characterization of the Nakoda as slaughterers of game was not only a grudging admission by some Euro-Canadians of the Nakoda's hunting skills but also an illustration of the view that First Nations' hunting and fishing practices were inimical to wildlife conservation. Subsistence practices that helped provide food throughout the year – such as fishing during spawning season, hunting birds and animals throughout the year, and gathering wild bird eggs – were all said to contribute to depletion. The sale of wild meat, a common practice among First Nations and Euro-Canadian hunters in some districts, was said to contribute as well. This was no doubt a valid criticism in an era of declining wildlife, and government game officials and fish and game associations were quick to point this out. In 1893 the Calgary Rod and Gun Club submitted a petition to the territorial government demanding that the game ordinance be enforced to stop First Nations people from gathering wild bird eggs, killing young birds and animals, and hunting during the breeding season. Similar petitions were presented by other associations, including those in Maple Creek, Edmonton, Lethbridge, Moose Jaw, and Red Deer.[37] First Nations hunters were also accused of being wasteful and wantonly killing game. F.J. Fitzpatrick, a Saskatchewan game guardian, claimed in 1914 that "Indians north and west of Prince Albert brought down tons of moose and elk," leaving the meat to rot and taking only the hides and teeth. Furthermore, he charged that the Department of Indian Affairs encouraged overhunting to save money because "the more game the Indians kill the less [government] rations the [Indian] agent will have to hand out to them." Showing the persistence of such views, Duck Unlimited (Canada) argued in 1939 that First Nations people in northern districts were "perhaps the worst enemies of waterfowl" because they gathered duck eggs and killed young ducks in the fall for food. As late as 1964, dire predictions that had been heard for decades were still being voiced; at the Manitoba Federation of Game and Fish Associations convention, a delegate from Flin Flon charged that "unless Indians are forced to abide by the same regulations that govern all hunters, Manitoba may be without any big game animals in 20 years."[38]

While this was the dominant view, some people argued that First Nations hunters behaved better or at least no worse than anyone else. This did not mean that they were without fault – both they and Euro-Canadian settlers hunted illegally and wasted game. In 1911 the game guardian in the Humboldt district east of Saskatoon observed that neither First Nations people nor settlers observed the game laws, and seven years later the depletion of big game east of Prince Albert was blamed on wolves, on Aboriginal hunters who killed female animals, and on young men who had moved north to trap and evade conscription and who were "an even greater menace to the game than the Indians."[39] In other instances, over-hunting by some farm settlers was said to be the result of similar First Nations hunting practices. In 1910 the Saskatchewan game branch argued that "the Indians are the chief violators of the provisions of The Game Act" and "white people, generally, observe the law fairly well, although there are some isolated settlers in the remote portions of the province who are almost as bad as the natives" because they believed they did not have to obey the law given that Aboriginal people did not do so. Others suspected that some complaints about Aboriginal hunting had more to do with competition for game and personal failure. One game guardian in northern Saskatchewan noted in 1915 that there was "a great tendency on the part of the hunters who do not have as good fortune as their brother sportsmen" to blame "the Indians for going up before the season opens and either making a big killing" or driving the game out of the district. Claiming that he personally knew First Nations hunters in his district who killed no more than twenty moose in the season, this was "nothing compared to the game killed by settlers in outlying areas" or by white hunters who went north before the season opened to set up their hunting camp and "who cannot wait for opening day."[40] Similarly, Andrew Holmes, the game guardian in the Prince Albert district wrote in 1921 that while First Nations people tended to kill game out of season, "I will say this for the Indian, that whatever he kills, he uses all and wastes nothing, in fact he is a gentleman compared with some of the white men in our country."[41]

Such defence of First Nations hunting practices was sometimes inspired by the view that Aboriginal people were "natural conservationists" and that if they over-hunted and wasted wildlife, it was because their traditional values had been corrupted by European contact and its commercial impetus. The notion of the Aboriginal as conservationist and as having a special relationship with wild animals and nature was perhaps a compensatory gesture in light of the feelings of guilt voiced by some commentators over the devastating impact of European settlement on North American wildlife. Certainly, part of Grey Owl's appeal in the 1930s was his deliberate cultivation of the notion of the Aboriginal hunter as a conservationist and the Euro-Canadian as a despoiler of nature. But as anthropologist Robert Brightman notes, it also reflected a belief commonly held in the European world that Aboriginal people conserved natural resources and consumed only what was needed for survival due to their ancient spiritual beliefs. While Brightman contends that there may now indeed be a commitment by many Aboriginal people to such practices, he questions "when these practices and understandings developed" and suggests that the evidence, though limited, points to their emergence, at least among Algonquian peoples, after contact.[42] However this view arose and whether it was an accurate description of Aboriginal practices or not, the Department of Indian Affairs became a strong proponent of this view during the interwar years when the department was trying to gain recognition of treaty hunting rights and ensure First Nations self-sufficiency. As the department's annual report stated in 1929, "by immemorial usage the Indians are conservationist of the game and fish."[43] And some department officials contended that this was indeed the reality of contemporary First Nations hunting practices. In 1935, as part of its campaign against the Wesley case, George Spargo of the Alberta Fish and Game Association wrote to the Department of Indian Affairs complaining of "the wasteful and destructiveness [*sic*] and unethical practices of the Indian population." Hard times had led them to "corrupt themselves" by peddling wild meat at lumber and mining camps. To deal with these problems, Spargo recommended that the federal government increase its subsidies to First Nations so that they would not need to hunt for income. Further, the government should manufacture pemmican from surplus bison in the national parks and distribute it to First Nations on the understanding that they would in return surrender their right to hunt for food under the NRTA and come under the game laws in "a like measure to the white hunter." In conclusion, he drew attention to a poem on the association's letterhead that claimed that conservation created a legacy for future generations. In a briefing memo to the department about Spargo's letter, M. Christianson, inspector of Indian agencies in Alberta, detailed hunting practices in each of the Alberta agencies to demonstrate that First Nations were not wasteful, did not sell meat, and hunted only to meet their immediate needs. In response to the association's haughtiness, he charged that First Nations hunters did not "kill game wantonly for sport" and would never surrender their treaty rights to satisfy the ambitions of sport hunters. Moreover, increased subsidies were out of the question, as was manufacturing pemmican to feed First Nations people. "I do not think pemmican would be any more acceptable to them," Christianson dryly noted, "than it would be to the members of the Alberta Fish and Game Association." As for the poem's message, Christianson commented, "I have been in this Western country for 48 years, and I believe I have never heard a White man refer to the purport of this poem, but I often hear the Indian express that very idea: That with the number of white people who hunt and trap now, there will be practically no game left for the generation to come."[44]

Figure 9.3 Among other things, Everett Soop, who was born on the Blood Reserve in southern Alberta in 1943, was a cartoonist. This is his wry take on the changing priorities of First Nations people that was published in the *Kainai News*, c. 1968–86.

IN PURSUIT OF PRODUCTION AND SOCIAL ORDER: REGISTERED TRAPPING

Trapping was an important part of Aboriginal life in northern and parkland districts in the region. The federal government, however, did not view trapping as a treaty right for First Nations peoples and the provinces of course agreed with this view. By the interwar years, both levels of government were confronting the depletion of fur-bearers and a breakdown in social order in many northern districts because of the competition for fur resources. Registered trapping emerged as the preferable way that the provinces could balance the need for social order, the conservation of resources, and long-term income for the state. While registered trapping was a management tool that in principle did not conflict with treaty rights, the way that it was implemented in each of the three Prairie provinces had an impact on First Nations peoples and their treaty rights.

Historically, northern districts in the three Prairie provinces were key fur-producing areas. During and after the 1890s these regions also saw increasing social and economic tension because of huge swings in the price of fur, improved transportation that brought Euro-Canadian and Métis trappers north, growing rivalry among traders, and increased competition between resident and non-resident trappers. Trapping served as an economic safety valve for some people in the south, and fringe-area settlers trapped to supplement low farm incomes. Boom prices (such as before the First World War and between 1915 and 1921) brought a flood of trappers into northern areas looking to make quick money, and many others continued to arrive as problems in the southern economy drove them north in search of income. While some Euro-Canadians became permanent residents in northern communities (as they had since the late 1800s), others stayed only long enough to strip an area of its fur. Some aggressively threatened local trappers and used poison to kill fur-bearers, and the lowest common denominator quickly became the rule. The Nelson House area in northern Manitoba, for example, had once been one of the best fur districts in the province, but white trappers had killed nearly all of the game in the area by 1930. Commenting on similar situations in other areas, T.R.L. MacInnes of the Department of Indian Affairs noted in 1936 that "if whites are allowed to deplete Indian hunting grounds, the Indians themselves will naturally take all they can, while they can, and there is a grave danger that such a situation may bring about intensive competition between whites and Indians, ending in the virtual extermination of valuable species. Indian families, in most cases, are permanent residents, and their hunting grounds [are] recognized among themselves, and handed down from one generation to another, whereas white trappers are frequently of the itinerant class, whose practice is to trap out an area and then move elsewhere."[45]

These conditions were in part the consequence of equal access to wildlife on Crown land. Anyone with a licence gained access to all fur resources and was not precluded from encroaching on another trapper's lines or from beginning to trap in a district that had been used, even for years, by others.[46] All the provinces licensed trapping. Manitoba, for example, instituted licensing in 1913, ostensibly to enhance its enforcement of game regulations and to collect statistics to improve fur management. But licensing was not used to control trapping practices and did little more than produce revenue for government. There was little other meaningful regulation; closed seasons were used to time trapping so that only prime animals were trapped, and hunting bans were sometimes declared for some endangered species, such as beavers and marten. But even these limited regulations were often not enforced, in part because traders wanted it that way. John Christian Shultz, then lieutenant governor of Manitoba, was told in 1893 that Manitoba's fur-bearer regulations were not enforced and that many species were becoming scarce. Better regulation was suggested, but Clarence Chipman, the trade commissioner of the Hudson's Bay Company and its most powerful representative in Canada, took "a very adverse view of the matter and put his foot down upon it" and vowed to fight any regulation of the trade. Governments were not eager for tighter regulation either; bag limits or other quotas would have interfered with higher production, which all the players – government, fur traders, and many trappers – wanted. This was still the case in Manitoba in 1930 when it was admitted that regulations were ineffective and serious depletion of fur-bearers had occurred in the previous five years because of "intensive trapping, unlawful use of poison by trappers, [and] hunting and trapping during close seasons."[47]

Some observers argued that the problem of over-trapping and the social stresses of excessive competition and depletion could be met by establishing trapping preserves in northern districts where only local Aboriginal people or long-term residents could trap.[48] This approach drew in part on the notion that Aboriginal people were "natural" conservationists, but it also recognized that conflict over fur-bearers was not just about resources but about a way of life. For many Aboriginal people, trapping was not a "job" in the Euro-Canadian sense of it, but a vocation and a subsistence activity that gave them a way to make a living.[49] Indian Affairs was a strong proponent of trapping preserves and a number were established in Quebec in 1928 and in Ontario slightly later, while policies in Yukon and the Northwest Territories paralleled this approach by using residency or Aboriginal status as criteria for granting trapping rights. None of the Prairie provinces employed approaches that recognized prior rights when they granted trapping licences. After briefly considering proposals for a trapping preserve in northern Alberta in the late 1920s, Alberta decided against its establishment because, as the minister responsible for the matter observed in 1926, "any rights granted to the Indians will also have to be granted to the white men."[50]

Instead of Aboriginal trapping preserves or regulations that favoured local residents, each of the Prairie provinces adopted the registered trapping system, which granted a form of private property rights to a trapping area. Such exclusive rights to exploit resources such as timber or minerals on Crown land were commonly granted by the state to individuals and corporations, and in keeping with this model, the registered trapping system gave long-term access to a trapline to those who registered the trapline in their name. This promised to rationalize and stabilize fur trapping and to maximize production within conservation goals by making each trapper a "game guardian" who protected his interests through long-range conservation of fur-bearers.[51] Such registered trapping also aimed to reverse the social breakdown occurring in northern areas. As C.F. Shirley, a trapper at Sekip Lake in northern Saskatchewan explained in 1942, "we cannot trap the way things are going on here now." Traps and fur were stolen from traplines and personal effects were stolen from cabins. "Lots of this work is done by young lads who just run around with dog and gun and kill every animal they can get a shot at or cripple them and leave them to die or suffer for days in the bush with broken limbs." A registered trapline system would help deal with such problems because whoever had a registered line would "naturally protect the fur on it and try his best to keep the worthless trapper out." As the Manitoba game branch phrased it in 1941, registered traplines would give trappers "legal protection of their individual trapping grounds," promote conservation by eliminating "the itinerant and predatory trapper," and encourage "the 'farming' of individual protected trap lines."[52] By imitating the model of the farm, it would bring a settled Euro-Canadian order in the wilds. As Leslie Sara of the *Calgary Herald* wrote in the mid-1930s, "turn a trapper into a fur rancher and he becomes a settler rather than a nomad" because security of tenure would prompt him "to build a decent cabin instead of a wilderness hide-out with all of its discomforts."[53]

In adopting the registered trapping system, prairie governments drew on the precedent of British Columbia, where it had been adopted in 1925. Saskatchewan created a short-lived lease system in about 1933 where trappers (usually Euro-Canadians, it appears) were given exclusive trapping leases and then implemented a comprehensive registered trapping system at the end of the decade.[54] In 1933 Alberta also began issuing registered trapping leases in some forest preserves on the eastern slopes of the Rockies, and in 1939 extended the system to the most northerly parts of the province. In 1941 Alberta broadened the system to include all unorganized districts north of the North Saskatchewan River (thus excluding most of the northern Euro-Canadian farming areas that had local governments, such as the Peace River Country). In 1939 Manitoba introduced the registered trapping system in an area north of The Pas, and the system was subsequently extended to some southern districts, including the Whiteshell area in eastern Manitoba.[55] In all cases, individuals could only register traplines by filling out an application form. In northern Alberta, this placed many Aboriginal trappers, who had a poor grasp of English and no skill at paperwork, at a disadvantage. Similarly, in Saskatchewan the most aggressive trappers, or at least those who had some knowledge of how to deal with the English-speaking bureaucracy in Regina, were the best positioned, and it was noted that "one of the main failings of the programme was that it enabled a few of the more aggressive white trappers to gain

Figure 9.4 This muskrat trapper was enjoying a social gathering with his daughter on his way to The Pas, Manitoba, in 1945.

control of the more productive fur bearing areas at the expense of the northern trapping population."[56]

None of these early systems took into account the use of wildlife by Aboriginal people in general or First Nations people in particular. Historian Arthur J. Ray notes that registered trapping "legitimized non-Native access to fur and game resources on undeveloped Crown land."[57] This began to change in 1944 when Manitoba and the federal government launched a joint study on extending the scale of registered trapping in northern Manitoba. The federal government wanted to see wild fur trapping reformed in order to reduce its long-term costs as well as its responsibilities for First Nations and to create social stability in northern communities. The next year, a ten-year dominion-provincial agreement on fur development was ratified. This cost-shared agreement extended the registered trapline system to all of northern Manitoba, provided for fur-bearer habitat rehabilitation in some areas, and encouraged the transplanting of beavers. Although non–First Nations trappers were not excluded, the system favoured long-term residents. The most significant aspect of the agreement was its modification of the province's registered trapping system to allow for what were called group lines (sometimes referred to as "areas" or "blocks") that could accommodate traditional Aboriginal common-use trapping practices. Under this provision, a district could be registered to a group, usually a family or a community, which then could collectively organize its use according to custom, prior rights, community, family or band use, or even on an intra-band basis. Such "groups" could later decide to allocate lines to individuals, and they could also even out income among members, since not every line produced the same income each year. Thus, when requested by a community, "larger groups were registered, such as a camp of several families who always hunted together in a definite locality."[58]

A similar approach was adopted in Saskatchewan. As early as 1939, Saskatchewan and the federal government were discussing a scheme whereby, "in return for the concession of all hunting, trapping and fishing rights to the resident population, and to the exclusion of the itinerant white trappers," the federal government would financially assist with water conservation programs to restore fur and game in the northern parts of the province. In addition, the Saskatchewan government had begun, at least on a limited basis, to restrict the granting of trapping licences to local residents only.[59] With the election of the left-wing CCF government in 1944, the province began to restrict all trapping to local residents. The CCF was uncomfortable with the self-interest assumptions that lay behind registered trapping and was more amenable to communal approaches such as those underlying the group trapping provisions in Manitoba's federal-provincial program. In 1946 it signed a similar ten-year agreement with the federal government whereby the whole of northern

Saskatchewan was declared a "Fur Conservation Block." This block was subdivided into smaller "areas," each of which was directed by a five-person council elected by that area's trappers. The council worked with the game branch and other provincial government officials to manage fur and game in the area and to assign traplines. Priority was given to residents of the area: while non-residents, which primarily meant Euro-Canadians, could have a registered line, the government established rules that made it difficult for them to obtain them. All fur was sold through the Saskatchewan Fur Marketing Service, a compulsory government marketing system set up in 1945. The marketing side of the policy was unsuccessful because prices were too volatile, and the marketing service was too small a player to influence prices. Moreover, it did not grant credit, which, along with the practice of tallying final returns only once all the fur was sold, created resentment and a lack of popular support. Quotas were set from year-to-year following a survey of fur-bearer populations, and trappers were allocated the number of pelts they could take so that 40 per cent of the species was left as breeding stock.[60]

Despite their limitations, these group systems in Saskatchewan and Manitoba at least attempted to grapple with the social and economic needs of Aboriginal and other northern people.[61] In contrast, the system established in Alberta allowed any individual to obtain a registered trapping line as long as citizenship qualifications were met and an application was filed. Traplines were registered on a first-come, first-serve basis. Prior use, residency, and First Nations status were irrelevant. This focus on liberal-individual rights sat poorly with some First Nations communities in the north, leading them to insist on group lines. While the Department of Indian Affairs saw some benefit in individual registered traplines because they would end the competitive scramble for fur that First Nations people tended to lose, it worried that "whites" would push First Nations aside by registering lines first. Indian Affairs therefore encouraged northern Alberta First Nations people to register their lines quickly to ensure that outsiders did not register traplines on lands close to settlements where First Nations had customarily trapped. As well, Indian Affairs tried to persuade the province to segregate First Nations and "white" lines and to accept prior use of land by First Nations as part of the criteria for granting the registered lines. The department also wanted the province to register group trapping lines, which it saw as a pragmatic solution to potential conflicts over lines in some parts of northern Alberta that were not mapped in sufficient detail to permit an equitable and accurate distribution of lines.[62]

Alberta refused to heed these requests. Euro-Canadian trappers and their supporters as well as some Métis lobbied against group lines and the recognition of prior rights for First Nations. Indeed, Alberta abolished trapping licences in 1943, which meant that only those with a registered trapline could now legally trap and sell furs. In 1944 it further required that First Nations trappers pay the annual licence fee for a registered trapping line, which meant that such trappers now had to pay for something than many of them believed was a treaty right. Because Alberta largely viewed the issue in terms of maximizing the economic value of furbearers as a natural resource, its registered trapping approach was part of a modernization agenda in which only liberal individualistic economic relationships with wildlife, not communal ones, had validity. While it seems that the Alberta game branch field men who laid out the lines were sympathetic to First Nations desires for group lines, official policy continued to stress the granting of individual lines. Group lines were only allowed in a few special cases when attempts to impose individually registered lines threatened to bring conflicts that the province feared it could not control. But these were granted only as a concession and not a right. In addition, Alberta viewed all registered traplines as land leases and the right of First Nations to hunt for food on unoccupied Crown lands under the provisions of the NRTA therefore did not apply to such "occupied" land.[63]

The contrast between Alberta's approach and that of Saskatchewan and Manitoba is striking and illustrates how human-animal relationships occur within specific social and political contexts. Alberta's view that everyone was "equal" and that First Nations therefore had no special rights was fundamental to its trapping policies. The province's refusal to work with the federal government to ameliorate deteriorating conditions in northern trapping districts was also a significant factor. While Alberta had long been more sensitive about protecting its provincial rights than Saskatchewan or Manitoba, its adversarial relationship with the federal government had intensified in the 1930s with federal disallowance of some of the province's Social Credit legislation. Alberta's approach to northern development also assumed that the north did not require any special consideration over any other part of the province and that its fur management policies were simply another element in broader attempts to maximize the value and production of public resources. Just as it had done in its pioneering efforts in petroleum regulation, the province aimed to professionalize and modernize trapping, which became an explicit goal in its fur-harvesting policies in the 1960s. Political scientist Peter Clancy's observations about conditions in the Northwest Territories apply to Alberta as well. As he argues, modernization and professionalization meant that efficient, better-capitalized, and high-volume white trappers could produce more fur because they were "not distracted by the social solidarities or subsistence imperatives that prevailed among Natives."[64]

While Alberta's approach to trapping was in some ways unique on the prairies, it broadly mirrored the commonly held view that the status of First Nations peoples and the promises made by the treaties did not reflect distinct historical, economic, and social needs or confer singular legal rights. Conflict over all of this had a long history, and Manitoba's insistence in the 1880s that hunting by First Nations was bound by provincial regulations was one of the earliest demonstrations of this view. The practical consequences of the rejection of these First Nations hunting rights and the loss of local and communal control that it entailed were made clear as state regulation of game and fur-bearing animals for the benefit of settlers, sport hunters, white trappers, and commercial fishers increasingly dominated state policy and trumped promises made in the treaties. This marginalization of First Nations needs was confirmed and entrenched by provincial policies following the transfer of natural resources to the provinces in 1930. While Euro-Canadians often couched their opposition to treaty hunting rights in terms of conservation, the essential issue was not about whether First Nations people were better or worse conservationists than Euro-Canadians, but about whether they had the legal right to define their own relationships with wildlife. Whether the honouring of treaty hunting rights (and the granting of Aboriginal hunting rights to Métis) would have made the lives of Aboriginal people better overall cannot be known, but it can reasonably be concluded that rights of access as promised in the treaties would have enhanced their economic and subsistence use of wildlife. Such rights might have also offered them a way to establish greater local control to meet the challenges of Euro-Canadian competition in trapping, fishing, and recreational hunting. While the economic and political disadvantages of First Nations people were the consequence of a complex range of policies, circumstances, and attitudes, the refusal of Euro-Canadians to honour the hunting rights promised in the treaties was a crucial factor. This was exacerbated by their belief that Aboriginal people should live like Euro- Canadians did, an assumption that, at its core, was based on their view of themselves as conquers of the land, its resources, and its indigenous peoples. Inherent in this was the view that Euro-Canadian standards were universal. Yet, much conflict could have been avoided, and improved social outcomes might have resulted, had Euro-Canadians admitted that their right to fish, trap, and hunt was different than that of First Nations peoples and that the use of wildlife needed to be negotiated as a social project. Rigid support for equal access to wildlife did not, in practice, create equal opportunity for all.

10

Economics and Nostalgia: Encounters with Fur-Bearers

> Other industries, when compared with the fur trade are mere children. They are noisy children, however, and sometimes attract so much attention that we are inclined to forget that Canada's pioneer industry, the fur trade, is still going strong.
>
> R.H. Macdonald, "Golden Fleece" (Part 2), *Western Producer*, 15 March 1951

Historically, the fur trade created important connections between Aboriginal and European people and made fur-bearing animals into crucial commercial commodities. Indeed, the fur trade validated the development of the country, and as one Manitoba government official commented in 1931, "the raw fur trade was the sole factor which gave the North American wilderness an economic value in the esteem of the European peoples. It was this trade that attracted the enterprise out of which the vast civilization and wealth of this continent had developed." While its role in the prairie economy declined after 1870 relative to the growth of other sectors, the fur trade remained an important economic activity that gave value to non-arable land and to wild animals not useful for sport or food. In the early 1920s the market value of Canadian furs was about $16.7 million annually, of which about $5.7 million came from the Prairie provinces.[1]

The money generated directly from the sale of fur was only one part of the wealth extracted from fur-bearing animals. The private sector benefited from merchandising, transportation, communications, and export trade activities, and although most Canadian fur was exported in its raw state, some processing and manufacturing provided further economic benefits. While Toronto and Montreal were the major fur centres in Canada, by 1950 some marketing, dressing, and dyeing of pelts and manufacturing of fur garments had also developed in Winnipeg. In addition, the trade provided revenue for government from royalties on the market price of wild fur. The price and desirability of different furs varied greatly over time. Muskrat, one of the lowest priced, had been important in the fur trade since the mid-1800s, and after 1900 it remained one of the most commonly harvested furs. In 1921–22 just over 3 million muskrat pelts, valued at about $4.7 million, were taken in Canada. By 1957–58 the Prairie provinces were producing about $5.5 million worth of wild fur, earned on about 3.65 million pelts, mostly muskrat.[2] But a wide range of other fur-bearers were also trapped in almost every part of the region. In Saskatchewan in 1929, for example, just over 1 million pelts of twenty different fur-bearing animals were taken in various areas of the province. The most valuable were mink, muskrat, and weasel, which were trapped throughout the province. Badgers and coyotes were caught mainly in southern districts, while foxes and beavers were primarily taken in the north.[3]

Making money from wild animals was not, of course, limited to the harvesting of fur-bearers. A significant freshwater commercial fishery developed in the late 1800s in Manitoba and later in Saskatchewan and Alberta, all based largely on exports to the United States. Market hunting was another form of commercial harvesting, although it had become illegal everywhere in the region by 1922. Consumption through the tourist's gaze provided other commercial connections with wild animals, as did less common activities such as the sale of animal parts for scientific purposes or as curios for tourists. But the commercial use of fur-bearing animals was the oldest, most widespread and socially diverse of these activities. As seen earlier, various political, economic, legal, and cultural factors influenced the adoption of registered trapping systems. Similarly, many factors had an influence on other commercial relationships that people had with fur-bearing animals, as this chapter will demonstrate through four case studies. First, the impact of growing ethical concerns are shown in an examination of the interconnections among the fashion industry, trapping methods, and a growing public demand for humane trapping after the 1940s. Second, the case of beaver conservation reveals a range of often contradictory environmental, economic, and cultural motives for bringing a fur-bearer of mythic stature back from the edge of extinction on the prairies. Third, concerns about efficiency, profitability, and the depletion of wild fur-bearers are shown in the growth of fur farming as an alternative to trapping wild animals. Finally, Manitoba's muskrat habitat rehabilitation projects, which began in the 1930s, demonstrate how changing views about fur management, habitat restoration, and economic development coalesced around one of the commonest fur-bearers on the prairies.

MARKETS AND THE ETHICS OF TRAPPING

Depletion of fur-bearers was ongoing after 1870. Beavers were nearly wiped out by 1900, and marten and fishers declined steadily and were almost extirpated in many districts on the prairies by the end of the Second World War. In what was cited as yet another example of how Europeans had upset the North American environment, it was commonly claimed by the 1920s that all fur-bearing species throughout North America, as in most other northern parts of the world, were under threat. Manitoba's case was typical; by 1937 badgers and marten were "dangerously near the point of extinction," and by 1946 fishers and lynx were said to be threatened as well.[4] Although steps were taken to control depletion, such as the implementation of various trapping regulations and the registered trapping system, the consumer fur market adapted relatively easily to the depletion of particular species. Muskrats, for example, filled the void created in the late nineteenth century by the decline of beavers. Muskrat is a handsome fur in its natural state, and Canadian pelts found a market advantage because they were darker and more lustrous than those from warmer climates such as the southern United States, one of the largest muskrat-producing areas in North America. Muskrat pelts, as well as other low-priced furs like mink, rabbit, and weasel, were also dyed and processed to add value and create demand.[5] Indeed, the many inventions of the fashion industry in the twentieth century so changed the nature of fur and the values ascribed to it that the existence of the animals themselves came into question. As one writer advised readers of the

Grain Growers' Guide in 1927, many furs "are modified in the course of manufacture, and are sold under names quite different from those which they bore in their natural state." Some of the names were fanciful inventions, and while these tricks could be uncovered, the writer confessed that "it took me a little while to get accustomed to the names used by the furriers but now I know what they mean when they say electric seal, Hudson seal, kolinsky and broadtail etc." Hudson seal was processed and dyed muskrat; electric seal was dyed rabbit; and American sable, sometimes called "Russian sable," was pine marten (a European species of marten). In other cases, there was straightforward fraud – dyed hare was sold as sable or fox, woodchuck was transformed into mink, skunk became sable, nutria was sold as seal or electric seal, and rabbit could "masquerade as seal or muskrat."[6]

While designers and the clothing trade invented, seemingly like magic, new names and fashions to create new markets for fur, supply was unstable because of depletion and because most fur-bearer populations have cycles of varying duration. The market popularity of particular furs also fluctuated. Some furs became irredeemably unfashionable and ceased to have much commercial value. By 1940, for example, wolverine no longer had much market value except in the Arctic, where it was used to trim parkas. Indeed, it was said that "not since 'horse and buggy' days" had wolverine "been in demand in civilized centres," and northern prairie trappers now saw wolverines only as pests that raided their traplines. Bear skins were no longer of any commercial interest either, and their only economic value came from tanning them as hunters' trophies. The furs that fell permanently out of fashion were often rough "frontier" types, and this was perhaps linked to a disdain for the everyday clothing made from the skins and leather of local wild animals that was most often associated with Aboriginal people. Even by the 1870s, such traditional indigenous clothing had come to be seen as something that distinguished Aboriginal from newcomer and as colourful, but somewhat eccentric, remnants of the past. William Taylor, a settler near Woodlands, northwest of Winnipeg, noted in 1875 that Aboriginal people still used the skins and leather of wild animals for clothing, and while he implied that these materials were admirably suited for the climate and the country, he did not feel that they were something that Euro-Canadians should adopt or had adopted.[7]

With the exception of wolverines and bears, furs came in and out of fashion. When trends swung in their favour, trappers could have a profitable year. In 1925, for example, "Dame Fashion has decreed that the longer haired varieties of fur should be the vogue," and while muskrat prices fell, those for wolves and foxes rose and even the "lowly badger has found himself in the company of the aristocrats among fur bearers." The fun-loving insouciance of wearing the fur of long-haired wild animals did not last, and by 1947 the tide had turned to favour the sleekness of furs such as mink, beaver, and muskrat and long-haired furs "found few or no buyers at any price." Predictably, sleekness was once again superseded by demand for long-haired fur, with wolf, fox, badger, and lynx earning premium prices. Yet, by 1958 all fur prices were low – less fur was needed for the now fashionably short fur jackets than was required for the previously fashionable full-length coats. The growing availability of synthetic fabrics and Russian competition further contributed to falling demand for Canadian fur. Under these ever-changing conditions, the popularity of particular furs depended on the social status and fashionableness attributed to them. As Chantal Nadeau, in her book *Fur Nation*, suggests, changes in fur fashions and their marketing reveal that no fur has had intrinsic value and that value has been contingent on changing attitudes about national identity, sexuality, nature, gender, and women.[8]

Using fur for clothing also increasingly referenced general concerns about trapping fur-bearers. In 1921 it was observed in *Saturday Night* magazine that while it was "impossible to conceive of a world of fashionably dressed, or even comfortably dressed, men and women without furs," there was an important concern about wearing fur. This was not, however, the familiar mid- and late twentieth-century concern about the ethics of killing animals for fashion. Rather, it centred on the view that the consumer had a responsibility to conserve wild fur-bearers by wearing only the fur of animals raised on fur farms, thus mitigating the rapacity of trappers that was leading to the decline of wild fur-bearing animals.[9] In contrast to such anxieties, official concerns largely centred on the methods used for securing wild fur, the goal being to ensure economic efficiency, enhance market value, and control depletion. Pelt quality was directly affected by how an animal was killed, and huge discounts levied against scarred pelts helped to minimize the use of spears, knives, hooks, and bullets. Nonetheless, shooting muskrats and beavers with

.22-calibre rifles (which caused less damage than shotguns) remained a common practice in many places, even though governments legislated against it to prevent waste and to ensure the highest possible value of the fur.[10]

Snares, which caught an animal in a wire noose that tightened around its neck or body, could also damage pelts, pulling the fur from the skin. Another significant problem with snares was that they indiscriminately killed all animals that encountered them. The use of snares was restricted by most game statutes to specific purposes, such as killing rabbits. They were widely used for this purpose by many Euro-Canadian farm and Aboriginal women in northern and fringe districts, but many people used them indiscriminately as well. Cheap to make, many were hung along wildlife pathways, and they killed or maimed any creature caught in them, including domestic animals. In 1938 Manitoba legislated that trappers could not possess materials for making snares, but they remained in common use. In 1940 one Manitoba game official argued that snares were responsible for "the steady decline in foxes and lynx in recent years."[11] Inexpensive thin wire was often used to make them, and while the wire was strong enough to tighten around the animal's neck, such a snare could break loose from its mounting as the animal struggled, allowing it to escape only to suffer a slow death by suffocation or infection. In some places, hundreds of snares were set out and never picked up, becoming a permanent menace.[12]

Poison was also used to kill fur-bearers, especially carnivores like lynx and wolves. Although its use for killing fur-bearers was restricted by game legislation from the beginning, strychnine and other poisons were nonetheless being widely used by the end of the 1800s and could be purchased at many pharmacies without a permit. It was argued in 1910, apparently without much evidence, that the pelts of poisoned animals were of lower quality, but most critics railed against the indiscriminate and wasteful slaughter that poisons caused and lamented their spread through an ecosystem. In 1934, for example, a trapper in northern Saskatchewan was charged with illegally setting snares and using poison to kill fur-bearing animals, and it was argued at his trial that meat baited with poison "might be eaten by a crow or raven which flies miles before dying. It, in turn, may be devoured by a mother fox and the entire litter be destroyed without one pelt being obtained." Such utilitarian concerns applied to the target animal as well, for "only a small percentage of the pelts can be secured as a dying animal often will travel some distance before succumbing."[13]

The conservation and environmental problems that came with poisons and snares, as well as the economic penalties associated with damaged fur, left leghold traps (or "legholds") as the most popular legal method of harvesting fur-bearers. Leghold traps snapped shut around an animal's foot when it stepped into the jawed trap. Although there were various types, steel legholds, available by the early 1800s, created a standard that remained in use for 150 years on the prairies. While they too were criticized as economically wasteful because animals could escape from them, leghold traps were increasingly censured for ethical reasons. Well before the Second World War, many people simply transposed their sentiments about the ethics of sport hunting to their judgments about legholds. One commentator claimed in 1930 that poison baits and snares were unsporting and brutal methods of killing wild animals, while the use of legholds was cowardly. A report that a golden eagle had been caught in a leghold near the Crowsnest Pass in 1929 prompted the *Coleman Journal* to ask its readers to "picture this noble bird, caught without a chance for its life, and you'll agree that the trap method of catching animals or birds is cruel indeed. Give them an even break, as all real sportsmen will."[14] Although trapping as a form of harvesting of course had entirely different objectives than sport hunting, the application of sport hunting ethics to leghold traps signalled emerging concerns about the cruelty inherent in their use.

The most direct way to put an end to the cruelty of the leghold trap was, of course, to convince people not to wear fur. As one reviewer in the American journal *Ecology* suggested in 1923, the extermination of some fur-bearers "can be prevented only through the creation of a sound sentiment and the education of the children and women, in the same manner in which the Audubon Societies have arrested the destruction of birds."[15] Another critic argued in a 1940 letter in the *Western Producer*'s Nature Lovers page that trapping was morally objectionable in principle and cruel as well because animals were killed only to serve the needs of fashion. Yet such thinking sat poorly with another correspondent who believed that trapping had built Canada and was no crueler than many other activities and "this world is not the bed of roses that we'd like it to be, either for man or beast. The theory 'Live and Let Live' isn't a practical or a practised one. Man has matched human wisdom against animal cunning and so far as I can see trapping is an even combat."[16] This letter seems to have reflected a widely held view, and since a ban on trapping was socially and politically impossible, efforts instead became more focused on humane trapping.

Humane trapping was promoted by the Association for the Protection of Fur-Bearing Animals, which was established in the mid-1940s. This association appears to have been quickly absorbed into the Canadian Association for Humane Trapping, whose goal was to "abolish the cruelty of the steel trap."[17] Humane societies endorsed this goal, and broader public discourse in the 1940s also reflected a greater awareness about humane methods of capturing fur-bearers. In a comment that would have been unusual in earlier years, an official with the Department of Indian Affairs explained to a correspondent in 1944 that muskrats and other fur-bearers in Canada were "caught, rather cruelly, in a steel trap. The trap must be frequently visited in order that the [musk]rat is put out of his misery as early as possible." Other critics were more categorical about this matter. In 1950 Tillie Richardson of Smeaton in northern Saskatchewan responded to an article in the *Western Producer* that had suggested that trappers "made the country safe from wild animals" and that animals caught in leghold traps only suffered for a short while. She had no patience with such claims, arguing that human sentience and animal sentience were parallel, if not identical. She asked, "[I]f you get your finger crushed or flesh torn does not a few minutes seem like hours? And what of the poor maimed animals?" By identifying herself as "we the anti-trap minded," she voiced the emerging sentiment of those who wanted "a law to abolish all unnecessary torture of any flesh." This did not, however, mean that killing coyotes or other wild animals that challenged human needs was wrong, but that "we are in favor of doing it as quickly and painlessly as possible."[18]

While Richardson was part of a minority, she was at the forefront of growing public discomfort about leghold traps. Yet, there was no real alternative other than to prohibit trapping altogether; alternate methods of

Figure 10.1 Ending the cruelty of the leghold trap became a priority for a growing number of people in the 1940s. In this 1945 photograph, a trapper in northern Manitoba holds up a leghold trap in which a muskrat had been caught.

trapping such as snares were even worse, and shooting was not a feasible option given the way that fur quality was judged and marketed. In the late 1940s, however, the dynamics of the debate were beginning to change because of the invention of a new type of body-grip trap. Frank Conibear, a trapper in Fort Smith, Northwest Territories, had observed that many animals escaped from legholds. Most importantly, he also saw that many of them chewed off their foot or leg to do so, which he found "inhumane and disturbing." As a trapper, Conibear recognized that a more humane trap had to be not only inexpensive but also lightweight, as trappers carried a lot of weight in the bush. He had developed a prototype of this new type of trap in 1929, but other events intervened and it was not until the 1940s that he designed another model. It was a square trap, "constructed with spring-powered steel jaws that snapped with tremendous force over the neck or chest of the animal as it passed through the square hoops, resulting in instant death." In 1958, in his nature column in the *Western Producer*, Doug Gilroy found reason to celebrate the appearance of the Conibear trap: "[T]here has always been much said of the cruelty involved in trapping wild animals in steel traps – the suffering they must endure while being held by a smashed leg or foot." In contrast, the Conibear trap killed instantly and was the "first completely revolutionary trap developed since the history of trapping began," and Gilroy hoped that trappers would adopt it.[19] By this time, the Canadian Association for Humane Trapping had financed the manufacture of fifty Conibear traps so they could be tested for commercial production, and the association also provided funds to promote their general adoption. It specifically promoted their use among trappers, although its attempt to approach Manitoba trappers directly was stymied when the provincial government claimed that it did not have usable lists of registered trappers that the association could consult. The traps were welcomed by other opponents of leghold trapping, and Conibear received an award for his invention from the American Humane Association in 1961. Critics of leghold traps had gained many supporters by the 1960s, and in response to growing public concern, the Manitoba government sponsored research on humane trapping methods in 1974.[20] By this time, the Saskatchewan government had conducted its own study, which showed that Conibear traps were the "most humane and effective" traps for small fur-bearing animals such as beavers, muskrat, mink, and squirrels. While a suitable humane trap was not available for large animals such as coyotes, foxes, or lynx, the study observed that Conibear traps killed instantly and caused "no damage to the pelt" when they were properly used for the appropriate size of animal. Although the Saskatchewan government was not willing to mandate humane trapping, it encouraged it, and in co-operation with the Canadian Association for Humane Trapping, it initiated a trap exchange program that enabled trappers to exchange their legholds for Conibear traps. The Saskatchewan government also promoted use of the Conibear by selling them at wholesale cost to trappers in the province.[21] While tentative, such developments revealed changing views about the human encounter with fur-bearers that had been emerging since the 1940s.

SAVING THE NATIONAL SYMBOL

The conservation of beavers was highly successful in contrast to that of other wild fur-bearers. Fearing their extinction, Manitoba banned beaver trapping in 1898, the NWT government did so in 1902, and the NWT's provision was carried over into legislation in Alberta and Saskatchewan at provincehood. While these bans were periodically lifted and certain districts – especially northern trapping ones – were sometimes exempted for economic reasons, permission to kill beavers was the exception and not the rule. In what was a model for the proactive preservation of a species, beaver conservation was achieved by balancing various economic interests with the need to ensure the survival of the species, demonstrating that depletion could be checked and controlled when there was sufficient political and administrative will.

Beavers live in colonies in areas where willow, birch, or aspen poplar grow alongside slow-moving streams that can be easily dammed to form ponds in which food can be stored and lodges constructed. Beavers were found in all parts of the prairie region except for the driest high plains, although they could be found in a few favoured spots there as well. People have long had a tradition of respect for and connection with beavers. Some Chipewyan people believed that beavers had been the rulers of the world and that humans had been their slaves until they finally became intelligent enough to throw off the beavers' tyranny.[22] Many non-Aboriginal people also held them in high regard and admired their social skills and intelligence. After observing beavers for three decades, Frank Conibear, for example, concluded that it was remarkable that the "basic domestic ways of beaver should so closely resemble those of mankind. And even more remarkable that in all probability the beaver reached their present peak of domesticity, involving the building of a dam, assembling of food and system of storage, the building of lodges with dormitories, nurseries, a separate room for eating, and even a privy, while man still lived in rude caves and slept on the ground."[23] Their ubiquity, historical economic importance, and character made them a national symbol, and no government was prepared to endure the humiliation of having this national symbol of industriousness (and, it can be observed, one with the ability to alter natural environments to meet its needs) become extinct under its watch.

The Englishman Archie Belaney, who posed as an Aboriginal named Grey Owl, was one of the most ardent advocates for beavers during the interwar years and tapped into national myths about them in his writing and lectures. His goal was double-edged; he wanted to preserve beavers from extinction as well as to stop the degradation of the environment from the constant expansion of mining, logging, trapping, and farm settlement. In his biography of Grey Owl, historian Donald Smith notes that Grey Owl used beavers as "the symbol of his crusade" for the preservation of Canada's wildlife and environment. Grey Owl contended that the environment and its wild animals had an inherent right to exist as "a living breathing reality," and he used, somewhat shamelessly, his tamed beavers, Jellyroll and Rawhide, for the greater cause of environmental protection. Grey Owl and his beavers were objects of immense public fascination that grew out of nostalgia for a heroic past. This fascination also drew strength from dreams of escape from the escalating economic crises of the interwar years.[24] His message was that white men were destroying the land and that, by killing beavers, they were not just exterminating a species but also setting in motion "a whole train of disastrous consequences. It was the beaver who created the ponds which nourished the forest, which in turn provided habitation for all the other creatures." But Grey Owl's relationship with beavers was also personal and direct. He loved them and treated them with more respect than he showed for many of the people in his life. Beavers were intelligent and had admirable social skills and an outstanding work ethic. As he told one school class, "Canada's National animal" had many enemies, such as otters, coyotes, and lynx, but among their enemies, humans were in a special class, one that killed thousands of them each year. By letting people see what beavers were "really like, how friendly, sociable, affectionate and industrious they are," Grey Owl hoped to make them "better liked and understood so they will not be killed in such large numbers."[25]

While such sentiments spoke of an enduring emotional connection with beavers, other more practical implications of beaver conservation had significant appeal. In 1887 the *Regina Leader* reprinted and endorsed an article from the *Edmonton Bulletin* that predicted that beavers would be extinct by the end of the century if the "alarming slaughter" did not end. The beaver's major enemies were humans, and, as in other parts of the country, beaver trapping had increased so much in northern Alberta that it was likely that they would be cleaned out "as completely as the buffalo have been and with greater loss to the country." The nub of this argument was that "although the beaver trade is not as valuable as that of the buffalo in its best days, the buffalo country can be turned to other and more valuable account, while a great part of the country which now yields hundreds of thousands of dollars worth of beaver annually, would if that produce were exhausted, yield next to nothing." Beavers were largely gone from the Edmonton district because of farm settlement and trapping, but vast areas, "empires in extent," that stretched far north and formed part of Edmonton's economic hinterland had "always been above all else beaver countries." The *Bulletin* urged a closed season in summer and the punishment of any trader buying unprime pelts. Eliminating a market for unprime pelts would be effective because "the Indian – unlike the white sportsman – does not kill for the mere sake of killing" and, without a market inducement, would hunt few beavers.[26]

Beavers, moreover, were understood to be valuable environmental stewards. It was commonly noted that beaver ponds created habitat for other animals and plants, while active and abandoned beaver ponds created firebreaks that reduced the intensity of forest fires. Beavers thus merited the most rigorous protection. As one dominion lands agent wrote in 1899, in comparison to the value of timber destroyed by fire, the value of beaver fur was "the merest bagatelle." The drought of the 1930s emphasized the importance of water conservation and confirmed the benefit of beaver conservation. Thomas Pugh, a field officer with the provincial Department of Natural Resources at Meadow Lake in northwest Saskatchewan, wrote in 1939 that he was convinced that the drought was intensified by land drainage and the disappearance of beaver ponds that had stored water and cooled the land. He believed that restoring beaver populations and protecting them in "this bush country" would solve many of the problems the province was confronting in conserving waterfowl, fur-bearing animals, game, and timber.[27] This widely held view was repeated in 1946 by an Alberta government official who noted that while their fur was valuable, beavers offered "a far greater value to man alive than dead." Fire was the greatest environmental threat in Alberta, and the "protection of the forest from fire must, therefore, always be regarded as the first and greatest of responsibilities in the life of this vast land." As water managers, beavers served this purpose, but their ponds had additional environmental benefits because they helped to prevent erosion by slowing spring runoff, offered habitat for fish, waterfowl, and big game, and became lush meadows when abandoned.[28]

Despite their many valuable attributes, the protection of beavers was not free of conflict. Beavers are great wanderers, and as Grey Owl phrased it, the beaver "is the imperialist of the animal world. He maintains a home and hearth, and from it he sends out every year a pair of emigrants who search far and wide for new fields to conquer; who explore, discover, occupy, and improve, to the benefit of all concerned." Not everyone agreed that the beaver's industriousness was beneficial, and as Gordon McDonald, a district game guardian in Saskatchewan, noted in 1915, "these original engineers are certainly detrimental to agriculture."[29] Beaver dams flooded farm pastures and croplands and reduced water flow for livestock in downstream areas. Livestock also became mired in beaver ponds or drowned in winter when they broke through the thin ice that formed on these ponds. Being herbivores that required fairly extensive tracts of land, beavers also destroyed valuable timber, fruit, and ornamental trees. In southern Manitoba in 1916, for example, beavers were "so daring" that they came into villages and cut down shade and fruit trees, and in Rapid City and Virden (both near Brandon) they even destroyed trees on a nursery farm. After only about fifteen years of total protection, they were appearing in parts of southern Saskatchewan where they had not been seen for years. Community reaction typically went from a warm welcome to hostility. As Saskatchewan's Department of Agriculture noted in 1916, when beavers first appeared in a district, they were "received very cordially" and were "an interesting topic of conversation for miles around." "Their industrious engineering feats" never failed to attract public attention, and people took "great pride in having such interesting animals in their districts." But trouble started when beaver dams flooded fields and the farmer decided "that after all he cannot afford to allow these gnawing rodents to usurp so much of his valuable land and expose his stock to the danger of being mired."[30]

The only recourse for farmers who were facing beaver damage but who wanted to live within the law was to demand remediation by provincial game authorities. While beaver protection was rigorously enforced, it was never pursued at the long-term cost of farmers. As Valentine Winkler, the minister in charge of Manitoba's game commission, observed in 1919, the province upheld farmers' property rights and "always" granted relief for beaver damage. Although each of the three provinces had slightly different approaches, the usual first step was to break open the dams so that the ponds would drain and encourage the beavers to move to a different location. If this failed – as it often did – the farmer could obtain a permit to trap the animals on his land. He was then entitled to keep the pelt but was usually required to sell it to the provincial government and pay a handling and processing fee. This naturally opened up an opportunity for people to earn extra cash, and while many complaints about damage by beavers were merited, it was not uncommon for beavers to be blamed for damage they did not cause. As was suggested in 1911 in Saskatchewan, farmers found it tempting to gain "a free hand" to trap beavers legally and to sell the pelts.[31]

Because beavers are prolific, it took little time for populations to rebound after the trapping bans of the late 1800s and early 1900s. They once again became common in many parts of the region, allowing governments to amend their trapping bans. In 1917 the Saskatchewan game branch declared an open season on beavers to counter growing complaints about the damage they caused. While the branch claimed that there were "many good reasons, both sentimental and economical" for beaver conservation, it noted that "all will agree that property rights should have the first consideration." It was soon apparent, however, that beavers were being trapped out even in districts where they were not a nuisance. Despite the protests of the game commissioner, the open season was maintained until 1922 when it was finally admitted that beavers had once again been almost entirely extirpated in many parts of the

Figure 10.2 Since killing beavers was strictly controlled, people attempted to mitigate any damage beavers caused to property. This often involved destroying their dams to persuade the beavers to move away, as shown in this 1950 photograph from Alberta.

province. Indeed, the future looked so bleak that one trapper told the game guardian responsible for the northern part of the province that he had collected a piece of poplar cut by a beaver so that he could someday show his children how the beaver procured food, since "that would be all the children would see of the beaver unless they were protected now."[32]

With the abolition of open seasons and the restoration of the permit system in Saskatchewan in 1922, nuisance beavers could again only be trapped in certain circumstances and beaver populations resumed their expansion. By 1938 Saskatchewan was trying to deal with public complaints by live-trapping nuisance animals and transplanting them to areas where they would not interfere with farming.[33] In conjunction with permit trapping, Manitoba and Alberta launched similar transplanting programs in the next few years. Beavers were also live-trapped in the national parks where populations had exploded because of total protection from trapping, and in co-operation with provincial authorities, these beavers were moved to suitable provincial lands. The Department of Indian Affairs participated in these live-trapping programs to establish fur-bearers in depleted areas in order to create economic opportunities for northern Aboriginal communities. Similarly, Saskatchewan and Manitoba made live trapping part of their northern development policies. In Saskatchewan, beavers were transplanted to fringe areas as well as to places further north, such as Île-à-la-Crosse and along the Churchill River where they had been trapped out. In 1946 it was said that Aboriginal teenagers in some northern districts, which a century before had been famous fur-producing regions, had "never in their lives seen a live beaver."[34] Typically, closed seasons were imposed to protect the beavers until populations had grown sufficiently to allow trapping. Ducks Unlimited (Canada) also supported beaver transplanting, arguing that the collapse of beaver populations in northern regions had led to a major environmental impact in the form of less water, excessive runoff, erosion, and forest fires that, in turn, had brought declining soil fertility and loss of waterfowl, big game, and fish.[35]

While these live-trapping programs involved a relatively small number of animals at first, 3,500 "seed" beavers were transplanted to Saskatchewan's fringe areas and further north in the decade after 1945. These undertakings had strong public appeal, but they were especially well received in northern communities, where they brought an emotional reconnection with the past for some people. In 1945 six beavers were taken by aircraft to Cross Lake in northern Manitoba, where there had been no beavers for many years. Phil Reader, the game guardian who took them north, reported that "on arrival at Cross Lake, we received the kind of welcome generally accorded movie stars. As we circled the post, a red

Figure 10.3 Live-trapping beavers controlled them in populated districts and helped re-establish populations to assist with water control and increase fur production. Here, a live-trapped beaver in the Edmonton district in 1950 is ready to be moved to a new home.

cloth was frantically waved and the dock was crowded as we taxied in." When the beavers were unloaded, "the older Indians were really overcome, as I don't think they ever believed such a miracle could happen. However, they had held a meeting the previous night and decided on the planting locations. These locations seem quite satisfactory, all being near to Cross Lake to allow close protection," even though it was "very doubtful if we need to worry on that score. These beaver are sacred, they even came from the sky, and mark a turning point in the Indians' life."[36]

Live trapping succeeded in re-establishing beaver populations in northern districts. Only 473 beavers had been trapped in northern Saskatchewan in 1944, but populations grew so quickly because of the proliferation of the protected seed populations that 34,000 pelts were taken in 1954.[37] Beaver populations also continued to expand in southern districts. In 1943, for example, they were increasing so quickly in some districts, such as Cypress Hills, that they could not be controlled by live trapping. Similarly, by 1949 applications for special permits in Manitoba had grown to the point where government officials could not keep up with the demand. Even though it was costly, as many beavers as possible were still live-trapped and moved to northern districts, but complaints continued to mount.[38] By then, the theory of carrying capacity was being applied to beavers, as it was in deer management strategies. In the mid-1950s a Saskatchewan government pamphlet about beaver conservation argued that an (unidentified) case on the eastern slopes of the Rocky Mountains in the United States showed that overprotection could be counterproductive. It claimed that the beaver ponds and thinned stands of trees at first "turned the foothills into a splendid natural garden where all forms of 'beneficial' wildlife thrived and all the 'harmful' predators were reduced to a minimum." But "nature had a cruel trick in store for those who tried to deny her law that the creatures of the wild must kill and be killed." Soon the beavers destroyed all the vegetation, bringing disease, flooding, and erosion. "Just a couple of summers of heedless over-protection had turned the gentle useful beaver population into a starving mob capable of making a paradise into a desert in a few short months." If Saskatchewan did not manage and harvest its beaver populations appropriately, the pamphlet's authors warned that the same would happen there. This reasoning also took hold in Alberta. The province began to allow trapping in 1951 because its game branch could no longer meet the demands for live trapping. By 1960 it was employing trappers to cull beaver populations in some areas and was studying the value of using repellents that could be sprayed to keep beavers away from certain areas.[39] This combination of deterrent and lethal methods was consistent with a century of attempts to accommodate beaver conservation to agrarian settlement and spoke to old relationships and ambiguities in the encounter with this mythic animal.

RAISING WILD FUR-BEARERS

Trapping wild animals was labour-intensive and production was uneven. Bad weather could prevent trappers from getting into the field, and predators often killed fur-bearers or ate the ones caught in traps before trappers could get to them. With the exception of beavers, whose cycles are longer and less pronounced, fur-bearers have sharp and relatively frequent population cycles. Beavers have thus historically offered a fairly reliable income base for trappers because they could be trapped when other animals were scarce. But as beavers disappeared, trappers were forced to rely on animals with pronounced cycles, such as muskrat and lynx. This boom and bust tied the lives of trappers ever more tightly to the population cycles of wild animals. In 1906, for example, rabbits and their predators were plentiful and trappers reportedly brought about 5,000 lynx pelts to sell in the town of Athabasca, north of Edmonton. Three years later, the rabbit population and the lynx population along with it had crashed and only 10 lynx pelts were brought in.[40] Such difficulties led people to treat fur production as a type of farming. Raising fur-bearing animals in cages began in the late nineteenth century in North America, and raising foxes on farms was pioneered in Canada in Prince Edward Island in the 1890s. By the First World War, fur farms had appeared in all three Prairie provinces, and as in the rest of North America, they almost exclusively raised mink and foxes in cages. It was less expensive to raise fur on a farm than to trap it in the wild, and economies of scale made it profitable to raise prolific animals such as mink. Population cycles were not a problem with farmed mink and foxes, and both could be bred to create distinctive market niches. Although farmed mink usually sold for less than wild fur, sophisticated breeding created colour mutations that did not occur in the wild, and these could be marketed as new fashion trends at higher prices. In the 1930s and 1940s, for example, Alberta mink breeders were finding market success with pastels and whites, while Manitoba mink farmers in the early 1960s were noted for producing sapphire and blue mink.[41]

Canada's climate, land, and resources offered important advantages for fur farming. Motivated by hopes of diversifying the economy, prairie governments provided incentives to the industry by sponsoring research at provincial fur experimental stations to improve husbandry and breeding stock. They also encouraged fur farmers to feed farmed mink and foxes with fish that was not wanted for human consumption, which gave "coarse fish" such as suckers a commercial value. For example, in 1939, in response to a petition by fur farmers, governments extended the season for catching rough fish on Lesser Slave Lake in northern Alberta by two weeks, and the minister of agriculture noted that the province "hoped to allow as much latitude" as was compatible with fish conservation. By 1943–44, Alberta mink farmers, most of whom were located in the northern part of the province, used about 1.4 million kilograms of such fish, which the province welcomed because this gave market value to all fish and there was now no "wastage." Showing a general pattern, most Saskatchewan fur farms were also located in northern areas near fish supplies.[42] Fur farmers only paid an annual licence fee, and domestic fur was exempt from fur royalties. Not surprisingly, the combination of government incentives and the depletion of wild fur-bearing animals led to the rapid growth of fur farming in the interwar years. While fur-farming production had been negligible ten years before, 23 per cent of all income from fur in Manitoba came from fur farms in 1936. And the sector continued to expand; by 1960, farmed fur made up 64 per cent of the raw fur sold in Manitoba, and ranched mink made up over half of the fur sold in Canada by value.[43]

The popularity of fur farming in the interwar years led to efforts in different parts of North America to expand fur farming beyond mink and foxes to counter the general depletion of wild fur-bearers. In 1933 the Manitoba game branch advised that closed seasons should be set "on all wild fur bearers showing sharp decline" until it was shown "that they can

Figure 10.4 Fur farms on the prairies primarily raised mink and foxes. The annual fur show in Edmonton in the late 1940s was a chance for fur farmers to compete for prizes, sell pelts, and network with colleagues and industry figures. This woman was modelling prize-winning platinum fox pelts at the 1948 show.

be raised consistently and profitably in captivity."[44] Most of this specialized interest on the prairies was directed at marten and fishers, both high-value and increasingly rare animals. It was noted in 1934 that if these animals "could be raised on the same scale as mink and fox," Manitoba would substantially increase the value of its annual fur production.[45] By the early 1930s the Manitoba Experimental Fur Farm was trying to breed strains of marten and fishers that could survive in captivity. Overtrapping was a problem with both animals, as fishers bred slowly and marten were threatened because they had "little natural instinct for self-preservation" in the face of forest fires. In both cases, the fur farm appeared to be "the best refuge" for their continued existence in Manitoba.[46] While trials with fishers were abandoned relatively quickly, experiments with marten lasted until the 1940s but were not successful.[47]

Although the fur farm industry often depended on commercially captive-bred animals, it nonetheless had a direct impact on wild animal populations. At Buffalo Narrows in northern Saskatchewan, domesticated mink escaped from fur farms in the 1940s and interbred with local wild ones, "leading to an amazing array of mink pelt colors" among the wild population.[48] While references to similar occurrences are rare, it would be

surprising if this did not happen elsewhere as well. A more significant impact on prairie wild fur-bearers came from the frequent use of wild foxes as farm stock. Fox pups were dug out of their dens in the spring (on the prairies they are usually born in late April) and were sold to local fur farms as well as eastern Canadian and American ones. The pups were raised in cages for around eight months and were then pelted, although some were kept for breeding and domestication trials. Trade in live pups was often brisk. For example, prices for live pups temporarily fell in 1913 in Edmonton when the market was flooded with pups that had been captured in northern areas.[49] But the market soon firmed, and pup hunting then "reached almost a boom stage," driven by high prices and the precedent of the Prince Edward Island fox industry, which was reputed to have made some Islanders rich with little effort.[50]

Mortality of caged wild fox pups was high because of poor care and farmers' lack of knowledge about raising them. In 1914 there were 148 "fur farms" with 803 foxes in Alberta, which suggests that many of these "farms" consisted of little more than a few animals penned in the farmyard. About 30 per cent of these animals were captured from the wild, while the rest were farmed breeds.[51] Because different colours can occur within a single litter of foxes, the fox trade became something of a lottery. Fox pelts of rare colour could make a trapper's winter very profitable; in Fort McMurray in 1910 (a year of low prices) such pelts brought an average of $75 each at a time when a muskrat pelt sold for 25 cents.[52] Such prospects animated pup hunters, and reports of the fabulous prices that were paid for rare crosses fuelled the pup-hunt mentality. Fox farmers naturally wanted only highly prized colours, and in the early 1920s they purchased "only silver, black and best crosses" from the pup hunters who threw away all others, including reds, which were most common. It was a harsh business. The writer Gill Shark went pup hunting as a boy and later in life recalled how on one occasion a male fox watched with "mute anguish" as Shark dug out the den. Later, Shark watched him go "over to the mound of earth, scraping hopefully with his forepaw as if he could not believe the whelps were gone. His mate followed me most of the way home." At the time, Shark looked forward to the coming winter when he might be able to trap the adults as well. He raised the three pups and sold their pelts, but he never caught the adults, which had moved away to "escape the relentless pursuit of men." But, as Shark recalled, "that was a good while ago, and it seems that through the years" a new "understanding had come to me concerning the great wilderness heritage that is ours. In those green years of my youth, no thrill could have equalled the digging up of that little fox family and the trapping of the older ones. Now, each time I recall it, it is always with a little keener regret."[53]

MANAGING HABITATS FOR MUSKRATS

For some observers, raising fur-bearers in captivity was only another chapter in a long story of human–animal relations. In 1927 the *Western Producer* claimed that the domestication of animals had slowed dramatically in the previous 1,000 years, and it contended that it was time to resume the "exploitation of dumb animals by human bipeds" by turning people's attention to muskrats.[54] Muskrats were suitable candidates because their pelts were in high demand and they were thought to be prolific and non-threatening and, unlike beavers, did not cause trouble for farmers or landowners. They were also thought by some to offer value in habitat management by keeping down coarse marsh rushes, which allowed wild rice to grow and provide food for wild birds. While there were complaints that muskrats raided gardens, their advocates claimed that this was un-

Figure 10.5 Muskrat ranching was popular during the interwar years. Small operations, such as this one in Alberta, were superseded by much larger and more complex government-sponsored undertakings in northern Manitoba and Saskatchewan.

usual because they rarely left water environments.[55] A number of efforts were made on the prairies to raise muskrats in captivity. Henry Tadei, a tinsmith in the town of Radisson northwest of Saskatoon, for example, started raising muskrats around the end of the First World War. He began by live-trapping muskrats and kept them in pens where he hoped they would breed. However, things quickly went wrong. Tadei knew almost nothing about muskrats; the penned animals froze to death in winter and were scalded by the sun in summer, and the older males sometimes killed the other muskrats in their pen. Even so, Tadei persisted and soon concluded that he had to alter his approach because he could not change the muskrats' behaviour.[56]

Muskrats require marshy land with relatively stable water levels about 1 to 1.5 metres deep. This water must not freeze to the bottom so that the muskrats can find food in the winter. As protection against predators and the weather, they usually construct lodges out of cattails and other vegetation, although they also dig riverbank burrows in some cases. Tadei met many of these needs when he transferred his operation to a nearby slough. It had a constant water level that did not freeze to the bottom in winter, it had intake and outlet streams that kept the water fresh, and it had plentiful and suitable food. He stocked the slough with muskrats that he live-trapped and fenced the area to keep out predators and prevent the muskrats from wandering. As one newspaper reporter commented, Tadei was raising muskrats "practically in the wild." But Tadei knew little about muskrat population cycles (their cycle can be up to ten years long) or their reproduction (at six months of age they can begin producing litters of four to nine kits up to three times a year in a cold climate such as Saskatchewan's), nor did he understand the carrying capacity of land for muskrats (which can reach a density as high as fifteen to eighteen animals per acre if there is no predation or trapping). Tadei learned only through observation and trial, admitting that he made mistakes at first because he "had no examples to follow." Nonetheless, he found a way to raise them successfully; the muskrats on his farm increased rapidly to about 2,200 animals, and he began pelting them about three years after starting his operation.[57]

Living and working in a natural environment was a popular ideal in the interwar years. Caught up in a romantic fantasy about wilderness as well as in a quest for solitude, a retreat from the world, and a personal engagement with wildlife, some Canadians dreamed of having a beaver or muskrat "farm." In the 1920s the Department of the Interior received hundreds of requests from would-be fur farmers for leases of swampy Crown lands in northern Canada where they could raise muskrats and other fur-bearers.[58] While not unsympathetic in principle – such projects would make non-arable land economically productive – government officials suspected that some applicants simply wanted to make an already well-stocked natural area into a private trapping ground. As well, officials were concerned that this might lead to exclusive and monopolistic conditions that could become precedents for challenging public access rights to wildlife on Crown land. Complicating things further, these requests were entangled in competing federal-provincial interests because natural resources on the prairies were still under federal control.[59] By the late 1920s, however, federal and provincial officials had worked out policies and procedures to permit the leasing of Crown land, and a number of muskrat habitat projects were soon established in the prairies.[60] On the basis of evidence that the breeding of muskrats in captivity was unworkable, it was concluded that such projects had to manage natural habitats in order to be successful. This was further confirmed in the late 1930s and 1940s by growing evidence that wildlife management needed to calibrate human activity with animal rates of reproduction and biology in conjunction with habitat conservation. These projects formed a part of broader northern economic development initiatives that in part were stimulated by the province-building possibilities created by the NRTA. As Manitoba's deputy minister of natural resources noted in 1935, the province was "not unmindful of the resources of the North country" and had given "much thought and consideration" to the development of fur, fish, timber, and minerals. However, until responsibility for natural resources was transferred to the provinces, this lack of jurisdiction had been somewhat of an impediment, and, equally important, weak transportation networks, the competing priority of southern development, and small, scattered, and impoverished populations in the north, as well as uneven markets for fur and fish, had held them back.[61]

One of the first muskrat habitat projects was set up in 1932 in northern Manitoba in the Saskatchewan River Delta at Moose Lake (east of The Pas) by Tom Lamb, a local fur trader and businessman. Lamb knew that while muskrat food was abundant in the Saskatchewan River delta, low water levels meant that the animals would not survive the winters. In the area he surveyed in 1931, Lamb found only 40 muskrat lodges containing about 300 animals, but he speculated that if water levels and trapping were controlled, populations would rebound. The next year, he secured a lease on 54,000 acres of delta marshlands from the federal government, and with the supervision and technical assistance of the Manitoba Department of Mines and Natural Resources, he built dams and dykes to control water levels in the marsh. By 1934 there were an estimated 65,000 muskrats on Lamb's lease, and the next year he trapped almost 24,000 animals in an area that four years earlier had been almost totally barren. Luck was also on his side; prices were high, and he sold his furs for an average of $1.45 a pelt.[62] In a parallel development, the HBC undertook a similar project at Steeprock, west of Lake Winnipegosis near the Saskatchewan border. Once a rich muskrat-producing district, it had been entirely trapped out by the early 1930s, and falling water levels ensured that muskrats could not re-establish themselves. In about 1932, an area of about 2,000 acres was set aside and a dam was built to raise water levels. While the HBC officially called it a "muskrat ranch," it was, like Tom Lamb's operation, a habitat management project and not a farm. By 1937 limited harvesting had begun, and the next year about 3,500 pelts were taken. A similar project was set up at Cumberland House in northern Saskatchewan. The HBC had established its first inland post there in 1774, and it had been an important trade centre at the heart of a rich muskrat-trapping region for over 150 years. The drought of the 1930s, however, lowered water levels in the Saskatchewan River, exacerbating ongoing geological changes that were causing the river to cut a new channel. Both factors reduced water in the Cumberland marshes and dramatically reduced muskrat populations that were already low because of over-trapping. The marshes caught fire in the dry summers, killing many animals, and froze to the bottom in winter, killing yet more. In response, the Saskatchewan government and the HBC established a community lease in 1936 on thirty-two townships, which together essentially served as a trapping preserve on which trapping was restricted to allow the muskrat population to increase. In 1938 a 300,000-acre parcel was removed from this community lease and granted as a private one to the HBC. By cutting new channels in the Saskatchewan River, the company brought water to re-flood the marshes and built rock-cribbed dams on tributary rivers to control water levels.[63] This infrastructure created conditions that restored muskrat trapping as an element in the community's economy.

These projects, especially the HBC Steeprock project and Lamb's project at Moose Lake, were models for more extensive ones subsequently established by the Manitoba government. They had proved that if suitable food and sufficient water were available, a substantial increase in muskrats would follow, leading the Manitoba government to anticipate that such projects would create jobs in northern and other areas and increase fur royalties from areas that were "totally unfit for any other purpose." In 1947 Manitoba premier Stuart Garson told the historian and social planner Frank Underhill that "the doctrinaire free enterpriser" doubtless saw the muskrat habitat projects undertaken by the Manitoba government as "paternalistic, communistic, totalitarian regimentation," but he believed that they helped rationalize the fur industry, maximized the province's fur wealth, and distributed its wealth more equitably.[64] Moreover, the projects coincided with increasingly intense interest in marshland management in North America. Drought conditions in the 1920s and especially the 1930s stimulated concern about water management, and demands by hunters and organizations such as DUC for habitat protection for waterfowl strengthened support for the marshland rehabilitation that the muskrat programs required.[65] While Steeprock had been called a "ranch," it was one of the last projects referred to in that way; instead, "preserve" and "habitat rehabilitation area" were increasingly used, indicating the reorientation that was taking place. Wild muskrats were now to be managed as units of production, not only as a part of habitat rehabilitation, but as a part of an integrated wildlife management system that would create new economic capacity by increasing animal populations through scientific planning and centralized management and marketing.

Both Saskatchewan and Manitoba established muskrat habitat rehabilitation projects after 1935, and the variety, longevity, and scale of Manitoba's projects illustrate the operation of such state-run muskrat habitat projects on the prairies.[66] The first Manitoba project, initially known as the Saskatchewan River Delta project and later as the Summerberry Project, initially consisted of several discrete areas on the shores of Cedar Lake (just north of Lake Winnipegosis) totalling about 135,000 acres. Construction began in 1935, and the addition of more districts, including a large area southwest of The Pas in the Pasquia and Saskeram districts, increased its size to about 1.25 million acres by 1954. Water levels were stabilized by the construction of dikes and dams, and ditches were dug to break up solid stands of vegetation to create open water.[67] The construction phase was completed in 1936, and trapping was banned for the next four years to allow the muskrats to reproduce. The first crop, totalling 126,600 muskrats, was taken in the spring of 1940 and sold for $160,000. By then, other projects had been started and by 1944 included, among others, Netley Marsh, a 45,000-acre project at the south end of Lake Winnipeg; Delta Marsh, a 35,000-acre project at the south end of Lake Manitoba; and the Fisher River Fur Development Project on Lake Winnipeg, which was 520,000 acres in size. By 1954 Manitoba's entire fur rehabilitation system totalled over 2 million acres.[68] All the projects (like those in Saskatchewan) had multiple partners. Given its commitment to rehabilitating marshland to increase waterfowl for sport hunters, DUC gladly partnered in some of the projects in the Cumberland House area in Saskatchewan as well as in those at the Netley and Delta marshes. It also constructed the main dam and contributed to other costs in the

Pasquia and Saskeram district projects (which were administered as part of the Summerberry network) to support their development as waterfowl areas.[69] This involvement was logical; neither activity interfered with the other because duck hunting took place in the fall and muskrats were usually trapped from early winter to spring. Local governments were also involved in some projects, such as the Big Grass Marsh project near Gladstone, northwest of Winnipeg, where 30,000 acres had been drained between 1909 and 1916 as an agricultural reclamation project. This land proved to be unsuitable for farming and by the 1930s was a wasteland of eroded boggy land. Through an alliance among DUC, local municipalities, and the provincial government, the area was rehabilitated as marshland for muskrats and waterfowl.[70]

Despite the involvement of DUC and other agencies, the federal government was the most important collaborator in these projects. Most of the muskrat habitat projects in Manitoba and Saskatchewan were carried out under ten-year federal-provincial agreements, the first of which was signed with Manitoba in 1936. The federal government paid engineering and construction costs, which, for example, amounted to $90,000 in the first phase of Summerberry, while the province administered and maintained the projects. Ottawa hoped that the projects would provide income for First Nations and reduce its long-term financial responsibilities for them. Construction of water-control and other infrastructure provided work relief schemes, and in the operating phase, the projects offered First Nations people the chance to earn income from trapping and the ongoing maintenance of the marshes. The federal government insisted that local First Nations have priority to trap in the rehabilitation areas, and the governments of Manitoba and Saskatchewan agreed. Alberta, out of its commitment to equal public rights of access to wildlife, refused any stipulations that gave First Nations priority in trapping. As a result, no federal-provincial muskrat habitat projects were undertaken in Alberta.[71]

These rehabilitation projects brought economic stimulus in northern Manitoba and Saskatchewan, and the federal government calculated that its initial costs for the Summerberry project were offset by the fact that "1,800 families then on relief have been restored to a self-sustaining situation and many more assisted to a higher standard of living."[72] Provincial officials identified who could trap on the basis of need and location so as to spread the benefits among all the communities in the fur preserve. While priority was given to local First Nations men (women were not given the right to trap), others (some of whom no doubt were Métis) also received trapping permits. In the first trapping season at Summerberry, 753 trappers participated and each was assigned a quota of pelts that he could take. Together, the Manitoba projects produced an estimated $4 million in revenue between 1940, when the first crop was taken at Summerberry, and 1947. In this seven-year period, just over 1.7 million muskrats were harvested in all of the Manitoba projects. A significant portion of the total revenue generated left the local communities in the form of auction fees (totalling almost $143,000), provincial government fur royalties and administrative charges (amounting to just over $868,000), and fees to merchants and landowners (totalling almost $394,000). Nonetheless, almost $2.6 million was paid to trappers.[73]

These muskrat habitat projects were tightly controlled and managed. Typical month-to-month activities in 1946 included regular patrols of the marshes to prevent poaching (a particular problem at the Fisher River project). Changes in water and food conditions were also recorded to help plan the harvest, and an annual fall survey was conducted to calculate probable winter survival rates given existing food and water conditions. When it was judged that winter mortality would be high, trapping in the fall was allowed to harvest the number of animals that it was anticipated would die over the winter. Specimens were also regularly collected for biological surveys and studies, predation was calculated, and "the cropping programme, selection of trappers to participate and removal of the crop" were organized.[74] Social and economic relations were also tightly managed. Incomes varied each year depending on fur prices, and trappers were paid under what was admitted to be an "extremely paternalistic" arrangement. Earnings were pooled and averaged among all the participating trappers and were then paid out in monthly installments throughout the year. Some proceeds were also withheld and placed in a fund that could be drawn on when prices were low or muskrats scarce. Further reflecting the paternalism of the system, payment from this fund was not paid to trappers if other employment was available.[75] Credit, the lifeblood of the old fur-trade system, was not part of the scheme, and while this may have made sense to urban professionals, it caused difficulties for people who needed credit to survive in a cyclical economy. As the Manitoba Fur Advisory Committee was told in The Pas in 1944, under former conditions the muskrat harvest had been "much less sure" than it was under the habitat rehabilitation system, but years of high production, albeit "sometimes at wide intervals," had provided "sufficient cash to enable trappers to replace important items of capital equipment such as canoes, out-board motors and firearms. It was also pointed out that during such periods the young folk frequently obtained the means for setting up their own domestic establishments."[76]

As project manager, the provincial government set policies and procedures in consultation with federal officials. Some community input was allowed through the Fur Advisory Committee, which was made up of local Euro-Canadian traders and merchants and federal and provincial government officials. A few trappers were on the committee, but they were outnumbered and largely powerless. Resource use was very tightly controlled, and predators that killed muskrats (such as mink, otters, weasels, and coyotes) were trapped.[77] The right to trap these predators was granted to local trappers by the block administrators. While broad environmental controls were achieved by regulating water levels, specific environmental activities were also sometimes necessary to ensure production. Cattails and wild rice, for example, were transplanted in 1946 to Netley Marsh to replenish muskrat food supplies.[78] While the management of fur-bearers in these preserves was theoretically grounded in scientific practice, relatively little was known about muskrat biology, especially when these projects first started. In 1943, for example, it was admitted that the intricacies of a muskrat's seasonal diet were still not clearly understood, nor was muskrat reproduction, including muskrat birth cycles or even how many litters were born.[79] Because they were of central importance for management, animal population cycles were plotted, but answers to the mystery of what governed them remained elusive.

Whatever such administrative and management routines achieved or hoped to achieve, they could not counter wider economic developments and changing policy considerations. By the mid-1950s, greater wage employment opportunities and low fur prices were drawing people away from trapping in northern Manitoba. Nonetheless, fur production from the fur rehabilitation areas remained satisfactory and fewer active trappers meant an improved catch for those who remained.[80] But far more devastating and permanent consequences were brought about by broader environmental and economic changes, such as the Saskatchewan Power Corporation's construction of the E.B. Campbell Dam (then called the Squaw Rapids Dam) on the Saskatchewan River, a short distance upstream from Cumberland House. Construction of the dam began in 1960 and was completed in 1962, and almost from the start water levels in the Cumberland House marshes fell, damaging the water-control systems in the muskrat habitat areas there and further downstream in the Summerberry projects as well. Far more damaging for Summerberry was the completion of the Grand Rapids Dam between Cedar Lake and Lake Winnipeg in 1962. By 1964 the dam had raised the water levels of Cedar Lake, Moose Lake, and other water bodies in the lower delta, flooding huge areas of land, some of which had been rehabilitated only twenty-five years earlier as muskrat habitat areas. Other lands that had been used for hunting moose, waterfowl, and other animals for food were also lost. Environmental studies and internal government discussions reveal that the long-term social and economic costs for local Aboriginal communities and the negative impact on wildlife were clearly understood, but the Manitoba government's growing fixation on hydro development led it to claim that such costs were an unavoidable part of projects that benefited the industrial development of the province, by which it meant its southern region. In the provincial government's defence, it was widely assumed that such projects would also contribute to the modernization of the northern economy. What this meant in reality, however, was that many Aboriginal northerners lost their livelihoods, and sometimes their homes and communities, and found that the promised new economy only marginalized them even further. Moreover, they faced these challenges on their own since the federal government was reluctant to stand in the way of hydro development and challenge the province in defence of their interests and needs.[81]

It was not only hydro development that created environmental changes that affected the muskrat habitat projects. In the Pasquia and Saskeram blocks southwest of The Pas, the habitats engineered for muskrats were increasingly damaged by agricultural development in the area. In a familiar pattern, the needs of the farm sector were the main priority of government. Beginning in 1946, farmers in the Pasquia area demanded that water levels be regulated to meet farm needs, not for the "interests of the muskrats and muskrat habitat." Even though this had been a highly productive fur-producing area, farmers had their way and trapping decreased in the face of competition from farming needs. By 1952 Pasquia and neighbouring districts were "almost out of the picture" as muskrat-producing areas, and depletion of muskrat populations intensified further as water management policy continued to focus on farmers' needs. By 1960 the Pasquia muskrat habitat rehabilitation projects were largely a memory.[82] Such developments revealed that the management and exploitation of wild animals could only help solve northern economic and social problems as long as they did not impair competing metropolitan priorities. Moreover, these developments also revealed that the fur sector could not create sufficient wealth to compete with conflicting industrial expansion. Harvesting wild fur-bearers remained a low-status, low-income occupation, vulnerable to outside markets and volatile prices, and forced always to operate within the interstices of an economy dominated by industrial and agrarian ideals and interests. In such an environment, wild animals could not sustain northern communities.

11

The Spectacle of Nature: Wildlife and the National Parks

Part of this destruction [of nature] is of course inevitable. As settlement progresses, plant and animal life is bound to disappear. At our present rate of development, even in Canada, the time is not far distant when it will be only in the national parks that the ordinary traveller will be able to find a perfect picture of primitive conditions as they were before the advent of man.

J.B. Harkin, 1917[1]

National parks were historically the most prominent wildlife sanctuaries in Canada, a system of public spaces that began in 1887 with the creation of Banff National Park in the Rocky Mountains in what is now the province of Alberta. The establishment of such wildlife sanctuaries helped to reconcile conflicts among conservation, settlement, and resource development by creating areas where wildlife could be protected without limiting economic growth. These sanctuaries also expressed various emotive and cultural relationships with wildlife. While wildlife conservation was not a reason for creating Banff National Park, it quickly became an important objective and created a model that was emulated in the national parks that were established later.[2] Management of wild animals in the national parks aimed, at least in part, to enhance the tourist experience by treating wild animals as an integral part of the parks and as elements in the cultural meaning of their landscapes. This function was expanded with the creation of parks for the specific purpose of preserving certain charismatic endangered species, notably bison, elk, and pronghorns.

The same year that the federal government established the park at Banff, it set aside 2,500 acres at the north end of Last Mountain Lake, north of Regina, to protect migrating marsh and water birds. This land was said to be the first bird sanctuary in the Western Hemisphere, and historian W.F. Lothian sees its establishment as linked to the conservation sentiments stimulated by the creation of Banff National Park. While federal authorities did not take steps to prevent hunting inside the sanctuary, keeping this land from the breaking plough was in and of itself an accomplishment. No other federal government bird sanctuaries were created until the 1916 Migratory Birds Convention mandated their creation, after which a number were established on the prairies. Usually set up in consultation with the provinces, these sanctuaries were significant for migratory bird conservation even though enforcement of protective regulations was often patchy.[3] Further opportunities for wildlife protection arose in 1894 when the federal government began establishing timber reserves to protect forests for commercial use. By the early 1900s, these also served activities such as grazing, watershed protection, recreation, and wildlife conservation, and further refinements in 1911 provided that they could be made into national parks.[4]

While the Prairie provinces lacked control over public lands before 1930 and were thus limited in their ability to create protected areas for wildlife, they in any event tended to see the federal government's system of forestry reserves, national parks, and migratory bird sanctuaries as sufficient.[5] After they gained this control, all three were primarily concerned with the creation of parks for outdoor recreational activities like camping, fishing, and swimming, interests that dominated provincial parks policy until the 1960s. Saskatchewan's first provincial parks were planned as "recreation and pleasure grounds," although natural features and vegetation were usually preserved as well. While it varied by park, game was also protected, although hunting and trapping were periodically allowed in order to control populations and meet the demands of local residents. Alberta was more ambitious when it set up its parks system in 1930, promising recreational opportunities as well as the protection of wildlife, wild plants, and objects of geological, ethnological, historical, and scientific interest. In the longer term, however, recreational objectives dominated its agenda, and by 1953 Alberta described its parks as serving scenic, historical, and recreational values.[6] While provincial parks were more concerned with recreation than with wildlife protection, they nonetheless established public spaces where people could pursue their interest in wildlife. Fishing was popular at all provincial parks, and as the naturalist

Doug Gilroy noted, some Saskatchewan provincial parks were also fine venues for watching wildlife. As he observed in 1960, "on a short trip" along the roads at Greenwater Provincial Park in the eastern part of the province "it is not uncommon to see deer, black bears, elk and moose as well as smaller animals such as beaver, muskrat and mink." It remained, however, that the parks and the few game preserves operated by the Prairie provinces before the 1960s did not have the size, stability, scale of operation, or historical depth of the federal system. By the 1960s, when the national parks administration was revamping its policies in an attempt to rein in commercial development and restating its commitment to the conservation of the natural integrity of the parks, the *Winnipeg Tribune* editorialized that "for the sake of future generations of Canadians," prairie provincial governments should copy the new federal parks model and expand the mandate of their parks beyond recreation.[7]

While all national parks served multiple uses, including recreation, economic development, and wildlife and landscape conservation, by the early 1900s they were broadly of two types. The oldest consisted of large scenic parks whose role as wildlife sanctuaries contributed to their identity and appeal as tourist spaces. Banff was the earliest of these; the federal government reserved land around the hot springs in 1885 and created Rocky Mountains National Park (now Banff National Park) at the site two years later. Other scenic parks in the eastern part of the Rockies included Waterton (established as a forest reserve in 1895 and made a national park in 1911) and Jasper (established in 1907). Prince Albert National Park was also established in 1927 in a scenic area in northern Saskatchewan between the boreal forest and the parklands. Although its creation respected the parks branch's goal of having a national park in every province, it more precisely reflected Prime Minister Mackenzie King's desire to reward the community of Prince Albert that elected him after his defeat in York North in 1925.[8] While they were not reluctant to make tourism a part of their operations, the second type of national parks aimed primarily to preserve threatened species. The groundwork for the first of these parks was laid in 1903 when a forest reserve east of Edmonton was declared an elk sanctuary. In 1913 it became Elk Island National Park. During this period, land near Wainwright in east-central Alberta was reserved as part of the management of Canada's bison recovery program, and it became Buffalo National Park in 1913. Nemiskam, Wawaskesy, and Menissawok parks were established in southern Alberta and Saskatchewan to protect pronghorns, and in 1922 Wood Buffalo National Park in the Peace Athabasca Delta in northern Alberta was established to protect wood bison. In 1929 the Riding Mountain Forest Reserve in western Manitoba, dedicated to protecting elk and the development of a local tourist industry, became a national park.

One characteristic that these national parks had in common with wildlife sanctuaries in general was that they did not allow hunting within their boundaries. This policy was first applied at Banff in 1889 and reflected the late nineteenth-century tenet that protection of wildlife depended on sanctuaries where hunting was banned, habitat was protected to some extent to save remnant populations of wild animals, and predators were controlled to allow desirable animal populations to increase. In keeping with broader thinking, most of these parks and sanctuaries were located in areas without agricultural or industrial potential, thus bringing value to land that was otherwise thought of as useless. As W.N. Millar in his report on big game in the Rocky Mountains commented in 1915, the mountain range had no agricultural value but its potential for tourism gave its game "a value as a national resource that can scarcely be exaggerated, and to let any of its numerous species be exterminated would be an act of extreme folly."[9] In any case, the national parks had by this time been promoting tourist development for two decades and wildlife policies were partly constructed around these tourism needs. But watching wildlife and being in the wilderness were also said to contribute to Canadian spiritual and social improvement and to express one understanding of national identity and history, especially in the case of parks that protected threatened species. The confinement of these animals to small areas did not threaten agricultural and industrial activities and respected the belief that all arable land should be devoted to economic production. Yet all the national parks encountered difficult problems in managing wildlife populations within restricted spaces, and their response to these challenges brought contradictory polices because of the broader demands of economic development, political and social views and prejudices, and changing approaches to wildlife management.

WILD ANIMALS, TOURISM, AND THE NATIONAL PARKS

As the birthplace of the national parks system, Banff National Park pioneered many of the system's policies. While the federal government's immediate concern in creating the park was to preserve an area of natural wonder and keep the hot springs in public hands, the broader goal was to develop tourism to benefit the Canadian Pacific Railway and the national economy. Other commercial objectives came into it, too, and some land in Banff (and in other parks as well) was leased for grazing, lumbering, and mining, a practice that continued in some parks until the 1960s. Such factors have led some scholars to argue that the origin of Canada's parks system owes more to a concern for public and corporate profit than to a desire to protect wilderness and wildlife for their intrinsic value.[10]

While concern for wildlife played little part in the rationale for the establishment of the park at Banff, it quickly became part of its development agenda. In 1886 William F. Whitcher, former dominion fisheries commissioner, set out a wildlife management plan for the area that was formative in future thinking about the place of wild animals in national parks. From the outset, he warned that once plentiful wildlife was "now scattered and comparatively scarce" because of habitat change, hunting, trapping, and natural predation. It was therefore vital to conserve the plants and animals that remained because their disappearance would "deprive the National Park of something of its many wild attractions; whilst plenteousness will be a source of profit and pleasure." Elk were "foremost among the game of the park" and added to the picturesqueness of the landscape. These "noble" and "magnificent" animals were increasingly rare, and he believed they were "admirably adapted by form and habits to the park-like woodlands which fringe the glades and cover the dark green slopes of the foothills country." The "lesser deer" added further to the picturesque scene. While living in remote and often inaccessible areas, mountain sheep and mountain goats were also worthy of protection, while bears completed Whitcher's roster of desirable large mammals. Forming a backdrop and adding richness of colour and character to the scene, fur-bearers like beavers and otters deserved protection as well because of their economic value for residents. Despite his emphasis on charismatic animals, Whitcher also advised protection of animals such as rabbits, squirrels, and marmots, which along with the "countless innocent and gay plumaged birds" were "living ornaments interesting to visitors." While predatory birds such as owls, hawks, falcons, and eagles should not be allowed to become "too numerous," he argued that almost all birds should be preserved because of their specific value. Vultures, pelicans, ravens, and crows were useful scavengers; waterfowl and upland game birds were attractive as well as valuable for hunting; and song and insectivorous birds would "enliven and beautify the Park." In contrast, Whitcher classified all animals with "destructive habits or peculiar ferocity" as "noxious." Thus, "wolves, coyotes, foxes, lynxes, skunks, weasels, wild cats, porcupines and badgers should be destroyed." While the destruction of predatory animals would help preserve the charismatic animals he had identified as the park's attractions, he also hinted that their destruction would help confirm the park's pastoral character by softening and reshaping the coarseness and violence of nature. Thus, for example, bears, as "vegetarian feeders," did not threaten desirable animals or upset the peace of nature and did not need to be "wantonly killed nor dealt with as we should do with the lupine, vulpine and feline vermin that prey upon furred and feathered game with savage impartiality."[11]

Whitcher's plans placed wildlife within a hierarchy of values conditioned by an interest in creating an attractive environment for tourists. Whitcher recommended that unlike Yellowstone National Park in the United States, where all hunting was banned, it should be allowed in Banff as a tourist amenity. He did not believe, however, that First Nations people should have the same right as Euro-Canadians and tourists to hunt in the park because it would entice them away from their homes to sell wild meat, furs, hides and horns in the Banff townsite, at local mines, and at the railway station. In contrast, Euro-Canadian hunting could be controlled by licensing it for "legitimate sport and domestic use" and restricting it to open seasons in the fall. Along with control of big game predators, a supply of game for sport hunters would thus be assured, while surplus big game would flow from the park into surrounding areas. This was an important justification for the park, since "the tendency of animal existence will naturally be towards the sheltered and protected central conservatory, especially during periods of reproduction." In a view that would dominate national parks wildlife policy in much of the next century, Whitcher observed that while use of public money for the park was still a novel expenditure, it would gain support as the public realized the park's value as a wildlife refuge whose overflow would provide food and sport to surrounding communities and as a legacy for the future. It was "unfortunately the case," Whitcher noted, that only naturalists and sport hunters seemed "to appreciate the intrinsic worth of the feral stock until increasing waste outstrips reproductive capacity" and decline became irreversible. Thus, with a "precautionary outlay on a liberal and enterprising scale, the remnant of game and fish still existing may be speedily saved and abundantly increased. The Canadian National Park would then be regarded for this, among other praiseworthy reasons, as a model reserve, exemplifying the public spirit, the wise foresight and the progressive views of modern Canadians."[12]

As a former fisheries official, Whitcher was also concerned with the connection between fish management and tourist development. Many species were depleted because of overfishing and the popular practice of dynamiting a lake in order to kill fish. The familiar villain, "the improvidence of Indian fishing," was also blamed. Whitcher therefore proposed that fishing be allowed at Banff "solely for sport and private use" from June to September and only by hook and line. Habitat could be improved by sowing wild rice in the lakes and streams near the townsite to replace "rank weeds and wiry grasses" and provide food and habitat for wild geese and ducks. This would also "make a cleaner and more extensive area in which to cultivate superior kinds of fish, with which its waters should be freely stocked" to replace indigenous fish such as whitefish, catfish, and suckers. He also advised that wing dams should be built on parts of the Bow River to raise water levels. This would improve fish habitat, contribute to scenic beauty by submerging fallen tangles of timber along the banks, create a firebreak for the town, and "drown out" the foxes and coyotes that lived in the swampy areas.[13]

As historian Janet Foster remarks, Whitcher's focus on the "profit, utility and commercial potential" of wildlife was representative of attitudes at the time.[14] Like a farm, the park's environment would be transformed by rooting out unwanted elements to allow favoured ones to flourish. While Whitcher's belief that wildlife should be at the centre of tourist development was only one view of how Banff should develop, it had the greatest public appeal, and many of his ideas continued to resonate in national parks wildlife policy for decades.[15] This emphasis on the tourism potential of parks was enhanced by the widespread view in the 1890s that little or no game remained in most parts of the United States and that American tourists would travel north to hunt or observe it.[16] And Banff had other benefits to offer tourists. As park superintendent Howard Douglas observed in 1901, it had no dangerous animals, reptiles, or poisonous snakes and a man could "sleep in perfect security and safety under his blankets anywhere." This did not, however, as Douglas stressed the next year, lessen the need for effective management to ensure that wildlife continued to offer "an incentive to recreation and health" and draw tourists to Canada.[17] Thus, despite Whitcher's recommendation, all hunting (other than that of predators by wardens) was banned in 1889 in order to protect all wildlife for the tourist gaze. This included hunting by First Nations peoples, particularly the Stoney (Nakoda), whose reserve was nearby. The ban was difficult to enforce against the Stoney, and park officials continued to demand that the reserve's Indian agent persuade them not to hunt in the park. While popular arguments today hold that the widespread North American practice of excluding Aboriginal people from national and provincial parks was part of an effort to make the parks an empty wilderness worthy of preservation, historians Ted Binnema and Melanie Niemi argue that the Stoney were excluded from Banff in order to protect the wild animals for tourism, to conserve them in nearby areas for sport hunters, and to contribute to the assimilation of the Stoney by forcing them to abandon subsistence hunting.[18]

Banning hunting in the park was only one of the policies that aimed to enhance the experience that tourists had with wildlife. Superintendent Douglas recommended that no further industrial permits be granted in the park because "large camps of men invariably lead to trapping and snaring and, in fact, to almost every possible breach" of the game laws.[19] While this recommendation was not followed, a requirement that guns had to be sealed when carried in the park was put in place, and by 1910 full-time wardens were on staff and patrol cabins had been built to allow tighter supervision of game regulations and to control predators. Trails to popular sights and areas inhabited by wild animals were also constructed to give tourists access to wildlife and the park's natural features. As Douglas wrote in 1902, "none of nature's landscapes are ugly so long as they are wild," and trails were needed to give people better access to them. Dogs were also prohibited from running at large because even small dogs would drive game to spots inaccessible to "average tourists." And as Douglas further argued, since game was "one of the most interesting features of the park to visitors," it was desirable that it should be disturbed as little as possible."[20] As discussed in chapter 7, new fish species were also introduced to enhance game fishing opportunities for tourists in the park.

Collectively, these and other policies treated national parks as sites of display where visitors could watch and learn from the spectacle of nature. Although "watching wildlife" is usually defined as a passive use of wildlife, animals and the natural world in the national parks were nonetheless manipulated to make them accessible and viewable – and perhaps more comprehensible as well. As historians Nicolas Jardine and Emma Spary argue, "even so-called nature reserves contain not untrammelled nature, but a managed, culled, restricted nature" where visitors are guided and access is controlled, which makes these reserves "'hybrids' between the social and natural." Yet many people who visited Banff and other national parks would have found this view remote from their own desires and perceptions and their definition of what was "natural." In 1946 the newspaper in Blairmore in southwest Alberta approvingly editorialized that "it is one thing to see a wild creature thwarted and restless in a cage, and quite another to observe it [in Canada's national parks] under natural conditions and behaving in a natural manner." The photograph was the most popular medium of memory for these encounters and "often each of these snapshots tells its own story to the wildlife enthusiast and brings back to mind years later, unforgettable encounters with primitive nature." As the editorial concluded, "wildlife adds materially to the pleasure of the park visitor. However charming the scenery, its beauty and interest are increased a hundredfold by the sight of wild creatures," making such encounters "an essential part of the national park idea."[21]

Such views were widespread. Many Canadians saw "natural" and "unnatural" landscapes in terms of a relatively easy dichotomy in which an unnatural landscape or environment was simply one that was farmed, mined, or built upon. As Commissioner of National Parks J.B. Harkin argued in 1917, natural spaces were "as yet untouched by private ownership" where "the wilderness awaits the people."[22] Yet such natural spaces did not mean an entirely unaltered landscape. As will be seen in the next chapter, in 1907 a zoo was opened as part of the natural history museum at the Banff park, and the park also had a paddock where bison and other animals were displayed, a feature that was subsequently added at other parks as well. But the animals in the open spaces of the parks were most important in this exhibitionary complex, and their presence and proximity to tourists were encouraged and enhanced. Fire was methodically suppressed in all parks to prevent loss of timber and wildlife and to regenerate what was believed to be the natural landscape that had been disrupted by railways and other developments. In 1903 it was noted that tourists were now commonly encountering mountain sheep and deer along the roads and trails at Banff, and by 1912 Howard Douglas was claiming that he was "convinced that the sense of the wild animal in the parks has told them that they are protected" and "all game is now less timid of the approaches of men and less fearful of natural enemies in the form of dogs and guns."[23] Historian Jim Taylor notes that official park reports about the increase of game and its tameness should be treated with caution as they were likely exaggerated to demonstrate park officials' success. Nonetheless, by 1916 it was clear that deer had become common around the Banff townsite, and while they were "somewhat of a nuisance during the early fall" when they invaded people's gardens, this was excused because they were such "a great source of interest to visitors." Elk transplanted from Yellowstone in 1917 and 1920 also supplemented small local populations at Jasper and Banff. Mountain sheep, formerly found only in remote areas, were so common and tame near the Banff townsite that tourists in 1917 were able to take fine photographs "at close range," and by 1919 sheep were so numerous along the highway in the park that cars had to slow down to get by them.[24] Rock salt was put out near the Waterton townsite to lure mountain sheep closer to visitors, and the increase in beavers in the park created "considerable interest amongst the tourists who are never tired of watching these busy little animals."[25] At Jasper, waterfowl apparently sought out the park "at the opening of the hunting season realising that within its border they were safe," and beavers had become "so plentiful that rabid utilitarians speak of them as a nuisance, not recognizing the pleasure the sight of these little animals at work gives to those who have never seen beavers and considered them extinct."[26] And enhancement of the visitor experience could have an impact

Figure 11.1 The chance for visitors to experience wildlife in the national parks was encouraged by drawing animals closer to townsites and accessible areas. In this undated photograph, mountain sheep were eating their daily feed of oats at Banff National Park.

on even the smallest creatures. Since mosquito season coincided "with the height of the tourist season," their control was important to tourist comfort. By the early 1920s the common practice of spreading kerosene or oil over breeding areas was undertaken around the Banff townsite to kill mosquito larvae.[27] All of these efforts contributed to the official portrayal of the parks as what Jim Taylor calls "a utopian sanctuary, a Garden of Eden where good animals – classified as game – were protected from the evils of the corrupt modern world including humans."[28]

THE DOUBLE MANDATE

In 1911 the national parks branch was created within the Department of the Interior with J.B. Harkin as commissioner. The legislation creating the parks branch foresaw a system of "resource reserves" where timber, minerals, watersheds, scenery, and wildlife would be conserved for rational exploitation.[29] As part of national parks policy, Harkin adopted the approach used by Superintendent Douglas in Banff where wildlife was conserved in order to enhance the park's attraction for tourists. In his first report, in 1912, Harkin stated that the function of the parks as engines of economic development was "of equal importance" to the conservation of scenic beauty, forests, wildlife, and fish. As historian Ted Hart notes, this dual goal guaranteed that "generations of future park administrators would face the so-called 'double mandate,' the challenge of navigating the delicate balance of development and preservation." Hart points out that it is unlikely that Harkin "fully realized the future import" of his statement because he initially saw parks largely in terms of the play movement and the belief that outdoor activity would promote physical and mental health and counter the degeneration brought on by urban life. Wildlife was thus only incidental in Harkin's thinking, but his views shifted quickly and by 1916 he was claiming that wildlife protection was "one of the most important features of parks work." Hart credits this change to Harkin's reading of William Hornaday's 1913 book *Our Vanishing Wildlife* as well as Ernest Thompson Seton's writings and his 1914 reports for the parks branch on the preservation of pronghorns.[30] It was thus not surprising that Harkin was soon expressing worries about the possible extermination of wild species and echoing Seton's and Hornaday's arguments about the value of sanctuaries for wildlife conservation. As he wrote in 1917, undisturbed wild areas in Canada were decreasing rapidly except in the national parks, where "the original balance of nature" was protected and where wild animals found "their natural food supply undisturbed, and free from all fear of man, they live and breed under the most favourable conditions." Appealing to the argument that sanctuaries created an overflow of game, he further contended that the whole Rocky Mountain area would soon be restocked with big game from the national parks. But, above all, wildlife was one of "the most important features of the parks. It is of no less consequence than the scenery itself, in fact it may almost be said to be part of the scenery." While some travellers found the Rockies lonely in comparison to the Swiss Alps with their villages and farms, Harkin noted that the unique wildlife of the Rockies animated the natural world and offered "something of the same interest which human habitation supplies in the Alps and gives many people even greater pleasure" because they could renew an "ancient relationship with wild animals, a relationship as old as man himself." In 1922 he broadened this argument by linking it to the decline of big game in North America, making the "abundance and fearlessness" of wildlife in the national parks a unique attraction for tourists, an advantage that promised to increase as depletion proceeded apace. Quoting Henry Osborne, president of the American Museum of Natural History, Harkin noted that the "entire vertebrate kingdom" was under threat because of the fur, bone, and hide trade and that people were living at "the close of the age of mammals." Other distinguished scientists also believed that wild game would soon be seen "only in museums and picture books," and Harkin concluded that "in view of these gloomy predictions, it is doubly gratifying to witness the steady increase of wild life in the parks and to note that the natural overflow is gradually re-stocking the adjacent districts."[31]

The presumption that the protection of part of the natural world from destruction could also serve economic and social needs helped to shape parks policies. Harkin followed the same path as the American preservationist John Muir, who saw increased visitation to national parks as a way to spread his message. In principle, this view is similar to the claims now often made for eco-tourism and sustainable tourism – that giving the natural world or particular species a monetary value makes their preservation achievable. Ted Hart argues that the combining of commercial and conservation goals in the double mandate was similarly necessary to gain political support for the parks system. Yet these goals also met the expectations of a growing segment of the public that cared about wildlife. Either way, tourism remained a powerful force in shaping relationships with wildlife in national parks. Banff's and Jasper's location on the transcontinental rail lines made the parks accessible to national and international travellers, and by the early 1920s cars were allowed in the national parks, which also expanded visitation. The number of visitors to Banff grew from about 8,200 in 1901 to almost 80,000 in 1922–23, and this helped to tie the tourist relationship with wildlife to the experience of road trips. Tourism was further encouraged at many parks with the construction of golf courses, swimming facilities, and sports fields to help fulfill their purpose as "playgrounds for the people." Under Harkin's leadership, the parks branch assiduously promoted the parks as tourist sites through advertisements, publications, films, press releases, and other means.[32] At the same time, policies were shifting somewhat to include the concept that the parks should preserve outstanding natural landscapes. In 1930 the mandate of the parks system expanded when new national parks legislation included support for their integrity as nature preserves. This goal was layered on top of older notions that linked conservation with tourism and economic development as well as with the view that parks were spaces for national social regeneration. As historian Jim Taylor remarks, this did little to resolve the "conflicting aims of use and preservation inherent in the idea of national parks."[33]

These conflicting aims were demonstrated in the appointment in 1931 of Grey Owl, the famous nature writer, to the staff at Riding Mountain National Park in Manitoba. Appointed as "caretaker of park animals," Grey Owl was also mandated to set up a beaver sanctuary at the park. Historian Donald Smith observes that Harkin saw Grey Owl as being ideally qualified for this work; the people who read his books also travelled and would come to visit him, thus increasing tourism at the park. Consistent with the double mandate, Harkin noted that Grey Owl's pet beavers, his personality, and his commitment would equally advance the cause of conservation.[34] While the relationship soon became tense, both sides wanted it to continue and Grey Owl moved to Lake Ajawaan at Prince Albert National Park in northern Saskatchewan later that year. The parks branch built a cabin there to his specifications and produced a film about him and his beavers to promote his work and the allure of the park. As one official in the parks branch concluded, Grey Owl had "the greatest publicity value of anything we have done," and during his seven years at Lake Ajawaan, Grey Owl wrote three books, gave public lectures about beavers and wilderness conservation, and received visitors who trekked out to see him.[35] His beavers were star personalities whose lives and performances reconciled conservation with economic development through tourism. Not unmindful of the contradictions, Grey Owl confessed in 1931 that he felt "a little queer at times putting the beavers through their paces before a crowd, but [I] never lose sight of the ultimate object; seed may be sown in a number of instances where little is suspected."[36] Grey Owl would, of course, be exposed after his death as a poseur. And as historian John Sandlos notes, his stint at Riding Mountain was doubly ironic because of the efforts to force out the Ojibway whose reserve lay within the park's boundaries near the public area on the shores of Clear Lake. These attempts began in 1931 (the same year Grey Owl was hired by the park) but, because of administrative complications, were not achieved until 1936. Motives for dispossessing the Ojibway were varied: the parks branch wanted them removed to ensure they did not hunt elk in the park; local tourism promoters and the parks branch agreed that the Ojibway's expulsion would eliminate what was said to be an incompatible land use in the park and would make the scenic areas along Clear Lake available for tourist amenities; and the Department of Indian Affairs supported the expulsion because it would encourage assimilation as band members would have to take up farming if they no longer hunted.[37]

While hunting was not allowed in the parks, their function as wildlife breeding areas whose surplus populations overflowed into neighbouring areas made Jasper, Banff, and Waterton important bases for sport hunting in the foothills and Rocky Mountains. The overflow of big game from Jasper alone was calculated in 1920 to have brought $2 million in sportsmen's spending to the area. As early as 1911, Alberta had approvingly noted that the overflow of big game into the foothills created sport-hunting opportunities, and the provincial game branch reported that mountain sheep and goat populations were increasing because of the protection provided by the mountain national parks. Towns like Pincher Creek on the eastern slopes of the Rockies midway between Banff and Waterton National Parks also profited, and the local board of trade advertised the town in 1933 as a "hunter's haven" with easy access to big game, upland game birds, and trout, some of which was overflow from the parks. Fishing was the only consumptive use of wild animals permitted in the parks, and as a "selling feature to attract tourists," fishing licences were not required and fish stocking continued in the effort to guarantee an adequate supply of game fish.[38]

Bear management in the parks also fell within the tourist matrix, and their control and contribution to the visitor experience was an ongoing preoccupation for parks officials. In 1886 William Whitcher had argued that bears were not threatening animals, and they continued to have many defenders. At a meeting of the Commission of Conservation in 1915, for example, a case where bears had destroyed the clubhouse of the Alpine Club in the mountains at Banff was discussed. Edward Prince, the dominion commissioner of fisheries, came to the bears' defence, arguing that while property damage should be prevented, the Alpine Club was at fault for having left food around that attracted the bears. And J.W. Robertson, another member of the commission, went even further, remarking that bears in the national parks did not constitute a threat. "They seem to have some boy-like Hallowe'en proclivities," he said, "but that is not a shootable offence." In 1916, recognizing their attraction for tourists, Harkin directed that grizzlies and black bears were to be protected but that nuisance bears could be shot. This policy was still in place in the late 1920s, and officials were instructed to see to it that there were sufficient "non-dangerous bears to provide the thrill that the tourists get from seeing live bears in the open."[39]

Visitor–bear relationships were acted out in a number of parks, but their history at Jasper provides a good example of how bear management was conditioned and legitimized by visitor demands. Bears became part of Jasper's identity and an important element in its branding as a tourist destination. The park played on the popular view of bears as cute clowns, but as Jim Taylor notes, there were "many instances of bear maulings," several deaths, and much property damage. Yet authorities were slow to confront this issue. While nuisance bears were shot, bears were never hunted down like wolves, cougars, or coyotes because, as Taylor concludes, "people liked them. They were seen as 'good' animals, sometimes bad 'good' animals, but nonetheless … an attractive species and symbolic of the wildness of the park." Park officials therefore tolerated practices that kept bears close to visitors, and by the early 1920s park visitors were even touring the town dump to photograph bears while they fed on garbage. As a result, people and bears were learning to get close to one another as the bears became dependent on garbage and human food. By 1950 the bears of Jasper had become national celebrities through the cartoon strip "Jasper," which started in *Maclean's* magazine in 1948 and ran for twenty-four years. As Jim Taylor notes, Jasper the bear, who became the town mascot, was "a likeable and friendly observer of park life, happy to share the park with the tourist visitors while sometimes running afoul of the park ranger." The message was a clear one: the park was Jasper's, and he legitimized the presence of tourists through his friendship with them, while they identified with the freedom he symbolized.[40]

Figure 11.2 *Right* Watching bears was a popular activity at many national parks, where they were seen as cute and authentic exemplars of wilderness. Here a young bear was photographed in 1914 along the road to Maligne Canyon in Jasper National Park.

Figure 11.3 *Below* From an early point, parks officials tolerated and even encouraged the habituation of bears to human contact. Here a visitor feeds a bear from her car on the Banff-Windermere Highway (Kootenay National Park) in 1927.

Yet such visitor–bear relationships were more precarious than this portrayal suggested. What Taylor calls "foolish luck" allowed the image of "the friendly harmless bear" to last into the 1950s. In the backcountry, damage to patrol cabins by bears searching for food was ongoing but tragedies were rare. The number of visitors to Jasper was also still relatively low and most people came in groups, while backcountry travel was usually organized by outfitters who knew how to behave when encountering bears. While signs instructing people not to feed bears were posted in 1938, a mauling by a grizzly at Lake Louise in 1939 and another at Yoho in 1943 forced the parks branch to recognize that it was facing deteriorating bear–human relationships in the parks. Even so, authorities were reluctant to alienate visitors, and as the number of visitors increased after the Second World War, so did the feeding of bears, especially by visitors in their cars along the highway in and around the Jasper townsite. While park authorities stepped up the educational campaign to discourage such practices, they continued to argue that only a small number of bears were a problem and that these bears could be relocated or destroyed if necessary. Although the public continued to find the bears feeding at the garbage dump entertaining and worthy of photographing, some observers were beginning to find the spectacle degrading. As the naturalist Dan McGowan wrote in 1948, many bears in the parks made "a good living by posing for pictures – the usual charge being a few cookies or a handful of candies." But bears were not toys and he thought it shameful that people's experience of them took place in garbage dumps, where "in a setting of tin cans and other assorted rubbish," even the handsomest bear was seen as little more than "a four-footed scavenger." In any event, human safety, not the bears' dignity, became the deciding factor. Jasper Park began to burn its garbage, and in 1951 new regulations made it illegal to feed bears. These rules, however, were not enforced until a seven-year-old girl died after being mauled by a bear in 1958. Public attitudes slowly changed and people became more respectful, or at least more wary, of bears. But, as Jim Taylor concludes, "the bears paid a bigger price. In 1959 and 1960, fifty-five black and twelve grizzly bears were destroyed in Jasper National Park" to ensure public safety.[41]

Figure 11.4 A bear on the Jasper National Park golf course. Photo not dated.

DEALING WITH PREDATORS IN THE NATIONAL PARKS

The treatment of bears in the national parks illustrated many of the contradictions that arose from the pairing of conservation with economic development. Contradictions also arose in regard to wildlife conservation and predator control. It will be recalled that predator control was a prominent concern in agricultural and sport-hunting communities, and the national parks also reflected similar concerns and attitudes about predatory animals. Big game and other favoured animals were protected in the parks by the routine killing of predators such as wolves, weasels, mountain lions, and wolverines. In 1917, for example, increased numbers of big game at Waterton were welcomed, and park officials, worried about the threat from wolves and coyotes, promised "strenuous steps" to deal with them. Other parks took action as well. At Jasper, wardens routinely hunted wolves, lynx, and coyotes as these animals moved into the park. When wolverine tracks were noticed during an inspection trip in the Mount Robson area in 1920, the park superintendent assured Ottawa that "all possible steps will be taken to eradicate this dangerous predatory animal." So, too, when two bands of wolves settled in Waterton Park in 1921, "a war of extermination commenced," and when a cougar was sighted, an "expert hunter" was hired to kill it.[42] And when wolves from Waterton were said to be killing cattle on nearby ranches in 1918–19, the parks branch – in an irony that went unremarked – hired Stoney hunters to kill them.[43]

While such predator-control practices continued, thinking about big game and predators was by this time also beginning to respond to evolving thinking about ecology and the role of predators. By the early 1920s, overpopulation of some big game animals was a growing problem. The number of elk at Jasper grew quickly after their introduction, and in little more than a decade it was feared that they would soon run out of food. At Banff the golf links were fenced to protect golfers from elk, and some elk were either shot or removed from the area. Deer were also coming into conflict with people in the town. Some critics were arguing that predators could help control the number of big game animals and maintain the vitality of their populations. Signalling the influence of such emergent thinking, wardens were instructed in 1924 to remove cougars, weasels, and lynx from the list of predators they could kill, which left wolves, coyotes, wolverines, and some birds of prey. And in early 1925, Commissioner Harkin issued a formal statement to guide predator control in the parks. Drafted by Hoyes Lloyd, head of the parks branch's Wild Life Division, it asserted that predation was indispensible for the balance of nature and that the branch now intended "to have wild life in the national parks remain as natural as possible." Quoting from the scientific work of E.H. Forbush, an American ornithologist, Harkin asserted that a species was kept "in its proper place" by the countervailing pressure of other species. As he argued, "any species of animal or plant life would become a plague if permitted to increase unchecked. It must be checked either by human agencies (costly and often inefficient) or by control exercised by its natural enemies (inexpensive and often marvellously efficient). Excessive reduction in numbers of the predatory animals and birds which act as controls will usually result in trouble." Intimating that predators now formed part of the parks branch's wildlife management tool kit, Harkin concluded that "what is most universally desired is the grand and beautiful mingling of indigenous species which nature itself provides."[44]

As part of this changing attitude, some long-standing park policies were also now invoked as a reason to protect some predatory animals. Harkin, for example, banned the elimination of eagles and ospreys because "the attractiveness of these showy species to tourists more than compensates for the small amount of damage they do" to game fish and various birds. Similarly, fur-bearers were to be protected because their overflow benefited trappers outside the park.[45] A changed direction, however qualified, was now in the air. When A.W. Cross, a member of an influential Calgary ranching and business family, protested the ban on killing cougars because this would endanger his cattle, Harkin replied that science was beginning to show that "predatory animals have a real place in nature, and because they feed upon animals in which we are interested" was "not a valid excuse" for killing them. In an attempt to reduce the number of predators killed, Harkin announced in 1928 that wardens could no longer keep the pelts of predators that they shot in the parks. A short time later, he also banned the use of traps to kill predators because many fur-bearers were accidentally killed as well. These changes in policy ended a money-making sideline for wardens because they were no longer able to sell the pelts of the predators they shot or of the fur-bearers inadvertently killed in traps. Ignoring the obvious potential conflict of interest, wardens saw these changes as an attack on their privileges and income, and for years they criticized Harkin and the parks branch in an effort to have them reversed.[46]

The complexity and range of goals and constituents, together with the ability of different actors at the centre and margins to shape the ways that state policy is applied, rarely allow any policy to find entirely consistent expression. Ted Hart calls Harkin's policy on predators a tightrope act in which he had to deal with "an uncomprehending public, an often openly hostile field service, and the opposition of influential game associations." He was, Hart notes, thus "constantly under attack and frequently [put] in a compromised position."[47] Within the parks branch, Hoyes Lloyd continued to advocate that predators had a legitimate place in the parks. When E.J. Duncan, the editor of the Banff newspaper, complained about predatory animals in the park and asserted that "the weasel family are of no benefit whatever to our visitors" and "should be trapped out" to prevent them from killing grouse, Lloyd replied that human relationships with wildlife in North America had not been very successful and new approaches were needed. Further, he argued, the parks were not preserves for only a few animals, but places "where all the wild things can exist as they have existed in their natural conditions."[48] This thinking was echoed by some field staff. In 1933 public complaints that wolves and coyotes from Riding Mountain National Park were attacking farm stock led one official to counter that unless predators became "exceedingly numerous," they should not be trapped because they were "Nature's method of getting rid of weaklings of all classes" and of controlling game populations."[49] And in 1937 the parks branch distributed a paper at the Provincial-Dominion Wild Life Conference that asserted that "predation is normal, not exceptional," and when people meddled with it, they started "a chain of events the results of which cannot be foretold and which will almost certainly be surprising and injurious to man's best interests." Citing the cases of deer in the Kaibab forest in Arizona, of moose on Isle Royale in Lake Superior, and of the environmental disaster caused by the acclimatization of deer in New Zealand, the paper argued that evidence of the importance of predation in controlling wildlife populations was increasingly clear. While admitting that scientific knowledge was still developing, it concluded that, "except for direct defence of domestic stock, the attitude of the public towards predators is based generally upon prejudice and lack of knowledge" and it was "the duty of those who have studied the problem, even though they have not fathomed all its intricacies," to educate the public about what was known "in the hope that a saner and more philosophical view will be accepted."[50] Frank Williamson, who replaced Harkin as commissioner in 1936, agreed in theory with this view. As historian Alan MacEachern argues, while predator control continued, Williamson responded to findings in ecological research and used cases in which the eradication of predators had caused irruptions in prey populations as "cautionary tales against tinkering with nature."[51]

Although the National Parks Act of 1930 initiated a new respect for the inviolability of park flora, fauna, and geology and shifted policy somewhat towards a stronger preservation stance, this new direction was only slowly realized as the new relationship with wildlife that it proposed was worked out. The friends and foes dichotomy continued to shape attitudes about useful and threatening animals. Historian Thomas Dunlap argues that the policies and practices of Harkin and the parks branch seemed contradictory because no scientific model about predation had been confirmed and new thinking about predation was only just emerging. Moreover, a policy that valued predators but still tried to keep them in balance was an ambiguous one that was easily compromised by political and public pressure and popular prejudice against predators. Indeed, pragmatism usually ruled. As Jim Taylor phrases it, both Williamson and Harkin "were conscious of public opinion and sought limited culls [of predators] while paying lip-service to ecological principles."[52] While Harkin had been willing to argue ecological principles when taking cougars off the predator list in 1924, it was easy to do so because cougars were rare at the time. However, when their numbers began to increase in the mountain parks in the 1930s, local outfitters, field staff, and others demanded they be controlled. Harkin responded by authorizing the purchase of hounds to hunt them at Jasper, and a professional cougar hunter was brought in from Vancouver Island to Banff in 1935. Harkin portrayed this change as one that was consistent with parks predator policy, which reveals how fluid the policy was.[53]

Parks branch policy on wolves and coyotes showed the same trends. By the early 1930s the official line was that they were to be killed only when their numbers were excessive and the "balance of nature" was being upset. But judging what was the proper balance of nature and how many wolves were too many was subjective and could rarely be separated from the prejudice and fear that shaped attitudes towards them. Moreover, it was clouded by the view that the parks were sanctuaries that would produce a never-ending crop of game. Most parks authorities remained committed to the assumption that their success was measured by the number of big game animals a park possessed, even if the animals were malnourished, and that any loss thus had to be prevented. Wolf control was further complicated by the view that the only animals that should overflow from a park were those that were acceptable to its neighbours. These assumptions were so widespread that they usually trumped contradictory emerging scientific evidence. Studies by biologists of deer and elk at Jasper and Banff in the early 1940s, for example, showed that overpopulation was causing starvation, overgrazing, and serious deterioration of land and vegetation. Low levels of predation (in part due to the killing of predators in the parks), the failure to cull big game populations, and ongoing fire suppression were also identified as causes. Other studies showed similar conditions with bighorn sheep at Jasper. But, reinforced by local mistrust of the scientific findings of outsiders, officials in Jasper continued to overreact to predators, all the while "ignoring or minimizing evidence that they were beneficial to the park's ecosystem." Claiming that big game was declining because of wolves and coyotes, efforts to eradicate them continued until the 1960s, when parks policy began to shift towards different management techniques.[54]

Despite official rhetoric about the value of restricting predator control, the reality was that ongoing predator-control activities kept wolves, coyotes, and cougars at extremely low levels in the national parks. Nonetheless, prairie provincial governments and other observers viewed the parks branch's predator policy as too lenient. Clearly, this criticism pointed to a belief that there should be no predators at all. W. Fisher of the Alberta Fish and Game Association argued in 1939 that the parks branch had "the wrong policy" when it viewed the parks as game sanctuaries where it was necessary to have "enough predators such as cougars *to kill off the increase in game*" because this only resulted in the spread of predators throughout the foothills, where they killed "thousands of valuable game" animals. The minister responsible for Alberta's game branch agreed, noting that the federal government "has the idea that nature should be allowed to take it's [*sic*] course; whereas, a system of good modern game management calls for a proper control of predators."[55] Others claimed that fur-bearers were also threatened by national parks policies, and in 1935 the Prince Albert Board of Trade passed a resolution urging the extermination of timber wolves in Prince Albert National Park and throughout northern Saskatchewan to protect fur and game animals. Harkin dismissed these claims, noting that park wardens had seen only four wolves in the previous year, but if the board had "concrete evidence to substantiate the alleged wolf menace," the department would be pleased to investigate it.[56] Others claimed that farmers, too, were victims of federal parks policy. In 1944, for example, W.J. Ward, the MP for the district encompassing Riding Mountain National Park, informed the parks branch that he "had received many letters" from his constituents complaining that coyotes in the park had "reached almost an epidemic stage" and were invading farm areas and killing sheep and poultry. Seeing the complaints as little more than an attempt by locals to obtain the right to trap coyotes in the park (coyote pelts were worth around $20 that year), the parks branch noted that it had no evidence of such problems, although it admitted that coyote numbers were high but understood they would soon fall because rabbits had recently passed the peak of their cycle.[57] Another popular argument maintained that federal policies on wolves and coyotes needed to mimic provincial ones, which, as seen earlier, to all intents and purposes favoured extermination. In 1950, when hysteria about coyotes was high in farming districts, Alberta's minister of agriculture claimed that the national parks and Indian reservations were "ideal breeding grounds for coyotes and it was futile for us to try and get rid of them elsewhere when they breed in parks and reserves and forage outside their boundaries." Alberta thus wanted the federal government to step up the killing of coyotes in the parks and on reserves or else to allow the province to enter the parks and reserves and do the job itself.[58]

NATIONAL PARKS AND ENDANGERED SPECIES: BISON

Relationships with wildlife found unique expression in parks set up to preserve plains and wood bison, elk, and pronghorns, all charismatic animals that were popularly associated with the romance of the bison days and the Old West. Individuals on the prairies were collecting bison from an early point. One such collection was a small herd owned by the former Winnipeg fur trader James McKay, who bought the animals as calves in 1873. While his motive for collecting them is unclear, he sold them about five years later to Samuel Bedson, the warden of Stony Mountain penitentiary near Winnipeg. To help finance the purchase, Bedson borrowed funds from HBC executive and co-founder of the CPR Donald Smith (later Lord Strathcona). Bedson was a collector of historical and natural artifacts, and while the bison purchase no doubt reflected these interests, Bedson also wanted to develop hybrids of domestic cattle and bison for sale. Such efforts had been underway at least since 1815 in North America. In 1886 Bedson claimed that the hybrids were hardier than domestic cattle and could winter in the open. They also had more finely grained meat, which he was able to market in the Chicago luxury food trade.

Figure 11.5 Developing from the private donation of captive plains bison, the bison herd in Banff National Park helped preserve the species and became a draw for tourists. Photo not dated.

Adding further value, mounted heads and hides prepared as robes had a ready market. In 1886 Bedson had sixty-eight wild bison and eighteen crossbred animals (although the males were apparently sterile). Showing a more general interest in hybridization, he also had plans to cross moose and domestic cattle, although it is unclear if these experiments took place. In 1888 Bedson sold about a hundred of his animals for $50,000 to Charles "Buffalo" Jones, an American bison promoter and breeder who had plans for a commercial bison meat and hide operation. As part of this transaction, Donald Smith received a few animals, which he kept at Silver Heights, his estate near Winnipeg. This collection became one of the town's attractions, and while visiting Silver Heights in 1892, Ernest Thompson Seton saw twelve bison as well as what he described as "a rough melancholy beast" that was a cross of a bison and a highland cow.[59]

In 1897 Smith's bison became part of the history of state efforts to preserve bison when Smith donated a number of them to Banff National Park. Their story mirrored many of the cross-currents in late nineteenth-century relationships with bison. As well as their potential for interbreeding, their iconic historical status and their threatened extinction made them worthy of being collected, exhibited, and preserved. Such preservation was thought to reveal human progress and to help reverse the negative impact of the European conquest on the continent's fauna. The poet Charles Mair expressed such sentiments in 1890 when he noted the value of bison for developing hardy hybrid cattle suitable for raising stock in northern districts. Arguing that only government could offer the continuity necessary for a long-term breeding program, he advised the creation of a national park as a bison sanctuary. Indicative of his understanding of the biological nature of what made wild animals wild, he believed that a national park would give the animals space to roam, which was vital, for if confined, "they would probably become sterile." Mair also argued that "no reasonable effort should be spared to rescue an animal from destruction which has been of great service on our continent, which is ultimately associated with its history, and whose extinction would be a disgrace to civilized man."[60]

As will be seen in chapter 13, the Banff herd that developed from Smith's and other donations was justified on the grounds that it preserved an endangered species and provided an opportunity for tourists to see animals central to prairie history and identity. While the exhibition of bison in paddocks became a spectacle for the modern age at most prairie national parks during the interwar years, bison conservation in national parks designed especially for this purpose began in 1907 when the federal government purchased the largest herd of wild plains bison left in North America from Pablo Michel at the Flathead Indian Reservation in Montana. Michel had tried to sell the animals to the US government, but the negotiations went badly and an embittered Michel was eager to deal when Canada expressed interest. The Pablo herd was initially estimated at about 300 animals, but the roundup took years and the herd proved much larger than anyone had anticipated. In 1912, when the last shipment took place, 716 animals had been sent north at a total cost of around $175,000. At the start of this project, the federal government reserved about 440 square kilometres of rolling prairie near Wainwright in eastern Alberta for the animals. The Battle River and Ribstone Creek that flowed through the park gave the bison shelter and water, while extensive hay lands promised that they could be kept "at the lowest possible cost to the government." It was estimated that the area would support 5,000 to 7,000 bison, and the reserve was given park status in

1913 as Buffalo National Park.[61] In her history of this park, Jennifer Brower argues that the government was driven by "sentiment for the nearly extinct species and a wish to show up the Americans by purchasing the herd ... rather than an interest in actually saving the species." Further, she suggests that these "misguided motives" overshadowed "any preservationist considerations for the bison." As evidence, she points to the site chosen for the park and the lack of knowledge about bison or their reproductive rates – problems that were exacerbated by drought in the 1920s and the Depression in the 1930s. This interpretation ignores, however, the fact that the melding of preservation ideals with commerce, nationalism, and agricultural needs was at the time not seen as incongruous. Historian Andrew Isenberg's conclusion that early bison preservation efforts in the United States should not be confused as a "precursor to the environmental movement" was germane to the Canadian case as well. American bison preservation was not, he argues, "an ideologically consistent continuum of wilderness protection" but contained "complicated motivations for preservationism that were specific to the culture and economy" of the times.[62] In these terms, the new regional economy accommodated bison preservation on the Canadian prairies through fenced sanctuaries that protected the animals and at the same time protected surrounding communities from the animals.

With only 117 kilometres of fencing completed at the reserve at Wainwright in 1907, the first shipments of bison were kept at the elk preserve that had been established near Edmonton that year and later became Elk Island National Park. In 1909 most of these bison were shipped to the reserve at Wainwright, and later shipments of animals from the United States went there directly. Most of the bison at Banff were transferred to the Wainwright reserve in 1909, leaving only a small exhibition herd in the mountains. About fifty bison were left at Elk Island; while it was said that these animals could not be caught for transfer to Wainwright, rumours suggested that the Elk Island authorities were lax in rounding them up because they wanted to keep some of them as a feature there. This was not an unreasonable supposition because many communities wanted bison. In 1907, for example, the Saskatchewan Game Protective Association recommended that some of the Pablo herd be placed in Saskatchewan to give the province its own bison park and game preserve. A decade later, believing that bison would be "a great drawing card" for tourists, John Turriff, a Saskatchewan MP, lobbied for a buffalo park in his constituency at the summer resort in the Moose Mountains in southern Saskatchewan.[63]

Despite such suggestions, preservation of plains bison remained focused in Alberta at Buffalo National Park and Elk Island National Park. Bison became an increasingly important focus at the latter park, and it was enlarged in 1922 to accommodate the growing number of bison and other big game.[64] Prudent budgeting was always a virtue in the parks branch, and the herds' existence was constantly justified by the revenue that would be earned by selling surplus animals. Predators were rigorously controlled at both parks. This was deemed necessary because while coyotes, for example, could not kill bison, they could damage hides and heads, thus reducing their market value. Coyotes and other predators were also controlled because it was said that they used the parks as a home base and, from there, raided nearby farms, drawing public criticism of the parks.[65]

Tourism was promoted at both Buffalo National Park and Elk Island Park, although not to the same extent as in the scenic mountain parks. Because Elk Island Park was only a "pleasant run by motor" from Edmonton, by 1913 it had already become a local attraction for camping and picnicking and to watch "the buffalo, moose, wapiti, and deer grazing amongst their natural surroundings."[66] A golf course and other tourist amenities were built there in the 1930s. As historian Graham Macdonald notes, while Commissioner Harkin directed that tourist infrastructure should be kept at a modest level so as not to sacrifice the park's role as a wildlife sanctuary, "the notion that the primary purpose of the park, that of a wildlife sanctuary, was compatible with a certain amount of unrelated public uses, was not seriously questioned except by a few."[67] Although the Wainwright board of trade had recommended in 1912 that Buffalo National Park be closed and "the ancient roamers of the plains" be moved elsewhere so the land could be farmed, tourism soon emerged as a local benefit of this park as well. About 19,000 people visited in 1929 and brought some additional business to the town. A paddock for a few elk, pronghorns, moose, and bison, as well as nineteen yaks transferred from the Banff paddock, gave people a chance to watch and feed the animals. As was remarked in 1930, "the chance of seeing such interesting animals at close range is now attracting tourists from all parts of the dominion."[68]

Tourists were also allowed to see the cattalo, the cattle-bison hybrid bred at Buffalo National Park. In 1916 the federal Department of Agriculture and the parks branch began collaborating on crossbreeding experiments to produce an animal hardy enough for the harsh prairie climate. Historian Jennifer Brower notes that while Commissioner Harkin argued that the public had a "natural desire to preserve specimens of the original dominant animal of the plains," he feared that this commitment would not last and that the cattalo experiments would more likely earn long-term public support by showing the practical value of bison preservation. Like earlier experiments, however, these ones failed to produce fertile male hybrids.[69] Maintaining the vigour of the plains bison was a related breeding concern. Maxwell Graham of the parks branch argued in 1912 that the government's plains bison herds could be improved if they were bred with wood bison. Frank Williamson, also of the parks branch, observed in 1915 that while "the danger is far from imminent, there is a tendency wherever wild animals are restricted in range and where inbreeding necessarily occurs, for the type to deteriorate." Thus the herd at Wainwright would need the "infusion of new and vigorous blood" to prevent

Figure 11.6 Buffalo National Park was established for the bison purchased in the United States by the Canadian government. The first animals came north in 1907, and by 1919, when this photograph was taken, the herd was expanding rapidly.

deterioration. The only source of "pure, vigorous, unrelated stock" was the larger, darker, and hardier wood bison, which Williamson thought "probably represents the finest species of bison now existent."[70]

Because of the substantial financial and emotional investment associated with the plains bison herds, they were the government's first priority. Nonetheless, concern about the status of wood bison was also long-standing. Wood bison populations had declined quickly after the 1860s, and in 1889 land surveyor William Ogilvie reported seeing them only in the Peace-Athabasca Delta near Fort Chipewyan and Fort Smith and advised that they be protected to guarantee a local meat supply.[71] While the preservation of wood bison continued to be justified for local economic and food needs at least until the 1960s, concerns about preserving them because they were an endangered species and the last free-ranging bison herd in the world were already being expressed by the late 1800s. The 1894 Unorganised Territories Game Act prohibited the killing of bison, and the North-West Mounted Police later became responsible for enforcing this ban. In 1907 Ernest Thompson Seton and E.A. Prebble (a naturalist with the United States Biological Survey) accompanied NWMP Inspector Jarvis to report on the herds. While wolves were blamed for the decline of wood bison, Aboriginal hunting was also said to be a factor, and Jarvis recommended that a national park be created to preserve the wood bison. Discussions about setting up a preserve were underway by 1911. The forestry branch favoured a fenced sanctuary, but wrangling between it and the parks branch over jurisdiction and the opposition of Indian Affairs, which objected to restrictions on First Nations hunting bison for food, stalemated the matter. Only in 1922 was Wood Buffalo National Park established to protect the approximately 1,500 wood bison in the area. Totalling about 27,000 square kilometres, the park spanned the Alberta–Northwest Territories border.[72]

Wood Buffalo National Park contained important traditional hunting and trapping districts, and all Treaty 8 First Nations (regardless of where they lived) were at first allowed to hunt and trap in the park and to cut hay and gather salt at the salt flats near Fort Smith. White trappers were excluded from the park, but local Métis were allowed to trap and hunt there until 1923, when they were also excluded. No one, however, was allowed to hunt bison in the park at any time. While the overflow of game from national parks was usually held up as a benefit for local hunters, this did not happen at Wood Buffalo when the bison expanded their range southward beyond the boundary of the park. By 1926 this additional land had been annexed to the park so as to exclude local hunters from the area. By this time, access to some parts of the park had been further limited, allowing only those First Nations who had previously used the area. This generally gave access to Cree First Nations trappers from Fort Chipewyan but only to a few Chipewyan First Nations people from the same community. These complicated rules of entitlement, as well as the extension of the park's boundaries, created economic inequalities, intensified pressure on wildlife resources outside the park, and led to conflict between the parks branch and local people. In 1926, for example, two Cree men were sentenced to the Fort Saskatchewan jail for killing bison. The trial in Fort Chipewyan drew a large crowd and

> the Chiefs and Headmen responded that they wanted "a written contract" to confirm that "their hunting rights would not be interfered with in the New Park area." When the Justice of the Peace replied that this was the responsibility of Indian Affairs, they retorted that despite "sweet words at other meetings," their hunting rights had been ignored. They considered "this Country their Country, as they were here before the Whitemen and that they never asked the Whitemen to come and give them Treaty. They also claimed the Buffalo were here before the Whitemen and this was their Country also" and that the government was "taking up more of their hunting grounds to feed more buffalo being brought into their Country without their permission."[73]

The argument that bison had inherent rights of occupation was one that few Euro-Canadians would have understood, and the concern about more bison was a clear reference to the moving of plains bison to Wood Buffalo from Buffalo National Park. As early as 1919 the parks branch noted that bison at Elk Island and at Buffalo National Park had increased by 500 per cent in ten years and that the herds needed to be culled.[74] This was not done, and by 1921 the bison population at Buffalo National Park exceeded the carrying capacity of the park. Problems were made worse by a series of dry years, and the park's resources were further drained by a large number of mule deer and elk that it was supporting. What had begun as the rescue of a magnificent animal from extinction had by the early 1920s become a crisis, with malnourished, overcrowded, and unhealthy bison.[75]

To solve overcrowding, the parks branch announced in 1920 that it would slaughter surplus bulls.[76] While it anticipated a good profit from the sale of mounted heads, robes, leather, and meat, this cull did not take place until 1922–23. This delay in part arose from fear of public criticism; historian W.F. Lothian suggests that this worry helped justify the commissioning of a film in 1923, *The Last Frontier*, to soften public opinion on the matter. Made at Buffalo National Park, the film celebrated the romance of bison hunting in the Old West and reminded the public that killing bison was a traditional and legitimate plains custom. Produced by a Hollywood company, it played to stereotypes about the West and Aboriginal hunters. In a dramatic stampede, thirty-four park bison were killed during the filming. Newspaper reports that the bison were killed with arrows (they were in fact shot with rifles concealed off-camera) sparked a public backlash, and similar negative publicity grew later in 1923 when the culling of the surplus bison began.[77] In this cull and in the one the next year, just over 2,100 bison were slaughtered under federal veterinary inspection, and bison meat that was safe to eat was sold. As part of its marketing scheme, the Department of the Interior arranged for some of the meat to be made into pemmican by "a number of Indians" using an "original Indian recipe." This was not merely a nostalgic reproduction of authenticity as a market for the product apparently existed in "outlying parts of Canada."[78] Missing from this story was news that the examination of the slaughtered animals had proven that the herd was very heavily infected with tuberculosis. The failure to publicly admit the existence of this problem was not a new stance for parks officials. The presence of tuberculosis in the Canadian plains bison herds had been known to parks officials for some time, but they had downplayed and even suppressed this information. In any case, public reaction to the cull was negative; many people were proud of the animals and opposed slaughtering them for any reason.[79] What was not clear was how the public expected officials to manage the problem of plains bison overpopulation.

The public relations difficulties associated with culling surplus animals exacerbated the serious problems that parks authorities were facing with plains bison overpopulation. Establishing a new park to absorb surplus animals was ruled out as too costly, and the parks branch rejected requests in 1921 and 1926 from groups in Saskatchewan who wanted a bison park established there as a tourist attraction.[80] Instead, bison were given away to parks and zoos; in 1923 alone, requests for the loan or donation of bison were received from Ireland, the United States, Scotland, and various parts of Canada. Such requests were honoured if the requesting party paid the shipping costs.[81] But such measures were inadequate and shipping surplus bison to Wood Buffalo National Park gradually came to be seen as the best solution. This proposal raised significant scientific opposition in Canada and the United States because of worries that crossbreeding would destroy the identity of the wood bison as a subspecies. There was also concern that the shipment would spread southern bovine diseases to the wood bison. Hoyes Lloyd in the parks branch voiced strong concerns about the proposal, but his colleague Maxwell Graham argued that the differences between the wood and plains bison were environmental (in other words, not biological) and that the wood and plains bison would in any event be geographically isolated from each other in the park. Commissioner Harkin agreed that animals should be shipped north, Hoyes Lloyd was censored, and the branch closed ranks, arguing that its staff was "better qualified to judge the policy, because of experience and practice, than are zoologists from a distance." The shipments began in 1925 and 6,673 plains bison had been moved north by 1928.[82]

The transplanting of plains bison to the north was an inexcusable error. There was adequate knowledge to foretell its consequences, and as Thomas Childs, the Canadian veterinary director-general argued in 1956, bovine tuberculosis was sufficiently well understood by the 1920s to have ruled out such a transfer from an infected herd. Not unexpectedly, Childs noted, the slaughter of bison in Wood Buffalo in the 1950s showed that tuberculosis had been transferred to the northern herd. And as predicted, the plains animals interbred with the wood bison and by the early 1950s it appeared that the wood bison subspecies had disappeared. Indeed, crossbreeding began almost immediately, and Dewey Soper, who carried out extensive studies on wood bison in 1932–34, wrote in 1933 that "the

Wainwrights are spreading everywhere and soon the Wood Bison will be only a memory as a separate and distinct entity of the Dominion fauna; this is just what the mammalogists of North America feared" when they protested the transfer. And Soper predicted that although a few pure wood bison might be left in the northern end of the park, it was only a matter of time "until the Wainwrights will have covered the whole ground while the Wood Bison giants will have completely lost their identity."[83]

Shipping surplus bison north was of course only a temporary fix for the overpopulated southern parks. The slaughtering of surplus animals resumed in 1927 at Buffalo National Park, and surplus bison were also slaughtered at Elk Island in 1929 and 1935.[84] The controversies over the slaughter of bison and the shipment of surplus animals north revealed a range of opinions and motivations. Historian John Sandlos argues that the shipments north reflected the contradictions of Canadian wildlife management in the 1920s in its combination of preservationist sentiment and quasi-agricultural approaches. This was even more evident in the case of general bison management; as historian Jennifer Brower notes, it was assumed from the start that bison management was little different than raising domestic livestock. Commissioner Harkin retroactively justified the plains bison transfer on the grounds that it provided food for northern residents and promoted northern development, which made it a simple business decision. Most importantly, however, the transfer offered Harkin and the parks branch an immediate solution to a political and public relations problem. In contrast, while scientific opinion admitted that there was a lack of knowledge of all particulars, it correctly predicted the risks of introducing disease to the north and the loss of wood bison through crossbreeding.[85]

Unlike their negative reaction to the bison cullings, the public's concern about moving plains bison north seems to have been minimal. Clearly, killing bison was the issue that reflected the public's emotional ties and admiration for these charismatic animals, even though most people would have only known them from photographs or paintings. This was apparent in one account of a vacation taken to Buffalo National Park in 1918. While the visitors found the cattalos novel, their most memorable encounters were with bison and elk. Numerous stories were recounted – of bison bulls attacking the wardens, of animals stampeding ("the earth fairly trembled with the thunder of their hoofs"), and of elk bulls fighting to the death. But when the visitors came face to face with them, the animals immediately became awesome living creatures. Driving with a park official, the visitors came upon a bison bull "not fifty feet away. To describe him is almost beyond me. Huge beyond belief with great shaggy fore-shoulders, an immense head, small beady eyes, all enclosed in a heavy growth of long brown-black hair." The visitors "gazed at him awe-stricken. No pictures of buffalos that we had ever seen even began to do him justice. He was much more massive than anything our imagination had created." He blocked the road, then moved off to the side, but "regarded us with lowered head, his tail switching angrily, then with his fore-paw he dug up a handful of sandy sod and tossed it high in the air. Next he snorted and pawed and backed up, scratching and bowing and tossing sand and dirt in all directions. That rite performed, he drove one horn into the ground and plowed up a huge furrow." Then "he got down and rolled over and then back again. We deemed that a good time to move on," and as the park official threw the car in gear, the party "swung into the open and beat a hasty retreat." An elk was later encountered as well, and "again did the reality surpass all pictures of the imagination. That graceful head balanced on the superbly arched neck and topped with those enormous widely branched horns beggars all descriptions."[86]

While the charisma of bison made their management complicated, it also ensured their perpetuation in the national parks system. This was not enough, however, to save Buffalo National Park. With war approaching, the land was turned over to the military in 1939. About 3,000 bison, 1,800 elk, and several hundred deer and moose were shot and the meat was processed and sold. The Alberta naturalist and writer Grant MacEwan later observed that the government offered three reasons for the closure, "none of them very convincing: the range, it was said, was overgrazed; there was tuberculosis in the wild herd and, finally, the range was needed for military purposes." The animals involved in the cattalo experiment were moved to the Range Experimental Station at Manyberries in southern Alberta, and experiments continued (unsuccessfully) until they were abandoned in 1964.[87] Clearly, though, the public retained an emotional affection for bison, and news of the closure of Buffalo National Park "resulted in a deluge of letters and telegrams to the department protesting the action." It seems that people believed that the park would be restored at the end of the war, and the news in 1947 that the military had received a ten-year lease on the land prompted the Alberta Fish and Game League to urge the government to restore the sanctuary.[88] The government responded by pointing out that bison were still preserved at Wood Buffalo and Elk Island parks and that the latter was being expanded as Canada's plains bison park. It also argued that bison preservation had been successful and that the species was no longer endangered. Indeed, demonstrating this success, descendants of the Pablo herd now lived in the paddocks at Banff, Waterton, Prince Albert, and Riding Mountain National parks.[89] Even so, control of bison populations in confined sanctuary settings remained an important management problem. Outcast old bison males were culled at Wood Buffalo as early as 1928–29 (at the same time that local Aboriginal hunters were prosecuted for hunting bison in the park), and the management approaches that had earlier guided the bison parks in the south were replicated at Wood Buffalo in 1954 when its management plan institutionalized the slaughter of surplus animals. Designed to control disease and commercialize the market value of the herd, the plan involved the processing of more than 4,000 animals between 1951 and 1967. The meat was sold locally to other government agencies like Indian Affairs as well as in southern markets.[90]

NATIONAL PARKS AND ENDANGERED SPECIES: ELK

Elk were once found across a wide swath of North America but had been pushed into forested and partly wooded areas in the prairie region by the late 1800s. In 1915 Edward Prince, a member of the Commission of Conservation, recalled that in Manitoba in the late 1800s there were elk trails "just like the old buffalo trails, beaten by the great herds of elk, and, now, there remains hardly a single head."[91] They were popular game animals and their meat, horns, and teeth were commercially valuable. Elk are unique in having two canine teeth. Euro-Canadians prized these "tusks" for watch fobs, mementos, and decoration, and some First Nations used them extensively for decorating clothing and as jewellery. The fraternal Order of Elks used tusks as part of its regalia but dropped the practice when it was criticized for contributing to the animals' extinction. Indeed, it was commonly said by the late 1890s that elk would soon meet the same fate as bison. They were gone almost everywhere in the prairies by 1915, when it was contended that "only very drastic measures taken at once will save the elk from total extermination."[92] Reflecting such concerns, the Manitoba government argued in 1911 that Riding Mountain Forest Reserve near Dauphin, where "one of the last important bands of the animals in their natural state in North America" was found, should be made a game preserve. The federal government subsequently designated it as a preserve, and in 1917 Manitoba banned all elk hunting and made the possession of elk meat illegal.[93] Elk protection was further extended in the province in 1929 when the reserve became Riding Mountain National Park.

While elk preservation was an important goal, the creation of Riding Mountain National Park also reflected the influence of well-connected political lobbyists who argued that the park would bring the development of a tourist industry and create local recreational amenities.[94] Of all the parks dedicated to preserving endangered species, Riding Mountain had the greatest recreational role, although elk preservation remained an important function. In 1936 the nature writer Tony Lascelles commented that many Manitobans were proud that the province had the largest herd of wild elk in Canada and that they could "view their heritage in a natural environment" rather than behind the bars of a cage. Reflecting the common belief that wild animals in national parks were a tourist draw, the park advertised itself as "the land where Wild Life is permanently protected. Elk and Moose, monarchs of the woods, stand and watch as you pass," while bison, "fenced for your protection," roamed the park's prairies and bluffs.[95]

Concerns about elk preservation were acted upon even earlier in Alberta, where the killing of elk was banned in 1908. Concern about their survival led the federal government to create an elk sanctuary (which later became Elk Island National Park) a short distance from Edmonton. In 1903 W.H. Cooper, a territorial game warden in Edmonton, reported that a small herd of seventy-five elk east of the city would likely be wiped out by hunters. A public petition demanded their preservation, and the Department of the Interior set aside about forty-two square kilometres as a sanctuary. In 1906 people from Edmonton and nearby Fort Saskatchewan began fencing the land, the area was designated as an "elk park" the next year, and twenty-five elk and thirty-five mule deer were enclosed there by 1908.[96] Elk were also found in other parks. A few elk at Buffalo National Park had come to the park as part of the Pablo purchase as well as from other sources, and their steady population increase gave the park a substantial herd. Elk numbers also grew rapidly in the mountain parks after about sixty were transplanted to Banff from Yellowstone National Park in 1917. Yellowstone was reducing its burgeoning elk herds by giving them away, and Commissioner Harkin was enthusiastic about acquiring some because they were rare in the Rockies and he believed they would be popular with tourists. Banff obtained 200 more from Yellowstone in 1920 and 100 arrived later that year at Jasper.[97] By 1923 elk populations located in the national parks were the only ones left in Alberta except for one herd in the foothills northwest of Rocky Mountain House.[98]

Elk were transplanted to the mountain parks even though officials had little understanding of their behaviour, reproduction, or the parks' ability to support them. By 1931 it was recognized that elk at Jasper had increased to the point where they were competing for food with moose and deer. The park's policy of killing wolves did nothing to help trim the population; instead of relying on natural predation, elk were slaughtered annually at Jasper between 1945 and 1970. Unlike the culling of bison, this seems to have raised little public comment, perhaps because elk were so clearly identified as big game. Like bison, elk were also given away; a carload from Buffalo National Park, for example, was sent in 1938 to the Cypress Hills Park and Game Preserve in Saskatchewan and another carload was released in the settlement fringe in northern Saskatchewan the following year.[99] Even so, elk populations expanded in the parks, as did conflicts with neighbouring communities. In 1946 at Waterton, for example, storms periodically drove elk from the park onto farmlands where they damaged feed and crops, prompting the province to declare an extended open hunting season that year.[100]

Such conflicts with elk were most intense at Manitoba's Riding Mountain National Park. By the late 1920s areas around the park were filling up with farms, and farmers were demanding an open season on elk to limit crop damage. The Manitoba Game Commission endorsed these demands in 1934, observing that there were too many elk for the food supply in the park, but the province refused to relax its ban on elk hunting.[101] This policy became untenable as the elk population in the park exploded. From 3,500 elk in 1933, the population reached 6,000 in 1941 and 16,800 by the fall of 1946. While farmers' complaints intensified, the situation was not clear-cut. Oral histories indicate that farmers were already killing elk illegally or with the connivance of park wardens, who in the 1940s turned a blind eye to local hunting as long as only a few animals were taken and the meat was not sold. The parks branch also allowed farmers to graze cattle in the park (a practice that only ended in 1970), which led some observers to conclude that this reduced the food supply for elk and pushed them into farm districts. Further, even though it was clear that Riding Mountain was overpopulated and that damage to surrounding communities was a genuine problem, one 1947 study indicated that the damage was sometimes exaggerated.[102]

Over the next decade, various solutions were proposed to address farmer-elk conflict. Some observers recommended that Manitoba pay compensation to farmers, but the province refused to do so, fearing the precedent this would create. Reducing the herd in the park by slaughtering surplus animals and allowing hunting in the park was also proposed, as was moving surplus elk to other national parks or to other parts of the province. In 1948 the province allowed an open season on elk in the districts around Riding Mountain, and open seasons were declared from time to time throughout the 1950s. Attempts to herd elk back into the park with helicopters and by building "elk-proof" and electric fences were undertaken in the early 1950s as well.[103] By 1959 wardens were authorized to undertake "herd reduction" by shooting surplus animals. The pragmatism of this management approach was effective, and after 1963 elk populations in the park ranged between 2,000 and 6,000 animals.[104] As in the national parks in Alberta, elk conservation in Manitoba had in just a few decades been transformed from a project to protect an endangered species to one that also had to manage that species to prevent damage to surrounding communities.

Figure 11.7 Elk were widely said to embody a virile wild beauty. Once very common on the prairies, they had been almost exterminated by the First World War. Efforts were made to protect the few that remained in national parks in the region. These elk were photographed in Banff National Park. Photo not dated.

NATIONAL PARKS AND ENDANGERED SPECIES: PRONGHORNS

Pronghorns were probably once as common as bison on the plains and were associated in the public mind with the mythic qualities of the Old West. They declined steadily after the 1880s because of hunting and the loss of habitat. The increasing use of barbwire fencing also posed a threat because pronghorns did not jump vertically (although there is evidence that they are now developing this ability), which caused them to become entangled or cut by the wire. By 1908 the fencing of the CPR mainline was completed, which caused additional difficulties for pronghorns by creating a long barrier to their north-south migration. Using historical accounts and calculations of carrying capacity, Ernest Thompson Seton argued in 1909 that the North American pronghorn population had been 20 million at its height but that he would be "agreeably surprised" if 100,000 remained. Half were in Mexico, and in Canada they were found only in Saskatchewan and Alberta. None had been seen in their historical range in Manitoba for decades.[105] Many of those in Canada had died in the winter of 1906–07 when heavy snow prevented them from reaching food, and two years later, populations had still not recovered.[106]

Banning hunting was one obvious way to preserve the species, but Benjamin Lawton, Alberta's game commissioner, also advocated placing some animals in parks where they could breed and be protected. Superintendent Douglas at Banff purchased some young animals for the park's paddock, and there were also a few at Buffalo National Park. However, these captives did poorly, and it was reasoned that pronghorns had a "delicate, capricious and easily upset" nature and those that did not die of the

Figure 11.8 Pronghorns, like bison and elk, were charismatic animals threatened by habitat change and hunting. Paralleling the models used for the conservation of bison and elk, national parks were created specifically to shelter them. Their appeal is evident in this photograph of a pronghorn in the Alberta badlands in 1944.

shock of capture remained psychologically and physically weakened. Further, it was thought that those that survived capture often died because of an unknown nutritional imbalance. These observations indicated that a preserve with natural range was the only way to save them. On the recommendation of the Saskatchewan game commissioner, several areas were set aside in 1910 by the Department of the Interior as temporary preserves until permanent sites could be selected. Despite ongoing recommendations for the creation of such permanent preserves, the federal government moved slowly.[107] The parks branch finally hired Ernest Thompson Seton in 1914 to select suitable sites, and he and the branch's Maxwell Graham toured southern Alberta and southern Saskatchewan and identified three appropriate locations. Two were in Saskatchewan – one was on the north slope of the Cypress Hills near Maple Creek and the other was near Old Wives Lake southwest of Moose Jaw. The Alberta location was north of Medicine Hat and was the largest of the three at 140 square kilometres. Again, progress on implementing the plan was slow, but the government was forced to act in the winter of 1915 when 200 starving pronghorns were found feeding on a ranch in the Nemiskam district south of Medicine Hat. Workers started to build a fence around the herd as it grazed so that it could be contained with minimal disturbance, but the herd moved on before the fencing was complete. Despite delays in obtaining funding approval for more fencing, another nearby herd of 42 animals was subsequently fenced in on land owned by Edgar McHugh. McHugh had a special fondness for pronghorns and a deep concern about their fate. As part of the parks branch's plan to establish a preserve, an arrangement was struck with McHugh whereby he deeded five sections of his land to the branch in return for land elsewhere. This land (and apparently some adjacent federal land) was declared Nemiskam National Antelope Park in 1915. Totalling about 5,000 acres, it was the first such preserve in North America. McHugh agreed to serve as warden, which he did without pay until 1919 when he was put on a part-time salary. By this point, the population of the park was nearing 100 animals. At about the same time, two of the sites selected by Seton were made into preserves and in 1922 were designated as national parks. The site north of Medicine Hat (which is now at the edge of Canadian Forces Base [CFB] Suffield) was named Wawaskesy (Cree for "pronghorn") National Park and the one south of Maple Creek became Menissawok (Cree for "common property") Antelope National Park. The site selected at Old Wives Lake was abandoned. In addition to these public initiatives, there were a few private ones as well. Charlie Blazier, a former wolf hunter who raised fawns on what was called the only "antelope farm in the world," near Brooks in south-central Alberta, believed that "it would be a shame" to see pronghorns wiped out. By the mid-1920s he had a successful record in supplying pronghorns to reserves and other sites in the United States. Although little is known about his endeavours, a Mr Lloyd of Davidson, in south-central Saskatchewan, also had a few pronghorns on his farm but had uneven success in raising them.[108]

The parks branch justified its network of pronghorn parks by noting that Nemiskam had proved that pronghorns could thrive in a park environment and that protection was essential because there were only about 250 wild pronghorns left in Saskatchewan and only about 1,000 in Alberta. The parks branch forecast that without a sanctuary the species would disappear from Canada because of illegal hunting.[109] Yet of these parks, only Nemiskam was successful. The Alberta writer and photographer Jim Gibson contends that "Edgar McHugh's enthusiasm made Nemiskam work. Certainly, the other reserves and parks, which had no such local boosters, never amounted to much." McHugh drilled wells, constructed dams, completed the fencing of the park, monitored the animals carefully, and added new ones to keep the herd healthy. In contrast to the way in which the relentless growth of bison and elk populations in other national parks was dealt with, McHugh released surplus pronghorns into the wild to keep the park's population manageable. This also helped build up wild pronghorn populations elsewhere in the region. Although the park's population of 42 animals in 1915 had increased to 405 by 1927, it usually stood at around 300. The animals were healthy and disease free, and in 1921 there were more pronghorns at Nemiskam than in the two larger reserves in the United States.[110]

While the parks branch deserved credit for designating and protecting physical spaces for pronghorns and for making a small annual expenditure for salaries and supplies, the real debt was owed to Edgar McHugh, whose work probably saved the pronghorns of the Canadian plains. This was demonstrated after McHugh's death in 1938 when the park began to deteriorate. The practice of selecting new stock and releasing surplus animals was abandoned, fences fell into disrepair, and Nemiskam had only

50 animals by 1941. Because the park had lost such a strong local defender, ranchers in the area began to cast envious eyes on the high-quality grazing lands it contained and to pressure the government to close it and lease it for cattle grazing. While some MPs protested its closure, local ranchers got their wish in 1944 when the park was turned over for cattle grazing and the remaining pronghorns were released. In 1947 the federal government transferred the land to the province in return for land needed for the expansion of Elk Island National Park. Alberta turned Nemiskam into a community pasture, and by this time the other pronghorn parks were no longer operating as well. Menissawok had been a failure and was closed in 1930 as part of the rationalization of the parks network under the new national parks legislation. The land was handed over to Saskatchewan, which apparently leased it as grazing land. Wawaskesy had an equally undistinguished history. The land was not fenced, and rather than looking after the few pronghorns it held, "the part-time warden was kept busy dealing with local landowners who held grazing rights to park lands" and who "complained constantly that park pronghorns were damaging grazing lands outside the park." The Alberta government supported the landowners, and in 1938 the federal government turned the land over to the province.[111]

The parks branch justified its abandonment of the pronghorn sanctuaries on the grounds that there were now sufficient pronghorns in the wild to guarantee the future of the species. A pronghorn census in 1945 claimed there were 30,000 pronghorns on the prairies, and while this estimate was probably high, it was clear that the species had been saved. While author Jim Gibson argues that the establishment of Nemiskam in 1915 made "Canadians aware of the importance of pronghorns as part of their heritage," this did not mean that the threat to them had disappeared. Even when the animals were clearly still endangered during the 1920s, they were hunted illegally.[112] And by the 1930s, when pronghorns were increasing, farmers and ranchers saw the growing wild herds as a "nuisance" because they caused crop damage and competed with livestock for food. As a result, both Saskatchewan and Alberta declared short open seasons to reduce their numbers, and some of the hunting that followed showed that bloody-mindedness was still a force to be reckoned with.[113] The *Calgary Herald* estimated that in 1934 during a ten-day open season 10 per cent of wild pronghorns in Alberta were legally and illegally killed, often in barbarous ways. As the *Herald* editorialized, new and better relationships were needed, and it appealed to the familiar argument that hunters needed to treat hunting as an ethical contest. Implying that the province had allowed hunting because of rancher complaints, the paper noted that if pronghorns had to be reduced because they were a nuisance, "it should be done in an orderly and humane manner. But to chase them with high powered cars until exhausted, herding them into fence corners to be riddled with buckshot or mutilated by contact with barbed wire is not sportsmanship. It is simple slaughter, pandering to a blood lust far removed from decency."[114] It was graphic evidence of why the use of national parks as dedicated sanctuaries had been critical for the survival of pronghorns on the Canadian prairies.

A COMPLICATED LEGACY

The establishment of national parks on the prairies mirrored significant public support for preserving certain species as a public undertaking. Public agencies took pride in bringing bison, elk, and pronghorns back from the brink of extinction, a view that continued to be exhibited in new contexts. In 2006, for example, plains bison, most of them sharing ancestry with the Pablo herd, were reintroduced from the Elk Island herd to Grasslands National Park in southern Saskatchewan, an area where bison had not been seen for more than a century. Yet wildlife conservation through national parks before the 1960s was selective. Not all wild animals were equal – charismatic and useful ones earned positive attention, but those thought to be threatening or without merit earned enmity or neglect at best. This reflected a general attitude that defined the value of wildlife in terms of economic and social needs, the prejudices of tradition, and immediate political and social priorities. Bison, elk, and pronghorns merited special attention as unique animals, but ones that could no longer find space in an agrarian economy. Support for their preservation endorsed a broader quest for identity and nostalgia for a way of life that was no longer compatible with or even desired in the new west. But if preservation of these charismatic animals in geographically restricted parks satisfied public needs, it did not solve the conundrum of how to manage such confined populations. Associated with this conundrum were technical and herd management problems as well as the challenges inherent in the double mandate's attempt to combine conservation and economic development goals. These latter goals framed such policies as the disastrous shipment of plains bison to Wood Buffalo National Park, the exclusion of Aboriginal people from their traditional hunting areas, and conflicts with neighbouring communities as elk and other animals moved from the parks into farming districts.

While these problems were ongoing, some were mitigated through good luck. The discovery in 1959 of an isolated herd of healthy, pure-bred wood bison at Wood Buffalo National Park serves as one example of this. In 1962 these animals were shipped to Elk Island where they were segregated to rebuild a disease-free wood bison gene pool. Yet, as was the case in all the national parks on the prairies, the legacy of the double mandate remained at Elk Island. As historian Graham MacDonald argues, the park's identity had been marked by "a fundamental ambiguity" from the start and "the emergence of the wood bison issue in the mid-1960s served to compound the dilemma." The idea that park visitors "should be able to 'see' animals was a longstanding one, as it was in many other National Parks. The need to reconcile the game sanctuary idea with visitor beliefs that they were in some kind of 'zoo' was a more or less constant policy question." This conflict was intensified by landscape change in the park; in earlier times, "when the forest was less dense, and travel more leisurely, the visitor would normally have his wishes satisfied. This was not the case in the late 1950s however. Fire suppression had aided in bringing on a forest in which the animals were often well hidden." Thus a paddock was developed in 1963 to show people the animals they had come to see.[115]

Beginning in the 1960s the national parks system spent considerable effort in reshaping its approaches as new public expectations were expressed and new theories about wildlife management and knowledge about ecosystems became more broadly accepted. These developments marked the 1960s as a transitional period in which approaches to wildlife and public participation in the parks' environment were changing. Regular slaughter as a solution to overpopulation began to fall out of favour when different wildlife management methods to control populations (but which still often involved culling) were adopted across the national parks system. As Graham MacDonald notes, "public pressure was mounting on parks officials that wild animals should not have to be routinely slaughtered. The pursuit of more sophisticated wildlife controls and methods increasingly became the response, and this took the form of a more or less sustained commitment to habitat research (such as browse studies) and experimentation with selective culls so as to induce preferred population sex-rations." Elk Island, for example, gradually phased out feeding the animals and ended hay production in 1967. Similarly, by 1959, when bison had spread beyond the boundaries of Wood Buffalo National Park, sport hunting through licensed outfitters was allowed, thus providing another method of population control. In 1967 commercial slaughter was also abandoned at Wood Buffalo.[116]

The perception of the parks as national playgrounds also began to change in the 1960s. In 1959 Parks Canada created an Education and Interpretation Section and began developing interpretive programs on ecology, zoology, geology, and history to educate visitors about the parks.[117] This, at least in part, tempered the view that wild animals were an entertainment feature and gave them a role in public education. Debates about the purpose of the national parks took on a new dimension in the mid-1960s with increasing environmental concerns. As the writer and naturalist Andy Russell remarked at the time, "if a firm hand is not taken with commercial developments in our national parks, they will ultimately be turned into oversized and trampled picnic grounds." In light of such demands, the federal government in 1964 confirmed the inviolability of the national parks and declared that commercial exploitation of park natural resources would be curtailed to let Canadians "gain the greatest long-term recreational" and educational benefits from them. Commercial entertainment culture would be restricted and confined to park townsites, and park planning would now prioritize the enjoyment of the outdoors and natural environments and would control and preserve "wilderness" areas for hiking, canoeing, and such uses. While development pressures remained high and the double mandate that had bedevilled parks wildlife and nature conservation remained, for some this new approach was a welcome halt to what they saw as the ever-increasing commercialization of Canada's national parks. Change would not, of course, prove to be easy, and the working out of these polices and philosophies would be a source of ongoing contention and debate in the next decades.[118]

12

Rounding Out a Full Life: Traditions of Natural History

We live in a fascinating world – a good naturalist knows just how fascinating. Much of its fascination stems from its mystery. While we know much about it, there is much we do not know.

Frank Brazier, Saskatchewan Natural History Society, 1955[1]

Interest in the natural history of prairie Canada has deep roots. Soon after its founding in 1670, the Hudson's Bay Company conducted meteorological studies and collected specimens of plants, animals, and minerals in northern North America for the Royal Society. During the next 250 years other scientific agencies in Britain and North America enlisted HBC personnel for this purpose. Company officials such as James Isham and Andrew Graham and explorers such as Samuel Hearne (who the Saskatchewan naturalists Stuart and Mary Houston called "the most talented of the early naturalists on this continent") made significant contributions to the collection and description of animals in western and northern Canada.[2] This collecting was compatible with the Enlightenment ambition to catalogue and know the entire natural world, a quest that endured in nineteenth-century explorations. The Franklin expeditions, for example, collected a wide variety of flora and fauna, and catalogues of these specimens were published between 1829 and 1837. The British Palliser Expedition of 1857–60 (especially the work of its scientist and physician, James Hector) and the Canadian Dawson-Hind expedition of 1857 added to this knowledge. Beginning in 1872, the Geological Survey of Canada conducted extensive geological field studies as well as ones on the region's flora and fauna, which included work by A.R.C. Selwyn and the botanist John Macoun, who conducted research and surveys from 1872 to 1881 in the prairies and other parts of the country. The Geological Survey published many of the studies done by its staff and others, and by 1898 it had published 655 papers on various topics. Additional studies also appeared as stand-alone titles or as articles in the transactions of the Royal Society of Canada or in scientific journals.[3]

This interest was often rooted in people's relationships with wildlife based on needs for food, commerce, sport hunting, protection of farming, and the stimulation of tourism. But it also grew from a fascination and curiosity about wild creatures and nature. Nature columns in newspapers reflected a personal interest in wildlife on the part of writers and readers alike. Natural history societies also reflected this dynamic, although like the personal passion for birds that led to the creation of private bird sanctuaries, they incorporated various utilitarian agendas as well. In addition to such institution-oriented expressions, a wider, more diffuse interest in watching, photographing, and interacting with wildlife was frequently apparent. While unified by intellectual curiosity and often by personal affection, the concerns of scientists, naturalists, wildlife enthusiasts, and people simply interested in wildlife did not occur in a vacuum but reflected a loose set of attitudes and commitments with deep historical roots. The appeal to nature as an antidote to the soullessness, poverty, and ghastly living conditions of urban life after the Industrial Revolution informed one part of the origins of public interest in natural history. So, too, the expansion of science and the revolutionary thinking of Darwin and others spurred interest in nature study, while nineteenth-century travel, exploration, and adventure narratives that described encounters with strange places and wild animals often presumed a regret for the loss of former times when people had supposedly felt closer to nature.

APPRECIATING WILDLIFE

When Fred Bradshaw, Saskatchewan's game commissioner, argued in 1913 that people who did not hunt had as much right to a say about game policy as hunters, he was arguing on behalf of naturalists and other people who were formally or casually interested in watching, studying, or appreciating wildlife.[4] Although many people related to wild animals in these ways, they were at times marginalized, especially during the settlement era. In an eight-part series about common prairie birds in the *Grain Growers' Guide* in 1915, Dr H.M. Speechley, a bird enthusiast at Pilot Mound, southwest of Winnipeg, noted that when he arrived in the district in 1901, he had asked people in the town about local birds. Their replies had been uniform: "everything that wasn't a hawk, owl, prairie chicken, goose, duck, meadow lark or robin was a 'little brown bird.'" Speechley perhaps envisioned more specialized knowledge than it was reasonable to expect, but he did note that "observation has sharpened of late years and eyes have been opened." Even so, he admitted that people still sometimes "passed by birds and flowers and such things so absorbed in wheat, cattle or machinery that they have simply been blind to them, just as a color-blind man cannot see the brilliant orange lily of our prairies. They do not see any value in noticing these wonderful gifts of the Almighty, and talk of people who do take notice as 'star gazers' or 'bug hunters,' or, as in my own case, the 'canary and weed man,' sometimes playfully, sometimes more or less scornfully." Speechley was not, however, "ashamed or apologetic" about his fascination with birds. As he asserted, "The old Book with supreme wisdom says: 'Man doth not live by bread alone,' and that 'A man's life does not consist in the abundance of things that he possesses.'" Yet he was eager to argue that birds were useful – which made them even more amazing – and he appealed to the utilitarian sentiments of his compatriots by noting that birds had great economic value for farm and town folk alike by controlling insects, weeds, and rodents.[5]

Two decades later, Neil Gilmour, a Saskatchewan game guardian, also confronted apathy about birds in Saskatchewan. In a speech to a teachers' convention in 1920 in Shaunavon, southwest of Regina, he noted that unlike people in other parts of Canada or the eastern United States, people in Saskatchewan had little interest in wild birds. Most settlers were poor and "came here with the hope of bettering their condition," and in their obsession with business and work, many things "deserving of attention, and that should have well repaid attention, were overlooked." One of these was "the value of our bird life," which like all beautiful things in art and nature exerted an "uplifting" and "refining influence" on human character. And in "all the realm of nature, there is nothing so beautiful as our wild birds," all beautiful in form and motion, and many beautiful in colour and song. Anyone watching and studying them would "surely have a finer nature because of that acquaintance."[6] And others saw the matter in more directly personal ways. As one correspondent to the *Western Producer*'s Nature Lovers column wrote in 1928, she could not understand how people could not pay attention to the birds and animals around them. "Doesn't it make you wonder what you could do to wake such people up to see the magnificent beauty that ever surrounded them? Doesn't it make you feel like shaking them?"[7] But some did pay attention and found their interests supported and entrenched through natural history societies, bird and other sanctuaries, and articles about wild animals in newspapers and magazines. This appreciation could be intensely personal and rewarding, even in a society that often seemed not to value it. Tommie Shepherd, a farmer and naturalist who lived at West Plains in southwest Saskatchewan observed in 1941 that his hobby of "natural history" gave him "a better, and I might say, greater, understanding of life as a whole." Working in the fields was monotonous, and watching wildlife gave him something important to think about. "When the average person sees a small bird, if they notice it at all, it is just a bird. Whereas, if a person has only a very limited knowledge of bird life, each small bird is someone you know – an old friend to meet again. And that adds interest to the day's work." He had passed this knowledge on to his children and it all helped "to round out a full life."[8]

While watching or studying wildlife was a pastime or a hobby for most of these enthusiasts, some were able to build careers and develop a public persona around it. A few, such as William Rowan at the University of Alberta and his student Dewey Soper at the Canadian Wildlife Service, found full-time professional careers in the field. Others, such as Carl Rungius, a painter who specialized in charismatic wildlife (especially moose and elk); Angus Shortt, an artist technician at the Manitoba Museum who specialized in waterfowl and did work on commission for Ducks Unlimited and as a book illustrator; and Clarence Tillenius, who wrote for and illustrated magazines, sold paintings, and painted museum dioramas, found popular success (although not academic favour) as wildlife artists.[9] Other careers were based on the commercial collection of specimens. Until the early 1900s, European and North American museums and collectors commonly purchased specimens of North American flora

Figure 12.1 The study of natural history was said to enhance powers of observation and attention to detail. This 1915 photograph shows children at the Presbyterian [Ukrainian] Mission Boys' School in Teulon, Manitoba, on a nature study outing.

and fauna, and as a collecting ground, Canada was an exporter of natural history specimens, just as it was an exporter of minerals, timber, and fur. Until high prices and the growing capacity of large institutions to collect their own materials curtailed (but did not eliminate) the trade, commercial collectors helped meet this demand.[10] Percy Gregson, one of the early figures in the North-West Entomological Society, who farmed at Blackfalds near Red Deer in central Alberta, collected fleas, a collecting passion he shared with Baron Charles Rothschild of the wealthy European banking family. While it is unknown how Gregson and Rothschild made contact, Gregson was collecting fleas on his farm for the Rothschild collection as early as 1895.[11] More conventional commercial collecting was carried out by Ernest Thompson Seton, who collected birds, eggs, and nests for the Smithsonian Museum. He sent specimens to Washington throughout most of the 1880s; in June 1888, for example, 23 specimens consisting of 15 bird species earned him $13.45, a decent sum at the time. Seton carried out similar work for other institutions; in 1892, for example, the Manitoba government paid him $100 to collect and mount birds for the province's exhibit at the Chicago World's Fair.[12]

Ernest Thompson Seton fit the conventional mould of the late nineteenth-century adventurer-writer-collector and became wealthy and famous through his consulting, lecture tours, books, and stories. Although they did not enjoy the wealth or international fame that Seton did, others on the prairies crafted successful writing careers with regional magazines and newspapers. As was seen in earlier chapters, fiction was part of this regional output, but signalling a growing public demand and interest during the interwar years, several publications began to run an increasing number of non-fiction articles and features about wildlife and natural history. Some examples include contributions by Lawrence Potter, a bird enthusiast at East End in southwest Saskatchewan who provided notes on birds to the *Family Herald and Weekly Star* in the 1920s, and those by A.G. Lawrence, who began a column in 1921 called "Chickadee Notes" in the *Winnipeg Free Press* that featured reports about birds found throughout Manitoba, in the other prairie provinces, and in adjacent parts of the United States. In a demonstration of the column's popularity, the *Free Press* distributed, at no charge, 70,000 coloured posters about birds and their migration routes. In 1940 the column celebrated its one-thousandth appearance. In 1928 B.W. Cartwright began an equally popular and long-running column in the *Winnipeg Tribune* called "Wild Wings." In the 1930s Leslie Sara wrote a column for the *Calgary Herald* entitled "What Nature Shows Us" in which he reported on wildlife topics for the paper. Sara also delivered radio broadcasts on nature topics on a local radio station. V.W. Jackson of the Brandon agricultural college wrote extensively in various prairie magazines about nature, and beginning in 1923 he delivered radio talks over radio station CKY in Winnipeg and "had thousands of people seeing things" about the flora and fauna around them that "they had never noticed before." By 1946 he had delivered almost 700 talks.[13] Dick Bird, later president of the Regina Natural History Society, began hosting a Regina radio program in 1936 called *Camera Trails* that featured talks about bird sightings, plants, and other nature subjects. The program showed "that good nature subjects could be found near Regina," and listeners wrote in with stories about what they had seen.[14] In 1945 Bird toured the west giving talks at Canadian Clubs that "thrilled audiences" with stories about wildlife "in all the beauty of its natural setting." His talk in Edmonton, for example, featured "a complete sequence of the habits of white pelicans that inhabit the west and whose characteristics are little known to the average person."[15]

After the Second World War, the reach of some of these writers widened and new ones emerged. While some, such as Andy Russell, became prominent nationally, most writers continued to appeal to regional audiences. Elizabeth Roley Cruickshank, a well-known Saskatchewan naturalist, wrote about nature topics in the Regina *Leader-Post*'s weekly Nature Notes column under the pen name Liz Roley. Beginning in 1956, she wrote more than 1,500 original stories over the next thirty years.[16] Although his most important public work in regard to wildlife was only beginning at the end of the period under study here, the historian and agricultural scientist Grant MacEwan shared his concern about wildlife and especially broader environmental issues through his many writings, including his 1966 book, *Entrusted to My Care*. In addition to his painting, Clarence Tillenius wrote and illustrated regular columns about wild animals in the *Country Guide*, where he expressed his concern that "wilderness, something wild and free that finds response deep in man's innermost soul" – and which was essential for a full life – needed to be preserved. And among others, Stuart and Mary Houston of Yorkton in southeast Saskatchewan wrote extensively about natural history topics in addition to doing conservation work, such as bird banding.[17]

While the concerns expressed in postwar popular writing about wildlife were as varied as the authors, Kerry Wood stands out as one example of writers who reached readers throughout the region over a long period of time. Wood was from Red Deer, Alberta, and wrote for a number of magazines, but most consistently for the *Country Guide* and the *Farm and Ranch Review*. His writing career began in 1923 when he was sixteen, and he went on to write hundreds of articles and commentaries for prairie magazines and newspapers as well as for some publications further afield. He also wrote a number of books and delivered numerous radio broadcasts and talks. His early curiosity about nature prompted Grant MacEwan to wonder if Wood's desire to write had drawn him to nature or if "his discovery of Nature's charm gave him something about which to write." In 1925 he became the Red Deer correspondent for the *Edmonton Journal* and continued to freelance as well. Although he believed that steel traps were inhumane, animal rights were not part of his world, his time, or his thinking, and much of his output consisted of essays about the habits and lives of particular animals. Yet he was not shy about taking on popular prejudices and discussing wider issues. While he agreed with the conventional harsh attitudes towards starlings, house sparrows, and magpies, he argued in the early 1950s that coyotes were unfairly maligned, which was perhaps the most unpopular thing a prairie wildlife writer could say at the time. Even more controversial was his 1950 article in which he argued that .22 rifles, found on every farm for "pest control" and regularly used by many young boys, were a "menace" to wildlife and public safety and should be licensed. And in 1955, when Alberta was celebrating its fiftieth anniversary as a province and the air was full of talk about the glory of its pioneers, Wood began an article by listing the species that had vanished or were endangered because of Euro-Canadian settlement. He then proposed that "we should hold a memorial service for Nature's Heritage. And at it, we should dedicate ourselves to cherish and protect the good things left in our keeping in this jubilee year of '55."[18] While he was aware of the mindset of many in his community and often argued the case for wildlife on economic grounds, he did not hesitate to put forward affective and aesthetic arguments as well. As he wrote in 1960, "we urgently need the birds, the green coverts that shelter them, and nature's police force of predators to patrol the good earth. [But] it isn't only the dollars and cents worth of our wild allies. Without their delightful melodies and bright plumes, this would be a drab and sorry land." Wood's commitment to wild things was strong, and he often despaired about the unfailing eagerness of farmers to kill hawks, owls, weasels, coyotes, and other animals that actually served their interests. In 1957, when a nature sanctuary that he had helped set up near Red Deer was vandalized, its birdhouses and other features smashed, and as many birds killed as the vandals could get at, he was brought to despair by the "brutish contempt for the love of wild creatures" that this act showed. While it made him lament the time he had spent on "lost causes," he nonetheless concluded that "the world's history is scarred with human destroyers" and that when hearing the sound of some nearby birds, "somehow, we became happy again, and willing to go on fighting for the wild things we love."[19]

Doug Gilroy, who made his living farming in the Boggy Creek Valley at Lumsden in the Qu'Appelle Valley near Regina, was another writer who reached a wide audience in prairie Canada. He had the same commitment to wild things as Kerry Wood, and like Wood, he tried to secure broader public support for wildlife protection by pointing out how such protection made good economic sense for farmers. Yet his commitment went well beyond utilitarian arguments. In 1954 he began writing a weekly nature column in the *Western Producer*, which at this time had an approximate circulation of 160,000. By the end of 1965 he had written over 600 columns, and when he retired thirty-five years later, he had written a thousand more. By then, he had also published four books on prairie birds. He set out his personal philosophy early in the life of his column, stating that "when we share with others what we see, it is then that we all learn one more fact or habit of something, be it birds, mammal, bug or flower. If it were possible to live till we reached the age of 200 we still would not know half of nature's secrets. But if the facts we do know are shared and recorded we can leave them behind for the benefit of those that follow in our footsteps."[20] The column was open to discussion about all flora and fauna, but birds were his passion, although he confessed in 1954 that "I sometimes think nature students spend so much time studying birds that we neglect paying attention to our wild animals." He wrote about what he saw while working in the fields or when travelling around the region. Although he referenced new and useful books about wild animals, his columns generally consisted of a narration of everyday life in which wildlife played an integral role. He was not a scientist, and he used unassuming language and illustrated his columns with his own photographs or ones sent in by readers. His interest in what he saw drew in others, and the columns became a focus for a community of wildlife watchers and enthusiasts. They commonly featured questions, observations, and reports by people throughout the prairies. These dealt with insects, mammals, and plants but were most often about bird identification, and when Gilroy could not answer a question, he invited his readers to help. His columns had a broad appeal. John Hicke, a schoolteacher in Grayson in southeast Saskatchewan, used the column as a teaching aid, and as he told Gilroy in 1955, "each week these articles provide the most interesting nature science lessons. It further creates interest among the children to seek more information about the birds and animals mentioned in Prairie Wildlife."[21]

While interested in all matters relating to wildlife, Gilroy only rarely devoted space to hunting. In 1958 he noted that "Prairie Wildlife is a nature column as you know, and up to now we have never told of anyone's hunting experience with a gun, but this week are going to break away from the rule" with a two-part story about a big game hunter's experiences north of Prince Albert.[22] At least in the next decade, reports about hunting were not repeated, and while Gilroy did not seem to be hostile to hunting in principle, he clearly found photography a more personally rewarding way of interacting with wild animals. To meet an increasing demand, in 1961 he and Joyce Drew of the extension division of the Saskatchewan Museum of Natural History offered a ten-week series of classes on wildlife photography. As he had argued in 1954, photography was "the only possible means of preserving bits of the present in its own true light. Memories change and fade but pictures seldom do."[23] Gilroy also consistently reported on the activities of natural history societies, promoting them as vital agencies for encouraging knowledge about and appreciation of wildlife and nature. Regional natural history society publications, annual meetings, proceedings, and lectures found frequent reference in his columns. He also reported on such subjects further afield, especially the Federation of Ontario Naturalists, whose activities he saw as a model of proactive natural history education and programming.[24]

NATURAL HISTORY SOCIETIES

The term "natural history" is a broad but serviceable one that has retained its currency to describe popular research and inquiry about wild flora and fauna based on field observation. An encyclopaedic study of the natural world, natural history was first popularized during the Enlightenment. In Europe, natural history societies drew together academics, civil servants, hunters, and a broad spectrum of other people interested in nature. Activities rarely involved laboratory research but focused primarily on collecting, describing, and classifying flora, fauna and geology.[25] The study of natural history was said to bring people aesthetic pleasure and to stimulate in them the growth of reason through the exercise of classifying plants and animals into categories (taxonomy) and linking this taxonomic knowledge to a broader understanding of the natural world. As a hobby, it was a useful pleasure, for it brought improved powers of observation and transmitted aesthetic, moral, and social values. Percy Gregson of the Territorial Natural History Society made this argument in 1902 when he stated that natural history taught mental discipline by teaching people how to keep "a definite purpose in view" and heightened their powers of observation.[26] It also served religious sentiments, since the study of natural history was said by some to reveal the hand of God in the working of the world. Natural history societies that promoted and focused this inquiry were found throughout the Canadian colonies by the mid-1800s. Historian Carl Berger notes that in their "encyclopaedic study of all objects in nature," these societies "incorporated approaches that would in time become compartmentalized into botany and zoology, ethnology and meteorology."[27] In the nineteenth century, natural historians were beginning to approach the field through the lens of the disciplinary interests increasingly seen in the scholarly world. Thus, one group of members might study birds, another plants, and yet another geology, making many natural history societies into umbrella groups that expressed a social identification and a shared general interest in the natural world.

The creation of the Historical and Scientific Society of Manitoba in 1879 was the earliest formal expression on the prairies of such approaches to studying nature. From its founding, the society reflected a mixture of personal interest in natural history and economic and social goals. Reverend W.A. Burman, the president of the society, argued in 1900 that the most important reasons for nature study were intellectual and spiritual. He claimed that the study of the "enchanted land" of nature had given him "sincere pleasure and genuine profit" but, more importantly, it had brought "quickened powers of observation, with a keen sense of the beautiful; and of the all-pervading presence of Him 'who hath made all thing beautiful in His time.'" Burman knew "of no recreation that will so truly *recreate,* and yield such unalloyed and elevating pleasure, as can be found in the pursuit of some branch of Nature-study." Yet knowledge and appreciation of nature also helped conquer it. As Burman argued, "we must either woo the forces of Nature, making use of them for the maintenance of life, the upbuilding of an empire," and increasing prosperity, or "we must prepare to contest with the various adverse developments of those forces, for the right to live in, and to rule and enjoy, this portion of the world."[28] This view was widely held. Robert Bell of the Geological Survey argued in 1907 that nature study, in which he included biology, geology, and astronomy, "cultivates accuracy of observation and truthfulness of description, and helps us to be more logical and concise in our conclusions." It also disabused people of the "all too prevalent" idea that "the pursuit of gold is the one great object in life." But the transcendence of material wealth by knowledge was not a questioning of human power over the earth. Indeed, it was a messianic phase of human history, for "the time is at hand when scientific knowledge is going to subdue the earth, when man shall come to a conscious realization of his power over nature, and achieve that dominion promised him."[29]

This intersection of science, knowledge, and power led historian Gerald Friesen to note the ways in which the Historical and Scientific Society of Manitoba's promotion of scientific knowledge fit with ambitions for Winnipeg's economic development and its status as the most important city on the prairies. As he observes, the society aimed "to set out the truth – and the promise – of the region" by delineating its resources and their economic potential.[30] This accords with the views of other historians who have charted how the study of natural history formed part of European expansionism. Studying natural history helped identify the existence of natural resources as well as of natural barriers to commerce. It also provided useful information for the creation of commercial and military networks and helped to make some European cities repositories of knowledge about the world. So, too, by the mid-nineteenth century the study of the mechanisms that were thought to govern the dispersal of plants and animals across the globe, coupled with a growing tendency to characterize animal and plant life as a competition for resources and a struggle for survival, had shaped intellectual assumptions about the natural world that were used to justify the legitimacy of European adventurism and capitalism. As Karl Marx bluntly declared, the study of nature and its "laws" became "a ruse" for subjecting nature "to the needs of man, whether as an object of consumption or as a means of production."[31] In a similar vein, historian Suzanne Zeller argues that the expansion and cataloguing of knowledge about resources, land, and climate strengthened the popular desire of late nineteenth-century Canadians to build a transcontinental nation by providing them not only with "the practical means to dominate their physical surroundings" but also with "an ideological framework within which to comprehend the experience of doing so."[32]

Its affinity with expansionist ambitions ensured that natural history was a socially respectable and useful pastime for middle- and upper-class people in Victorian Canada. Not surprisingly, the twenty-six founding members of the Historical and Scientific Society of Manitoba included members of Winnipeg's elite.[33] Like other natural history societies in Canada, the Manitoba society was divided into branches that specialized in meteorology (which Reverend Burman asserted in 1900 was "a most important subject to an agricultural country"), history and ethnology, geology, botany, entomology, zoology, and ornithology. The society gradually built up a library and a collection of natural history specimens, collated the fieldwork of its members and correspondents, and maintained contact with individuals and other societies outside Manitoba. An 1884 report in the *Regina Leader* noted, for example, that specimens of parasitic hair worms found in locusts in Regina had been sent to Winnipeg for the society's collection.[34] While its library and natural history collection were the core attractions, the society also sponsored public lectures and publications on the history of Red River, ethnology, and natural history. By 1889 it had published thirty-three papers that had been read at its meetings. As well as disseminating knowledge through publications and talks, the society saw itself as setting research agendas and priorities in natural history. In his inaugural address as president of the society in 1889, Charles Bell celebrated the society's accomplishments but warned that much remained to be discovered about the west. Research was needed in geology (including palaeontology) and meteorology, botany remained "almost a virgin" field on the prairies, and entomology needed attention because of its relevance to agriculture. It was also important to encourage the study of wild animals to capture knowledge of a vanishing world, as "the advance of civilization is surely driving away animals abundant but a few years ago," such as grizzly bears and bison. Indeed, Bell contended that research on almost all wild animals was needed. There had been no study of reptiles or molluscs in the northwest, and while some research had been done on birds, many gaps in knowledge about them remained. And economic needs clearly inspired his other recommendations. While the eight- to ten-year cycle of rabbit populations was shown in the HBC fur returns of lynx, more needed to be discovered about this cycle. As well, the role of gophers and badgers and moles in soil aeration and soil quality had often been commented upon but not studied. So, too, the society had published nothing about fish, even though research was urgently needed because there was much confusion about the names of different species and no knowledge about fish behaviour or population cycles.[35]

Notwithstanding such lofty goals, it was the library – with over 10,000 volumes by the late 1880s – that bound the public to the Historic and Scientific Society of Manitoba. In 1884, for example, the society had 269 members, most of whom did not contribute to its scientific or historical work and were primarily interested in having library privileges. In fact, in the 1880s, only between 30 and 60 members were actively involved in research on scientific and historical topics. But the library, even after a librarian was hired, was an onerous responsibility for a voluntary organization, and when the chance came in 1905 to transfer it to Manitoba College and the new Carnegie public library in Winnipeg, the society seized the opportunity, apparently seeing it as a way to pursue its scientific and historical objectives unhindered. At the same time, the geology specimens in the natural history collection were transferred to the University of Manitoba and the rest of the collection was put in storage. Despite hopes that this devolution would offer greater freedom, it instead sealed the fate of an organization that was already faltering because of social change. By 1900 Winnipeg was becoming more socially and intellectually complex with an expanding range of institutions and activities that claimed its citizens' attention. Moreover, as Gerald Friesen notes, the society's function as a "public relations and scientific research institute had been taken over by a dozen agencies including the government experimental farm, the University, the many local newspapers and other journals," as well as by public schools and libraries. Without the focus of its library and collection, the society declined rapidly. By 1910 it was moribund and an attempt to revive it in 1913 failed.[36]

The First World War was disruptive for all social and cultural activity, including natural history. Although a branch of the American Audubon Society was established in Winnipeg in 1915, it did not succeed. After the war, there was renewed activity in natural history with the founding of new societies in Manitoba and Saskatchewan. These societies shared many of the traits of earlier ones, with members interested in a wide variety of topics and having a strong curiosity about the natural world. In other respects, however, they showed different orientations and emphases. Collecting was no longer so central an activity, although it still went on. Publication and dissemination of research remained central concerns, but research and observation were now less personal and had a stronger collective character. As illustrated in the widespread adoption of bird banding as an ongoing project of natural history societies, activities that contributed to regional, national, and even continental collective knowledge became more pronounced. These changes approximated those taking place elsewhere, especially in Britain, where the "new ornithology" of the interwar years brought about what historical geographer Mark Toogood calls "a minor revolution in the underlying ontology and purposes of nature observation." In this "minor revolution," "the local, normally amateur, naturalist moved away from a supposedly meandering search for stimulation" through the collecting, displaying, and sorting of wild specimens. In its place, observations were ideally recorded in consistent format using standardized field surveys. Observations were collated and archived at a national level, making local observation part of collectively formed national knowledge. This was not so much a clear break from earlier practices as it was a "building upon a 'critical spirit' to approach the study of nature in an apparently serious way, unburdened by what self-styled 'modern' naturalists saw as the ignorance and selfish concerns of latter-day Edwardian naturalists for specimens, lists and number of records for personal use." The goal was "to encourage individual observation of, reflection and attention to the living animal *in situ* as well as collective and systematic endeavour" as part of a progressive national undertaking.[37]

Although it found a more limited application in practice, the change taking place in British ornithology was reflected in natural history societies on the prairies during the interwar years. The Natural History Society of Manitoba (renamed the Manitoba Naturalists Society in 1971 and known as Nature Manitoba since 2009) was founded in 1920 to foster research and appreciation of the natural environment. In the traditional pattern, it was divided into disciplines, including ornithology, zoology, botany, entomology, and geology. As in the older style of natural history, these disciplinary boundaries reflected individual interests, but they now also demonstrated the increasingly firm lines that divided various disciplines in academic science. The society reconciled these divisions by admitting their importance and using them to benefit collective effort and efficiency within a community of practice. The recognition of disciplinary boundaries made the society stronger by preventing conflict among members and inefficiencies caused by "the competition of one set of naturalists against another." Reflecting what was evidently seen as a more progressive social context for natural history, the new society no longer drew the majority of its members from Winnipeg's commercial and professional elites. Holding their meetings at the University of Manitoba, the society's membership varied between 125 and 250 between 1920 and 1940, and as the society's historian remarked in 1940, its members represented a mix of "real scientists with amateurs," all of whom were superior people, for "true naturalists are simple plain determined people of all ages, not 'Society' folk and not selfish folk governed by superficial aims in life." The society included many women members among the university professors, civil servants, and others, all of whom were committed to attaining "high scientific ideals."[38]

The Natural History Society of Manitoba was active on a number of fronts. It published monographs and newsletters and organized public talks, special events, and field trips. It did not eschew collecting but no longer saw it as so central in its mandate, valuing to a greater degree the study of wildlife, geology, and landforms in the field. In keeping with international trends in natural history, there was a growing commitment to a broader community. Beginning in the early 1920s, the society organized and coordinated bird banding as part of the international effort to track the movement of birds. It was also involved with applied research at the community level. In 1927, for example, it established and coordinated the Winnipeg Anti-Mosquito Campaign. Nothing could have been of greater public service; with twenty-two species native at the forks of the Red and Assiniboine rivers, the city was infamous for its plagues of mosquitoes, and the expertise of society members helped create control programs to make city life more tolerable in summer. Members contributed to scientific knowledge in other ways as well, especially in ornithology, which was a strong interest of many society members. Through their work, for example, the Manitoba bird list of known species grew from about 250 to about 325 species by 1940, creating a fuller record of bird species in the province. Members also compiled data about avian biology, habitat, and migration, and some members individually carried out research on specific species.[39]

While the study of natural history on the prairies had the strongest historical tradition in Manitoba, there was also a relatively long record of natural history societies in Saskatchewan and Alberta. Under the patronage of the territorial government, the North-West Entomological Society was founded in 1898–99 by Dr Henry George of Innisfail in what is now central Alberta. Showing a high degree of specialization and the particular enthusiasm of its members, the society aimed to increase the knowledge of prairie insects beneficial or injurious to farming. Despite its economic importance, this focus gave the society only limited public appeal.[40] It was therefore reorganized in 1902 as the Territorial Natural History Society, with a more conventional focus on botany, entomology, and ornithology.[41] In 1904 Dr George noted that the principle object of the society was to discover "the farmers' friends and foes, not only in the insect and vegetable worlds, but also amongst the animals and birds," and to campaign for the protection of economically valuable birds.[42] With the formation of Alberta and Saskatchewan as provinces in 1905, the organization became known as the Alberta Natural History Society. It continued on the same path as its predecessor, but it failed to draw a substantial membership and the departure of Dr George for British Columbia in 1923 seems to have dealt a death blow to the organization.[43]

In Saskatchewan, an organization with natural history interests came about when forty-eight enthusiasts formed the Regina Natural History Society in 1933. It was set up with the conventional natural history divisions of ornithology, botany, entomology, and geology, but, showing an important shift from earlier concerns in natural history, it included a division focused on conservation.[44] Demonstrating concern for community service and the social relevance of natural history, the society pledged "to maintain an open mind to grasp the meaning of conditions new and old affecting the life of our province" and to "devise measures for the protection and conservation of wild life." Consistent with this goal, the society submitted nine resolutions to the provincial government in 1935, asking for "nature sanctuaries, water conservation, protection of water fowl, special attention to the prairie chicken population, protection of beaver for its water conservation practices, prohibition of shooting female moose and more educational literature for hunters and [the] public." In about 1946 it leased land in the Qu'Appelle Valley north of Regina and built an interpretive trail. Field trips to other sites and talks by society members and others about natural history rounded out its activities. The latter included fund-raising speeches by notable figures such as Ernest Thompson Seton and the Banff naturalist Dan McGowan. Indicating the popularity of such speakers, McGowan drew 300 people to his presentation.[45]

A society with goals much like those of the Regina Natural History Society was formed in Yorkton in eastern Saskatchewan. While local birdwatchers had organized field trips as early as 1936, Isabel Priestly and Stuart Houston formed the Yorkton Natural History Society in 1943. Its most notable undertaking was the publication of a magazine, *The Blue Jay*.[46] In 1949 natural history activities in Saskatchewan were reorganized with the formation of the Saskatchewan Natural History Society with chapters in Regina and Yorkton. The goal of the provincial body was to promote a "better understanding and utilization of our wildlife and natural resources; this heritage once lost can never be regained." The society took over responsibility for *The Blue Jay* so that it could help build a province-wide community of interest and disseminate its members' research. By 1951 the magazine had 1,000 subscribers and was being published with the assistance of the Saskatchewan Museum of Natural History. This was only one example of the close and productive relationship between the museum and the society. Society members helped with museum programs and special projects, including the publication in 1958 of *A Guide to Saskatchewan Mammals* by W.H. Beck of the University of Saskatchewan's biology department. An update of W.A. Fuller's 1944 book, the volume upheld enduring traditions in natural history by including a section on collecting specimens.[47] While birdwatchers formed the core of the Saskatchewan Natural History Society, the society maintained a commitment to broader wildlife conservation issues and to the formation of a community of practice in natural history. In 1953, for example, it protested the Saskatchewan Fish and Game League's magpie-, crow-, and hawk-control programs, and the next year it expressed concerns about the rapid clearing of woodlands and its effect on wildlife and soil erosion. The society also recommended a closed season for sharp-tailed grouse and all large white birds (as part of rescue efforts for whooping cranes) as well as greater emphasis on nature study in schools.[48] By the mid-1950s the Regina chapter had purchased 320 acres in the Qu' Appelle Valley for a wildlife refuge where wild animals could be watched in their natural setting and, the chapter hoped, "wild things will work out their own lives without our help or hindrance."[49]

A PASSION FOR BIRDS

The widespread fascination with birds among naturalists and natural history society members was shared by many other people. Indeed, in 1934 sixteen-year-old Ina Watson wrote in the *Western Producer* that birds were "the most remarkable of God's creatures." Their ability to fly was "a power that even man envies," and she so admired it that she was "almost willing" to trade her "intelligence" for it. In 1928 another young correspondent had written in the same column that birds made "a very interesting study," which involved identifying them and learning to recognize their sounds.[50] Adults were not immune to the charms of the bird world either. Fred Bradshaw, a former Saskatchewan game commissioner and by then the director of the provincial museum, argued in 1930 that the study of all phases of bird life and activity was a "sheer delight that only those who have tasted it can appreciate." And the nature writer Dan McGowan wrote in 1924 that Franklin's gulls were the farmer's friends because they ate grubs and worms as they followed the plough through the fields but, most importantly, they were beautiful and for "many a lone tiller of the soil weaving his way to and fro across the prairie wheatfields," the gulls that followed "in his wake give a sense of companionship which is of no small value."[51]

The empathetic relationships that people like McGowan articulated were demonstrated in various ways. In 1920 the Herbert Hughes family of Dolard in southwest Saskatchewan hung bones in trees (then a common practice) near the front door of their home and kept scraps of meat on a window ledge for chickadees, magpies, blue jays, and other birds to feed on in winter. And in 1928 fifteen-year-old Leonard Burke of Aberley in southeast Saskatchewan wrote that he found birds so interesting that he did what he could to save them by moving their nests when he was ploughing the fields.[52] Other people followed their own cultural traditions. As a Saskatchewan correspondent in the *Western Producer* remarked in 1930, "during holidays I noticed a man had a large oat-sheaf placed on a tall pole so the snowbirds and others could come there for food. The sheaf was covered with our feathered friends. This is an old Norwegian custom which is done at Xmas time." While such personal relationships with birds were often not documented, B.W. Cartwright wrote in the *Winnipeg Tribune* in 1933 that "the past few years have witnessed a wonderful growth in public interest in birds." Indeed, he pointed out that a greater number of observers were keeping notes on bird arrivals and departures, providing important information for understanding migration.[53] Even though birdwatching was an informal hobby for most people, this suggested the emergence of a more broadly based community in which local observation was gaining a wider context and application. Birdwatchers had always shared information, but largely on an irregular and individual basis. Ernest Thompson Seton, for example, had appealed in 1885 to individuals in Ontario, Manitoba, and the NWT to cross-check his lists of Manitoba birds.[54] By the interwar years such networks had developed substantially. For example, the clipping files and correspondence of Dewey Soper, the federal government's chief migratory bird officer, include newspaper columns written by local birders that Soper referenced to help him understand the distribution, habits, and population of bird species. In the 1930s and 1940s he kept in close touch with birdwatchers and ornithologists across the country, such as William Rowan, Lawrence Potter, and Frank Farley. Farley, a bird enthusiast and hunter from Camrose in central Alberta, wrote *Birds of the Battle River Region* and planned to write (but did not complete) a book on all Alberta birds. He too had ongoing correspondence with other bird enthusiasts throughout Alberta and outside the province as well. In 1938 Farley was receiving more or less regular reports about birds from about thirty people.[55]

The popularity of birdwatching grew after the Second World War and drew on the long tradition of official and popular advice on building birdhouses and feeding stations to encourage insectivorous birds for the benefit of agriculture. While such utilitarian objectives did not disappear, postwar articles, even in the farm press, increasingly justified birdwatching for its aesthetic and educational value and its personal enjoyment.[56] A request by the *Farm and Ranch Review* in 1950 for reports of birds doing unusual things drew so many responses that the paper was "snowed under" with submissions, and in 1959 the *Regina Leader-Post* contended that birdwatching was increasing in popularity because of more leisure time and the "increased volume of reference material within easy reach of the amateur." It was also a socially respectable hobby. The *Leader-Post* informed its readers that while she had been "interested in birds for a number of years" and had watched them in her "tree-lined back yard," Mrs T.C. Douglas, the wife of Saskatchewan's premier, participated in the Regina Natural History Society's birdwatching outings. Group events like these helped to make birdwatching a collective endeavour and to create a cumulative record of observations. While no doubt encouraged by changes in the international culture of birdwatching in the interwar years, such collective efforts had been a feature of birdwatching in North America since about 1900 when the Audubon Society in the United States began to encourage a Christmas bird count to collate bird sightings on the same day across the continent. Although they may have been undertaken earlier, Christmas bird counts were widely conducted on the prairies after the Second World War. One began in Calgary in 1951, another in Saskatoon in 1954, and by 1963 the bird count in Winnipeg (organized for a number of years on Boxing Day by the Natural History Society of Manitoba) had become a well-established tradition.[57]

Public interest in birds during the interwar years found further expression through the development of sanctuaries. As is noted in chapter 11, a bird sanctuary established in 1887 by the federal government at Last Mountain Lake north of Regina was the first public bird sanctuary in Canada. In 1921 it became a sanctuary under the migratory birds convention legislation, and a number of other public sanctuaries were similarly established to protect migratory birds.[58] Additional sanctuaries were created through private initiatives. One of the most famous in Canada was Jack Miner's bird sanctuary at Kingsville in southwest Ontario. By the end of the First World War it was one of the largest and most important waterfowl sanctuaries in North America. A number of other sanctuaries, some inspired by Miner's example, were also established on the prairies. The Inglewood Bird Sanctuary in Calgary, for example, was established in the late 1920s on the estate of Colonel James and Euphemia Walker, located on the Bow River. Ducks had wintered there for decades, but by the 1920s they rarely stayed long because they were increasingly shot by hunters. W.J. Selby Walker, Colonel and Euphemia Walker's son, was interested in wilderness conservation, having been involved in the formation of the Canadian National Parks Association in 1923 to stop the Spray Lakes hydro project in Banff National Park. In about 1925 he was joined by George Pickering, a former mountain guide and "friend of animals," who began to manage part of the estate as an informal bird sanctuary. Pickering and his wife, Nina, lived on the site and tended to the needs of the birds. In 1926 the Pickerings fed 15 mallards that had overwintered there, and by 1928 this number had increased to 200. It was said that Euphemia Walker, who died that year, had known Calgary in the days when it was "a birds' paradise" and had wanted the land used as a bird refuge. Her hope was posthumously respected the following year when Selby Walker succeeded in having the site designed as a Dominion Bird Sanctuary. At fifty-nine acres, it was the largest bird sanctuary located inside an urban area in Canada.[59]

While sanctuary status prevented hunting at the Inglewood site, hunters soon set up across the river and shot the birds as they flew to and from the sanctuary. To address this problem, 350 more acres of land were added to the sanctuary, and it was enlarged again in 1932. By 1935 between 4,000 and 5,000 ducks were overwintering there. The sanctuary was a local attraction, inspiring Alberta poet Elizabeth Garbutt to write a poem describing how the birds came "home" to Inglewood at dusk on a winter's day and then fell silent as they settled in for the night. And like the birds at Inglewood, Garbutt hoped that "Uncompassed, lost mankind … Shall sanctuary find." More prosaically, Leslie Sara of the *Calgary Herald* valued the sanctuary because it saved ducks from being shot by American hunters during migration. Moreover, he contended that the birds that overwintered there produced stronger offspring earlier in the spring than those that migrated. There were other benefits as well. George Pickering (whom Sara called "Alberta's Jack Miner") attempted to crossbreed domestic ducks and geese with wild ones to produce hybrids with better meat. He also gave talks to and performed demonstrations for visiting school groups, offering "more knowledge of the ways of waterfowl than they would likely acquire by mere book study." And as Sara further noted, of the 6,000 visitors to the sanctuary in the previous two years, many were tourists who had been drawn to the city by the birds.[60]

In addition to its other benefits, the Inglewood Bird Sanctuary was of value to hunters and farmers. As Selby Walker pointed out to Premier Brownlee when asking the province for financial support, the sanctuary assured sport hunters "a reasonable supply of ducks" during open season as well as "a certain amount of flight shooting as these birds enter and leave the Sanctuary almost daily." And because the birds were fed in the refuge, farmers benefited because the birds caused less damage to crops. But feeding thousands of birds over the winter was costly. In the winter of 1931–32 eleven tons of wheat and screenings were consumed, mostly "donated by the elevator and grain men through the Calgary Fish and Game Association," which was "very much interested" in keeping ducks at home for Canadian hunters.[61] In 1936 Walker confessed that the sanctuary was having a "rough ride" financially because the money it had earlier received from the Department of the Interior for banding birds had been cut and it did not receive any financial support from the provincial government. The sanctuary earned some income selling seed and bread to visitors to feed to the birds, and in 1936 it stepped up its school programming "in the hope that these children would in turn bring their parents to the Sanctuary, who would possibly lessen the operating expenses through the sale of scrap bread."[62] But reliance on public attendance for support was double-edged. The public wanted charismatic birds, and as Walker told Dewey Soper in 1935, "'Have you any swans?' is asked very often by those accustomed to European parks." Since Inglewood had a collectors' permit under the Migratory Birds Convention Act for four whistling swans (tundra swans), Walker wanted to collect some to make the sanctuary more attractive to the public. "We need tame ones to call

Figure 12.2 As a tourist destination, the Inglewood Bird Sanctuary in Calgary was featured on this 1929 postcard.

them [migrating whistling swans] down to feed as they pass over, then we could get wonderful publicity and possibly some private support if a flock of swans could be seen and photographed in the Sanctuary." All that was needed was a way "to catch them."[63] But doing so was not easy, and although discussions between Soper and Walker stretched on, the difficulty and cost of the task ensured that it was never undertaken.[64] The sanctuary continued to have a hand-to-mouth existence, continually trying to find financial support. In 1936, when the Miner foundation was trying to raise money, the *Calgary Herald* editorialized that Inglewood should have the "first call" on Calgarians' generosity. Later that year, the local Gyro Club took on support of the sanctuary as a club project, and by 1938 the government of Alberta was paying George Pickering's salary. But the sanctuary's financial problems continued. In 1940 Selby Walker was again asking, still unsuccessfully, that the province help to fund the sanctuary. After Walker's death in 1953, Inglewood was purchased by the Alberta Fish and Game Association, but it gradually fell into disrepair. Interest in the sanctuary did not revive until 1970 when the city of Calgary bought some of the land and rehabilitated the sanctuary.[65]

Another urban bird sanctuary was established in Regina when the Saskatchewan Natural History Society and the Saskatchewan Museum of Natural History, along with a number of corporate and public agencies, collaborated to create the Regina Waterfowl Park within Wascana Park in the centre of the city. The project was driven by the vision and enthusiasm of Fred Lahrman, who worked for the museum. In 1953 Lahrman obtained a breeding pair of Canada geese with the goal of reintroducing the then scarce birds in southern Saskatchewan. Beginning with this single pair of Canada geese, the park was sheltering ninety-one geese by 1957, the product of natural increase and artificial incubation, as well as a variety of other waterfowl and marsh birds. By 1960 the goose population had increased sufficiently that the transplanting of birds to other parts of southern Saskatchewan could begin. When the park was declared a federal bird sanctuary in 1956, authorities gained greater control over public behaviour at the site. Until then, vandals had damaged the site, people had dumped dirt and garbage on the grounds and used them as a place to wash their cars, and dogs had run loose and killed the birds. The Saskatchewan Natural History Society lobbied successfully to have a warden appointed to police the marsh and to feed the birds in winter with grain provided by the society. Conditions quickly improved; by 1958 the Saskatchewan Museum of Natural History was able to use the sanctuary for school field trips and the park was said to provide a "unique opportunity for public study and enjoyment of waterfowl and other native species at the doorstep of Regina."[66]

Wild birds were also protected in an array of other sanctuaries that were inspired by varied motives, reflecting a sometimes interconnected range of relationships with wild birds. Some of these grew from family needs or expressed emotional connections with wild birds, while others promoted conservation to assist sport hunting and to respond to the crises faced by prairie waterfowl in the interwar years. Still others grew from a commitment to rescue and protect injured wild birds, and some even came to have commercial motives.

Unique in demonstrating family and personal needs was a sanctuary run by Reuben Lloyd at Davidson, southeast of Saskatoon. As the *Regina*

Leader-Post editorialized in 1942, Lloyd might not have developed his keen interest in birds "had his son not been in poor health when young. In an effort to keep the lad in the open air as much as possible, the father sought to interest him in the birds that were so plentiful about the farm. The plan worked quite well, and Mr. Lloyd soon found himself becoming as devoted as the boy to this new hobby." Everyone benefited: the son went on to become a naturalist at the Pittsburgh Museum, and Lloyd built a highly successful sanctuary, where he banded many of the birds. Personal fulfillment of a more abstract sort motivated R.W. Burley, who established a sanctuary on his farm near Cadillac in southwest Saskatchewan. After he flooded about twenty acres of dry slough bottom to create a habitat for ducks, Burley found "his whole life enriched by the presence of these handsome Mallards, Pintails and Shovellers" that had come to live on his place.[67]

Other sanctuaries, such as the one established by Ralph Stueck at Kerry Farm, east of Regina near Abernathy, combined a fondness for birds with the goal of increasing the numbers of waterfowl for sport hunters. In 1927 Stueck rescued "a beautiful young goose" that he "didn't have the heart" to kill after tip-winging it while hunting. Stueck saw his sanctuary as a way to increase goose populations on the prairies (in part to benefit hunters), and his success in the endeavour was further demonstrated by the fact that he supplied the breeding pair that formed the nucleus of the goose population at the Regina Waterfowl Park.[68] A similar mixture of emotion and pragmatism motivated Alfred A. Hole, who lived east of Winnipeg. In 1938 Hole, an "ardent hunter," was given some goslings after their parents were killed. Finding the Canada goose to be "the embodiment of courage, devotion and wisdom," he established a sanctuary (later known as the Alfred A. Hole Wild Goose Breeding Sanctuary) as a breeding site for migratory geese, most of which Hole was able to band. By the mid-1950s it sheltered 300 geese, but Hole was growing old and was worried about the future of his birds. He persuaded the Manitoba government, with the assistance of Ducks Unlimited (Canada), to take responsibility for the sanctuary as a permanent wild goose breeding site.[69]

Humane motives lay behind the establishment of other sanctuaries. Inspired by Jack Miner's work with birds, D.H. Bendick near Leduc, south of Edmonton, created a sanctuary for rescued birds as well as for birds he hatched from eggs he gathered from abandoned or damaged nests.

This commitment was gradually combined with commercial motives, and Bendick began raising wild waterfowl and importing Hungarian partridges from Europe for sale. By 1927 he was selling birds to game parks and conservation agencies around the world.[70] Rescuing birds was also the goal of the three Gebauer brothers, who farmed near Morris, just south of Winnipeg. When the brothers dug out a 55- by 36-metre pond for their livestock, they also levelled the banks, fenced the pond, and gave the cattle only one access point so that the pond was suitable for waterfowl as well. Their goal was to rescue injured birds, and they "let it be known around the district that they would like to have any sick or injured geese that were left behind after the spring and fall migrations." The sanctuary quickly attracted healthy birds as well, and the brothers soon had a variety of geese and ducks and a pair of whistling swans. They then dug a second pond, almost double the size of the first, to make room for more birds. A third pond was also dug, and a fourth was under construction in about 1938 to accommodate the approximately 10,000 geese that were gathering at the farm each year. "The entire available water surface was covered with geese, and thousands congregated on the surrounding fields awaiting their turn to use the water."[71] While many of the sanctuaries set up to rescue wild birds were in rural areas, some were located in urban areas as well. For example, Christopher Ridley opened his first sanctuary in Winnipeg in 1930, and over the next thirty years he cared for a variety of rescued animals at various locations in the city. By 1962 he was caring for 400 birds and mammals. Some were domestic animals, but Ridley mostly looked after wild creatures, including numerous ducks and geese that had been wounded by hunters and brought in by individuals or the Humane Society. It was a hand-to-mouth enterprise, and Ridley offered veterinary services and pet boarding to keep the doors of his sanctuary open.[72]

Creating nesting opportunities was another way that people demonstrated their fondness for wild birds. One popular project was the establishment of "bluebird trails," where a series of nest boxes for bluebirds were placed on fence posts and trees around farms or along roads. Bluebirds were rapidly disappearing on the prairies by the early 1950s because of habitat loss, and increased numbers of starlings and house sparrows compounded this decline because they took over suitable nesting sites and pushed out bluebirds. One way of dealing with this problem was to build nesting boxes with appropriately sized entrance holes to encourage more bluebirds and other native songbirds to nest in the region. This was the approach adopted by Charlie and Winnie Ellis, a brother and sister team who farmed near Red Deer in central Alberta. Noticing in 1955 how rare bluebirds had become at their farm, Charlie used instructions from a farm magazine to build a nesting box. It attracted a pair of tree swallows, but they were driven out by house sparrows. Determined to protect native songbirds, the Ellis siblings trapped and killed sparrows and starlings, built more nest boxes, and attached them to fence posts, creating a 300-box "trail" around the farm. This project was only one of a number in the region; in 1959 Dr John Lane of Brandon undertook a similar project, mounting nest boxes on fence posts along Manitoba roads. As other enthusiasts joined in, this "trail" of boxes formed a long network that by 1976 stretched from west of Winnipeg to Regina, south to the United

Figure 12.3 There were a number of privately run bird sanctuaries on the prairies. Among these was the Bendick Bird Sanctuary near Leduc, Alberta, which was inspired by the Jack Miner Sanctuary in Ontario, c. 1930.

States border, and then north to North Battleford. Branch lines off the main trail extended the reach of this bluebird trail. A wide range of community groups got involved in mounting the boxes and cleaning them each fall in preparation for the arrival of bluebirds in the spring, and by the end of the 1970s bluebirds could be found in areas where they had never been or had not been seen for years.[73]

On the whole, the bluebird nesting boxes, like private bird sanctuaries, were individual initiatives that received little state support. With the exception of the North-West Entomological Society, which received a small grant from the territorial government to help it promote an understanding of threats to farming, natural history societies before about 1960 were largely self-financed. Nor did the concerns of naturalists and bird lovers earn much official government support or endorsement. And, as will be seen in the next chapter, with the exception of support from the federal government for the museum at Banff and from the Saskatchewan government for its provincial museum, the collecting and display of natural history specimens in museums earned only scant government support on the prairies before the late 1960s. Government concern with wildlife was focused on economic considerations and sporting values and needs; indeed, the support extended to hunters and their organizations, to farmers and ranchers, and to the fur and fish industries stands in sharp contrast to that extended to naturalists and other enthusiasts. And these attitudes remained, and remain, powerful. Nonetheless, there was an emerging public demand that wildlife policy reflect other relationships with wildlife as well. Recognizing this, the Saskatchewan government stated in 1958 that its wildlife policy needed to extend to more than the needs of trappers, hunters, anglers, and residents of northern and fringe areas who relied on game for food and income. Also important was "the nature lover who places a high aesthetic value on our game." To meet the needs of this constituency, the province developed a public information program about its natural resources, including wildlife, but it still maintained that wildlife did not possess intrinsic value. "No object of nature has value by the mere act of its physical existence," the government announced, although it admitted that wildlife and other natural resources made "significant contributions" to "the material and spiritual well being of the people to whom they belong." While the province seems not to have changed its interpretation of the intrinsic value of wild animals, it continued to broaden its definition of legitimate users. In 1963 the Saskatchewan wildlife branch noted that "wildlife resources are not just 'hunter' resources and it must be remembered that there are other groups and individuals who place high value on them," such as naturalists and campers. In keeping with this developing perspective, the branch defined the groups with a stake in wildlife as sport hunters, trappers, naturalists, campers, northern residents using wildlife for food, and landowners on whose land wildlife lived.[74]

Such views were in part a response to a new wave of environmentalism that was emerging, one stimulated in part by fears about the adverse impact of chemical insecticides and pesticides on wildlife and the environment. Older organizations were reinvigorated (for example, the Audubon Society in the United States and the Royal Society for the Protection of Birds in Britain), and many new organizations (for example, Greenpeace in 1970), representing a huge range of strategies and beliefs, emerged to protect wildlife and wilderness areas. These developments were even more prevalent at the community level. In Alberta, for example, the Federation of Alberta Naturalists (now Nature Alberta) was formed in 1970 to represent six natural history clubs in various parts of the province, only one of which had existed before the Second World War. And the federation's growth has continued; by 2014 it represented forty clubs with varied character and objectives. While this unprecedented growth in organizations interested in natural history in some ways justifies the contemporary environmental movement's belief that it constitutes a historical vanguard, this belief ignores the fact that similar, albeit less socially powerful, concerns had formed a diverse set of relationships with wildlife through the organized study of natural history and personal commitment for over a century on the prairies.[75]

13

Displaying Wild Animals

> The value of the museum depends, not upon the large number of knick-knacks that may be documented, but upon its usefulness to the public as an educative factor in developing a love of nature.
>
> *Annual Report, Saskatchewan Department of Agriculture, 1914*, 236

In the century after 1870 wildlife was considered worthy of collection and display as a part of regional heritage. Collecting specimens added to the public's understanding of the value of natural wealth and created knowledge that aided its commercial exploitation and conservation. Such collections were most often displayed in natural history museums, institutions that were thought to mark social advancement and to demonstrate progress in a developing community. Of the museums on the prairies that displayed natural history specimens, the Saskatchewan Museum of Natural History was the most successful before the 1960s. Like many other regional institutions, it reflected interwoven efforts to increase knowledge about natural history, satisfy people's fascination with the natural world, conserve wildlife for economic as well as emotive reasons, and provide useful recreation for residents and tourists. Like the collections assembled by natural history enthusiasts and their organizations, natural history museums preserved the physical remains of animals in order to classify and describe them and to add to current and future knowledge. This has indeed proven to be the case, as technologies unimagined by past collectors can now be used to extract new information from specimens collected many decades before.

The conventional use of the term museum includes zoos, many of which have had motivations and goals in common with natural history museums. The Assiniboine Park Zoo in Winnipeg, a zoo-museum complex at Banff National Park, and the Calgary Zoo demonstrate this commonality; they all collected, preserved, educated, and entertained. They also played an important role in shaping the way that wild animals were represented in public discourse. A growing concern in animal studies argues that much of "human understanding of animals is shaped by representations rather than by direct experience of them."[1] Representation finds various expressions – books, illustrations, photographs and film, museum displays, and taxidermied specimens, all of which have their own context and significance. When animal bodies are mounted and preserved through taxidermy, for example, they are turned into artifacts whose verisimilitude cannot hide the fact that the viewer is looking at a representation created from the actual body of the animal. Whether they are intended for educational, entertainment, or aesthetic purposes, such artifacts have the potential to be used in ways that the animal living in nature cannot: they can be handled, moved and stored, traded, bought and sold, and they can be arranged and displayed in combinations that reflect the ambitions or priorities of the exhibitor.

People's experiences of live animals in zoos share some of these characteristics. The caged animal is removed from its natural context and is commodified in various ways. It can be traded, it can be bought and sold, and it can become an entertainment spectacle or an educational or aesthetic object. Its display in a constructed environment, even one that appears to be more or less natural, further confirms the animal as a representation, albeit a living one. Most significantly, the human encounter with the animal in the zoo is predicated upon its captivity. This becomes apparent when the encounter is compared to one between humans and non-captive animals in the outdoors, even in the constructed environment of national parks. While these outdoor environments in national parks are manipulated through various management policies to frame and enable the tourist gaze, the encounter remains radically different from that in a zoo. Crucially, the animal in a national park retains the power to escape the human gaze or to reciprocate it. In an influential essay, British cultural critic John Berger contends that the "public purpose of zoos is to offer visitors the opportunity of looking at animals. Yet nowhere in a zoo can a stranger encounter the look of an animal. At the most, the animal's eye flickers and passes on," or they look sideways or scan mechanically. "They have been immunised to encounter, because nothing can any more occupy a *central* place in their attention." Thus Berger contends that the zoo has eliminated "the 'look between animal and man," which he speculates was likely crucial in human development. In her history of the Adelaide Zoo, geographer Kay Anderson further argues that animals, in the artifice of the zoo, were "confined and subdued," interpreted and classified, and boundaries were drawn between human and animal as well as between animals, which asserted "a cultural sense of distance from that loosely defined realm that has come to be called 'nature.'" The zoo also facilitated the use of animals as commodities that could be bought, sold, or traded and which the observer paid admission to see.[2] Similarly, the classification, display, and interpretation of an animal's remains in a museum strongly influenced how an observer perceived the animal, although the observer's relationship to the display was mediated by his or her personal attitudes and needs, which were in turn shaped by broader social and community attitudes, concerns, and expectations. This is a dimension of what museum theorist Tony Bennett has called the exhibitionary complex: a set of beliefs and attitudes that arose in the nineteenth century and presumed that the museum (like the zoo) was a place where the observer was invited to share in displays of power and to identify with that power and its precepts.[3] Through the hegemony of the museum or the zoo, the observer was implicated in a relationship with wild animals that explicitly and implicitly asserted a quest for knowledge about them at the same time that it asserted human power over them.

MUSEUMS

One of the earliest collections of natural history specimens on the prairies was displayed in a museum opened in 1893 by Dr Henry George in Innisfail, a small town north of Calgary.[4] The collection was the result of Dr George's efforts as well as of those of local volunteers and members of the natural history societies he was later involved with. Reflecting the conventions of natural history collecting, the collection – as described by historian Karen Wonders – included "mounted animals and birds, shells, butterflies, moths, birds' eggs, insects, pressed plants and geological samples, as well as a substantial collection of heads and horns" of local big game animals. Dr George's collection had a peripatetic history; it followed him in 1907 when he moved from Innisfail to Red Deer, where he built a two-storey addition to his residence to house it, and when the Georges left for Victoria in 1923, the collection was moved to Calgary. By then it had become a ragtag assortment of materials, including natural history specimens as well as some early newspapers, guns, swords, coins, stamps, souvenirs, and 3,000 Canadian Army badges. The Georges kept the stamps and coins and a few personal items, and what remained was given to the Calgary Natural History and Arts Museum. This museum had been established in 1911 by local natural history enthusiasts and held a disparate collection: geological and paleontological specimens, 130 bird and some big game specimens, an excellent butterfly collection, and some fine art. It relied on civic charity for its space and was first located in the public library and then in the basement of the courthouse. It closed in 1935 because of financial difficulties, and much of the collection had been lost or destroyed by the time of the Second World War.[5]

Community commitment to public collections of natural history was stronger in Saskatchewan than in Alberta. The first public collection of natural history specimens in the province was formed in 1906 when the provincial government assembled an exhibit for the Dominion Fair in Halifax. It included small birds and mammals, a few fish, amphibians and reptiles, and mounted heads of larger mammals. On its return to Saskatchewan, the collection became the nucleus of a provincial natural history collection. At first, agriculture department staff responsible for game legislation cared for it in their offices. In 1908 the province intimated that it intended to build a new building to house a provincial museum, a prospect welcomed by the *Regina Morning Leader*, which remarked that "no capital would be complete without its museum" and it was therefore natural that "Regina should have a museum too."[6] The hope for such a museum was unfulfilled for almost fifty years, and the collection was instead moved into a room in the new Saskatchewan Legislative Building in 1912. The move was an unfortunate one; a cyclone hit Regina that year and blew out the windows of the room, destroying most of the collection.

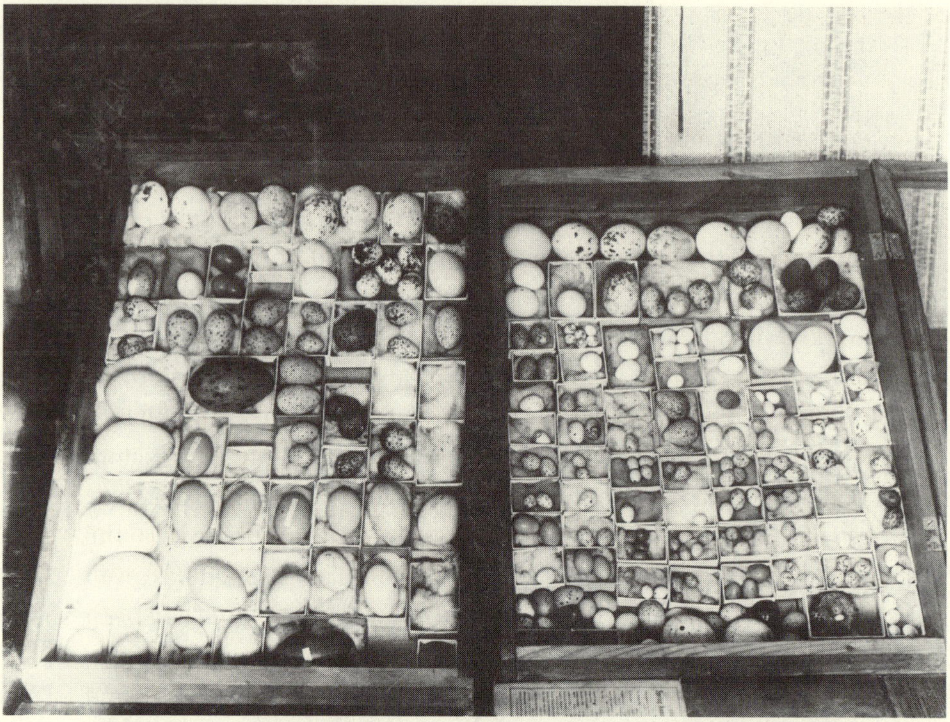

The next year, Horace H. Mitchell, formerly a taxidermist at the provincial museum in Toronto, was hired as Saskatchewan's provincial taxidermist and the museum's first full-time employee. His first task was to repair what he could of the damaged collection, but he also announced that he would begin collecting in the field as soon as possible because he found that Saskatchewan was "full of interest to the naturalist and nature student" and that there was "no lack of material" in the province for a "museum builder." Although the existing collection had about 350 animal specimens, a few geological ones, and some Aboriginal artifacts, Mitchell wanted to build a representative collection of the province's fauna. The collecting of specimens of endangered animals, in which he included whooping and sandhill cranes and pronghorns, was a priority.[7]

The museum was the responsibility of the game branch, and Game Commissioner Fred Bradshaw took an active interest in its development and management. The museum received a small annual appropriation, but it did not have sufficient status to retain coveted space in the Legislative Building and was relocated in 1916 to the basement of the nearby Normal School, a teacher-training facility. Bradshaw found the move "very regrettable," again recommending that the museum have its own building. He noted that the new space was cramped and the lighting poor, and its only advantage was that it gave future teachers exposure to the collection and that "the knowledge thus obtained will be radiated throughout the rural districts where these students will eventually labour."[8] While the museum was not the first – nor would it be the last – to operate in inadequate physical space, it was not shy to apply new ideas in exhibit design. The dominant exhibition method at the time was to display mounted specimens on shelves and in glass cases where they could be closely examined. As historian Lianne McTavish notes in her history of visual culture in New Brunswick, this method of display was "a Baconian

Figure 13.1 *Opposite top* The principle of "productive looking" used in museums in the late nineteenth and early twentieth centuries presumed that the museum visitor would carefully study assemblages of specimens. This principle inspired what many visitors must have found to be dauntingly detailed displays, such as this one of bird eggs at Dr George's museum in Red Deer, Alberta, in 1908.

Figure 13.2 *Opposite bottom* This grouping of owl specimens at Dr George's museum in 1908 did not provide visitors with any interpretative assistance except for a very broad grouping of specimens by their taxonomic order.

approach to ordering the products of nature, emphasizing the inventory, description and methodical observation of natural history." Through what McTavish calls "productive looking," it was assumed that "objects could educate the eye of the average person." This approach had wide appeal, and "many nineteenth century museums were likewise filled by what historian Stephen Conn calls 'naked eye science,' which affirmed that the study of natural history rested on the close observation of specimens."[9]

Yet close observation did not happen on its own – people needed to be taught how to observe by knowledgeable guides, through labels or publications, and by methods of display. While most of the Saskatchewan collection was shown in the customary fashion as individual specimens on shelves and in cases, Horace Mitchell strongly endorsed what was called group work, a display method in which animals were mounted "in lifelike surroundings, showing to the best advantage the history or habits of the given species." Exhibits using this relatively new approach were time-consuming and costly to prepare and required high-quality, specifically prepared specimens. The Field Museum in Chicago used this method extensively, but the only example in Canada was an exhibit that Mitchell had prepared in Toronto. In his first year on the job in Regina, Mitchell built four group exhibits, thus putting the Saskatchewan museum in the forefront of contemporary museum practice. One group exhibit showed two northern flickers posed on either side of a log that was cut away to show the nest and eggs inside. Another exhibited a male and a female yellow-headed blackbird alongside a nest built in reeds. The third group exhibit displayed red-winged blackbirds, and the final one showed various species of terns and gulls.[10] By 1915 Mitchell was amplifying the effect of these exhibits by adding a painted background against which each grouped specimen was displayed, in effect making them into small dioramas. This work was ongoing, and in 1921 he developed five new exhibits that employed this principle, including ones that featured the upland plover, the American goldfinch, the cedar waxwing, and the tiger salamander. Showing the museum's growing interest in ethnography, he also developed a reproduction of an Aboriginal grave.[11]

The group technique and the narratives that it presented were seen as particularly useful for promoting public education. Game Commissioner Fred Bradshaw praised the technique as "far more valuable from an artistic and educational standpoint than the usual run of individual specimens" because it gave the viewer an impression "which is more apt to stay with him than if the specimens were simply mounted on stands without any evidence of their interesting ways and habits." But he also made it clear that whatever display technique was employed, the museum's collections were elements in a strategy to help "stimulate an interest in the wild life of our province and win for it many friends who will find not only health and pleasure in studying its habits in the open, but will advocate its protection at all times." Horace Mitchell agreed that one of the important contributions that his craft made to society was to encourage conservation. As he observed in 1913, effective display would help with the protection of birds. While the intricacies of their lives and habits made them popular and fascinating subjects, they, most importantly, were essential for human life because without them "the ravages of other life that they prey upon would make the earth *uninhabitable*." While modern taxidermy could thus help the cause of conservation, it was also clear that professional pride and demonstration of skill were also motivations. As Mitchell speculated in 1913, "supposing a group of four or five buffalo, mounted amidst their natural surroundings in a great case, suitably built for the purpose, and a large painting of the same subject, both masterpieces by artists in their own line, were placed side by side in a museum or elsewhere, I would venture to say the reproduction by the taxidermist would 'draw the crowd' every time." And as he further noted in 1920, taxidermy and the creative mounting of animals using group techniques was a modern art and a science and showed "the difference between present-day methods and the old time 'bird stuffing.' Birds and animals are not 'stuffed,' a term abhorrent to the scientific taxidermist – but are modelled and mounted."[12] In other words, the animals' bodies had become representations of the living animal. While museums did not fully mount all their specimens – the great majority of bird specimens, for example, were preserved as study skins stored in drawers – fully mounted specimens were indeed a crucial part of the public face of the modern museum. Historian Paul Faber has noted that modern ornithology did not emerge until the late eighteenth century when the long-term preservation of museum specimens had been perfected, allowing comparative study and the objective verification of species by a number of scholars over a period of years. For museum visitors, however, taxidermy animated the bodies of dead animals, allowing them to be posed to suggest motion and life. Anthropologist Jane Desmond concludes that this animation truly transformed the animal into a representation that depended not only on its death and the erasure of "all marks of killing," but on "the complete dismemberment and rearticulation" of its body into a particular pose as well.[13]

By 1920 the museum's collection contained a range of materials, including some Aboriginal artifacts as well as geological and plant specimens and moths and butterflies. Nonetheless, mammals and birds were its primary focus in these early years. Mitchell actively collected specimens in the field and took pains to counter any assumption that collecting was a mere romp in the outdoors. Instead, it was a muscular activity requiring self-control, endurance, and persistence. The collector of specimens had to "have great patience at times and be able to put up with a certain amount of hardship, for collecting is not all fun by any means, at times wading in icy water, tramping many weary miles, or almost [being] devoured by swarms of mosquitoes." In 1915 Mitchell's collecting was temporarily supplemented when the museum commissioned a Mr Buchanan to collect specimens in northern Saskatchewan. As a result, 209 specimens of birds and eggs, as well as a caribou, white timber wolves, and, among others, arctic and cross foxes, were added to the collection. The museum also solicited public donations to the collection, and Mitchell drew up instructions for the care of specimens in the field and their preparation for mailing to the museum, where they could be mounted for display or prepared as study specimens.[14]

In a society that valued the accumulation of material things, collecting was rarely controversial, but Mitchell vehemently objected to the popular boys' hobby of collecting wild bird eggs and nests. Not only was it illegal under the migratory birds legislation, Mitchell noted, it was a wasteful and anti-social pastime. "No boy," he stated, "should be allowed to collect birds' eggs or nests. The accumulation of such specimens is of no real value, and tends to create a desire on the part of the boy to destroy other useful things."[15] However, the widespread collecting of eggs for scientific study was acceptable, as was killing birds for museum collections. The physical examination of a bird was seen in the early 1900s as providing the only authoritative identification of a species, a practice that led some to wonder how people who supposedly appreciated and loved wild creatures could go on collecting expeditions that ended with the death of so many of them. Mitchell countered that collecting was foundational for displays and for the education of the public. So, too, only through physical collecting could "study skins" be accumulated for research or scientific study undertaken on such things as diet, which required the examination of a bird's crop or stomach. Nonetheless, while Mitchell accepted that death was an inevitable part of forming collections, he further noted that many students of nature had now discarded guns, going out "armed with nothing more destructive than a camera and an observation tent," which was making research in the field entirely harmless.[16]

Zoological field research was limited in the museum's first forty years when the priority was to increase and display its collections. Nonetheless, by 1929 the museum had apparently sponsored research projects on sage grouse, prairie dogs, and whooping cranes. This work was most likely undertaken to support exhibit development, and research that connected to a broader scientific context only began to develop after the Second World War. By 1950 the museum was sponsoring bird-banding projects and assisting with wildlife population surveys, including counts of woodland caribou, whose populations had fallen dramatically in northern Saskatchewan.[17] Community education, another accepted museum function, was only slightly more advanced. By the First World War the museum had prepared duplicate mounted specimens of birds to lend to student teachers to assist them in preparing their practice teaching lessons. In 1913 Horace Mitchell advised teachers that the complexity of bird species required careful introduction in the schoolroom. "For the purpose of holding the pupils' interest in the subject," he advised, "first learning to name a bird and the extent of good or harm (if any) it does, may be better than going into the more technical part at first, such as learning names of the parts of a bird." More difficult topics, such as migration and changes in plumage, should be "deferred until at least most of our commonest birds can be named at sight by the pupils." By 1937 the museum was supplementing these mounted specimens with photographs. The museum and the game branch also lent lantern slides free of cost for public talks. A typical set consisted of thirty-five slides, and the museum provided a written syllabus so that the lecturer could quickly "acquaint himself with the subject to give an intelligent talk." Nineteen talks were delivered in 1913, all in the Regina and Moose Jaw districts. One of the talks in Moose Jaw drew 300 people, but audiences were usually smaller. In 1919–20, forty-two such talks were delivered, including some by game branch staff.[18]

The creation of the Regina Natural History Society in 1933 gave the museum an opportunity to expand its educational outreach. This was not unexpected; Fred Bradshaw was one of the founders of the society, and after his career as Saskatchewan's game commissioner ended in 1928, he became director of the museum, a post he held until 1935. Society volunteers collected, labelled, and identified botanical, entomological, and geological specimens and gave lectures. They also acted as guides, a task that expanded in 1935 when the museum began to open on Sundays. On

the first open Sunday, 150 people braved the cold to see the exhibits. Activity slowed during the Second World War, but by 1949 the museum had begun to further develop its educational programming, even making a film, *The Pelicans of Last Mountain*, which was released the next year. Even so, it was reported in 1953 that "the program of education has not been as broad as it could be, although it is anticipated that illustrated lectures, guided tours and other activities normally associated with such a program will be inaugurated in the next two or three years."[19]

The museum thus primarily collected and prepared exhibits, and it continued, other than for a brief period during the Second World War, to be housed in the Normal School building. Inadequate lighting was an ongoing problem, and at only about 930 square metres (plus 45 linear metres in a corridor lined with exhibits in display cases), the space was cramped and inefficient. Even so, attendance was high and in the early 1950s averaged 25,000 to 30,000 visitors per year. In 1947 Fred Bard became director, and the museum entered a long period of revitalization under his leadership.[20] Group exhibits (now called "assembly groups") continued to be used, but by 1949 Bard was encouraging the development of more immersive exhibits as far as resources and space permitted. Cases that held large panels with photographs, mounted specimens, graphics, and text began to be developed. In an exhibit about wrens, for example, images and specimens were posed on a large relief map of

North and Central America to show the birds' summer and winter ranges. More didactically, an exhibit about badgers explained the economic value of wildlife conservation. Two mounted badgers were posed on either side of a case, one of them holding a gopher in its mouth. A mounted gopher and a sheaf of wheat completed the three-dimensional elements. Background graphics showed a badger, a sheaf of wheat, and a gopher all connected by arrows and text that stated, "EACH GOPHER EATS ONE BUSHEL OF GRAIN A YEAR." The arrow connecting the gopher and badger was labelled "BADGERS EAT GOPHERS, RENDERING A VALUABLE SERVICE." The museum also soon began to develop twenty dioramas to replace its older exhibit cases. While large dioramas had been used in leading and well-funded North American museums for decades, they remained an innovative and challenging technique for smaller institutions. In these dioramas, specimens were posed in large exhibit cases with realistic backgrounds painted on a curved wall. Specimens were posed in the foreground in arrangements that integrated them with a skilfully painted background, creating the illusion that living animals were being viewed in their actual habitat. Verisimilitude was enhanced by adding dried plants and rocks from the location being portrayed. The museum's long-term plans included the development of exhibits about each part of the province so that residents could see a display of an animal and a landscape that was typical of their "home locality." Demonstrating the public's fascination with whooping cranes as well as Fred Bard's commitment to their preservation, a "whooping crane memorial habitat case" was developed. A wounded crane that had died shortly after its discovery was used in the exhibit and was displayed as if it was living in its last known Saskatchewan nesting site near Kerrobert in the west-central part of the province.[21]

Work on these new exhibits had just begun when the province announced that it would construct a new natural history museum to open in 1955, the fiftieth anniversary of the province. It was a community effort, and the Saskatchewan Natural History Society raised about $1 million for the construction project. Located in a prime location across Wascana Lake from the Legislative Building, it promised to be "a lasting tribute to the pioneers of the province." The celebration of pioneering was mostly a political gambit, and though it could never be admitted, the ethos of pioneering had been (and in some respects remained) a major threat to the province's wildlife. In any case, the province made it clear that the new museum was to be an instrument for promoting the future conservation of wildlife by "creating a deeper appreciation of the aesthetic and practical values" of the province's natural heritage through exhibits and extension programs. This approach was based on a recognition that conservation could not be achieved simply through legislation and enforcement, but could only be "truly effective" when it was "understood and appreciated by every citizen" who would then make a personal commitment to it. The museum would be the mechanism for creating this new attitude.[22]

At just over 5,000 square metres, the new museum was one of the largest natural history museums in Canada. Exhibits were mounted on two floors. Displays on geology, Aboriginal history, and fish, as well as a series of displays on "ecological relationships of one group of animals with another," were located on the lower floor. The ecology displays presented different habitats along "with general information on the effects of various environmental controls" on wildlife populations. This information was intended to demonstrate "the management programs currently in effect" and to give the visitor "a clearer understanding of the why and wherefore of Saskatchewan's fish and game laws." Large display units – 4.2 metres wide, about 2.5 metres high, and about 2.5 metres deep – occupied the second floor. These typically employed a diorama technique. The model developed earlier at the Normal School location, in which each case dealt with one dominant species in a particular part of the province, was maintained, and some of the cases, "built while the museum was in its former, less spacious location," were "simply dismantled and transplanted to the new building" with plans to enlarge them in the future. There were, however, a number of new installations, including one for children. Based on *Bambi*, the Walt Disney movie, this exhibit contained "a number of mounted young animals and birds grouped around the foot of a tree upon which 'the wise old owl'" was perched. With the assistance of museum staff, children learned the story that the owl was telling about the young animals.[23]

The new museum's dioramas usually featured one species, set in its natural habitat. Habitat was primarily illustrated by a painted backdrop that portrayed the landforms of the district. The accuracy of the backdrop was ensured by reference to photographs of the district being depicted. While some ecological relationships among plants, land, and animals were shown, the relationships that the animals had with one another were not usually depicted, although it appears that they were explained by interpretive staff. Showing a more direct treatment of ecological issues, one exhibit that focused on a forest area north of Hudson Bay Junction in eastern Saskatchewan portrayed "the food chains that become established among animals co-existing in a common area."[24] At first, museum staff concentrated on finishing such exhibits; the opening of an elk display in 1957, for example, marked the completion of the large dioramas in the upper gallery. Although new displays were also planned, "a number of minor changes" were made to other exhibits that had been "prepared on a somewhat temporary basis." New large dioramas were subsequently developed, including one on sandhill cranes. Showing the birds at Last Mountain Lake Bird Sanctuary, it was designed to direct attention "to the problem of crop depredation as against the spectacular sight of the great flocks of these birds during fall migration." A display on coyotes was also opened. Demonstrating that the long-held animosity towards these animals was dissipating, it offered the visitor "some understanding of the much maligned coyote."[25]

Given the mandate of the museum, public education was the other immediate concern. The new museum had a theatre for public lectures and films and planned to develop a variety of extension programs, including a mobile museum to reach communities throughout the province. The latter plan seems not to have come to fruition, but other outreach plans were being implemented by the late 1950s. In 1956–57 the museum set up twenty temporary exhibits at various venues in co-operation with other government departments, agencies such as the Prairie Farm Rehabilitation Administration (PFRA), and private organizations.[26] By 1959 the museum, as part of the "public information" function of the Department of Natural Resources, was using lectures, publications, films, slide sets, film strips, and press, radio, and television releases to encourage communication between "the scientist and the layman" and to make "the outdoors more meaningful to the general public." In addition, educational programs at the museum were tied to the school curriculum to help educators enrich their teaching of biology, and the museum helped community museums create an interest in nature and its conservation. As it had for almost thirty years, the Saskatchewan Natural History Society remained an important collaborator in museum outreach programming and benefited by using museum facilities for its meetings and public events. By this point as well, the museum had developed a more active research program, reporting in 1959 that "the heritage of knowledge is increasing as a result of museum explorations." Zoology remained an important element in this research program, although fieldwork in archaeology and palaeontology was also undertaken. While this research activity was no doubt driven at least in part by the need to obtain detailed information for its dioramas and exhibits, by 1960 the museum was also providing support to a network of scientific communities. This support included whooping crane counts in association with the US Fish and Wildlife Service, assistance to the National Museum of Canada's reptile and amphibian collecting program in Saskatchewan, and assistance to individual scientists at various North American universities.[27]

While the museum had previously drawn a respectable number of visitors, attendance increased dramatically with the opening of the new building. In its first ten months, the new museum attracted almost 182,000 visitors, and numbers continued to be strong. At the end of the 1950s the museum was drawing between 221,000 and 225,000 visitors a year. These were high numbers relative to the population and attested to strong public interest in the museum's programming. This interest was also shown by ongoing donations of specimens collected by the public. And while the museum's contributions to public life could be measured by knowledge gained and lessons taught, its impact on personal well-being was important too. The museum recognized this in its 1958 annual report, stating that while disseminating information about conservation of natural resources was an important goal, the museum also wanted to give the public an opportunity to become familiar with and to simply enjoy and admire the natural environment. And some visitors took advantage of that. In 1957, for example, one Regina resident told nature columnist Doug Gilroy that "sometimes in winter I get to feeling blue and downhearted. When this happens I head straight for the Museum of Natural History" and sit down "before some of the splendid wildlife exhibits with marvellously painted backgrounds and drink in the beauty of the scene. And when I come away I feel like a new and better man."[28]

The Saskatchewan Museum of Natural History until the 1960s was the most highly developed, best-funded, and professionally run museum in the prairies. While there were a number of small museums in all three provinces, often run privately or by volunteers, neither Manitoba nor Alberta had provincial facilities similar in quality to Saskatchewan's.[29] Alberta only began construction of a provincial museum in Edmonton in 1963 as a Canadian centennial project. Until it opened in 1967, a few mounted specimens were displayed in a room in the Legislative Building, but it was a desultory effort and little more than a gesture to show that the province realized that it should do something to encourage knowledge of its natural history. The situation in Manitoba was only marginally better. The Manitoba government acquired an important ornithological collection when it accepted the gift of a collection of bird skins that had been assembled in the 1920s by Manitoba ornithologist and collector Cyril Guy Harrold. Shortly after his death in 1928, Harrold's family donated the collection to the province, and it was kept in the Legislative Building on an interim basis until a provincial museum was established.[30] This did not happen until 1932 when the Manitoba provincial museum opened in "temporary" quarters in the provincial auditorium building in downtown Winnipeg. Thirty years later, the Manitoba Museum was still in these temporary quarters, where it tended a collection of Aboriginal and early settler artifacts, some botanical, geological, and paleontological materials, as well as a zoology collection that included "all of Manitoba's present-day animal life – shellfish, insects, fishes, amphibians, reptiles, birds, and mammals." With only 510 square metres split between the ground and third floor spaces, the museum had very limited storage and preparatory space and little room for exhibits. Given its layout, it was only open when the auditorium was in use. Moreover, materials had to be carried up three flights of stairs, and visitors had to climb them as well, causing a "constant source of complaint." The museum was also chronically underfunded; the Winnipeg Board of Trade at first provided some financial assistance, and the natural history society contributed what it could. While the province provided space, it made no financial contribution until 1944 when it gave the museum $950, but not as a permanent appropriation. By 1963 the museum's $15,900 annual budget was made up of a $10,000 grant from the city, $5,000 from the province, and $900 from donations, which barely covered the salaries of the museum's four staff members. In contrast, the Saskatchewan provincial museum received $80,000 a year from the government.[31] This museum served Manitobans until 1970, when a new Manitoba museum opened as part of the museum-building euphoria stimulated by the celebration of Canada's centennial.

COLLECTING AND DISPLAYING WILDLIFE AT BANFF

In his 1886 planning report for Banff, William Whitcher supported the creation of a museum of natural history where the remains of "animal species weeded out" of the park could be displayed for the education of visitors. Despite Whitcher's recommendation, planning was soon underway for a museum and library about geology. While it was assumed that such facilities would interest visitors, the primary goal was to make the park a centre for the mineralogical exploration of the mountains. But tourist tastes ran more to wild animals than to geology, and in 1895 the Rocky Mountains Park Museum was established with an emphasis on botany and zoology as well as geology. Under the direction of Norman Sanson, a naturalist interested in botany, entomology, and meteorology, it featured a number of specimens collected in the Rocky Mountains in the late 1800s by John Macoun of the Geological Survey.[32] This museum drew few visitors; it was out of the way and was housed in a poor building. In 1903 a new building was opened on the banks of the Bow River near the downtown. To emphasize its concern with nature, the museum featured a "rustic design" and was finished inside with Douglas fir. It had a reading room and exhibition space on two floors where minerals, fossils, and specimens of insects, birds, and animals were displayed. It apparently also had a few Aboriginal artifacts because, as Banff's park superintendent recommended, like the flora and fauna being preserved in museums as evidence of vanished times, "it would be well to get together a representative collection of Indian work (typical) before the day for these things is gone." In its first year, the museum welcomed 3,000 visitors.[33]

The museum planned to collect specimens of all the animals living in the park and complete its collection of minerals in order to represent all of the park's resources. Plans were also still being mooted for the further

Figure 13.3 Constructed in 1903, the Banff National Park natural history museum was finished with Douglas fir and lit by wall and clerestory windows. Most of the specimens were displayed in glass-fronted cases. Others were artfully placed throughout the exhibition space. This photograph was taken in 1943, by which time the trend in museum display was moving towards more immersive approaches.

collection of Aboriginal materials, since "the day for procuring these things will soon be past." The museum's collection grew incrementally – in 1906, 150 bird specimens as well as other animal specimens were added and acquisitions in 1908 included a caribou head, bird eggs, and insects. Because museums often find it hard to reject donations, the museum accepted gifts of sponge corals, shells from the Bahamas, and garnets from Alaska. Further specimens from the park were added when a bull bison and an elk that had died in the paddock were taxidermied for the museum. Other specimens from the park were added from time to time, and such collecting was sometimes opportunistic. In 1920, it was noted that the museum's collection of bird study skins and mounted specimens was being augmented with birds killed in the park when they flew into windows or telephone wires.[34] As a form of educational programming, the museum curator helped visitors to identify plants they had collected in the park, and the museum published a guidebook in 1914 with descriptions and general background information on the collection's specimens to help people better understand the displays.[35]

Park visitors could also view collections of live animals in a number of park venues. These displays evolved over a decade, having begun in 1897 with the acquisition of bison, which were displayed in a paddock. An aviary was constructed in 1904 on the museum grounds, and a zoo was added in 1907. While this combination of venues was unusual at the time, it was not unique. Such a model had been proposed by the American writer and zoologist William Hornaday for a zoological park at the Smithsonian Institution in Washington in 1889, but plans for it were not realized. Ten years later, Hornaday used the same concept to exhibit animals at the New York Zoological Park in the Bronx. Native ungulates could graze in a field in the park, and some of the caged animals were kept in

"naturalistic settings – water, logs and plantings for reptiles; forest trees for birds." Although Hornaday had proposed that the Bronx's buildings and landscaping should be subordinated "to the needs of the specimens," historian Helen Horowitz concludes that this proposal was ultimately subverted in favour of monumental buildings with caged animals that better asserted Anglo-American control of nature and provided a more direct model for teaching ethnic and racial minorities, who were said to threaten the goals of game conservation.[36]

While the Bronx experiments idealized the "natural" confinement of wild animals in a variety of spaces, the display of animals in cages in the Banff zoo and in the open space of the paddock seems to have developed on an *ad hoc* basis, largely because of the actions of elite donors and local administrative decisions to encourage tourism. Banff's paddock collection began in 1897 when businessman Thomas Blackstock of Toronto offered the park a bison bull and two cows that he had purchased in Texas. This offer was made on the condition that the Department of the Interior establish a bison herd, and this commitment was honoured the next year when Donald Smith (Lord Strathcona) – who, it will be recalled, had been involved with bison breeding experiments at Winnipeg – donated thirteen bison that he had kept on his estate at Silver Heights in St James, near Winnipeg. While Smith had planned to give his entire herd to the park, Winnipeggers protested the loss of all the animals and so several were kept in the city.[37] Smith's donation was welcomed by parks officials. The bison were lodged at the site of the abandoned NWMP detachment a few kilometres from the Banff townsite. The site was treated as a pasture and was fenced to keep the animals safe and to prevent them from wandering, while the former police barrack building was used as a barn. It was hoped that the protection of these "rare specimens" in the paddock would be a way to perpetuate an "almost extinct race."[38]

Visitors found the bison fascinating, which whetted the collecting appetite of Park Superintendent Howard Douglas, who recommended in 1899 that the park acquire more indigenous animals "such as deer, moose, elk, mountain sheep, musk ox etc." This plan was implemented the next year with the acquisition of five elk and twelve young pronghorns.[39] Subsequently, more animals were purchased, captured in the wild, or obtained through trading with other zoos. In 1904 three black bears caught in the park were added along with three red and two cross foxes. The collection now also contained moose, bison, mule and white-tailed deer, a coyote, a timber wolf, cougars, a badger, and two great horned owls. That year, the paddock welcomed 8,000 visitors, which indicates that almost everyone who visited the park also went to the paddock. Eight years later, the paddock had expanded to include mountain sheep and a mountain goat. Despite the park's intention to showcase regional fauna, anomalies had also crept in. By 1904 some angora goats and Persian fat-tailed sheep were on display, and in 1910 Professor Bell of Nova Scotia donated three Zulu sheep. In 1912 six yaks, donated by the Brandon experimental farm, arrived; first given to Canada by the Duke of Bedford, their "quaint appearance" made them "objects of much interest to visitors." In 1919 a Siberian pony was donated to the paddock by the 31st Battalion. It was a war trophy, having been recaptured from the Germans by the Canadian Army at Amiens in 1918, and was displayed as a patriotic object.[40]

The completion of a good road around the paddock in about 1922 allowed tourists to drive around the whole enclosure to look at the animals from their cars. Not surprisingly, the number of visitors to the paddock increased. By then, the paddock only held large mammals.[41] Birds had been transferred from the paddock to an aviary on the museum grounds in 1904, and smaller mammals and ones that needed more secure confinement were moved in 1907 to a zoo located near the aviary. These sites

Figure 13.4 The Banff paddock held a variety of large mammals, including moose, separated from the public with fences. Photo not dated.

offered the chance for closer encounters with the birds and mammals. The aviary was an octagonal building with separate compartments. Although it was dominated by game birds, it also held several golden eagles and some owls. Two vultures, which had been captured near Edmonton, were added in 1910. By 1920 the aviary housed waterfowl, ospreys, brown pelicans, and a peacock. It was a low-key attraction, and in 1910 it was reported that the birds were "doing as well as a flock of well-cared-for domestic fowl."[42]

The attractions in the aviary were complemented by the smaller wild animals housed in the adjacent zoo. As early as 1900 Park Superintendent Douglas had proposed that visitors would enjoy seeing an assortment of caged animals, but this was impractical at the time owing to a lack of civic infrastructure. A zoo on the museum grounds only became feasible with the installation of water and sewage facilities in the town in about 1905. In preparation, animals were acquired in 1906 and kept in cages at the paddock until the zoo was ready.[43] The zoo consisted of iron cages with cement floors that could be kept clean and free of odours. Walls were built from local "sulphur rock," which gave them an "artistic rustic appearance" and created an appropriate local context for the display. Each cage had water and sewer connections and a cement tank for "the animals to bathe in." At the end of its first year, the zoo's collection consisted largely of regional animals – a mountain lion, three bears, three timber wolves, three porcupines, and a pair each of coyotes, foxes, lynx, racoons, badgers, and marmots. Given the public appeal of rare and aggressive animals, park authorities wanted to add wolverines and grizzly bears to the collection. It seems that wolverines were not added until the end of the First World War, but the zoo had more immediate success with grizzlies when two cubs were purchased from an Aboriginal hunter in 1912. They were "large militant representative specimens of this ferocious species" and grew so quickly that their cage needed to be reinforced.[44] That year, the zoo also obtained a polar bear from an undisclosed party, trading two moose for it, and this addition soon became one of the zoo's prize animals. In 1922 the Royal Canadian Mounted Police (RCMP) at Chesterfield Inlet secured a polar bear cub for the zoo, and it immediately became the zoo's biggest attraction.[45] By 1920 the collection was augmented with pine marten and mink. While most of the animals in the zoo were local or at least from Canada, a few exotic animals had been acquired to enhance the tourist experience. In 1910, for example, the zoo purchased two rhesus monkeys, which proved "a great attraction" to visitors, and in 1912 two ring-tailed monkeys were added.[46]

The appeal of the live animal collections in the aviary, zoo, and paddock reflected a range of public attitudes towards wild animals and their display. Current studies of zoo visitors reveal that most people today want to watch newly born animals, see animals engaged in activity, and have some interaction with them.[47] Historically, visitors at the Banff zoo showed the same preferences. They also found the rarity and novelty of the animals appealing and seemed to enjoy being near caged dangerous animals. While the bars that separated visitors from the animals created mental as well as physical barriers, zoo visitors did not want to see animals simply lying in cages. This was most easily dealt with by turning feeding time into a spectacle. As was noted in 1923, the crowd at the Banff zoo was so thick around the cages when the animals were being fed that it was difficult to get near them. While reports do not indicate whether the

Figure 13.5 As is apparent in this undated photograph of the bear pen at the Banff Park zoo, public security was a high priority in the display of these animals.

public was allowed to feed the caged animals, it is likely that they were because this was then a standard feature at zoos. As historian Harriet Ritvo notes, feeding and touching wild animals was a popular activity in zoos in the nineteenth century and symbolized proprietorship and domination. This interaction was analogous to the relationship people had with domestic animals and may have been part of the attraction of some of the animals at the zoo and paddock. By 1910, for example, the pronghorns at the paddock were "so tame as to come in answer to a call and be fed by hand," and it was reported the next year that the mountain sheep ate from the keeper's hand. Animals that were cute or vulnerable were also a draw. The popularity of the zoo's polar bear cub was one example of this, and when a moose was born in the paddock in 1902, it for a time became "the most interesting feature in the animal collection."[48]

The public also wanted to see animals that were rare or not commonly seen in captivity. In 1901 the paddock proudly announced that it had obtained a mountain goat, the only one in captivity in Canada and one of only three in captivity in the world, the other two being in London and New York. It was thought that the draw would be even greater if the zoo had more than one, and thus Stoney hunters – who were otherwise banned from hunting in the park – were hired "to secure a second one and also a pair of mountain sheep." Mountain sheep did not appear in the collection until 1909, when their acquisition led park officials to say that "we can safely boast of having the only full grown Rocky Mountain sheep in captivity." Indeed, it was said that visitors expressly came from the United States to see them, and in 1910 it was claimed that the mountain sheep enclosure was "constantly lined with visitors and photographers" who found them even more attractive than the bison.[49] While this suggests that bison had now become familiar and accepted, park visitors nonetheless still had an enduring affection for these rare and powerful animals. From the start, paddock visitors revelled in nostalgia for the vanished way of life that the bison represented. This sentiment was demonstrated by the history of one of the bison donated by Donald Smith, a bull born in about 1870. Paddock officials named him Sir Donald, which perpetuated the memory of his former owner's munificence and at the same time placed the bull firmly within the tradition of naming domesticated farm animals. Sir Donald developed a familiar relationship with his human keepers and the visitors who gazed at him. Although he had been in captivity for almost all of his life, his wildness was unimpaired in the eyes of his keepers and he was said to be the last wild bison, all others having been born in captivity. His life during thirty-seven years of captivity was one of "romantic interest to thousands of people as the sole survivor of a noble type of animals that in the wild state have become only a memory." In 1908 Sir Donald – then thirty-eight years of age and the oldest bison at the Banff paddock – was killed by younger bulls. Although it had long been planned that he would be taxidermied and shipped to Ottawa for the new national museum, the young bulls had trampled Sir Donald's carcass so badly that only the head was useable. Mounted as a trophy (now in the Banff Park Museum), Sir Donald's head was said to be "a memento of the last and noblest specimens of the buffalo of by-gone days."[50]

Despite their popularity with visitors, the bison posed management problems for park staff in much the same way they would later do at Elk Island, Wood Buffalo, and Buffalo National parks. As early as 1901 it was recognized that a more varied genetic pool was necessary to keep the herd healthy, and efforts were made to exchange bulls with some from the Yellowstone National Park herd. Population increase was another problem, although not an intractable one since surplus animals from the paddock were simply shipped to other parks. By 1910 the bison paddock was overcrowded and the animals were doing poorly. Consequently, seventy-seven animals were shipped to Wainwright and replaced with seven or eight bulls from the Pablo herd in order to enrich the Banff herd's bloodline. As the parks branch noted, it was important that the bison be managed properly because they were one of Banff's "chief drawing cards, [and] too much cannot be done to give those we have here every assistance to thrive" and to keep them "in the pink of condition."[51]

Such herd management techniques were not usually applied to other animals. For example, the Angora goats were prolific, and while they were considered "pretty interesting little animals," the herd was culled as early as 1902 to keep it a manageable size.[52] Nonetheless, keeping all of the animals healthy was an important ongoing concern. Pronghorns were a particular worry; in the winter of 1901–02 all of them died, and it was speculated that they had been too young when they were captured and had never recovered from the shock of being shipped to the park. Park

Figure 13.6 This commemorative photograph of Sir Donald titled "The Last of a Noble Race" was likely displayed in a public space. Part of its caption read: "This is 'Sir Donald.' He used to be the boss of the herd, but in a fight he got licked – got the far side horn broken off. After that he would not associate with the rest of the herd." Photo not dated.

staff had only marginally better luck with the pronghorns that replaced them, and by 1911 all were developing throat and neck ulcers. The cause was unknown, although it was thought that this might be due to an inappropriate diet or to the mountain climate. Some were sent to Buffalo National Park in Wainwright, and by 1912 the park had stopped trying to rear pronghorns and sent the surviving ones to Wainwright.[53]

While pronghorns were difficult to raise in captivity and no zoos had much success with them, other problems at the paddock, zoo, and aviary demonstrated poor husbandry and a lack of knowledge about the management of confined wild animals. When elk were introduced to the paddock in about 1901, for example, they were simply mixed in with the bison, which then numbered about forty animals. The bison killed two of the elk, which indicated that separate enclosures were needed because it was "not safe to have any other animals kept with the buffalo as they are very treacherous and will charge unexpectedly at any time."[54] In other cases, problems stemmed from the lack of even the simplest precautions. In 1906 the snow was so high around the paddock that coyotes were able to get in and kill seven deer. The lesson, however, seems not to have been learned, and in 1911 coyotes killed two pronghorns, one deer, and an angora goat.[55] Problems occurred at the aviary as well. In 1904 one of the golden eagles killed the other and another specimen had to be found. Despite a new eagle cage that was built in 1906 – said to be one of the finest in North America – this happened again in 1911 when another eagle died while "fighting with its companion." That same year, one of the vultures recently added to the collection also died.[56] The zoo had similar problems. Although it was reported in 1911 that the animals had done "unusually well this winter," it was clear that the observation was only relative. A wolf pup had been killed by coyotes housed in the neighbouring cage, two of three young racoons had been killed by the adult male, and the third had only been saved in the nick of time, even though it "had one foot eaten off." The kit foxes and the badger had also been failing but had been improving since earthen floors were put in for them to dig in.[57]

While it is unclear whether the public was aware of or concerned about these continuing deaths and other difficulties, the Banff zoo, paddock, and aviary were popular attractions. In 1908 it was claimed that the museum and the zoo were of great interest to visitors, one of whom referred to these features of the park as "the University of the Hills." Although people were undoubtedly educated by seeing live wild animals that they would otherwise never have encountered, the zoo had no interpretive programs to enhance their experiences and its display of live animals served only to entertain and to provide whatever knowledge could be gained through personal observation. While paddocks proved to be a long-standing feature of almost all national parks in the region, including Banff, the parks branch came to see the Banff zoo and aviary as incompatible with its goals and philosophy. In keeping with the spirit of the new National Parks Act in 1930, the branch began to wind down the zoo and aviary and gradually transferred the animals to the Calgary Zoo. By 1937 both the zoo and the aviary had closed down completely. Although Norman Sanson was not replaced when he retired as director in

1932, the museum remained open until well after the Second World War, but by then it was in need of repair and rejuvenation. The museum has now been restored as a historic site and features many of the specimens that were displayed there in its heyday.[58]

CIVIC ZOOS

The closure of the Banff zoo – apparently because it had come to be seen as incompatible with environmental stewardship – suggested changing attitudes about the caging of wild animals. While the parks branch saw no ethical problem with using animals in the wild as tourist attractions, the justification for confining wild animals in cages or pens for these purposes had become less clear. Certainly, there was a growing minority who saw it as unethical. But unlike the current argument that confining animals for display is wrong because it degrades the inherent integrity of the animal and obscures its natural context, critics on the prairies before 1950 generally criticized the practice because it deprived an animal of its freedom. Seeing wild animals as symbols of freedom and liberty, Ernest Thompson Seton had written in 1898 that there was a huge difference between "wild animal prisoners, perfectly useless and untameable to the last," and animals in the wild. Such criticism grew in the interwar years. One correspondent in the *Western Producer*'s Nature Lovers column wrote in 1928 that she thought that capturing wild birds and caging them was "an equal, if not a greater sin than killing them" because their liberty was part of their character. And in 1934 Beresford Richards, an Alberta poet, wrote a poem entitled "To a Little Wild Duck Captive Born" in which he described holding a captive bird in his hand. Richards took no pleasure in the bird happily snuggling into his hand, but rather reflected on "The wrong I've done to you / Pinching the soul of thousand leagues of sky Into that starless cell, captivity." In 1949 concern for this loss of freedom was also expressed by the prairie writer Gill Shark, who commented that "no matter how we delude ourselves, few animals generally take to captivity. Go to the zoo some day and watch how the coyotes stalk around their cages, how the captive badger tries to make himself happy by digging in the dirty sand."[59] Yet many people felt no discomfort with such scenes, and some argued that the zoo expressed a humane relationship with wild animals. In 1929 Dr O.H. Patrick of the Calgary Zoological Society noted that "while a number of persons agitated against taking away freedom from the animals," he thought keeping some of them in zoos was a humane gesture because they were fed and protected there in contrast to the wild where they were "continually in danger and in want of food."[60]

While captive wild animals were most commonly encountered in zoos, there were some private collections, such as one at the Albion Hotel in Winnipeg, which in 1892 had three moose, two elk, and a couple of deer "running loose in the yard." It also had five caged bears, all different colours, from cinnamon to black.[61] These animals were likely kept, at least in part, to advertise and draw attention to the hotel. Similar commercial motives led to the establishment of a collection of wild animals at Winnipeg's River Park in 1900. River Park was originally developed as

an amusement park by a street railway company in 1891 to promote traffic on its lines during off-hours and weekends. The collection of wild animals added another attraction to the carousel, racetrack, roller rink, and other park features. Perhaps reflecting that their commercial value had become limited, the animals were replaced by a children's entertainment area in 1927. The animals were transferred to the Winnipeg zoo at Assiniboine Park, a city facility that illustrated the practice of establishing zoos as a civic amenity to enrich visitor experiences at public parks and gardens.

Before the 1960s such civic collections were menageries where animals were kept in cages or large pens, a practice that created barriers between visitors and the animals and symbolically represented a boundary between humans and animals. Their confinement in cages kept the animals visible for the visitors' gaze by preventing their natural tendency to hide. There was no pretence of creating an environment that mimicked the animals' native habitat or of meeting what was understood to be their psychological needs. Most zoos also lacked interpretive programs, which in some ways belied the claim that their displays helped to educate people about the wildlife of the region or the nation. This claim was also contradicted by the view that a zoo's success was measured by the number of unfamiliar and exotic animals it displayed. This view – exemplified by the ring-tailed and rhesus monkeys at the Banff Park Zoo – was prevalent everywhere and became even more pronounced after the Second World War. While many visitors found the native animals intriguing at first, they soon demanded more variety and novelty. This quest encompassed an implicit devaluation of native and now familiar wildlife in favour of the unknown and exotic, creating tension between the commonplace and the exotic in the entertainment objectives of zoos.

The Assiniboine Park Zoo, which took the animals no longer wanted by River Park, was established in 1904 when the Winnipeg Parks Board purchased a number of native animals, including deer, bison, and elk, for display in its largest civic park. In 1908 cages were built for bears, and by the next year the zoo held 116 animals representing 19 species.[62] Most of these animals were native to Manitoba, and some of the collection consisted of wild animals confiscated by the province's game branch. By 1905 the branch had distributed seven moose, deer, and elk among the Assiniboine Park and River Park zoos in Winnipeg and the Brandon Agricultural and Arts Association Zoo. The next year a further eight animals were similarly apportioned among them. Indicating that this was an ongoing practice, the province consigned a beaver to the Assiniboine Park Zoo in 1912.[63] By 1927 Assiniboine Park Zoo occupied twenty-two acres and had fifteen species of birds and seventeen species of mammals, including bison, elk, deer, bears, coyotes, beavers, and porcupines. All of the mammals were native except for some monkeys and guinea pigs, while the bird collection had a larger proportion of foreign species. The birds included a variety of waterfowl, some unusual domesticated species, a few exotic birds, and some game birds. The birds and the monkeys had winter quarters, but most of the other animals stayed in the same enclosure year-round. By this time, the collection had grown to the point where the zoo was offering to sell to other zoos an elk, several bears, and some racoons and pheasants and sixteen peafowls (including four peacocks, or male peafowls).[64]

Ungulates were kept at the Assiniboine Park Zoo in pastures about 91 by 123 metres behind a 2.5-metre fence. The bears were held in a pen about 25 metres square surrounded by a 1.2-metre cement wall topped by 2.5-metre-high bars. Small animals such as badgers, skunks (deodorized), and porcupines were kept in pens about 3.6 by 9 metres, while the monkeys lived in a cage about 1.8 by 2.5 metres square and 2.5 metres

Figure 13.7 The display of wild animals aimed to expand the value of civic parks as social and recreational community spaces. In 1925 swans and ducks added beauty and interest in one of the open spaces at Assiniboine Park in Winnipeg. This goal was furthered by the animals housed in the zoo located in the park.

high. In the summer, beavers were confined to a round cage almost 8 metres in diameter. This cage was enclosed by a 1.2-metre-high wall and had a pond about 3.6 metres in diameter. In winter, the beavers were moved into a smaller pen. The cost of keeping these animals ranged from a high of forty cents per day for bison to seven cents per day for bears and a couple of cents per day for most birds.[65]

Because the animals were basically warehoused in cages for display, some people believed that some of them were not well cared for. In 1931 rumours reached Grey Owl at Riding Mountain National Park (where he had just been appointed to the parks staff) that the beavers in the Assiniboine Park Zoo were "on their last legs." Grey Owl offered to take care of them until "suitable accommodations" could be arranged, which the Winnipeg press interpreted as an indication that he wanted to remove them from the zoo. Since he was already planning to travel to Winnipeg, Grey Owl arranged a meeting with city officials to discuss this accusation. The acting superintendent of Riding Mountain National Park subsequently reported to Ottawa that he believed that Grey Owl had "impressed the

Mayor and some of the high dignitaries," but that the man in charge of the zoo "was a little hostile." Most importantly, he noted that Grey Owl had "gained the point that better conditions are going to be created" for the beavers that the zoo had in captivity.[66]

Whether his suggestions were adopted is unclear, and despite any personal motives he may have had in promoting beaver conservation, Grey Owl's subsequent correspondence with the chair of the Winnipeg Parks Board offers insight into how one sympathetic individual viewed the humane treatment of captive animals. At the outset, Grey Owl argued that the beaver pen needed to be designed to respect the animals' habits and routines. They required a space that they could modify and adapt to suit their needs, like building a lodge with an underwater entrance. A water surface at least twice the size of what they currently had in the zoo would be preferable, and "as large a tank as possible" would allow zoo visitors to watch "the really remarkable aquatic exhibitions these animals give at intervals when living in congenial surroundings." Their psychological needs needed to be respected as well. The park was near the river, and while Grey Owl noted that a wooden wall separating the beavers from the river would not "deceive their instinctive knowledge of the water," a high-enough wall would show them that they could not escape and, "being of a mentality which desists absolutely from considering a project once found to be impossible, they would become more completely resigned and contented, and pursue their various interesting activities even in confinement." Even so, it was essential that the beavers' view of the river be blocked because if it was still in sight, they would "suffer at times intensely, and unnecessarily, resulting in long periods of listless apathy, which detract much from their value as an educational exhibit." As he also recommended, educational goals could be met with a small interpretive program. Grey Owl explained that beavers could be trained to come when called by someone familiar to them. The training was "simple and easy" and required only "patience and regular attention" from an attendant who would visit them throughout the year. Because beavers "soon get to recognise a voice," the attendant could train them to come to a call and reward them for their response with wholemeal bread, an apple (their "favourite delicacy"), or "some other trifle they show a taste for, till they got to depend on it." Once the beavers were trained to come to a call such as "Come here," the attendant could "easily call them out of the house at a stated hour" so that the public could see them and throw sticks into the water. The beavers would seize the sticks and "proceed to mount the house with them." They would stay out for a long time and would "work after receiving their treats." As Grey Owl noted, "I have established a personal contact with my beaver *without feeding* that enables me to call them from any part of the lake at any time after four o'clock, although I do at all times reward them."[67]

Grey Owl's advice to the zoo that it should respect the social and psychological needs of its animals was ahead of the times, as was his belief that interpretive activities could enhance the public's appreciation of the animals. But his view of animals as objects of display was in tune with current attitudes, which remained dominant in the treatment of zoo animals on the prairies. This was the case at the Assiniboine Park Zoo, where attention seems to have been concentrated on making new acquisitions to expand the collection. In 1935 a local Shriner's chapter donated a female lion, the zoo's first, and in 1939 the zoo obtained its first polar bear, an orphaned cub named Carmichael. The next year Carmichael received a partner, a female named Clementine, and a suggestion of romance intimated how emotion and sentimentality could still be used to create public interest. In 1951 the Assiniboine Park Zoo was given another bear, a one-year-old grizzly bear cub caught in the Yukon. The cub was named Minto, after the district where he was caught. The donation was brokered by the Calgary Zoo, and the crate in which the cub was shipped was labelled "I'm 'Minto' from the Yukon. I flew to Calgary in a Mannix Construction [company] plane. Watch me grow with a greater zoo for a Greater Winnipeg." While the boosterism surrounding Minto revealed one aspect of the zoo's role in civic culture, his acquisition was the first step in the zoo's new policy of building a collection of animals "once native to Manitoba." The last grizzly in the province had apparently been killed at Portage la Prairie in the winter of 1805–06, which made Minto the first one "to be a permanent inhabitant of Winnipeg for probably 200 years."[68]

The expansion of the zoo's programming was further enhanced in 1956 when the Zoological Society of Manitoba was formed to revamp the zoo and raise funds. One of its first projects was the development of a petting zoo, "Aunt Sally's Farm," which confirmed the zoo as a place for family outings where some of the animals could be used as playthings for children. And perhaps reflecting the level of entertainment expected by a postwar society with an increasingly international outlook, more wealth, and the desire for more varied leisure experiences, the zoo began to pursue a different role than it had in earlier years. As Gerald Ives, the director of the Montreal zoo, told the Zoological Society's members in 1959, modern zoos were not the menageries of old. They were conservation agencies concerned with preserving wild animals that were facing unprecedented threats at a global level. The modern zoo therefore needed to help "save remaining species from dying out" and to educate the public (especially youth) about the need for conservation throughout the world. Yet the public had always found exotic animals attractive, and the idealization of the modern zoo as a global player in conservation efforts matched a new worldliness where increasingly exotic animals were needed to offer maximum drama and excitement. Expansion in this direction at the Assiniboine Park Zoo was rapid; in 1964 it opened its gibbon and monkey house, snow leopards were added to the collection in 1966, and in 1968 construction began on the "tropical house." Notwithstanding these developments, wild animals with a closer tie to the province also received attention. While they had been a part of the collection for three decades, the zoo began to stress its interest in polar bears – animals that most Manitobans still found exotic, even though they were found in areas in the far north of their own province. As well, in 1969 construction began on a native animal exhibit that featured regional species. Located near the zoo's main entrance, it was one of the first exhibits visitors encountered.[69]

While the Assiniboine Park Zoo was an important institution in the history of prairie zookeeping, Calgary's zoo by the 1960s was the most successful in terms of visitation and the size of its collection. Like the Assiniboine Park Zoo (as well as a small zoo established at Borden Park in Edmonton in 1909), the Calgary Zoo began as an amenity provided by the parks department.[70] Located in a city park on St George's Island in the Bow River, the zoo was only one of several features of the island park, which included a dance hall, refreshment booths, and picnic sites. At some point, midway rides were also installed. William Reader, the superintendent of the parks department, reported in 1919 that, after "repeated recommendations," a pair of black-tailed deer had been acquired for the park. The next year, some turtles were donated by a local resident and a pool was constructed for them next to the deer pen. As Reader noted in 1921, the deer were a "never ending source of interest" for park visitors and he very much wanted "to see this feature extended by the addition of other native fauna." The next year, officials at Banff National Park tamed a deer "sufficiently so as to be fence broke" for the Calgary Zoo. Delighted with the acquisition, Reader noted that Calgary would be happy to have any "surplus animals (native fauna)" that Banff had. The collection expanded further in 1922 with the addition of a swan, several owls, and a kangaroo. Although Calgary officials knew nothing about how to care for the kangaroo, Reader claimed in 1926 that the animal had become "thoroughly acclimatized." Donated by the Vancouver parks department, it was an opportunistic acquisition rather than part of Reader's collecting plan, which, as he described it, was to establish "a worth-while exhibit of a more or less complete collection of the fauna of this Province. Such a collection could be gathered and maintained at comparatively small cost and would be of considerable educational value."[71]

The Calgary Zoo had a very limited budget, and specimens were acquired from a range of sources. In 1928 citizens donated coyotes, pheasants, ducks, and other birds. A goat from the city pound was also accessioned into the collection, Jasper National Park donated a bear, and some Muscovy ducks came from the federal Department of Agriculture. The Toronto Parks Department donated six black squirrels, which "were turned loose in the park and roamed to all parts of the city, two only remaining on the Island." The squirrels rapidly became part of Calgary's fauna and have continued to roam the city to this day. Using a permit under the provisions of the migratory birds legislation, staff also collected native birds. By 1930 the collection contained seventeen species of mammals and a variety of birds. With the exception of some guinea pigs (and some monkeys acquired the next year), the mammals were all native to the region and included bears, wolves, coyotes, a mountain sheep, and a variety of small mammals, such as ground squirrels, badgers, and porcupines. The zoo had a few exotic birds, but native birds such as mallards, Canada geese, and various hawks and owls made up most of the avian collection.[72]

While St George's Island Park was also a horticultural showcase, with rose and peony gardens, perennials, decorative shrubs, and trees, the zoo was the main draw for the public. It thus aimed "to have the animals and birds really tame and on view as much as possible." In 1930 Reader reported that "never in the history of this park were so many attracted to it as in the past year" when upwards of 500 cars were in the parking lot at one time. This popularity was "entirely due" to the zoo, and Reader noted that while the park had formerly been deserted in winter, visitors now came every day and there were "quite large crowds on fine Saturdays and Sundays." Indeed, one Sunday over Christmas, 2,500 people visited the zoo.[73] Even so, not all Calgarians were entranced. Although nothing apparently came of it, the city commissioner was told in 1931 that some residents of east Calgary were organizing a petition to have the zoo moved from their neighbourhood because of odours as well as noise during the night, apparently from the coyotes and wolves.[74]

Figure 13.8
This eagle at the Calgary Zoo in 1930 demonstrated the zoo's emphasis in its early years on displays of local and regional animals.

In 1929 the administration of the zoo broadened with the creation of the Calgary Zoological Society. Although Parks Superintendent Reader wanted assurances that the parks department would maintain control, the zoological society rapidly became the driving force in the zoo's operation. While the city retained ownership and met most of the maintenance costs, the society took on fundraising and soon developed social and financial connections with the community that would have been difficult for a city department to achieve. The society had elite connections: Prime Minister R.B. Bennett and local business leaders A.E. Cross and Patrick Burns were honorary patrons. By the early 1930s the society had also taken on most of the responsibility for acquisitions, which allowed it to meet a number of goals. The society believed that the zoo would "help make the City more interesting for our visitors and residents" and would give people the chance to "learn more of Western Canada's wild animals." These achievements would, in turn, help to develop public support for wildlife conservation and would further "assist our children with their nature studies." Moreover, they would encourage civil behaviour and respect. Assuming that violent treatment of wild animals arose from ignorance, the society argued that when people learned more about animals, they would be "less cruel to them." In an interesting amplification of this argument, the society contended in 1932 that wildness in animals was a result of human aggression. As a place that brought humans and animals together, the zoo would teach people that "these beautiful creatures are only wild because we have been cruel."[75]

In addition to its role in fundraising and acquisitions, the society took on several other functions. While the city was responsible for feeding the animals, the zoological society assisted by securing donations from local food processors and retailers of offal, screenings, and food that was unfit for human consumption. The animals also received "natural food" at least two days a week; depending on the species, this included "brush, leaves, berries, rats, mice, sparrows, crows, rabbits, etc.," which proved "a great factor in keeping them healthy." Particularly in its early years, the zoological society was also especially active in building infrastructure. An aviary, an enclosure for carnivores, and another for waterfowl were among its first contributions. It also helped restock and repair the zoo after the Bow River flooded in 1929. In 1932 the society erected a totem pole, installed fencing and "rustic seats" where visitors could sit and watch the animals, and planted 300 trees as well as shrubs and flowering plants. It also initiated the creation of an exhibition area with life-sized cement casts of dinosaurs and other prehistoric animals. When the first casts were put in place in 1936, the parks department landscaped the area to resemble as much as possible "the supposed appearance of the habitat of these animals."[76]

People in the surrounding district often approached the zoological society and the parks department with offers to sell animals that they had captured. These included bears, eagles, coyotes, and small mammals, and some of these animals were occasionally purchased. This was a practical way for the zoo to develop its collection, as it had neither the space nor the money to house and collect exotic animals.[77] Even so, public expectations indicated a craving for the exotic. Vastly exaggerating the extent of the collection, the *Calgary Herald* grandly reported in 1931 that "from the plateaus of Mexico, the snow crowned Rocky Mountains, the forests of South America, and the wastes of the Far North, the prairies of the West and the veldts of Africa, come the strange and engaging denizens who make Calgary's zoo a Mecca for children and a place of absorbing interest for their elders." Yet this aggrandizement was not entirely misplaced because the zoo eagerly obtained exotic animals when it could afford them. By 1933 it had acquired a baboon, a coati, and four types of monkeys, and in 1935 more monkeys as well as a peccary were added. By 1938 a polar bear, a yak, and other animals had arrived from the Banff zoo, which was being wound down. By 1939 the further acquisition of exotics had brought lions, leopards, mouflon sheep, and a sea lion, among others, into the collection.[78]

During the 1930s the zoo began to grapple more directly with reconciling its role as a park amenity with the needs of the animals it had collected. In early 1931 the zookeeper, Tom Baines, wrote to William Reader protesting plans to move back the fence around the deer and sheep pens to make more room for landscaping and walkways for park pedestrians. Implicitly arguing that the zoo should pay more attention to the welfare of the animals than to the upgrading of public access, Baines claimed that "already these poor creatures are penned in what in my opinion are terribly cramped quarters and yet now they are to have the most valuable strips of ground taken away. Fallow deer will have to live in a pen approximately 90′ by 40′ [12 by 27 metres]. Unless a shelter is built almost at once – that is before it rains – the Rocky Mountain Goat will die of Pneumonia in any case. For years deer have been kept here in inadequate quarters and with what result we know." As he concluded, "if the animals are to be kept under these conditions I feel we are not fulfilling any useful purpose." Reader did not respond directly to Baines's concerns about the welfare of the zoo's inhabitants but instead asserted that more space would better serve sightseers who "now flock" to the zoo. In other words, visitor needs trumped those of the animals. As Reader observed, wider spacing would help enhance the park-like aspects of St George's Island, and while this would create more work for the staff, whenever he had travelled he had found that "the practice is to arrange pens at comparatively wide intervals with considerable landscaping distributed amongst them." Notwithstanding Baines's criticism of the zoo's husbandry, the zoo typically reported low death rates among its animals. This did not, however, mean that the animals lived in humane conditions. Although the Calgary Zoo was likely no different than other prairie zoos, after a visit in 1939 the ornithologist Lawrence Potter wrote, "What a terrible dump! Most of the inmates looked cramped and unhappy, there was little or no shade for hot weather, and it *was* hot too. The two timber wolves were lame as if they had some foot affliction, the coyotes looked as if they craved a hole to creep into out of sight of gaping humans; to see two golden eagles and snowy owls cooped up like chickens made me sick."[79]

Potter's comments were in effect a broad criticism of the practice of keeping animals in small cages in a menagerie layout. Tom Baines was also increasingly dissatisfied with this style of zookeeping. In 1939 he noted that the animals were generally "housed in their correct zoological order" and most pens had labels giving the common and scientific names and habitat of the species. Yet, notwithstanding its popularity, Baines believed that the zoo needed to move beyond being "merely a menagerie" and become a modern zoo that was "educational as well as interesting and amusing." In part, this goal could be realized with an educational centre. This would put the Calgary Zoo on the map, since Baines claimed that no other zoo in western North America had such a facility. While school groups were given guided tours of the zoo – in 1938 ninety classes went through the zoo – a formal space with seating for up to eighty students was needed. This space could also be used for adult education and lectures. Further, Baines felt that the centre should have exhibits and teaching materials such as mounted specimens, bird nests and eggs, and maps and charts showing the location and habitat of each animal in the zoo. Promoting the view of wild animals as friends or foes, he recommended that one of the charts should explain the differences between "harmful and useful predatory birds" to show "why some should be reduced in number and others left that are real friends to farmers and gardeners."[80]

Despite Baines's desire for change, a "modern" zoo was slow to emerge. Following his attendance at the 1947 annual convention of the American Association of Zoological Gardens and Aquariums and inspired by the formal educational programs at the San Diego Zoo, Baines returned to Calgary full of enthusiasm for transforming the Calgary Zoo into an educational venue. Repeating the position he had taken nine years earlier, he argued that if a zoo was "to properly take its place in a community," it needed to have a more active educational role and transcend its character as "merely a menagerie where people gawk and throw peanuts to animals."[81] This was not, however, the course that the Calgary Zoo took. Rather, management was devoted to expanding the collection and entertainment facilities. Postwar petroleum development in Alberta brought an economic boom to Calgary, and with it, a population that expected and could afford leisure activities and community services such as the zoo.

While the ongoing development of the prehistoric animal garden and the expansion of picnic and entertainment facilities for family outings confirmed the zoo's role in community life, postwar growth at the zoo was most evident in its collection. This growth was at times almost frantic and matched the ethos of a society thoroughly committed to economic growth and material accumulation. By 1949 the zoo had 200 species in its collection, almost seventeen times the number in 1930. Growth continued throughout the 1950s, and by 1959 the zoo had 1,099 animals representing 353 species.[82] The zoo continued to accept private donations, in 1953 accessioning gifts of deer, bear cubs, and cougar kittens, among others. It also took animals that came into the hands of the provincial game department, including pronghorns, lynx, raccoons, bear cubs, and deer. Further, the zoo asked the province to give it additional beaver kits when available so that it could expand its beaver collection because "Canada's national animal" was one of the zoo's "key specimens." In 1956 the zoo mounted the first exhibit of fish at a Canadian zoo when it received a small display of freshwater tropical fish courtesy of the Calgary Aquarium Society.[83] And the zoo continued to expand its purchase of Canadian wild animals. It placed standing orders for animals through the Yukon's game and publicity branch, which acted as its agent with trappers who live-trapped lynx, wolverines, otters, and marten. The zoo preferred an equal number of each sex, paying $100 each for otters and wolverines and $50 for lynx.[84]

Exotic animals were also purchased. In 1953, for example, the Calgary Zoo became the only zoo in Canada to have a pair of Sumatran tigers, having purchased them (both born in captivity) from the Rotterdam zoo. In 1956 it purchased gibbons, and private funds – donated specifically for the purpose – were used to acquire polar and Kodiak bear cubs. And in 1959 the zoo acquired a tuatara from the New Zealand government. There were only six of these reptiles in captivity in the world, and the zoo relied on its influential supporters to facilitate this acquisition. In addition, the Calgary Tourist and Convention Association lent the zoo its support because it anticipated that the tuatara would enhance the zoo's appeal as a tourist attraction. Expansion of the collection continued in the early 1960s. In 1963, for example, through purchase or exchange, the zoo added giraffes, zebras, black leopards, bison, water buffalo, orangutans, and lemurs.[85]

The zoo sometimes sold animals, and in 1947 it reported that while currency restrictions had unfortunately ended sales, there had been a good market for North American animals in Europe, where zoos were rebuilding collections devastated by the war. The zoo also commonly exchanged or traded animals with other zoos. As examples of the international network of trade and exchange in animals that fuelled zookeeping, in 1962 the zoo obtained camels from Chicago; a giant condor from Pittsburgh; herons from Tarpon Springs, Florida; and a Barbary ape, a New Guinea wild dog, and a common marmot from San Diego. In return, the Calgary Zoo sent out white-tailed deer, a mouflon, swans, a dingo, a wolverine, and lynx to various zoos in the United States and Canada.[86]

And in 1963, when wild lynx began showing up in Calgary city limits, the zoo turned public service into institutional advantage when it helped to trap them and then used them as "exchange material," especially with zoos in Europe. A long-standing practice, such exchanges were important for the expansion and renewal of the collection. They also created a network among zoos and built up what Tom Baines in 1947 called "moral debts" that enhanced the long-term influence of the zoo. As well, these exchanges served civic identity by demonstrating that the Calgary Zoo was a player in the North American zoo community. Civic promotion was not neglected either; when four geese were sent to the zoo at Wheeling, West Virginia, in 1962, the crate was labelled "From the Children of Calgary, Oil Capital of the World to the Children of Wheeling, Park Capital of the World."[87]

The public was attracted to the Calgary Zoo out of a curiosity about animals, a search for entertainment, and a hope to meet their personal psychological and social needs. The zoo skilfully publicized the novelty and excitement of its collection and its newest acquisitions and often used the animals for its promotional needs. From at least 1959 zoo staff showed live animals on weekly appearances on the local children's television program, *Romper Room*. And beginning in 1953 the zoo began taking its animals into the community, using them as entertainment features at seasonal parties or other events at hospitals, health care facilities, and schools. Over the next decade, guinea pigs, rabbits, birds, and other animals, including very young ones, were taken to a total of 840 parties. These included "coloured chickens" for Easter, and a young deer dressed like Rudolf the Red-Nosed Reindeer was for many years a popular choice for Christmas tours at various facilities.[88] Such activities suggest that the public viewed the zoo's animals as playthings, which was also demonstrated by the zoo's practice of allowing the public to feed the animals, a practice now accepted as a serious violation of zoo husbandry. While many North American zoos were prohibiting it by the early 1960s, Calgary zoo officials recognized that the public wanted such interaction to continue. As was noted in a 1964 zoo newsletter, officials hoped that "drastic steps" would not be needed and that they would "never have to stop feeding by the public entirely." Nonetheless, the pubic was urged "to exercise care so the animals do not suffer, however innocently, from

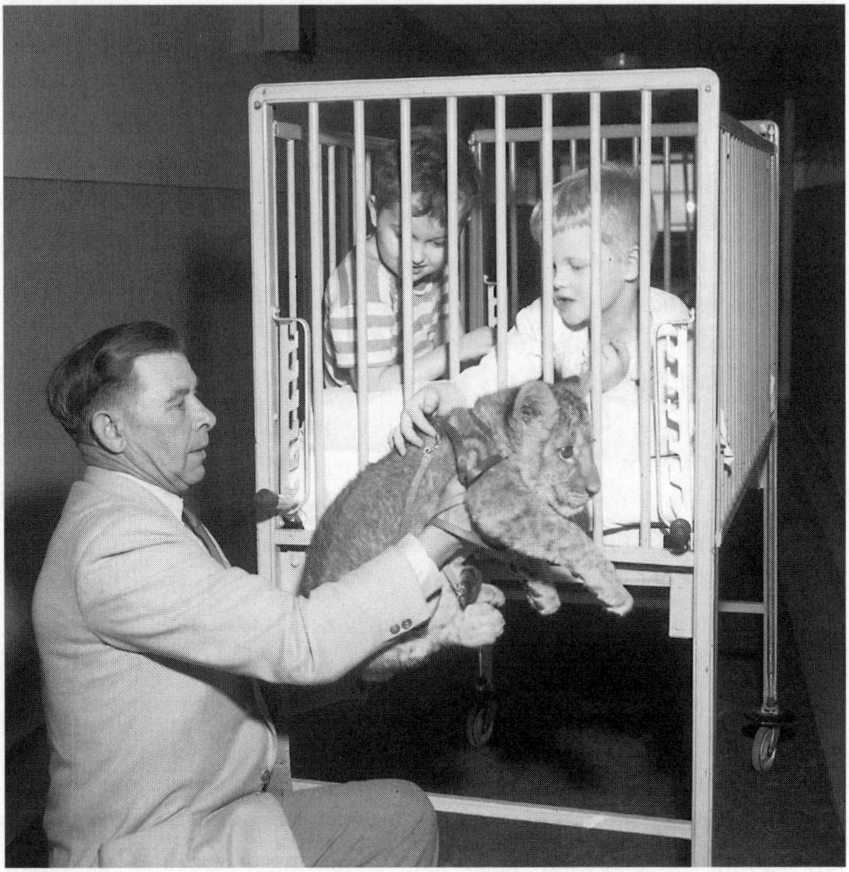

Figure 13.9 The Calgary Zoo used some of its animals as part of its community outreach in hospitals, schools, and other venues. As this 1955 photograph suggests, the use of zoo animals for social service objectives often treated them as toys or playthings.

feeding accidents." Thus the public was warned not to feed popsicles and lollipops to the animals because the sticks could seriously hurt them if swallowed and not to carry food in plastic bags when approaching animals because an animal could snatch the bag and eat it.[89]

The persistence of public feeding, despite its risks, was consistent with the tradition of treating zoo animals as objects meant to enhance visitor experiences. Modern formats of display confirmed these attitudes. In 1954, for example, the zoo opened its new small-mammal building, the first zoo building with "exhibits behind glass." Despite its gloss of modernity, as American cultural theorist Susan Willis argues in her critique of the modern zoo, the use of glass walls was really a "security shield" that attempted "to engineer intimacy" by giving the illusion of closeness at the same time that it gave observers complete safety and control and confirmed their dominant position over the animal.[90] As these practices revealed, the Calgary Zoo in the 1950s was in some respects still very similar in its outlook and approach to what it had been before the Second World War. This was further revealed by the state of interpretive programming at the zoo. Although Tom Baines had since the 1930s urged that more attention be given to public education, developments in this area were only slowly realized. Public appreciation was fostered more by ensuring that animals were kept visible and on display than by more deliberate interpretive devices and activities. Nonetheless, a few steps in this direction began in 1956, when large pictorial maps of the grounds were erected to guide tourists to the sights, and in 1957, when the zoo began to develop interpretive elements that went beyond the mere labelling of cages. An example of this was "a large electrically operated display" consisting of a map of the world with models of animals placed on it to show their distribution. Another such display featured showcases with models of North American animals at 1:20 scale together with maps and descriptive labels. A fifteen-metre long panorama sponsored by the local electrical utility company showing how fast various animals could run was also under construction.[91]

Such changing attitudes towards interpretation were further suggested by the development of a "children's zoo" which opened in 1957. As a facility specifically targeted at an important segment of zoo visitors, it asserted a new attitude to public education. But even within this new format, traditional values about the treatment of zoo animals endured. In a complex of linked buildings and spaces, young visitors began their tour by viewing an orientation panel that explained the facility's layout and the animals it displayed and where they came from. Once into the exhibits, visitors first encountered small mammals such as skunks and prairie dogs, a tapir, and some raptors penned behind plate glass windows. Nearby, a vending machine sold "correctly blended pellets suitable for any animals or birds." After moving upstairs to a viewing platform from which various animals could be watched, visitors passed through an open space that contained a large tortoise and several "orphaned Bambi Deer," some of which would "visit many Children's hospitals and Homes next Christmas." Next came a large building with infant gorillas named Ruff and Tuff (names chosen by a radio contest sponsored by the zoo) and rows of cages with other newborn animals and small creatures such as snakes, hedgehogs, and kinkajou. Outside, there was a "contact pen where small kiddies can play with young Deer, Rabbits, Guinea Pigs, Chickens, etc." as well as pens containing coyote and wolf pups, a young elephant (named Gyro in honour of its donor, the Calgary Gyro Club), and "a fine collection of Macaws, Cockatoo and Parrots." Further on, the young visitors encountered a newborn camel, monkeys, and gibbons, and on the way to the exit they passed by the "Goat Castle with its collection of Nubian Goats" and an area with "various animals running at liberty, including Lambs, Deer, Puppies, Rabbits, etc."[92]

The techniques and arrangements employed in the children's zoo conveyed to the upcoming generation a number of traditional attitudes about animals in North American popular culture. Animals were objects or playthings, cuteness was adored, humans both figuratively and literally dominated animals, and the exotic was privileged over the local. It is not clear how these views related to the zoo's often-stated goal of promoting wildlife conservation, but they were clearly not at odds with popular attitudes. During Easter week in 1962, the children's zoo had 25,000 visitors, 7,000 on Easter Sunday alone. The zoo in its entirety was by then attracting about 1 million visitors a year, and in 1963 it was estimated that it had 1.2 million visitors, making it the most popular entertainment venue in the city.[93] Tom Baines argued that while the zoo's success arose from effective promotion, the "real secret" to its growth was that it exhibited "healthy animals, in clean if not elaborate surroundings," and had staff members who "not only like animals but also people."[94] And the zoo was careful not to make its visitors uncomfortable. As it told its Yukon collectors in 1953, "an animal that shows damage or injury is of no value to us as it causes too much adverse criticism from spectators." And as it noted in 1956, when new specimens were received at the zoo, aged or injured ones were "excluded." This practice perhaps contributed to good survival rates at the zoo; in 1955 the loss of animals at the zoo was said to be half that at any other zoo in North America.[95] As prairie zoos had done for decades, the Calgary Zoo continued to feature animals as a prominent part of a broad entertainment package, thus meeting a range of public recreational demands. In 1962 the zoo added a roller coaster to complement its "other thrilling rides," which included a "Merry Go Round, Ferris Wheel, Boat Ride, Sky Fighter and our Famous Zooliner which is the largest and finest miniature railway in Canada."[96] Such activities persisted until the mid-1970s when the Calgary Zoo began a transition towards practices that were then becoming common in other North American zoos and that better reflected a changing social ethos that favoured giving more attention to the social and psychological needs of animals. Peter Karsten, who had worked at the zoo for a number of years, became director in 1974 and brought in a "ban the bars" program. As a result, habitat-based spaces replaced concrete pens and cement floors covered with wood shavings. The zoo also became more involved in conservation programs and toned down its use of animals as objects for community service and entertainment.[97]

Figure 13.10 Like other animals in the Calgary Zoo at the time, birds were exhibited in 1954 as objects in a cage. This began to change in the 1970s with approaches that paid greater attention to an animal's natural context.

After the Second World War the Calgary and Winnipeg zoos occupied the forefront of prairie zookeeping. No other regional zoos approached the extent of their collections, although the Alberta Game Farm, opened in 1959 near Edmonton by naturalist, wrestler, and filmmaker Al Oeming, rivalled them for a time in the 1970s with a very large facility featuring many exotic and native animals.[98] The Moose Jaw Wild Animal Park near the city of Moose Jaw in south-central Saskatchewan was a smaller but similar facility with a longer history. Located on several hundred acres, the park was established in 1929 after persistent lobbying by Frank McRitchie and other local "wildlife enthusiasts." Operated by the Moose Jaw Wild Animal Park Society, the park aimed to attract tourists to the city and enhance its reputation as an up-to-date and exciting place to visit. The park also hoped to stimulate public interest in wildlife conservation. Drawing on nostalgia for the natural abundance of the region before settlement, it preserved a piece of prairie landscape on which it was possible to "reproduce the actual scenes as the early settlers saw them," with "wild animals roaming on the hills and in the valleys." The park also had an animal welfare dimension and took in injured animals and added them to its collection after they were nursed back to health. It was also involved with hatching and releasing chukar partridges and pheasants and also carried out crossbreeding experiments with domestic and wild turkeys.[99]

On opening day the Moose Jaw Wild Animal Park had only a few specimens; visitors could see a mountain sheep as well as some elk, bison, and deer donated by Banff National Park and Buffalo National Park. The collection subsequently expanded, and by 1933 it contained 100 birds and mammals. While smaller mammals, birds, carnivores, and bears were kept in cages, ungulates were kept together in a single large open field.[100] The park claimed that it was located on the "former campsite of Chief Sitting Bull" and advertised this to attract visitors to the park. Whatever appeal this claim may have had, the wild animals were the primary attraction. As one visitor wrote in 1934, the animals in the park showed that "you don't have to go far away to some distant place to find beautiful and interesting things." Deer, elk, and bison could be seen in the paddock, and the visitor found the park's animal rescue work particularly appealing. The rescued animals included a fawn saved "from a fire near Midnight Lake [in east-central Saskatchewan] and shipped to the park by the Game Department," a white-tailed deer rescued as a fawn from a Manitoba forest fire and bottle-fed at the park until it matured, and a bear cub sent "from the north by a hunter who shot his mother." The opportunity to feed the animals was also an attraction. Visitors fed peanuts to the bear, and a newborn racoon washed the food people gave her – "no matter if it is a piece of lovely angel food cake you give her, she washed it just the same." Almost fifteen years later, a fourteen-year-old visitor reported many of the same features. He fed candy to the deer and watched people feeding popcorn to the bear, commenting, "They just loved it!" While the monkeys "looked as though they had been teased" and were aggressive, most of the birds were interesting, although the turkeys, chickens, and ducks "weren't very fascinating for me since I live on a farm." A lion had recently been added to the park, but the visitor lamented that she wasn't in her cage when they visited.[101]

In 1952 the Moose Jaw Wild Animal Park drew just over 70,000 visitors, mostly in the summer. The province gave it a small operating grant, reasoning that it offered people the chance to become familiar with the province's wildlife. Suggesting its marginal financial condition, in 1958 the local Lion's Club assumed responsibility for the park as a civic social service project. By the early 1960s the park had a cougar, lynx, black bear, silver fox, coyotes, wolves, racoons, elk, white-tailed deer, bison (kept in a drive-through paddock), a Barbary sheep, some monkeys, some lions (the "King of the Beasts"), and some exotic South American mammals. The park also featured a Kiddie's Zoo with budgie birds, a midget pony, and various newborn animals. The exhibits were built around nursery rhymes, and children could handle the animals under supervision.[102]

These practices were the same as those in place at other prairie zoos and suggest a shared culture in the use and display of confined wild animals. While the management of live animals in zoos posed more complex problems and cost more than did the management of taxidermied specimens in natural history museums, both forms of display used wild animals to enhance tourism and to provide entertainment and education. They also both reflected the public's curiosity about wild animals and their eagerness to see them, whether mounted in a museum or confined as live specimens in a zoo. In both venues, the representation of animals evolved, albeit slowly, through various phases that showed a growing

Figure 13.11 Feeding captive wild animals was a widespread practice in zoos and other venues. It asserted human control over the animal and offered an entertaining spectacle. In this 1948 photograph, a boy feeds ice cream to a bear at the fair in Camrose, Alberta.

public and professional desire that greater attention be given to the animals' natural context. The earliest museum displays had tended to organize individual specimens on shelves or in cases according to various criteria (of which zoological classification was the most popular), thus placing a heavy responsibility on the observer to learn through close examination of the specimens. Museums soon began to provide context for these specimens through different display methods, texts, and frameworks that facilitated learning. In prairie Canada before 1960, the large dioramas installed at the Saskatchewan museum provide the best example of this process. The region's zoos were slower to reflect these concerns. The zoos at Banff, Winnipeg, and Calgary long remained close to nineteenth-century precedents of the menagerie where visitors walked along a path and viewed animals confined in cages. Significant change only began in the 1970s. Perhaps encouraged by increasingly sophisticated photography of animals in the wild as well as by changing sentiments about ecology and the interconnections of living things, the public and professionals alike championed the redesign of zoos in order to display animals as if they were in the wild. Yet, as historian Gregg Mitman comments, the zoo without bars was still a statement about human domination of nature, albeit a more subtle one because it gave the impression that the animals were free and that the human–animal encounter in the zoo was "natural." Nonetheless, such change meant that more attention was paid to the social and psychological needs of the animals. And preservationist sentiments, which had motivated the saving of endangered plains bison, elk, and pronghorns in national parks, were increasingly recognized in zoos because of the growing acceptance of the belief that the confinement of some wild animals was ethically justified on the grounds that it preserved them as "samples of endangered gene pools held in trust for future generations."[103]

Conclusion: Fitting and Not Fitting Together

Wild animals are central in the natural heritage, history, and collective memory of the Canadian prairies, and the relationships between humans and wild animals that developed there have centred on determining how their lives fit and did not fit together. Until at least the 1960s, the utilitarian view that wild animals did not have intrinsic rights and that they could be used freely to promote economic and social needs dominated state wildlife policy and found support in a number of economic and social sectors. This view also fit with a popular and easy dichotomy in which wild animals were thought of as friends and foes – as useful or harmful, as competitors or companions. But usefulness referenced a number of values, and for some observers, wild animals also offered models of moral behaviour as well as beauty. Nor were they mere automatons; rather, in varying degrees, wild animals were sentient and could feel pain or pleasure, they had more or less complex cognitive capacities such as reason and memory, and they had emotional sensibilities such as feelings of sorrow, fear, and fondness. All this did not presume that they were morally, intellectually, or emotionally equal to humans. But while the divide between humans and animals remained firm, there was uncertainty among many about how wild animals should be treated and what respect they deserved. This was compounded by questions about the role of wild animals in nature and their contribution to human success.

A vision of nature beautiful and alive with sentient and intelligent creatures of ethical and aesthetic value and another vision in which they could be used freely as articles of trade or destroyed at will were at the extremes of views in popular culture about human relationships with wild animals. While the emotive vision had an imprecise and variable appeal from the end of the nineteenth century, it rubbed against the culture's social grain and economic priorities. Moreover, support for it was too individualized and personal, and perhaps it was too gentle a vision to reshape public and state preoccupations in the early settlement years and during the consolidation of settlement in the interwar years. Nor did it find support from a scientific community dominated by behaviourist methodologies and fearful of accusations of anthropomorphism. Yet while rigid economic utilitarianism had many supporters, it was too soulless and mechanistic to have universal appeal and it could not dispel doubts among many that the lives of wild animals might have more meaning than simply as objects of threat or commerce. Thus, in practice, affective and utilitarian views often overlapped and were sometimes uncertain and idiosyncratic, with elements of one sometimes appearing in the other, ostensibly with little contradiction. This overlapping accommodated multiple views and found different emphasis and expression over time. Beaver conservation, for example, offered economic and environmental benefits through water control and the production of a valuable fur at the same time that it expressed a fondness and admiration for an animal that was part of the nation's mythology, had remarkable engineering abilities, and was recognized by many to have sophisticated social organization and high intelligence. So, too, the double mandate of the national parks promoted economic interests such as tourism while simultaneously aiming to preserve charismatic animals such as plains and wood bison, pronghorns, and elk as a part of national and regional history and identity.

The themes that are apparent in the diverse attitudes, thoughts, needs, and expressions that shaped people's relationships with wild animals in prairie Canada between 1870 and the early 1960s reveal how encounters between people and wild animals were part of the broader history of the region. This was not a single or linear narrative, and like an ecological system, these themes were interconnected and best understood relative to one another. At the outset, relationships with wildlife were intertwined with the implanting of a new Euro-Canadian economic and social order in the prairie region. They thus were part of state formation, public policy, province-building ambitions, and Aboriginal and Euro-Canadian relations. They also reflected the newcomers' cultural assumptions about the appreciation of nature as well as its conquest and control. And blaming wild animals was often a convenient way to deal with personal failure, or to explain why life had not turned out as planned, or to excuse the negative environmental changes that Euro-Canadians themselves had caused. These relationships with wild animals were expressed and gained concrete reality in particular sites and activities – on farms, ranches, and traplines; in zoos, parks, and museums; and in hunting and nature outings. While these sites of connection commonly involved the economic and social utility of wildlife, they also, in varying degrees, symbolized hopes and anxieties and created further connections with landscapes and the natural world and the ineffable wonder of being-in-the-world. The diverse encounters with wild animals were also grounded in the environmental change created by Euro-Canadian settlement. While some species were eliminated or marginalized because of habitat change, others found new advantage because of it. This dynamic process was, and remains, ongoing, but it has occurred at different rates in different parts of the region – most rapidly and dramatically on the southern grasslands and more gradually in the parklands and northern areas. But throughout, change in one part of the environment brought change in others. This change was sometimes intentional and brought predictable consequences, but in other cases it resulted in unexpected and unwelcome developments. The extirpation of bison and wolves, for instance, was a deliberate strategy that aimed to free land for agriculture and to safeguard livestock. But the disappearance of these keystone species and a broad change in habitat also brought unexpected benefit to other species that Euro-Canadians often saw as a threat to farming, ranching, and sport hunting, creating new tensions and forcing people to grapple with the unforeseen consequences of the changes they had wrought.

Since wildlife was a public resource owned and regulated by the state, governments and centralized bureaucracies imposed rules about its use and treatment and asserted judgments about the value of particular species. More active management of wildlife by the state, a trend that began in the late nineteenth century and grew in strength in the early 1900s, sought to ensure equal public access; to conserve wildlife for food, commerce, and as recreational resources; and to produce government revenue from licence fees and royalties. The breadth of these goals was made clear by the disparate range of activities that fell within the ambit of conservation, which at various times included the preservation of certain species and the killing of others, the regulation of hunting and trapping, and the transplanting of new species of fish and game birds. One observer, looking back in 1947, quipped that while many of these strategies were "part of what is called conservation of natural resources," he thought the term was "a misnomer" because "what we are really doing is attempting to re-develop natural resources, after having once wasted them."[1] This conclusion had merit, but the drive to regulate and manage wildlife also expressed an evolving understanding that the management of natural resources and the control of nature were critical functions of the modern state. Such factors structured the creation of the "national commons" in the late nineteenth century, in which local resources were publicly regulated in favour of a wider community of users. This public use regulation sometimes reshaped local use of wildlife, such as when its use by settlers and especially Aboriginal people for food and subsistence was compromised in order to benefit outside sport hunters, trappers, or fishers.[2] While this reshaping of local use of wildlife in favour of a broad community of users reflected a conflict between the centres of power and the margins in some cases, the creation of a national commons in regard to the use of wildlife was more often about a geography of social and economic inequality, networks of influence, and conflicting valuations than about physical location. On the whole, the state met the needs of influential local Euro-Canadian interests and sectors, helping to focus and coordinate the control of nature and maximize its commercial value. Although the exemption from provincial game laws of northern districts at certain periods suggests respect for local subsistence needs as well as for the needs of resource exploration and farm settlement in these districts, the needs of farmers and ranchers, sport hunters and commercial harvesters of wildlife, wherever they were located, had a clear priority in state policy. And while they sometimes had conflicting needs, the state and private interests often co-operated to achieve commonly held goals. This was especially so in the case of sport hunting. The state increasingly privileged it (regardless of where the hunters lived) and typically co-operated with fish and game associations to remake and control nature to enhance sport hunting. This remaking of the natural world led both groups, independently and co-operatively, to attempt to eliminate wild predators that competed with hunters for game. This privileging was also seen in the transplanting of foreign game birds to mitigate the depletion of native upland game birds and to replace them with ones that had more desirable sporting features. By speaking for locally influential interests (often against less influential local ones), fish and game associations expressed local needs at the provincial level. Further, the fish and game advisory committees established by all prairie provincial governments in the 1940s institutionalized these influential local interests at the centre, integrating them more firmly as elements in wider provincial networks of power and social authority.

These networks easily accommodated the steadily expanding scale and scope of wildlife management activity. As was the case in Manitoba, management concerns in the 1870s had largely been expressed through the relatively simple devices of wolf bounties and the mandating of closed and open seasons to conserve game animals and fur-bearers. These strategies had broadened with respect to hunting by the 1890s to include bag limits as well as an increasingly complex array of rules that governed hunting practices and behaviour, which, ideally, would increase the chase and reduce the killing. Although the state's ability to enforce such rules and their subsequent iterations was limited, they were significant in setting standards and establishing models for legally acceptable and socially respectable relationships with game animals. Moreover, while it was well understood that habitat destruction posed a major threat to wildlife, it was equally clear that for political and economic reasons such destruction could not be controlled in a meaningful way and that all wild animals would continue to decline before an inevitably expanding human sphere. Therefore, seed populations and desirable animals were protected in bird sanctuaries and national parks, and wildlife conservation outside of these venues relied on regulating the people who used wildlife and, increasingly, on controlling predators. The latter would ideally stabilize the "natural balance" that remained or would create a new one. This presumed that a self-regulating balance of nature had once operated but had been dislocated by hunting, trapping, fishing, and habitat change. Proof of the upset was thought to be apparent everywhere in depleted game and fur-bearing animals and in imbalances between predators and prey. In the case of fish, stocking programs, bounties on predatory fish, and coarse fish removal aimed to stabilize or enhance commercial and sport fishing. This goal of remaking nature for human needs – from afresh if need be – was most bluntly expressed in lake rehabilitation projects where lakes were poisoned with rotenone and toxaphene to create an empty environment that could be repopulated with desirable game fish species. With strategies less radical in application but sharing the same intention, sport hunters and their associations aimed to eliminate coyotes, crows, and magpies, which they blamed for the falling numbers of waterfowl and up-land game birds, and targeted wolves and other high-level predators as threats to big game. As the Saskatchewan Fish and Game League frankly stated in the early 1950s, "man, the greatest predator of all, has an immediate interest in the [game] species being preyed upon and in order to derive more benefit from the desirable forms of wildlife must eliminate some of the competitors. Admittedly, every creature, no matter how shady its character, may at times have some good qualities and it is to assess the good with the bad that is required when formulating a control program."[3]

Stocking and predator-control programs, in conjunction with efforts to make hunters, trappers, and fishers personally responsible for conservation, sidestepped the need to regulate land use or to control habitat destruction in the interests of wildlife. Aside from bird sanctuaries (often established as part of the Migratory Birds Convention) and the often contradictory policies of the national parks branch, the only major attempts to manage habitat as a conservation technique were undertaken with respect to waterfowl and fur-bearers. The habitat conservation work of Ducks Unlimited (Canada) in the 1930s was important in respect to waterfowl. Coupled with predator control, DUC hoped that such conservation would increase waterfowl for sport hunters and, not inconsequentially, make the tighter regulation of sport hunting in a time of declining waterfowl populations unnecessary. Similar recognition of the importance of habitat was seen in beaver conservation efforts as well as in the rehabilitation of marshlands to increase the production of muskrats. But even when wildlife management shifted after the Second World War towards more complex interventions that drew upon changed thinking about the ecological complexity of natural systems, little was done to control land use or habitat destruction.

While wetland rehabilitation by DUC was significant in expressing a regional perspective about conservation, each province enjoyed almost total control over wildlife within its boundaries. This autonomy was reflected in an overall lack of interprovincial co-operation. As was noted in 1943, for example, conferences held by wildlife managers from the Prairie provinces had "functioned in a desultory manner for a few years" and then ended because of a lack of funds and disagreement over bag limits for migratory birds.[4] More importantly, provincial authority over wildlife was only marginally affected by national political life, thus confirming the constitutional status of wildlife as a local concern within the framework of Canadian federalism. Although the federal government showed initiative and commitment to wildlife issues in the national parks where it had exclusive jurisdiction, its attention to wildlife matters as expressed through the short-lived Commission of Conservation did not lead it to expand its involvement in this area of provincial jurisdiction. This general passivity left the provinces with almost unchallenged control.[5] While it gained responsibility for migratory birds under international treaty (which the Commission of Conservation promoted), it remained timid in the face of provincial jurisdiction and did not attempt to force conservation measures on the provinces, even during a time of collapsing waterfowl populations in the 1930s. The same reluctance was especially evident in the case of hunting rights promised in treaties with First Nations peoples. The Department of Indian Affairs was unable and ultimately unwilling to challenge provincial intransigence on the recognition of such hunting rights, nor did the federal government later show a strong commitment to the enforcement of First Nations hunting rights agreed to in the 1930 natural resources transfer agreements with the provinces. While this federal passivity perhaps indicated respect for provincial jurisdiction and the practical limits of forcing provincial compliance, it also indicated that the federal government was not willing to expend political capital on an issue that engendered a deep and sometimes vitriolic antagonism in many Euro-Canadians. In any case, this failure to act reflected the commonly held view that treaty rights stood in the way of Euro-Canadian manipulation and use of nature in pursuit of economic and social goals. This in turn was reinforced by the federal government's assimilationist policies and its attitude that Euro-Canadian patterns of use and management of wildlife were normative and universal.

Whatever provincial–federal sensitivities existed, the prairie region showed important shared patterns in the exploitation of wildlife. In all three provinces, wildlife was used for food, as well as for market hunting for a time, and sport hunting was widespread. Each province had similar economic activities that impacted wildlife – mixed and grain farming, sport and commercial fishing, tourism and trapping. To be sure, there were different emphases throughout the region. Commercial fishing, for example, took place everywhere but was most prominent in Manitoba, livestock ranching was largely found in southwest Saskatchewan and southern Alberta, and tourist use of wildlife occurred in all three provinces but most extensively in Alberta. Nonetheless, all of these activities shared cultural outlooks and goals that aimed to control and reshape wildlife and the natural environment in order to enhance economic production as part of the everyday need to make a living.

Such shared cultural outlooks and goals were particularly evident in the agricultural sector's impact on wildlife throughout the region. The use of modern technologies in the breaking and clearing of land for farming and the profound transformation of the region's flora, fauna, and land necessary for the creation of an industrialized farm economy were celebrated in prairie popular culture as critical accomplishments. In varying degrees, wild animals affected the profitability of the newly established agricultural sector, reframing an ancient and perhaps inherent competition with farmers and ranchers for land and resources. While the decimation of the bison was a portent of what was to come for some wild animals, the land breaking and clearing and the methods of field husbandry that destroyed the habitat of some wild animals inadvertently created new habitat for others, most importantly, coyotes, foxes, white-tailed deer, crows, magpies, and some species of ground squirrels. With the exception of white-tailed deer, which were welcomed by sport hunters and naturalists, the success of these newly advantaged animals was typically an unwelcome development that tarnished settlers' dreams of remaking the land for their own use. The reappearance of wolves on the plains along with cattle ranching was met with a virile hostility, and efforts to eradicate wolves in ranching areas were generally accepted as a sign of civilization and progress.[6] Such thinking was equally apparent when the hands of grain and mixed farmers were raised against ducks, coyotes, gophers, raptors, and other wild animals that threatened farm production. The extent of this hostility was demonstrated by the killing of almost the entire coyote population in Saskatchewan by the mid-1950s. While critics argued that people could only blame themselves after coyotes had been exterminated and fields and towns were then overrun by mice, such reasoning did not temper the violence of the coyote-control campaigns. Agrarian competition with wildlife was blunt and all-encompassing, made all the more aggressive because of the view that land was a commodity that owners had the right to use in its entirety for their private gain. While the often brutal treatment of wild animals by many farmers had no redeeming features, the modern industrial farm economy gave farmers little economic reason to coexist with them. Most begrudged the loss of any field or farmyard production to wildlife, and such losses were sharpened by the economic precariousness of many prairie farmers and their vulnerability to climatic and market conditions. As a cultural expression, agrarian antagonism to wild animals drew upon a rich array of myths and perceptions about human superiority over animals and about the control and use of land. It also redirected anxiety about economic marginality and confirmed a narrow vision of an environment tailored to agricultural production. For example, the early use of strychnine by farmers and ranchers to kill wild animals that stood in their way revealed almost no concern about the poison's spread through natural food chains and the damage it caused well beyond the target animals. Indeed, the broadcast spreading of strychnine on fields to kill gophers implied a future in which no living thing would exist unless it contributed to farm production.

Since agriculture was the most important economic sector and source of livelihood for most people in the region, the farm sector's adversarial relationship with wildlife was respected and enhanced in public policy and legislation. Game legislation allowed farmers and ranchers a free hand against most of their wild foes, and more directly, the state implemented numerous programs of extermination or, at least, extirpation for their benefit. Wolf bounties to protect an emerging livestock industry in Red River in the mid-1800s were the first public subsidies of the fight against wild competitors and similar subsidies were utilized again later in the century when an alliance of the territorial government and ranchers worked to eliminate wolves on the southern plains. And at various times from the late 1800s, all levels of government extended financial and other support to farmers in their conflict with gophers, magpies, and coyotes. While farmers believed they faced an excessive number of wild animal competitors, they nonetheless had friends in insectivorous birds, whose value was affirmed by provincial legislation in the 1890s and later by the Migratory Birds Convention Act. Among others, Manitoba bird enthusiasts George Atkinson and H.M. Speechley contended in the early 1900s that this protection should also extend to hawks and owls because the handful of farm chickens lost to them was a small price to pay for their control of rodents. Such ecologically oriented approaches made only slow progress against entrenched farmer (and hunter) hostility to raptors. And signalling that the acceptance of ecological views about insectivorous bird conservation was somewhat tenuous, concern for their protection lost its urgency for farmers after the Second World War. This perhaps was the result of the postwar chemical revolution that promised the total control of insects and weeds, thus making the protection of insectivorous birds less important. At the same time, a declining commitment to bird conservation was offset by a growing willingness on the part of some farmers to see some wild animals that preyed on small rodents, such as weasels and badgers, as beneficial to farming.

Such ideas tempered the adversarial mindset of some farmers as well as others and drew upon a view that held that while a natural world tilted in favour of human needs was a legitimate goal, it could be better and more economically achieved by working with and not against nature. For some, this sentiment was conjoined with a fascination about the world around them, and for others, it revealed a desire to be connected with the reality of the natural world and to express a sense of being-in-the-world. For yet others, concern and respect for the sentience and integrity of wild animals formed one element of a moral life or of witnessing the presence of God in the world. Such affective views had in varying degrees been part of European thinking for centuries. Clearly, many people, for example, derived personal pleasure from observing the birds they encountered in everyday life or saw off in the fall and welcomed back in the spring. As the ornithologist George Atkinson argued, birds were fascinating precisely because of their social complexity and cognitive abilities, which made watching and studying them a richer experience. Others contended that the lives of wild animals provided symbols and models of courage, freedom, maternal sentiments, domesticity, and joyfulness, as well as of good and evil, and still others reported finding spiritual grace or private solace in their relationships with wild animals. So, too, some people found a sense of belonging and connection to their environment and landscape through wild animals, and as the generous public support for the campaign to preserve whooping cranes in the 1940s and 1950s demonstrated, this relationship with wildlife sometimes created an engagement that expressed local identity. For still others, such relationships with wild animals reflected inherent human characteristics. As the Saskatchewan game commissioner Fred Bradshaw contended in 1918, programs such as bird protection appealed to innate human qualities, for there was "something in human nature," whether "in the tiny tot of five or the burly engineer of mature understanding, which readily responds to a reasonable appeal for the protection of beautiful things."[7]

While people's views of wild animals remained constant in some ways, they also shifted along with social and intellectual change. The casting of the life of wild animals as a manifestation of the working of God's hand in the world, for example, was strongest before the Second World War but then began to lose ground to the view that a greater appreciation of the value of all wild animals was important in helping people to cross the divides of biology, culture, and history and to understand the unity of the natural world. This confluence of old and new ideas characterized many of the forums through which people expressed their interest in wildlife. The preservation of bison, elk, and pronghorns in national parks, for example, drew on an emerging nostalgia for the vanished bison landscape of the Old West, but such longings remained firmly rooted in contemporary priorities and goals. The protected animals were segregated in places where they could not interfere with agriculture or industry, while less charismatic animals that had equally been part of the bison landscape, such as badgers and swift foxes, were given scant or only selective regard at best. And predators, such as wolves and coyotes, which had also been defining species in the bison landscape, were targeted in order to safeguard the protected animals. By the mid-1920s, however, some national parks officials had begun to argue that predatory animals could be used to "ecologically" or "naturally" control big game populations. This shift was, in theory, reflected in 1930 with the enactment of new national parks legislation that endorsed the preservation of the integrity of all elements in a park's environment. Although predator control was not entirely abandoned, concern for the integrity of park environments subsequently evolved and predation by wolves came to be popularly viewed as a demonstration of the grace and wholeness of a naturally balanced and authentic nature, not the horror of an untamed wilderness.

Other organizations and institutions also demonstrated this meeting of old and new ideas and values. Natural history organizations showed an early commitment to wildlife, and the first of these organizations appeared in Manitoba shortly after settlement. Such organizations continued to operate for the next century, helping to confirm the economic, social, and cultural value of wildlife by promoting and focusing public interest through talks, fieldwork, tours, and publications. Reflecting an evolving culture beyond the demands of pioneering, their scope broadened in class terms and they increasingly took a more global perspective by participating, for example, in continental bird counts and animal research projects. These natural history groups also participated in a widening public discourse about wildlife by means of the new medium of radio and, more conventionally, in newspapers and magazines. At the same time, every major newspaper and magazine in the region ran stories and reports about wild animals from time to time during the interwar years and increasingly after the Second World War. Many of them also featured regular nature columns that expressed a broad and evolving range of ideas and views about wildlife.

A similar evolution occurred at zoos and museums. While early museum methods exhibited the remains of wild animals as individual specimens in splendid isolation, these were soon supplemented by display methods that suggested an animal's physical setting, though not necessarily its ecological context. Such attention to ecology only became popular in the late 1950s when exhibits, most notably some of those at the Saskatchewan Museum of Natural History, increasingly attempted to enhance public understanding of wild animals as elements in interconnected environments. Zoos were much slower to adopt such practices. The view that zoos should entertain by displaying cute and exotic animals or provide the thrill of being close to threatening and powerful animals was prevalent and operated alongside the public's desire to see and learn about local and foreign animals. Public safety and the zoos' practical need to protect their collections helped sustain the use of what were essentially menagerie formats, and zoos only began to replicate more natural environments with the "ban the bars" movement in the 1970s. While this only better disguised the fact that the animals were captive, it did help to improve their living conditions and to a certain extent addressed growing concerns about what rights the animals had. It also paralleled a growing commitment by zoos to encourage conservation. While display practices changed in zoos as they had in museums, the exhibition of live animals inevitably made zoos more controversial than museums. Ethical concerns about depriving wild animals of their freedom had been expressed as early as the late nineteenth century and reflected the belief that animals possessed a desire for freedom and autonomy, just as humans did. Criticism of zoos for depriving wild animals of their freedom grew during the interwar years, and while this critique remained important, it was gradually broadened to include the belief that the confinement and display of wild animals violated their dignity and integrity.

Such criticism in some ways echoed ideas about the ethical treatment of wild animals that had been expressed for decades in Canada. At the end of the nineteenth century, writer Ernest Thompson Seton had asserted that the intelligence and emotional capacity of wild animals mandated their respect and decent treatment. Similarly, while the attention of the humane movement had at first been focused primarily on the prevention of cruelty to domestic animals, wild animals were increasingly included in such concerns by the end of the Second World War. Trapping had for many years been criticized in these terms, and it became the beachhead of concern about the humane treatment of wild animals after the 1940s. Although some critics argued that trapping was inherently cruel and should be banned, a more publicly and politically acceptable stance was that if animals were to be killed for their fur, it should be done humanely. This targeted leghold traps, and opposition to them had gained sufficient support by the 1960s that the use of the body-grip Conibear trap was officially encouraged, albeit not mandated, in Saskatchewan and was being studied in Manitoba.

While some of its supporters would have disagreed, the humane trapping movement generally did not challenge the principle of killing furbearers but instead focused on the ways they were killed. This reflected the very broad social attitude that wild animals, whatever their social or cognitive attributes, could be killed out of need – for food and recreation, to protect economic activity, and to ensure personal and family safety. This attitude was, after all, inherent in most publicly accepted definitions of the usefulness of wild animals. But for some, killing out of need had limits, and acceptable levels of violence towards wild animals were thought to mark and measure the ranking and valuation of animal life as well as what constituted ethical human behaviour. Thus, some people viewed the ferocity of the coyote-control campaigns during the interwar and postwar years as excessive and believed that the recruitment of children to kill thousands of gophers, crows, and magpies was unethical and served only to brutalize society and sanction cruelty to animals.

Justifying the killing of wild animals out of need required especially delicate manoeuvring in the case of sport hunting. While the need (as opposed to the desire) to hunt for sport rather than food could not be as easily defended, sport hunting was sanctioned by tradition and its ubiquity, and the social influence of some of its leading practitioners gave it added social respectability. Its social status was also confirmed by the fact that it was dominated by men. The relationship of hunter and prey was cast as a gendered contest that revealed male prowess and domination. This was echoed in the common trope that the killing or conquest of a wary or powerful wild animal marked a transition to adulthood, confirmed gender identity, and created self-confidence. More concretely, sport hunting was said to promote military preparedness and to bring economic benefit to the state through tourism and licensing revenue. Yet, despite these social and economic benefits, unease remained. Clearly, sport hunting was recreational and not an activity that one had to participate in, and it undoubtedly threatened the survival of some species of game animals. It was within such considerations that the state regulated sport hunting through rules that conflated conservation and social priorities. While often not honoured in practice, by the end of the 1800s the sportsman's code of conduct had articulated a justification for sport hunting by asserting that a hunter had an ethical duty to be fair, to kill modestly, cleanly and quickly, and to eat what he had killed. He was also obligated to track and kill injured animals to prevent their lingering and painful death. The rules of engagement set out by the sportman's code thus helped to confirm sport hunting as a socially useful and, most importantly, an ethical and even humane activity. This justified its violence and helped to obfuscate questions about the ethics of killing reasoning creatures for pleasure. While some people believed that an animal's wariness expressed an intelligent effort to avoid death, many others claimed it was merely an instinctual reaction. Yet, by the admission of many hunters, the appeal of sport hunting lay in the battle of wits with an intelligent animal. The sportsman's code circumvented this potential ethical dilemma by requiring that animals be pursued fairly and without guile.

Whatever ethical concerns it may have engendered, the record suggests that a majority of men did not participate in sport hunting (even though some of them would have hunted for food). The reasons for this are unclear. Even though hunting was said to be normative male behaviour, perhaps these men had other priorities for their time or were simply not interested in this outdoor activity. Or perhaps their non-participation reflected unease about killing animals for pleasure. While such views were not commonly articulated, this is not surprising; criticizing an activity that was portrayed as normative male behaviour and a measure of masculinity required strong commitment and self-confidence. It is also further apparent that a majority of the region's non-hunting population neither opposed nor supported sport hunting. Once again, the reasons for this stance are unclear, but perhaps, as W.N. Millar suggested in his 1915 report on game in the Rocky Mountains, most people found no reason to take a stand either way because they were satisfied that state regulation of sport hunting mitigated its contribution to wildlife depletion and created acceptable basic standards for appropriate sport hunting behaviour. Even so, a few people publicly stated that they had no taste for sport hunting and that they liked live animals better than dead ones. And others rejected the activity on moral grounds or as a result of personal experience. For example, after participating in the killing of a bear, the Peace River newspaper editor Charles Frederick declared that he was "done with hunting." Likewise, author Andy Russell abandoned his gun for a camera after shooting a bighorn sheep, and others, like nature columnist Doug Gilroy, also found the camera rather than the gun a more rewarding way to encounter wildlife.

The foundations for public acceptance, or at least tolerance, of sport hunting began to weaken after the Second World War. This indicated that a growing but still small segment of the population was drawing different conclusions about human–animal relationships. While some people throughout the twentieth century had argued that coyotes, wolves, gophers, crows, and other foes played a role in the balance of a self-regulating nature, their arguments had been ineffective in the face of public hostility towards these animals and the political and economic power of those advocating campaigns to control them. Yet by the 1940s research was showing, and public opinion was increasingly accepting, that all animals in an ecosystem had a positive role. In the early 1950s, for example, avian biologist William Rowan argued that wild animals could not be valued and defined as "bad," "good," "beneficial," "pests," or "varmits" because all had a place in nature.[8] This represented a profound change from the older duality of friends and foes, and although this duality endured in popular thinking, it was increasingly marginalized by the growing acceptance that ecological interdependency gave some value to all wild animals. Such loosening of older hierarchies of value found expression in more inclusive and holistic attitudes and interest about the natural world. And the appeal of this thinking sometimes had surprising scope. Carling Breweries, for example, ran a series of public service advertisements using wildlife in 1949. One, titled "Nature in Balance is Nature Unspoiled," featured pileated woodpeckers, formerly ignored on the prairies except by birdwatchers. Using the language of modern ecology, the ad described how "food chains help keep the right proportions amongst all plant and animal life" and that other interconnections, such as the use of abandoned pileated woodpecker nesting holes by goldeneye ducks, were equally significant.[9]

The anticipation that the end of the Second World War would provide an opportunity to improve social equality and to create a more progressive society seems to have inspired some people to claim that wild animals should be beneficiaries of such postwar advancement as well. While few would have claimed that concern for wildlife was equivalent to a quest for social justice, a changed sensitivity about wild animals gained broader acceptance. This change expanded upon an earlier view that the way that a society treated animals mirrored its moral, ethical, and social character. While traditional utilitarian economically driven relationships endured and participation in familiar pursuits such as sport hunting increased after the war because of greater wealth and mobility, a growing number of observers equally gained a new focus in urging that measures be taken to protect the intrinsic value of wild animals and the ecosystems in which they lived. Similarly, naturalists and their organizations argued that their interests deserved greater attention by the state, or at least a hearing equal to that routinely given to hunters and other consumptive users. While still on the defensive, a formerly divided and marginalized stream of thought thus began to find new expression, coherence, and confidence in the late 1940s and early 1950s. A decade later, the growing environmental movement of the 1960s would be able to build upon attitudes and inclinations that had been gaining focus and strength since the end of the Second World War.

***

Increasing concerns in the early 1960s about the growing impact of humans on global and local environments and about personal well-being in an age increasingly reliant on chemical pesticides and herbicides marked a transition in thinking about wildlife and natural systems. This transitional period can be said to have matured by 1970 when Earth Day became a formal annual recognition of environmental concerns. Yet, while environmental thinking in the last quarter of the twentieth century went through different phases, some observers have concluded that the hopes and promise of the environmental movement in the 1960s were not realized. Certainly, many environmental problems became worse after the 1960s and wildlife continued to be crowded out by expanding industrial, urban, and agricultural development. Historian Neil Forkey suggests that the promise of the 1960s and early 1970s was thwarted because Canadians refused to surrender economic growth to environmental protection.[10] And while prairie culture had long put a very high value on material gain as a measure of individual and collective success, the power of a money culture – in which economic growth and material well-being were the only measures of worth – indeed gained immense ground after the 1970s. At the same time, the period after the 1960s saw important developments in thinking about wildlife. A new emphasis on species diversity, for example, enriched ecological thinking. While legislation to support diversity was slow to develop, ongoing efforts were made after 1970 to preserve species at risk, which on the prairies included peregrine falcons, swift foxes, and burrowing owls. And while the mindset for them was still present even though much diminished, state-sponsored extermination campaigns on the scale of those mounted in the 1950s against coyotes had by the end of the twentieth century become almost impossible, given the political controversy they engendered. The concept of sustainable resource use was also reinvigorated and given a broader social context. In the early 1900s, sustainability (as articulated by the Commission of Conservation) meant maximizing the value of natural resources for the current generation without lessening their use by future ones. With some refinement, this approach retained its currency for the next half century and was echoed in postwar wildlife management. While the 1987 report of the United Nations Brundtland Commission, for example, did not question goals such as the efficient and sustainable exploitation of natural resources (including wildlife), it shifted the ground by defining sustainability in wider social and global terms. Thus it recommended that the concept of sustainability be expanded to include long-term thinking, the pursuit of "equity and justice in an interdependent world," and the use of integrated and holistic management. While the notion of wise use to conserve resources for the future would not have been foreign to conservationists of the early 1900s or to postwar wildlife managers, and although the latter group would have been comfortable with the idea of integrated management, neither would have seen the pursuit of global equity and justice as a relevant or warranted concern in conservation.[11]

Thinking after the 1960s about environment and wildlife thus departed from earlier thinking in some ways but built upon it in others. While popular thinking remained anthropocentric in many ways, the development of an inchoate sense that people are just particularly clever animals signalled a significant shift from the long-held notion of the otherness of humans. Ecotourism scholar David Fennell argues that, among other factors, attitudes in the early twenty-first century towards animals arise from the findings of ethology, the increase in pet keeping, the intimate portrayal of animals on television and film, and "green" sensitivities about the fragility of the earth. Fennell also argues that increased urbanization has changed humans' connections and encounters with wild animals.[12] Prairie Canada has shared this urbanizing trend: by 1961 around 64 per cent of the population of Manitoba and Alberta was urban. Saskatchewan's urban population reached this level in 2001, by which time just over 80 per cent of the population in Alberta and 72 per cent in Manitoba was urban.

There is little concrete evidence that urban people on the prairies in earlier times had thought differently about wild animals than did rural people. While there had been conflict about balancing user rights, such as those between farmers and sport hunters, such disputes were as much about access to land as they were about relationships with wildlife. Although the recreational use of wildlife (especially sport hunting) had likely been more significant to city and townsfolk than to farm people, urban and rural people nonetheless had shared the same dominant view – that wild animals should be seen as friends and foes and that nature should be controlled and managed according to economic values. But the standards and conditions of contemporary urban life and the scale and autonomy of the modern city in the early twenty-first century seem to have created a new context for people's interactions with wild animals. For most urban people, wild animals are now overwhelmingly encountered through their representation in various media and in zoos and parks. While representation had always been important in the human encounter with wild animals, the agrarian society of the prairies before 1960 had also had other tangible encounters with wild animals – as sources of food, income, and pleasure, as threats to personal safety, and as competitors for resources. Other than in urban zoos and parks, most contemporary prairie urban people now have only passing contact with a limited range of wildlife – magpies, crows, Canada geese, varying hares, squirrels, and the odd coyote. Some of these animals are adapting to the city as their ancestors did to the farm landscape. But as environmental critic Clive Hamilton phrases it, as people have become more remote and isolated from nature, it has become easier for them to romanticize and idealize it.[13] In this sense, the modern prairie city has formed a new, if imagined, site of connection with most wild animals. But this change has been more than simply situational and has occurred at the same time as broader changes in intellectual and community life. For one, scientific study of animal behaviour has become more broadly based, offering new ways to think about enduring questions such as what abilities wild animals have. Legitimacy has thus been conferred by science on the view, for example, that wild animals possess cognitive abilities like memory and reason or that some of them have complex social structures.

Changed circumstances and contemporary worries about the future of wild animals and the habitats they need to survive do not exist apart from past relationships between people and wildlife. Instead, they are elements in a continuum of historical attitudes, policies, and approaches. This history in prairie Canada casts a long shadow, and while it is very much a record of regret, it also offers some models for more positive ways of engaging with the world. The contemporary re-wilding movement in Europe is attempting, after a long and destructive environmental history, to reintroduce species or to allow those that remain to increase and disperse naturally. Significantly, its protagonists see this as a social project in which the conflicts that occur between people and wild animals can only be resolved through changed public attitudes and government incentives, restrictions, and encouragement to achieve coexistence in sharing the land.[14] This offers one model for the future, but as in the past, relationships with wildlife will be shaped by entrenched economic, social, and cultural attitudes and interests. Most people remain unwilling to sacrifice economic growth, personal benefit, or social priorities for the good of wildlife. Historically, changing such mindsets required changing people's attitudes, including their definitions of needs and wants. On the

Canadian prairies, killing and eating songbirds lost their appeal because of legislation and changing standards of respectability. Hunting became less a lust for slaughter and more a sport under similar legislative and social pressures. A sense that tolerance of or coexistence with predators was possible equally involved social change. Similarly, marshland rehabilitation for prairie waterfowl and muskrats led to the development of models that involved shared stewardship approaches to managing land and wildlife. Of course, wild animals continued to be seen as commercial assets, market hunting still went on, as did the slaughter of game, and predators continued to be needlessly killed. But all of this occurred at reduced levels and increasingly against the social grain. In part, our contemporary engagement with the ongoing loss of habitat and species diversity will be equally imperfect, but this is not an impediment to moving forward. Rather, it requires taking ownership of the often shameful history of relationships between people and many wild animals on the Canadian prairies, just as it requires recognition of the generally more positive relationships demonstrated with insectivorous birds, pronghorns, elk, beavers, and whooping cranes. Such successes contained contradictions, served multiple goals, and in some ways were incomplete; nor, frankly, did they ultimately require people to give up very much to achieve them. Relationships with wildlife will continue to be socially, culturally, economically, and politically complex and will no doubt continue to exhibit contradictions, conflicts, and multiple goals. Securing a future of increased coexistence will thus not be one-directional or uncontested. Indeed, it will require paying sober attention to which competing goals are given priority and what people are willing to surrender to give space and dignity to the wild animals with which they share the land.

Notes

PREFACE

1 Coleman, "Two by Two," 492.

2 This version of the legend is from Francis, Jones, and Smith, *Origins: Canadian History to Confederation*, 1–2.

3 Kollin, "The Wild, Wild North," 45; MacEachern, "Voices Crying in the Wilderness," 216; Steinberg, "Down to Earth," paras 3, 9.

4 Colpitts, *Game in the Garden*, 103–6, 118–24.

5 Glavin, *Waiting for the Macaws*, 177–9.

6 Cronon, "Cutting Loose or Running Aground?," 41. Cronon's article is a response to Demeritt, "Ecology, Objectivity and Critique in Writings on Nature and Human Societies." On this debate, see also Demeritt, "The Nature of Metaphors in Cultural Geography and Environmental History"; and Dombrowski, "Bears, Zoos, and Wilderness."

7 Ehrenreich, *Living with a Wild God*, 47–8.

8 Fennell, *Tourism and Animal Ethics*, 13–39.

9 Hewitt, *The Conservation of the Wild Life of Canada*, 2.

10 On some of the complexities of the term "wilderness," see Isenberg, "The Moral Ecology of Wildlife," 48.

11 On pickerel on the prairies, see Nelson and Paetz, *The Fishes of Alberta*, 369.

12 MacEachern, "Changing Ecologies: Preservation in Four National Parks," 361.

13 Worster, *Nature's Economy*, 1–98, 342–50.

14 Ritvo, "History and Animal Studies," 401.

15 Colpitts, *Game in the Garden*, 13.

CHAPTER ONE

1 Epp, "Henry Kelsey's Journals and Correspondence," 196–214.

2 In 1870 a tiny area centred on Winnipeg became the province of Manitoba. In 1881 Manitoba's boundaries were extended north to 53 degrees and west to the current Saskatchewan boundary. In 1912 they were further extended north to the 60th parallel. After the creation of Manitoba in 1870 the rest of the region was administered as the NWT, which was later divided into several administrative districts, including the unorganized districts in the northern parts of the region. In 1905 the provinces of

Saskatchewan and Alberta were carved out of the NWT and took in all the lands between the 49th and 60th parallels. After 1905 the rest of the region became the districts of Yukon and the Northwest Territories.

3 Beaudoin, "What They Saw," 36.

4 Ecozones Prairies, http://canadianbiodiversity.mcgill.ca/english/ecozones/prairies/prairies.htm.

5 For example, see Conrad, *Reading the Entrails*, chaps 4 and 9.

6 Butler, *The Great Lone Land*, 217.

7 "Report by Lachaln Kennedy," *Annual Report DI for Year Ended 30 June 1881*, 67.

8 Roe, "The Red River Hunt," 200–3.

9 Denny, *The Riders of the Plains*, 32–40; Brower, *Lost Tracks*, 4; *Report and Minutes of Evidence on Existing Natural Food Products [1887]*, 35, 50.

10 Hanner, "Government Response to the Buffalo Hide Trade," 241–3, 263–7.

11 Secretary to McCraight, 30 April 1940, SAB, "Various Materials on Bison," SHS82; Potyondi, "Loss and Substitution," 213–15. An argument that absolves the Red River Métis is Roe, "The Red River Hunt," 203–18. For the United States, see Flores, "Bison Ecology and Bison Diplomacy," 465–85.

12 *Report and Minutes of Evidence on Existing Natural Food Products [1887]*, 51. Forget gives the date as 1880, but the trip was in 1881.

13 *Annual Report Commissioner of the NWMP 1888*, 22.

14 Seton Journal, 3 August 1892 and 13 September 1904, AMNH, E-3, vol. 4; Atkinson, *A Review-History of the Passenger Pigeon*, 7–8. The last passenger pigeon died in the Cincinnati zoo in 1914.

15 Nelson, *The Last Refuge*, 128–43, 181–4.

16 *Report and Minutes of Evidence on Existing Natural Food Products [1887]*, 65, 125–6; Nelson, *The Last Refuge*, 170–81.

17 Extract from a report of exploration by Professor John Macoun, *Annual Report DI for Year Ended June 30, 1880*, 21–5.

18 "A Trip Northward," *Regina Leader*, 28 June 1883.

19 *Macleod Gazette*, 7 March 1889; *Calgary Herald*, 6 March 1902.

20 *Annual Report Saskatchewan DA 1905*, 60–1; Gillese, "The Wildlife Journals of William Rowan," 49.

21 *Annual Report Saskatchewan DA 1906* (reprinting 1905 report by H.K. Tod, US ornithologist), 129; *Annual Report Manitoba DAI for Year Ended 30 November 1916*, 113–14; *Annual Report Saskatchewan DA 1907*, 215.

22 Voisey, *A Preacher's Frontier*, 5, 8–9.

23 DUC, *Newsflight No. 55* (1939), 7–8 (copy in PAA, Premiers Papers, file 663B).

24 *Annual Report Saskatchewan DNR for Year Ended April 30, 1938*, 2.

25 *Saskatchewan Fish and Game League Year Book 1943*, 31, 35 (copy in SAB, R-E4531).

26 Houston and Bechard, "Decline of the Ferruginous Hawk in Saskatchewan," 166–70.

27 Soper, *Mammals of Alberta*, 44, 275–6. Soper called them kit foxes but they now are known as swift foxes. On their reintroduction, see http://www.sararegistry.gc.ca/species/speciesDetails_e.cfm?sid=140.

28 Seton Journal, 7, 9, 28 June and 4 July 1892, AMNH, E-3, vol. 4. The wolves he heard about probably moved south in winter from the Interlake district.

29 "Report of Exploratory Survey to Hudson's Bay 1884," *Annual Report DI 1884*, 15, 28.

30 *Annual Report NWT DA 1904*, 109; *Annual Report Saskatchewan DA 1905*, 60–1; Hewitt, *The Conservation of the Wild Life of Canada*, 28–31.

31 *Annual Report Saskatchewan DA for Year Ended 30 April 1921*, 113; *Annual Report Saskatchewan DA for Year Ended 30 April 1925*, 295–96; *Annual Report Manitoba DAI for Year Ended 30 April 1926*, 102; *Annual Report Alberta DA 1934*, 45.

32 Tony Lascelles, "The Mammals of Manitoba" (Part 1), *Winnipeg Free Press* [magazine], 18 April 1936; *Annual Report Saskatchewan DNR for Year Ended 31 March 1950*, 117.

33 *Annual Report Manitoba DMNR for Year Ended 31 March 1956*, 69.

34 Soper, *The Mammals of Alberta*, 295; *Annual Report Manitoba DAI for Year Ended 30 April 1930* (typescript) (n.p.).

35 "Natural Resources and the Department of Natural Resources," n.d. (c. 1931), SAB, Pamphlet File, "Natural Resources, Saskatchewan"; Radio talk on CKY-CKX, 11 December 1946, MA, GR1600/G4527/36.2.1.

36 *Sharp Tailed Grouse in Saskatchewan*, 10–11.

37 *Winnipeg Tribune*, 27 June 1964.

38 Nelson, *The Last Refuge*, 190–3; Hanner, "Government Response to the Buffalo Hide Trade," 268.

39 Kricher, *The Balance of Nature*, 91.

40 A.V. Mitchener, a University of Manitoba entomologist, noted that new insects had appeared with settlement: the alfalfa plant bug was first reported in Nova Scotia in 1917 and reached Manitoba in 1941 and Saskatchewan by 1956. European corn borers appeared on the Atlantic coast of North America in 1917 and reached Manitoba in 1948. Hessian flies appeared in North America in the late 1800s and in Manitoba in 1899 and southwest Alberta in 1916 (Mitchener, *Field Crop Insects and Their Control*, passim).

41 "Exotic Animals in Manitoba," at http://www.naturenorth.com/YOTF/Invasive_Species.html. In 1926 it was reported that "it was not so many years ago that there were no rats west of the Great Lakes" in farm districts. They established a presence in Manitoba and "steadily and in increasing

numbers" moved west. By 1926 the eastern parts of Saskatchewan were described as "rat infested" (*Western Producer*, 18 January 1926 and 19 January 1928). They reached the Saskatchewan–Alberta border in 1950, and the Alberta government was able to control their movement further west ("A History of Rat Control in Alberta," Alberta Agriculture, Agdex Publication 682-1, February 1989, 1). I do not discuss rats in this book since on the prairies they could rarely survive the winters except by moving into or adjacent to human settlements or buildings.

42 Mitchener, *Field Crop Insects and Their Control*, passim.

43 Seton, *Life-Histories of Northern Animals*, vol. 1, 113; *Annual Report Manitoba DMNR for Year Ended 30 April 1933* (typescript), 126.

44 *Annual Report Manitoba DMNR for Year Ended 30 April 1946* (typescript), 56; Soper, *Mammals of Alberta*, 350.

45 Leola Love (Ridgedale, SK), Nature Lovers page, *Western Producer*, 31 May 1928; *White Tailed Deer in Saskatchewan*, 2.

46 *Annual Report Alberta DA 1909*, 96; *Edmonton Journal*, 31 August 1929 (editorial); Soper, *The Mammals of Alberta*, 350; Ondrack, *Big Game Hunting in Alberta*, 150–9, 172, 183, 189.

47 Seton Diary, 14 August 1904–31 December 1904, AMNH, E-3, vol. 14; Soper, *The Mammals of Alberta*, 109–11.

48 *Winnipeg Tribune*, 4 August 1958; "Exotic Animals in Manitoba," at http://www.naturenorth.com/YOTF/Invasive_Species.html.

49 The popular and scientific determination of wolf subspecies was (and remains) complex. The term "buffalo wolf," for example, was used imprecisely and sometimes referred to coyotes. Soper used the term to describe the Great Plains wolf (*Canis lupus nubilus* Say), which he judged was very rare and found only in southwest Alberta by the 1920s and was extinct in Canada by the early 1960s (Soper, *The Mammals of Alberta*, 265–6).

50 Millar, "The Big Game of the Canadian Rockies," 108.

51 *Fish, Fur and Feathers*, 163–4.

52 Criddle, *The Habits and Economic Importance of Wolves*, 8, 12–13. In Canada in the 1920s, coyotes were found mainly in British Columbia and the prairies. Their spread into other areas would appear to be a complex phenomenon, and Criddle's environmental determinism has subsequently been disproved by their appearance throughout the country, including forest and mountain areas.

53 Kerry Wood, "The Yapping Yodler," *Country Guide*, February 1945; Phil Schwarz, "Predator Fox," *Country Guide*, February 1947.

54 Farley, *Birds of the Battle River Region*, 47–8.

55 *Annual Report Saskatchewan DA for Year Ended 30 April 1927*, 167.

56 Farley, *Birds of the Battle River Region*, 46–7; Alberta Natural History Society, PAA, 67.293.

57 *Annual Report Alberta DA for Year Ended 31 March 1939*, 47; *Winnipeg Tribune*, 22 November 1954.
58 John Patrick Gillese, "The Hunters," *Country Guide*, February 1952, 51–2.
59 Fowke, *Canadian Agricultural Policy*, 3.
60 *Annual Report NWT DA 1901*, 108.
61 Worster, "Transformations of the Earth," 1101.
62 Elofson, "Grassland Management in Southern Alberta," 153–9.
63 McCormack, "Transportation and Settlement," 1–18.
64 D.G. Wetherell and Associates, *Agriculture in Western Canada*, 136–7, 153–4.
65 Royal Commission on Canada's Economic Prospects, *Progress and Prospects of Canadian Agriculture*, 72.
66 Stead, "The Plow," *Songs of The Prairie*, 8–11.
67 *Calgary Herald*, 8 March 1907. On settlers' rights to wildlife in the United States, see Tober, *Who Owns the Wildlife?*, 19–20.
68 Isenberg, "The Moral Ecology of Wildlife," 52.
69 den Otter, "'The Wilderness Will Rejoice and Blossom Like the Crocus,'" 83, 96.
70 Thomas, "Alberta Perspectives, 1905," 1–3. The Minifie quote is in Harrison, *Unnamed Country*, 1.
71 For example: Letter to the editor 20 September 1883; "A Scotchman on the North-West," 26 April 1883; "Mr Carss on the Qu'Appelle Valley," 26 April 1883; "The Climate, Soil and Resources of the Turtle Mountain and Other Districts," 21 April 1885; and "More Misrepresentations of the North-West," 10 March 1885, *Regina Leader*.
72 Seton, *The Arctic Prairies*, 312–13.
73 Francis, *Images of the West*; Francis and Kitzan, *The Prairie West as Promised Land*.
74 *Regina Leader*, 29 March 1883; *Camrose Canadian*, 10 December 1903. On Robert Stead's critique of prairie materialism, see Karr, "Robert Stead's Search for an Agrarian Ideal," 41–4.
75 Wetherell and Kmet, *Useful Pleasures*, 304.
76 Kerry Wood, "The Farmer's Game Birds," *Country Guide*, September 1946.
77 *Farm and Ranch Review*, May 1905.

CHAPTER TWO

1 Leopold, *A Sand Country Almanac*, 117.
2 A comprehensive outline of historical thinking about animal cognition is "Animal Cognition," S. 2 "Foundational Issues," in Edward N. Zalta, ed., *The Stanford Encyclopedia of Philosophy (Summer 2011 Edition)*, http://plato.stanford.edu/archives/sum2011/entries/cognition-animal (accessed 12 November 2012).
3 Merchant, *The Death of Nature*, 99–100; Salisbury, *The Beast Within*. On

later Christian thinking, see Preece and Fraser, "The Status of Animals in Biblical and Christian Thought," 245–63.

4 Gray, "The Early Animal Behaviorists," 372–9.

5 Merchant, *The Death of Nature*, 99–103.

6 Thomas, *Man and the Natural World*, 129, 301–2; Worster, *Nature's Economy*, 77–97.

7 Ritvo, *The Animal Estate*, 36–41.

8 Berger, *Science, God, and Nature in Victorian Canada*, xiii and Lecture 3. On anti-Darwinist thinking in Canada, also see Waiser, *The Field Naturalist*, 8–11.

9 On the Cree in Quebec, see, for example, Scott, "Knowledge Construction among Cree Hunters," 193–208.

10 Brightman, *Grateful Prey*, 41, 51, 160–9, 177.

11 Ibid., 8, 2–3, 317.

12 Ibid., 76, 91–3, 187–8, 288.

13 Ibid., 95, 99, 168–9, 188. On the connection of dreaming and hunting among the Dunne-za, see Brody, *The Other Side of Eden*, 32–3, 258–61.

14 Brightman, *Grateful Prey*, 104, 106–10.

15 Ibid., 119, 188, 289.

16 Ibid., 103, 117, 188, 307–9.

17 Ibid., 203, 207.

18 Ibid., 104, 283, 287–8, 290–1.

19 Ray, *Indians in the Fur Trade*, 117–24, provides a good summary. Historian Calvin Martin argues that the slaughter of wild animals by some Algonquian people in the late eighteenth century was an act of revenge because they saw the devastation of European disease as evidence that animals had withdrawn their protection of humans. This slaughter was enhanced by the preaching of missionaries and the incentives of the fur trade (Martin, *Keepers of the Game*, esp. chap. 5). Brightman cites extensive literature that argues that the evidence for a belief among Algonquians that animals provided protection from disease is weak (Brightman, *Grateful Prey*, 285).

20 Brightman, *Grateful Prey*, 290, 301–2.

21 Ibid., 188–99.

22 Ibid., 196–202, 286–7.

23 Ibid., 16, 287, 303, 307–9. Over-hunting was one result of competition between the HBC and its rivals in some districts. With the amalgamation of the warring parties into a single company in 1821, efforts were made to rebuild populations, and in districts where it had sufficient power the company introduced a range of conservation measures between 1821 and 1850. In many districts, Aboriginal trappers resisted these measures, seeing them as an unwarranted restriction on their mobility and on access to fur resources (Ray, *Indians in the Fur Trade*, 117–24; Ray, "Some Conservation

Schemes of the Hudson's Bay Company," 49–68). Brightman suggests that in northern Manitoba these plans clashed with Rock Cree beliefs that animals reproduced infinitely and that hunter–prey relationships were regulated by the animals and the spirit beings, not by human manipulations of the environment.

24 Mandeville, *This Is What They Say*, 201; Patrol Report, NWMP Fort Chipewyan Detachment, 1 April 1913, LAC, RG 18, vol. 440, file 167; Brody, *The Other Side of Eden*, 117, 256–7.

25 Mandelbaum, *The Plains Cree*, 219–20, 308–9. On bears and the Rock Cree, see Brightman, *Grateful Prey*, 115, 205–7.

26 Hugh Dempsey, personal communication, September 2012. I am also grateful to Dempsey for pointing out to me Joseph Dion, *My Tribe, the Crees* (Calgary: Glenbow Museum, 1979), 56; and to Mandelbaum, *The Plains Cree*, 275, for additional details on the Bony Finger.

27 Brink, *Imagining Head-Smashed-In*, 155–60; *Report and Minutes of Evidence on Existing Natural Food Products [1887]*, 147.

28 Brightman, *Grateful Prey*, 36, 211–12.

29 Thorpe, *Learning and Instinct in Animals*, 14.

30 "Instinct," *Winston's Cumulative Loose-Leaf Encyclopedia: A Comprehensive Reference Work*, vol. 5 (Chicago, Philadelphia, and Toronto: 1912 [1922 printing]).

31 McClung, *Clearing in the West*, 144; *Nor'West Farmer*, April 1898; *Saturday Night*, 15, no. 31 (1902): 9.

32 For one view of Descartes on this matter, see Cottingham, "A Brute to the Brutes," 551–9.

33 W.A. Waiser, "Robert Bell," *Dictionary of Canadian Biography*, http://www.biographi.ca/009004-119.01-e.php?BioId=41323.

34 All references to this talk are from Robert Bell, "On the Intelligence of Animals," a paper read before the Natural History Society of Montreal, 31 January 1870, LAC, MG29 B15, vol. 1, file 1.35.

35 Taverner, "A Discussion of the Origin of Migration," 323.

36 Atkinson, *Insectivorous Birds of Manitoba*, 3–8.

37 Mair, "The American Bison," 98; *Regina Leader*, 7 January 1886.

38 On Roberts, see Coldwell, *Sir Charles God Damn*. A Nova Scotian, Saunders published under the name Marshall Saunders. Her most popular book was *Beautiful Joe: An Autobiography* (1893), a story told in the voice of an abused dog. It was the first Canadian book to sell more than a million copies (LAC, "Celebrating Women's Achievements, Margaret Marshall Saunders," http://www.collectionscanada.gc.ca/women/030001-1470-e.html). Thirty years after its publication, a correspondent wrote to the *Western Producer* in 1928 saying that reading *Beautiful Joe* had convinced him that animals had a language of their own (Sal's Pal, Nature Lovers page, *Western Producer*, 24 May 1928).

39 Atwood, *Survival*, particularly 69–86. Atwood was inspired in part by Northrop Fry's argument that Canadians feared the country's vastness and being swallowed up by it, which created a society that he called a "bush garden" where people turned inward to keep a threatening natural environment at bay. Critics of Atwood's argument include historian John Sandlos, who observes that she ignores the intellectual and historical context of writers such as Seton and Roberts and overlooks the fact "that the animal stories address primarily the philosophical implications of late-nineteenth and early-twentieth century natural science rather than those of postwar literary criticism." Writer Wayne Grady also points out that the eagerness with which many Canadian nature writers and scientists engaged with the natural world around them "can hardly be said to have been produced by writers who harboured deep-rooted psychological disturbances when confronted with the wilderness" (Sandlos, "From within Fur and Feathers," 79–81; Grady, *Chasing the Chinook*, 115).

40 Mighetto, "Science, Sentiment, and Anxiety," 33–50; Dunlap, "The Realistic Animal Story," 56; Sandlos, "From within Fur and Feathers," 79–81.

41 Nixon to Potter, 12 April 1941, SAB, SHS50; Lenny, Youth Page, *Western Producer*, 18 March 1956.

42 Keller, *Black Wolf*, 70–1, 93–113; Foreword to the journal, AMNH, E-3, vol. 1, Pt 2.

43 Anderson to Seton, 5 October 1925, LAC, MG25 D108, vol. 10, file 10.6; Arnason, Afterword, in Seton, *Wild Animals I Have Known*, 251. *Lives of Game Animals* came out in four volumes, the last appearing in 1928.

44 Other fiction that directly drew on his Manitoba years includes *The Trail of the Sandhill Stag* (1899), set in the wild hills near Carberry, and "The Winnipeg Wolf," published in *Animal Heroes* (1905).

45 Arnason, Afterword, in Seton, *Wild Animals I Have Known*, 249–52; Fiamengo, "Looking at Animals; Encountering Mystery," 36–59.

46 Seton, *Wild Animals I Have Known*, 13–14, 15, 143 (italics in original).

47 Ernest Thomson Seton, "The Minds of Animals" (lecture with miscellaneous notes) [n.d.], LAC, MG29 D108, vol. 3, file 3.47.

48 Seton, "The Springfield Fox,"149; "Redruff: The Story of the Don Valley Partridge," 223–33; "Lobo: The King of Currumpaw," 19–44; "Silverspot: The Story of a Crow," 54–8; "Raggylug: The Story of a Cottontail Rabbit," 70–87; in *Wild Animals I Have Known*. The theme of the intelligence of the fox appears in a different guise in a draft story, "How the Wolf Thought It Out," about a coyote at Carberry in 1882 that evaded Seton's attempts to shoot it. Seton notes that the animal "thought it all out carefully" (Seton, "How the Wolf Thought It Out" (n.d.), LAC, MG29 D108, vol. 7, file 7.61).

49 Seton, "The Springfield Fox," 135; "Silverspot: The Story of a Crow," 50–4; "Raggylug: The Story of a Cottontail Rabbit," 69, 93–6; "The Pacing Mustang, 186; in *Wild Animals I Have Known*.

50 Fiamengo, "Looking at Animals, Encountering Mystery," 44; Seton, "Raggylug: The Story of a Cottontail Rabbit," 96, 101; "Lobo: The King of Currumpaw," 19–33, 41, 44; "Silverspot: The Story of a Crow," 47–8, 65; "Redruff: The Story of the Don Valley Partridge," 234–8; "The Pacing Mustang,"159–86; in *Wild Animals I Have Known.*

51 Seton, "The Springfield Fox," 155; Silverspot: The Story of a Crow," 60; "Raggylug: The Story of a Cottontail Rabbit," 77–9; "Redruff: The Story of the Don Valley Partridge," 216; in *Wild Animals I Have Known.*

52 "The Intelligence of the Wolf," typescript, n.d. (c. 1900), LAC, MG29 D108, vol. 4, file 4.41 (italics in original).

53 "Early Anecdotal Methods," section 3.1, *Stanford Encyclopedia of Philosophy.*

54 Mighetto, "Science, Sentiment, and Anxiety," 40–4.

55 Seton, *Life-Histories of Northern Animals*, vol. 1, v–vi; Merriam to Seton, no other date, 1910, LAC, MG29 D108, vol. 9, file 9.30.

56 Anderson, *The Chief*, 122–8; Mitman, "When Nature *Is* the Zoo," 125–6.

57 Dunlap, "The Realistic Animal Story," 61.

58 Mitman and Burkhardt, "Struggling for Identity," 165.

59 For two contemporary popular critiques of this methodology, see Morell, "Minds of Their Own"; and Natterson-Horowitz and Bowers, *Zoobiquity*, 12, 96–7.

60 Ainley, *Restless Energy*, 93. Prairie universities were active, however, in entomology with a focus on insects and agriculture. Typically, it was taught in association with agricultural sciences.

61 Ibid., 71, 102–6, 123, 191–3.

62 Ibid., 133, 146, 176.

63 Ibid., 186–205; "The Memorable William Rowan," at http://www.ualberta.ca/ALUMNI/history/peoplep-z/93sumrowan.htm.

64 While it was understood that rabbit populations were cyclical, this was usually credited to unique causes such as the regular outbreak of some type of insect-borne disease or perhaps a virus. See, for example, *Report and Minutes of Evidence on Existing Natural Food Products [1887]*, 39; Correspondence with F. Bell [New Zealand High Commission London] to High Commissioner of Canada, 7 November 1886; and Clarke to the Lt Governor, 21 June 1887; SAB, R-192.4, file 20.

65 Rowan, *The Ten Year Cycle*, 9.

66 Ainley, *Restless Energy*, 241, 243.

67 Rowan, *The Ten Year Cycle*, 13, 15; Ainley, *Restless Energy*, 245.

68 Ainley, *Restless Energy*, 266–8; Soper, *The Mammals of Alberta*, 45–6.

69 "White Rats Taught Us Plenty about What Food to Eat," *Farm and Ranch Review*, April 1951.

70 For outlines and references to some of this literature, see Adams, "The Relation of General Ecology to Human Ecology," 325–6; and Egerton, "The History of Ecology, Part One," 296–8. The scope and concerns of

such research can be seen in Thorpe, *Learning and Instinct in Animals,* which provides reference to the research concerns after the 1920s and especially the 1950s.

71 Doyle, "Science, Literature and Revolution," 7, 10; Roger J. Moran, "Herbert Dyson Carter," *The Canadian Encyclopedia,* http://www.thecanadianencyclopedia.com/articles/herbert-dyson-carter.

72 Carter, "Can Animals Think and Feel?," *Saturday Night* 58, no. 7 (1942) (Part I): 26.

73 Ibid.

74 Carter, "An Ape at the Breakfast Table," *Saturday Night* 58, no. 9 (1942) (Part II): 4.

75 Kerry Wood, "Why Are Bluebirds So Stupid and Sparrows So Smart?," *Farm and Ranch Review,* June 1950; Soper, *The Mammals of Alberta,* 49; "Animals as Their Own Doctors," *Country Guide,* July 1956.

76 "The Instinct of Preservation," *Good Roads,* September 1923; *Regina Leader-Post,* 16 September 1933.

77 *White Tailed Deer in Saskatchewan,* 6; *Winnipeg Tribune,* 20 August 1964.

78 Mrs L.M. East, "Nature's Love Stories," *Western Producer,* 28 June 1934.

79 "Mating Time," *Country Guide,* May 1948; *Winnipeg Tribune,* 20 August 1948; Kerry Wood, "Qu'Yuk the Beautiful," *Country Guide,* November 1947.

80 Alexander Krenakerich (Hamton, SK), Nature Lovers page, *Western Producer,* 28 March 1928; V.W. Jackson, "Birds of a Different Feather," *Grain Growers' Guide,* 3 March 1926.

81 "Who Says Our Wild Life Has No Sense of Humour!," *Farm and Ranch Review,* April 1951; John Patrick Gillese, "The Hunters," *Country Guide,* February 1952.

82 Ralph Hedlin (Renown, SK), Nature Lovers page, *Western Producer* 25 January 1940. The Nature Lovers page was a weekly feature in the newspaper throughout the interwar years. The page was for writers up to twenty years of age. It seems to have been an important forum for many young people, including some who went on to have writing and scientific careers. In addition to Hedlin, Dick Beddoes, who became a columnist in Toronto, and Walter Schowalter, who became a famous rose breeder and naturalist, were correspondents at various times. After the mid-1940s Nature Lovers and a Bird Watchers column ran on alternate weeks on the Youth page.

83 "Animal Cognition, Experimental Methods," section 3.2, *Stanford Encyclopedia of Philosophy;* Midgley, "Against Humanism," 3.

84 Morell, "Minds of Their Own" (online).

85 *Social Learning in Animals,* a selection of essays edited by Celia Heyes and Bennett Galef, is among the pioneering academic works on the subject, but studies on crows, whales, and many other animals have extended the debate. As only one example of a steadily growing literature, see Balcombe,

Second Nature. For a popular press report, see Peter Christie, "Who Taught This Bird to Open a Milk Bottle?," *Globe and Mail*, 25 January 2005.

CHAPTER THREE

1 Taverner, *Bird Houses and Their Occupants*, 126.

2 Atkinson, *Insectivorous Birds of Manitoba*, 15.

3 Ritvo, *The Animal Estate*, 23.

4 Hewitt, *The Conservation of the Wild Life of Canada*, 15–16; *Annual Report Saskatchewan* DA *for Year Ended 30 April 1920*, 75.

5 *Annual Report Manitoba* DAI *for Year Ended 30 April 1927*, 1; *Annual Report Manitoba* DMNR *for Year Ended 30 April 1933* (typescript), 125; Dick Beddoes, Youth page, *Western Producer*, 23 March and 29 April, 1944.

6 Seton journal, 25 June 1892, AMNH, E-3, vol. 4.

7 Mildred Galloway (Melfort, SK), Nature Lovers page, 2 February 1928; Goldine (Ratcliffe, SK), Nature Lovers page, 1 June 1933; *Western Producer.*

8 *Western Producer*, 21 July 1927, and 6 and 27 September 1928.

9 Woodsworth, *Thirty Years in the Canadian North-West*, 29; Hugh Duddrige, "The Chickadee," *Western Producer*, 3 May 1928. James Woodsworth was the father of J.S. Woodsworth, the social-democratic theorist and politician.

10 Roy, *Enchantment and Sorrow*, 34; J. Arch. McLeod, "Progress," *Saskatchewan Poetry Book 1952–53*, 8.

11 "Birdlover" (Gravelbourg, SK), Youth page, *Western Producer*, 27 June 1946; Vina Chilton, "The Robins Came," *Saskatchewan Poetry Book, 1952–53*, 9; Dowson, "Flying Geese," in *Country Places and Other Poems*, 32. For other examples about bird migration, see A.E. Roland, "The Meadow Lark," *Western Producer*, 15 April 1926; N.E. Graham (Senlac, SK), *Western Producer*, 26 April 1945; Irene Louise Harrison, "Return of the Wild Geese," *Farm and Ranch Review*, March 1957.

12 Seton, "Raggylug," 101, and "Redruff," 238, *Wild Animals I Have Known*; Atkinson, *Insectivorous Birds of Manitoba*, 5; Nature Lovers page, *Western Producer*, 3 July 1930; "Bad Birds," *Saturday Night* 42, no. 4 (1926): 10.

13 "Pamela," Nature Lovers page, *Western Producer*, 13 September 1928.

14 Edna Jacques, "Birds," in *Beside Still Waters*, 78; Criddle, *The Birds of a Manitoba Garden*, 2–3; "Courageous Birds," *Farm and Ranch Review*, February 1953.

15 Gruelle's stories were also published in book form, including *The Meadow Folks Story Hour*, which appeared in 1921. The stories on which this paragraph is based are from the *Western Producer*, Nature Lovers page, as follows: "Doris Makes a Happy Discovery," 19 January 1928; "Hoppy Toad Settles a Dispute," 23 January 1928; "David Has a Chat with Bunny Rabbit's Cousin," 2 February 1928; "Doris Meets a Dainty Little Traveler," 15 March 1928; "Good Night Stories," 3 May 1928; "David Changes His Mind about

Freddy Frog," 10 May 1928; "Doris Meets Mama and Daddy Red Winged Blackbird," 17 May 1928; "How Miss Blue Bird Found Blue Feathers," 24 May 1928; "Doris Meets a Useful Little Neighbor," 7 June 1928; "Why Blue Bird Is Always Happy," 15 November 1928; and "Dicky Is Introduced to Mama and Daddy Catbird," 2 May 1929.

16 Ingram, "Beastly Measures," 231, 244; "Animals and the Criminal Law (Canada) (June 2008) Chapter 1, Overview," http://www.isthatlegal.ca/index.php?name=overview.animal_criminal_cruelty_law; Charles Delmer-Radcliffe, "Wild Animal Life in the British Empire," Speech to the Empire Club of Canada, 12 February 1931, http://speeches.empireclub.org/60156/data?n=3.

17 *Annual Report Saskatchewan DA 1914*, 212; A.C.H., "Making the Most of Our Environment," *Farm and Ranch Review*, 5 May 1922.

18 Edward McCourt, "Walk through the Valley," *Western Producer*, 12 September 1957.

19 Jennie Pringle, "Shooting Mountain Sheep – for Health," *Farm and Ranch Review*, 10 June 1927.

20 Connor, *The Man from Glengarry*, 15, 55–7; Andy Russell, "Johnny Get Your Gun," *Country Guide*, November 1961; "Coyote: A Story Based on a True Experience," *Country Guide*, May 1962.

21 Ralph Hedlin, "Aya-Oo," *Country Guide*, November 1956.

22 *Annual Report Saskatchewan DA 1913*, 255; Mac Coleman (Vanguard, SK), Nature Lovers page, *Western Producer*, 17 August 1933.

23 *Annual Report NWT DA 1901*, 65–6.

24 *Annual Report Saskatchewan DA for Year Ended 30 April 1921*, 89–93.

25 *Grain Growers' Guide*, 22 March 1922; V.W. Jackson, "The Balance of Nature: The Value of Birds to Agriculture," *Grain Growers' Guide*, 11 June 1924.

26 Nash, *The Birds of Ontario in Relation to Agriculture*, 30; Taverner, *Bird Houses and Their Occupants*, 119; *Western Producer*, 16 May 1935.

27 Merriman, *Attracting Birds with Food and Water*, 4, 14; Taverner, *Bird Houses and Their Occupants*, 119. See also Atkinson, *Insectivorous Birds of Manitoba*, 19; H.M. Speechley, "Common Prairie Birds," *Grain Growers' Guide*, 29 April and 19 May 1915; V.W. Jackson, "The Balance of Nature: The Value of Birds to Agriculture," *Grain Growers' Guide*, 11 June 1924.

28 Gwennie McWhirter (Redvers, SK), Nature Lovers page, *Western Producer*, 21 June 1928; Mary Mracek (Druid, SK), Nature Lovers page, *Western Producer*, 3 May 1928.

29 John Kyba, Nature Lovers page, *Western Producer*, 26 April 1928. See also "A Ray of Sunlight," Nature Lovers page, 10 May 1928, *Western Producer*.

30 William Rowan, "Bird Protection," University of Alberta *Press Bulletin*, 18 February 1921; *Annual Report NWT DA 1901*, 65–6; H.M. Speechley, "Common Prairie Birds," *Grain Growers' Guide* (Part I), 21 April 1915.

31 Atkinson, *Insectivorous Birds of Manitoba*, 15; Kerry Wood, "The Crow: Friend or Enemy?," *Farm and Ranch Review*, March 1953.

32 *Annual Report Saskatchewan* DA *1915*, 217, 227; *Annual Report Saskatchewan* DA *for Year Ended 30 April 1924*, 291–3.

33 See, for example, *Western Producer*, 8 April 1926.

34 Holliday, "Norman Criddle: Pioneer Entomologist," 6 (online version); Kalmbach, *The Crow in Its Relation to Agriculture*, 2, 18; Nature Lovers page, *Western Producer*, 16 June 1938.

35 L.B. Potter, "Let's Be Fair," *Western Producer*, 6 June 1940; Seton, "Silverspot: The Story of a Crow," in *Wild Animals I Have Known*, 57.

36 Svoboda [secretary treasurer, Saskatchewan Fish and Game League] to Buckle [minister of agriculture] 27 July 1932, SAB, NR.1/1, file G5000-S; *Annual Report Saskatchewan* DNR *for Year Ended 30 April 1937*, 58.

37 *Annual Report Saskatchewan* DA *for Year Ended 30 April 1927*, 166; Kerry Wood, "Feathered Outlaws," *Country Guide*, March 1945.

38 *Winnipeg Tribune*, 4 July 1955.

39 Among other animals, wolverines faced consistent hostility. According to trickster myths among the Rock Cree, wolverines were viewed with a mixture of respect for their intelligence and hostility for their aggression, strength, and plundering. Most Euro-Canadians believed they were inherently bad, and when describing wild animals found in northern Alberta, oil sands promoter Alfred von Hammerstein noted that wolverines were "called the 'devil'" in the west (Chambers, *Canada's Fertile Northland*, 38, Brightman, *Grateful Prey*, 42–3).

40 Blackfoot stories, http://www.glenbow.org/blackfoot/en/html/traditional_stories.htm#wolfTrail

41 Hampton, "Shark of the Plains," 4.

42 MacLachlan, "The Historical Development of Cattle Production in Canada," 10 (online).

43 *Annual Report* NWT DA *1898*, 85; *Prince Albert Herald*, 17 January 1935 (copy in SAB, NR 1/2, file P800C (3)).

44 McClung, *Clearing in the West*, 77; W.H. McKay, "Recollections as a Wolf Hunter," *Calgary Herald*, 15 January 1944; C.E. Craddock, "Nazis of the North," *Country Guide*, May 1946.

45 "A Night in the Woods," Nature Lovers page, *Western Producer*, 18 September 1930; Clarence Tellenius, "Sketch Pad Out-of-Doors, No. 11 in the Series," *Country Guide*, December 1952.

46 Criddle, *The Habits and Economic Importance of Wolves*, 6–9, 15; Harkin to H.C. Nolin, 28 March 1923, LAC, RG84, vol. 120, file U266, reel T-12898; *Winnipeg Tribune*, 1 July 1950 (editorial); Leslie Sara, "Ferocity of Wild Animals Said Myth," *Calgary Herald*, 14 March 1936.

47 Bill Yeo, letter to the editor, *History Now* (Historical Society of Alberta), no. 2 (April 2000): 5. Glenbow has now restored the title *The Wolves*.

48 Z.M. Hamilton, "The Big Buffalo Wolves on Our Cattle Range," *Yorkton Enterprise*, 8 July 1943 (copy in clipping file "Wolves," SAB, SHS231).

49 The family Canidae on the prairies includes wolves (*Canis lupus*), coyotes (*Canis latrans*), and foxes, mainly the red fox (*Vulpes vulpes*). See the Hinterland Who's Who factsheets on "Coyote," http://www.hww.ca/en/species/mammals/coyote.html, and "Wolf," http://www.hww.ca/en/species/mammals/wolf.html. Wolves weigh up to thirty-six kilograms and coyotes are usually about twenty-three, but both can be larger when food is plentiful. Although wolves are typically a grizzled grey, there is great variation in their colour, including black and white. Coyotes are usually a tawny grey with grey- and rust-tipped fur, and they are a lighter colour in winter than in summer.

50 See, for example, William Taylor, Woodlands [MB], letter to the editor, *Brampton Times*, 10 February 1875, p. 3 (typescript), MA, MG8/B68. On confusion about wolves and coyotes on the American Great Plains, see Hampton, "Shark of the Plains," 4.

51 Letter to the editor, *Farm and Ranch Review*, January 1949; Criddle, *The Habits and Economic Importance of Wolves*, 6–9, 15.

52 Hampton, "Shark of the Plains," 11.

53 Details about the rabies campaign can be found in *Within Our Borders* [Government of Alberta staff newsletter], 1 November 1952, 15 January, 1 February, and 1 March 1953, and 15 November 1954, as well as in *Annual Report Alberta DLF for Year Ended 31 March 1951*, 54–6 and *Annual Report Alberta DLF for Year Ended 31 March 1954*, 50.

54 *Annual Report Alberta DLF for Year Ended 31 March 1954*, 50; *Annual Report Alberta DLF for Year Ended 31 March 1956*, 56. In 1954 Saskatchewan also confronted problems with hyatid, a tapeworm that moved from a host carnivore to a deer when it consumed an infected carnivore's feces while browsing. The cycle was also perpetuated when a carnivore ate an infected deer. There was fear that hyatid could spread to humans if they ate undercooked deer meat or were exposed to the parasite during butchering (*Annual Report Saskatchewan DNR for Year Ended 31 March 1953*, 125; *Annual Report Saskatchewan DNR for Year Ended 31 March 1954*, 99).

55 Gill Shark, "Don't Tame Pets from the Wild," *Country Guide*, September 1949.

56 Moodie and Kaye, "Taming and Domesticating the Native Animals of Rupert's Land," 10–19.

57 "Peshew: The Lynx," *Saturday Night* 12, no. 26 (1899).

58 "Raise Ringtail Racoons for a Profitable Hobby," *Farm and Ranch Review*, December 1946; May Lassie, Nature Lovers page, *Western Producer*, 7 April 1938.

59 Russell, "The Wild Side of Animal Domestication," 286–7.

60 *Report and Minutes of Evidence on Existing Natural Food Products [1887]*, 124; Archie McKishnie, "The Alien," 18 February 1925, and "The Outcasts," 4 February 1925, *Grain Growers' Guide*.

61 Arthur G. Storey, "Honker," *Country Guide*, April 1962. The same theme, this time with a sparrow hawk, appeared in a story by Canadian author Vick Branden, "The Wind Climber," *Country Guide*, August 1966.

62 On pet keeping as an urban practice, see Olson and Husler, "Petropolis," 133–43.

63 "One of the Many Co-ops," Nature Lovers page, 15 October 1931; Gwennie McWhirter (Redvers, SK), Nature Lovers page, 19 January 1928; "Smiling Monty" (Verwood, SK), Nature Lovers page, 4 May 1933; L.H., Nature Lovers page, 14 September 1932 – all in *Western Producer*. See also "Pet Crows," *The Blue Jay*, 17, no. 3 (1959): 124.

64 Oeming to John Langford, June 1954, and Langford to Tom Baines, 14 June 1954, CCA, PRDF, series III, box 37, file 1953.

65 "Unky the Porcupine," *Country Guide*, September 1952; Criddle, *The Habits and Economic Importance of Wolves*, 11; *Western Producer*, 9 July 1953.

66 J.A. Capling, "Our Two Most Interesting Pets," *Country Guide*, May 1956; Kerry Wood, "A House-Broken Weasel at Home up a Pole," *Farm and Ranch Review*, March 1953.

67 Annie L. Gaetz, "Mickey the Pet Beaver," November 1945; Kerry Wood, "The Life Story of a Pet Beaver," April 1957; and Kerry Wood, "A Girl and a Beaver," April 1960 – all in *Farm and Ranch Review*. See also Wood to Ure, 13 September 1944, PAA, Premiers Papers, file 663B.

68 "A Partnership-Pioneer and Wildlife," *Saskatchewan Afield*, 1, no. 4 (April 1959): inside cover (copy in SAB, R.E4531).

69 "Circus Career Foiled for Manitoba Moose," *Western Producer*, 1 July 1976.

70 Berry Richards, "Gus Lindgren and His Moose Family," *Country Guide*, April 1957.

71 *White Tailed Deer in Saskatchewan*, 7; Gill Shark, "Don't Tame Pets from the Wild," *Country Guide*, September 1949. For the same view as Shark's, see Kerry Wood, "A Budgie for Bobby," *Farm and Ranch Review*, February 1959.

CHAPTER FOUR

1 This was later specified in legislation, likely to assert provincial rights. The Alberta game statute of 1942, for example, stated that all wild animals and birds belonged to the provincial crown (*An Act for the Protection of Game*, RSA 1942, chap. 70, sec. 3).

2 Foster, *Working for Wildlife*, 9–10; Donihee, *The Evolution of Wildlife Law in Canada*, 4–5. The federal government was responsible for inland fisheries and export standards, but administration of fisheries regulations was left to the provinces (http://www.gov.mb.ca/waterstewardship/fisheries/

regulations/leg.pdf). Historian Alan MacEachern notes that a perception that Canadians are less environmentally oriented than Americans has arisen because in Canada environmentalism "has been more provincially directed making it less visible to a national audience and less obvious in a national narrative" (MacEachern, "Voices Crying in the Wilderness," 220).

3 Donihee, *The Evolution of Wildlife Law in Canada*, 14–17.

4 Morris, *The Treaties of Canada with the Indians*, 228, 236.

5 Alexander Morris, Special Appendix F, 4 December 1876 (Departmental Report), xii–xiii, lxiii; Report by Dickieson, Special Appendix C, 7 October 1876, xxxv–xxxvi; *Annual Report DI for Year Ended 30 June 1876.*

6 McCandless, *Yukon Wildlife*, 17.

7 *Debates of the Senate*, Canada, 8 April 1878, 502; Nelson, *The Last Refuge*, 167.

8 *Debates of the House of Commons*, 26 March 1877, 994–5.

9 Ibid., 992–3.

10 "All Our Yesterdays," *Montreal Gazette*, 23 September 1946 (reprinting 1879 article), copy in SAB, clipping file, "Bison, American"; Isenberg, "The Returns of the Bison,"180–1.

11 "Report of Thomas Fawcett DTS on the Survey of the Township Outlines between the 2nd and 3rd Base Lines," *Annual Report DI 1881*, 66; Colpitts, *Game in the Garden*, 44–5.

12 McClung, *Clearing in the West*, 84; *Report and Minutes of Evidence on Existing Natural Food Products [1887]*, 179; Seton, *Life-Histories of Northern Animals*, vol. 1, 301.

13 Robert Christie, "Notes on the Mammals of Manitoba," *Natural History Journal and School Reporter*, 9, no. 76 (May 1885): 72 (copy in LAC, MG 29 D108, vol. 31, file 31/25).

14 Mair, *The American Bison*, 93; Albert Braz, "Wither the White Man" (online, not paginated); Voisey, *A Preacher's Frontier*, 45.

15 Clarence Tillenius, "Through Field and Wood No. 36," *Country Guide*, September 1961.

16 Crozier, "Buffalo Stone" in *A Saving Grace*, 67–8.

17 *Annual Report Manitoba DAI 1882*, 64; Friesen, "Grant Me Wherewith to Make My Living," 247.

18 Goatcher, *Manitoba Legislation Affecting Wildlife*, 2–4, 87–104; Canada Department of Agriculture, *Agricultural Gazette* 1, no. 6 (1914): 468. Manitoba's insectivorous birds legislation was incorporated into the Game Protection Act in 1931, and in 1944 the Wolf Bounty Act was repealed and its provisions folded into the Predator Control Act. In 1890 the Manitoba Game Protection Act was retitled *An Act for the Protection of Game and Fur Bearing Animals* to remove any confusion that fur-bearers were game. *An Ordinance for the Protection of Game* (Ordinances of the North-West Territories, no. 8

of 1883) was very close to Manitoba's 1876 Game Protection Act. Alberta combined the NWT's legislative provisions to protect useful birds and game animals that it had adopted in 1905 into a single game statute in 1907.

19 Loo, *States of Nature*, 15, 17–18; Tober, *Who Owns the Wildlife?* 14, 139.

20 Goatcher, *Manitoba Legislation Affecting Wildlife*, 11–16, 18, 73–4, 87-90; Bossenmaier, *100-Plus Years*, 4–5.

21 Hart, *J.B. Harkin*, 111; *Annual Report Saskatchewan* DA *1913*, 246–7.

22 Letter to the editor, *Regina Leader*, 11 August 1891; Seton, *Life-Histories of Northern Animals*, vol. 1, 46.

23 Isenberg, "The Moral Ecology of Wildlife,"49; Atkinson, *A Review-History of the Passenger Pigeon in Manitoba*," 1.

24 Reprinted in *Conservation* 1, no. 4 (June 1912): 3.

25 *Winnipeg Tribune*, 8 October 1953; Millar, "The Big Game of the Canadian Rockies," 109.

26 *Manitoba Game and Fish Association Bulletin*, no. 7, September 1938 (copy in MA, EC0016/GR/666/649); E.S. Russenholt, "Trails of Destiny: Address on Wildlife Conservation," March 1950, MA, P2993/9.

27 For early examples of this analysis, see Nash, *Wilderness and the American Mind*, and Hays, *Conservation and the Gospel of Efficiency*.

28 Foster, *Working for Wildlife*, 34–42; MacEachern, "Voices Crying in the Wilderness," 218; MacEachern, "The Conservation Movement," 1–2, 9–15; Worster, *Nature's Economy*, 266–7.

29 Foster, *Working for Wildlife*, 127–8, 210–16.

30 Hewitt, "The Conservation of Wild Life in Canada in 1917," 118–19; Colpitts, "Wildlife Promotion, Western Canadian Boosterism," 103–30.

31 "Provincial Revenue from Fisheries and Game," 236; Commission of Conservation Canada, *National Conference on Conservation of Game, Fur Bearing Animals and Other Wild Life*, 154.

32 *Annual Report Saskatchewan* DA *for Year Ended April 30, 1922*, 338; *Annual Report Manitoba* DAI *1921*, 57.

33 Wetherell and Kmet, *Alberta's North*, 372–3. On the inequality of the fur tax, see Grew to Acting Director, 12 January 1945, LAC, RG10, vol. 6732, file 420-1-5-3, reel c-8093.

34 *Annual Report Alberta* DA *1923*, 68; *Annual Report Alberta* DLF *for Year Ended 31 March 1963*, 61.

35 "Declare War on Crow" (Red Deer Board of Trade), *Blairmore Enterprise*, 30 May 1935; *Annual Report Saskatchewan DNR for Year Ended 31 March 1959*, p. Wi-12.

36 On the ways that sport hunting confirmed social status and the mastery of empire and nature, see MacKenzie, *The Empire of Nature*, 26–53; and Gillespie, "'I Was Much Pleased with Our Sport among the Buffalo,'" 555–84. Referring to published hunting narratives for the somewhat later period

1875–1914 in the Canadian west, historian Karen Wonders argues that masculinity, not empire, was the predominant theme (Wonders, "Hunting Narratives of the Age of Empire," 269–80).

37 *Annual Report Alberta DA 1908*, 104; *Annual Report [Parks] DI for Year Ending 31 March 1923*, 88.

38 Groff to Brownlee, 16 May 1932, PAA, Premiers Papers, file 13B; Wallace to Aberhart, 6 January 1939, PAA, Premiers Papers, file 663B; "Game and Sport Fishing Resources of Manitoba," [1 November 1941], MA, GR1600/G4527/21.1.1.

39 See, for example, Notes for Speech, Minister, Travel Bureau Conference, 11 November 1944, MA, GR1600/G4527, file 1.2.1.

40 "Parks Committee Preliminary Report April 30, 1945," MA, GR1600/4520/36.4.1.K.

41 *Annual Report Manitoba DMNR for Year Ended 30 April 1941* (typescript), 48; *Fisheries Research in Saskatchewan*, 1; Quiring, *CCF Colonialism in Northern Saskatchewan*, 179.

42 See, for example, Stevens to Beresford, 11 February 1944 (attachment), MA, GR1600/4519/36.3.

43 W.M. Baker, "A Recreation Report on the North-West Part of Saskatchewan with Recommendations for the Establishment of a Provincial Park on the Waterhen River and the Improvement and Development of Facilities in Other Places," December 1956 (typescript) (copy in SAB Pamphlet File "Recreation"); Vogel to Cowan, 26 August 1948, MA, GR1600/G4529/11.2.

44 See *Winnipeg Tribune*, 7 June 1949 and 24 September 1951; Saskatchewan Fish and Game League, Annual Convention [Minutes and Resolutions], 23–24 June, 1941 SAB, NR.1/2, file 115; and *Saskatchewan Fish and Game League Year Book 1943*, 11 (copy in SAB, R-E4531).

45 This interpretation is supported by research on the social and political impact of tourism. While tourist development sometimes offers local groups or outsiders a chance to challenge social and economic authority, it more typically supports the priorities of local elites. See Hal Rothman, *The Culture of Tourism*. On local elites and tourism, see Jasen, "Native People and the Tourist Industry," 5–27.

46 Saskatchewan Fish and Game League, Annual Convention [Minutes and Resolutions], 23–24 June 1941, SAB, NR.1/2 file 115. For criticism of American sport hunters, see Eben to Turner, 27 December 1946, LAC, RG10, vol. 6743, file 420-2-1-1, reel C8096.

47 Haskell, "Protection of Migratory Birds," 66–73; Hewitt, "Conservation of Birds and Mammals in Canada," 144.

48 Dorsey, "Scientists, Citizens and Statesmen," 426–7; Master, "Report of Committee on Fisheries, Game and Fur Bearing Animals," 254. Some critics in the United States challenged the constitutionality of the Migratory Birds Convention, but it was upheld in 1920.

49 Hoyes Lloyd, "Canada and the Migratory Birds Treaty," n.d. (c. 1925), LAC, MG 30 E441, vol. 10, no. 11; *Annual Report Saskatchewan DA 1916*, 231. The convention was implemented in Canada by the Migratory Birds Convention Act in 1917.

50 *Annual Report [Canada] DA for Year Ended 31 March 1918*, 6; "Migratory Birds Convention," 102–4; *Annual Report Alberta DA 1916*, 127.

51 Foster, *Working for Wildlife*, 161–2.

52 Burnett, *A Passion for Wildlife*, 4–15; *Annual Report [Parks] DI for Year Ending 31 March 1923*, 89.

53 J.B. Harkin, "Effects of the Migratory Birds Treaty Act," *Good Roads*, August 1923, 19. The first bird "banding" in Canada was performed by Ernest Thompson Seton in Manitoba in 1882 when he marked several snow geese with ink. This was more about curiosity than systematic study. Percy Taverner pioneered the use of standardized bands and central record keeping in 1904, which became the North American standard (*The North American Banders' Study Guide*, 3–4).

54 MacEachern, "Changing Ecologies: Preservation in Four National Parks," 365; Burnett, *A Passion for Wildlife*, 13, 20, 22; Macdonald, *Science and History at Elk Island*, 32. On the 1922 conference, see *Annual Report [Parks] DI for Year Ending 31 March 1923*, 88; and Foster, *Working for Wildlife*, 216–19. 523

55 Jack Miner, "The Unfinished Treaty" [24 January 1927] (typescript), LAC, MG 30 E441, vol. 12, file 1; "Rod and Gun in Alberta," *Good Roads Magazine*, March 1923, 23–4. Another example was trumpeter swans. Hunted for their feathers, they were believed to be extinct by 1900, but the population was rebuilt in the United States from tiny remnant populations discovered in 1919 (*The Trumpeter Swan*, Publication G3647, n.p., University of Wisconsin Extension Service, [1995], 2–3).

56 *Annual Report Saskatchewan DA for Year Ended 30 April 1925*, 290; Dunlap, "Organization and Wildlife Preservation," 202.

57 "The Whooping Crane Is Nearing Extinction," *Country Guide*, June 1945; Dunlap, "Organisation and Wildlife Preservation," 203–16; Burnett, *A Passion for Wildlife*, 230–6.

58 "Where the Whooping Cranes Nest," *Western Producer*, 22 July 1954; Dunlap, "Organisation and Wildlife Preservation," 197–221.

59 Editorial, *Regina Leader-Post*, 19 July 1955; Doug Gilroy, "Prairie Wildlife No. 36," *Western Producer*, 25 September 1954; Dorothy M. Powell, "Wings in the Wind," *Country Guide*, October 1958, 15, 38–40.

60 Loo, *States of Nature*, esp. chaps 1, 2, and 6.

61 Egerton, "Changing Concepts of the Balance of Nature," 324–5.

62 Robert Bell, "The Extinction of Useful Animals in Modern Times," lecture to Ottawa Field Naturalists Club, 1901, LAC, MG 29 B15, vol. 3, file 3.31; Robert Bell, *Evolution with Prevision*, Address Baccalaureate Sunday, Eighth Annual Commencement, Thomas Clarkson Memorial School

of Technology, Potsdam, New York, 1907 (reprinted from *Clarkson Bulletin*, vol. iv, no. 33 (July 1907): 3 (copy in LAC, MG 29 B15, file 3-52); *Annual Report Saskatchewan DA 1913*, 263–4.

63 Kricher, *The Balance of Nature*, 86, 91; V.W. Jackson, "The Balance of Nature: The Value of Birds to Agriculture," *Grain Growers' Guide*, 11 June 1924; Worster, *Nature's Economy*, 205–20.

64 Howard Fredeen, Nature Lovers page, *Western Producer*, 25 January 1940. This idea was an old one. British geologist Charles Lyell wrote in 1830 in *The Principles of Geology* that "every species which has spread itself from a small point over a wide area, must, in like manner, have marked its progress by the diminution, or the entire extirpation, of some other, and must maintain its ground by a successful struggle against the encroachments of other plants and animals" (quoted in Browne, "Biogeography and Empire," 319).

65 Hewitt, *The Conservation of the Wild Life of Canada*, 17–25. Although Hewitt defined "wild-life" as all fauna (including fish), he dealt only with birds and big game. This kept his book a manageable size and was also a logical approach, as the primary objective of state regulation was big game and birds (Hewitt, *The Conservation of the Wild Life of Canada*, 2–4).

66 Ibid., 22.

67 Worster, *Nature's Economy*, 280–1; *Calgary Herald*, 19 September 1931. See also "A Calgary Zoological Student" (letter to the editor), *Calgary Herald*, 11 January 1936.

68 Millar, "The Big Game of the Canadian Rockies," 109, 111, 115; Hewitt, *The Conservation of the Wild Life of Canada*, 235; Wood, *Draft of a Plan for Beginning Animal Sanctuaries in Labrador*, 5 (italics in original).

69 Goatcher, *Manitoba Legislation Affecting Wildlife*, 11–12, 16.

70 "Fisheries of Manitoba," 164–75; Dunlap, "Value for Varmits,"143.

71 Rowan to Lloyd, 23 February 1931, LAC, MG 30 E441, vol. 15, file 14.

72 On use of local reports, see *Bulletin No. 7* (September 1938), Manitoba Game and Fish Association (copy in MA, EC0016/GR1661/649). On the Alberta practice, see O.B. Johnson to Chief Game Guardian, 26 January 1926, GA, M1327, file 29.

73 Suchet, "'Totally Wild?,'" 141–57; *Report and Minutes of Evidence on Existing Natural Food Products [1887]*, 39.

74 *Annual Report Manitoba DA for Year Ended 30 April 1927*, 2.

75 Egerton, "History of Ecology, Part Two," 120–1; Isenberg, "The Moral Ecology of Wildlife," 53; Dunlap, "That Kaibab Myth," 61.

76 Dunlap, "Values for Varmits," 150; Dunlap, *Saving America's Wildlife*, 71.

77 Worster, *Nature's Economy*, 272–3, 279; *Fish, Fur and Feathers*, 35.

78 Bossenmaier, *100-Plus Years*, 13; *Annual Report Saskatchewan DNR for Year Ended 31 March 1947*, 78. See also *Annual Report Manitoba DMNR for Year Ended 31 March 1960*, 114.

79 G.W. Malaher, "Our Fur Resources," radio talk, 11 December 1946, MA, GR1600/G4527/36.2.1; "The History of the Northern Wildlife Conservation Programme," Fur Programme Report, January 1958, 58–9, SAB, R-E4287; *Annual Report Saskatchewan DNR for Year Ended 31 March 1956*, 22. On caribou programs, see Loo, *States of Nature*, 125–38.
80 *Annual Report Manitoba DMNR for Year Ended 31 March 1959*, 57–9, 65.
81 Loo, *States of Nature*, 145 (italics in original).
82 Dunlap, "That Kaibab Myth," 66.
83 *White Tail Deer in Saskatchewan*, 1, 11–12, 14, 16–17.
84 *Annual Report Manitoba DMNR for Year Ended 31 March 1950*, 59; *Annual Report Manitoba DMNR for Year Ended 31 March 1953*, 50; *Annual Report Manitoba DMNR for Year Ended 31 March 1960*, 93; *Winnipeg Tribune*, 23 November 1951 (editorial) and 17 August 1957.
85 "Big Game Report for the Winter of 1955/56," MA, P3609, item 2.
86 Ainley, *Restless Energy*, 253; Worster, *Nature's Economy*, 285–9.
87 Cotton, "From Game Refuges to Ecosystem Assessments," 41.
88 Worster, *Nature's Economy*, 283–4, 342–50; Jones, "*Never Cry Wolf*," 65–93.
89 Goatcher, *Manitoba Legislation Affecting Wildlife*, 70; Bossenmaier, *100-Plus Years*, 13–14; Donihee, *The Evolution of Wildlife Law in Canada*, 19–21, 27–8, 53.
90 Mallory, "Canada's Wildlife," 99–100.
91 *Country Guide*, August 1966, 42.
92 *Annual Report Manitoba DMNR for Year Ended 31 March 1962*, 54.

CHAPTER FIVE

1 At http://www.collectionscanada.gc.ca/cuisine/027006-119.01-e.php?&page_ecopy=nlc009578.U52.1&&&&&&PHPSESSID=h2efcoc2pr5dgm6gu2br2735k1 (accessed 2 November 2008).
2 Brightman, *Grateful Prey*, 344–9; *Report and Minutes of Evidence on Existing Natural Food Products [1887]*, 35.
3 On Aboriginal plains peoples' use of wild turnip (*Psoralea esculenta*), see Mandelbaum, *The Plains Cree*, 74–5.
4 *Calgary Herald*, 12 May 1911. The bias in favour of big game perhaps reflected a perception that, in the case of domestic animals, dark meat was the best. Clyde Campbell, a homesteader in the Peace River country in the 1920s, claimed that beef and pork provided the best nutrition because veal and other light-coloured meat had "no strength in it and a man can't work on it" (Moyles, *Challenge of the Homestead*, 289).
5 Rowles, "Bannock, Beans and Bacon," 6; F.A. Twilley, "Moose Hunt," *Farm and Ranch Review*, May 1958; Tony Lascelles, "The Mammals of Manitoba," *Winnipeg Free Press* (magazine Part 2), 9 May 1936; Isern, "Gopher Tales," 192; *Country Guide*, October 1964 (letters to the editor); *Winnipeg Free Press*, 24 January 1948.

6 *Good Roads Magazine*, September 1926; Doug Gilroy, Prairie Wildlife No. 124, *Western Producer*, 5 July 1956.

7 *Annual Report Manitoba DAI 1883*, 203; *Annual Report Manitoba DAI 1887*, 86; "Notebook re: Canadian Wildlife" (n.d.) (c. 1900–10), QUA, Collection 2026a, box 6, file 4; Moyles, *Challenge of the Homestead*, 255. Snow buntings were called snowbirds in the 1880s.

8 Spry, "Aboriginal Resource Use in the Nineteenth Century," 81–92. On theories about the stability of food supply among hunter gatherers, see Ray, "Periodic Shortages, Native Welfare, and the Hudson's Bay Company," 22–5.

9 Tough, *"As Their Natural Resources Fail,"* 176–8. As an example of the many comments by surveyors about local food supply, see "Extract from the Report of the Operations of the Western Section of the Standard Survey by Montagu Aldous, DTS," *Annual Report DI for Year Ended 30 June 1880*, 52.

10 *Annual Report DIA 1880*, 94; *Report and Minutes of Evidence on Existing Natural Food Products [1887]*, 4, 50.

11 *Report and Minutes of Evidence on Existing Natural Food Products [1887]*, 5; Senate. *Debates*, 18 May 1887, 74–8, 80–1.

12 *Report and Minutes of Evidence on Existing Natural Food Products [1887]*, 4, 16–17; Senate, *Debates*, 18 May 1887, 74, 78.

13 *Report and Minutes of Evidence on Existing Natural Food Products [1887]*, 4–5, 15–16, 25–8. Bell also mentioned that "Indian parsnip" was a traditional food. He was perhaps referring to cow parsnip (also called Indian celery).

14 *Report and Minutes of Evidence on Existing Natural Food Products [1887]*, 5, 35; Senate, *Debates*, 18 May 1887, 75.

15 *Report and Minutes of Evidence on Existing Natural Food Products [1887]*, 5, 18–19, 63–5.

16 *Annual Report DIA 1887*, xv. Planting wild rice remained in vogue for many years. In 1892 Ernest Thompson Seton purchased wild rice in Winnipeg and threw it out the window of the moving train near White Horse Plains (journal entry 9 June 1892, AMNH, E-3, vol. 4). During his Canadian tour in 1924, he stopped at Carberry and noted that "[i]n 1892 I sowed wild rice here. It is now a flourishing crop – great feed – resorted to by ducks and geese. The Blue Lake is now a sanctuary and full of ducks and geese" (journal entry 29 October 1924, AMNH, E-3, vol. 26).

17 Senate, *Debates*, 18 June 1887, 515, and 20 April 1888, 355; *Report and Minutes of Evidence on Existing Natural Food Products [1887]*, 6.

18 *Annual Report [Indian Commissioner] DIA 1888*, 125; *Annual Report DIA for Year Ended 31 March 1911*, 163.

19 Mandelbaum, *The Plains Cree*, 71–4; Brightman, *Grateful Prey*, 346–7.

20 "Report of P.R.A. Belanger, DLS, Survey of Trails and Settlements at Lac Ste Anne and Lac la Biche," *Annual Report DI 1889*, 41; Tough, *"As Their Natural Resources Fail,"* 176–8; Monthly Report, Lesser Slave Lake Detachment, August and October 1906, LAC, RG18, vol. 315, file 206.

21 Harrison Young, "The Northern Fur Trade," *The Nor'West Farmer*, 5 July 1902; Monthly Report, Lesser Slave Lake Detachment, August and October 1906, LAC, RG18, vol. 315, file 206; Field to OC, 25 August 1907, LAC, RG18, vol. 134, file 189.

22 Milne, *Trading for Milady's Furs*, 52.

23 *Annual Report Saskatchewan DNR for Year Ended 31 March 1959*, p. NA21; "Annual Report Alberta Fur Supervisor," 10 January 1949, LAC, RG10, vol. 6734, file 420-2-1-3, reel 8096.

24 *Report and Minutes of Evidence on Existing Natural Food Products [1887]*, 43; Rowles, "Bannock, Beans and Bacon," 7.

25 *Annual Report Manitoba DA 1882*, 65; *Annual Report [Deputy Minister] Department of Marine and Fisheries 1909–10*, 270; *Annual Report Alberta DLF for Year Ended 31 March 1954*, 48.

26 Rowles, "Bannock, Beans and Bacon," 7.

27 *Annual Report Alberta DLM for Year Ended 31 March 1948*, 79; *Annual Report Manitoba DMNR for Year Ended 30 April 1932*, 183; *Annual Report Manitoba DMNR for Year Ended 30 April 1933* (typescript), 151. Whitefish was the dominant commercial fish. The only regional specialty fish was goldeye, found in many spots in northern Canada but in commercial quantities mostly in Manitoba. It was found in the 1890s that smoking improved its flavour. Known as Winnipeg goldeye, the smoked fish became "an epicure's delight" on the trans-Canada railway dining cars in the interwar years. Despite its commercial value, almost nothing was known about goldeye when the Manitoba government and the National Research Council began studying its biology in the late 1930s (Kerry Wood, "A Creel-Full of Fish Facts," *Farm and Ranch Review*, August 1949; *Annual Report Manitoba DMNR for Year Ended 30 April 1939*, 62).

28 *Annual Report Manitoba DMNR for Year Ended 30 April 1932* (typescript), 183; *Annual Report Saskatchewan DNR for Year Ended 31 March 1953*, 80; *Annual Report Manitoba DMNR for Year Ending 30 April 1944*, 71.

29 *Annual Report Manitoba DMNR for Year Ended 31 March 1954*, 70; *Annual Report Saskatchewan DNR for Year Ended 31 March 1955*, 44.

30 *Annual Report Manitoba DMNR for Year Ended 31 March 1964*, 13; *Annual Report Saskatchewan DNR for Year Ended 31 March 1959*, p. F14.

31 Colpitts, *Game in the Garden*, 75–87; Weir, "Settlement in Southwest Manitoba," 60.

32 "Extract from a Report of Exploration by Professor John Macoun," *Annual Report* DI *for Year Ending 30 June 1880*, 21, 25; Farley, *Birds of the Battle River Region*, 28.

33 "North and West Homesteading at Rosthern,"171; Anderson and Niskala, "Finnish Settlements in Saskatchewan,"160; *Annual Report Manitoba* DA *1883*, 202. In 1911, for example, Manitoba gave "settlers" free permits to catch fish for home use (*Annual Report [Manitoba Appendix 8] of the Department of Marine and Fisheries 1910–11*, 285).

34 Kerry Wood, "The Farmer's Game Birds," *Country Guide*, September 1946; Spry, *Papers of the Palliser Expedition*, 20; *Report and Minutes of Evidence on Existing Natural Food Products [1887]*, 68, 74, 149, 169; *Report of the Canadian* DA *[Appendix 10, Winnipeg Agent, Table C], 1883*, 90–1.

35 "C.C. Fairchild Surveys in the Athabasca District," *Annual Report* DI *1901–02*, 73; *Conservation*, 7, no. 5 (May 1918): 20; Rowles, "Bannock, Beans and Bacon," 6.

36 *Annual Report Saskatchewan* DA *for Year Ended 30 April 1924*, 312; Kerry Wood, "The Farmer's Game Birds," *Country Guide*, September 1946.

37 See, for example, John Nelson, "Canada's Newest Inland Empire," *Maclean's*, 1 October 1924.

38 Kitto, *The Peace River Country*, 41; Griesbach, "Sheriden Lawrence," 21; *Edmonton Journal*, 1 October 1932.

39 "Ways to Prepare Game," *Grain Growers' Guide*, 8 October 1924.

40 *Annual Report Alberta* DLM *for Year Ended March 1932*, 60; "Peace River" (1928) (copy in PAA, 73.307/14); R.J. Dawson, *The Big River Survey: A Comparative Study of Natural Resources as an Aid to Improved Utilization* (Saskatoon: University of Saskatchewan, 1943), 22, 24 (copy in SAB Pamphlet file "Natural Resources, Big Game"); Moyles, *Challenge of the Homestead*, 35, 55, 82–3, 176, 184; Rowan, "Some Effects of Settlement on Wildlife in Alberta," 33.

41 Hewitt, "The Conservation of Wild Life in Canada in 1917: A Review," 135; Andrews to Craig, October 25, 1935, PAA, Premier's Papers, file 15; Barnett to Bence, 8 June 1933, SAB, NR 1/1, file G500S.

42 Colpitts, *Game in the Garden*, 77–80; Goatcher, *Manitoba Legislation Affecting Wildlife*, 16. By the 1920s the daily limit in Manitoba had been reduced somewhat, but the total limits for the entire hunting season remained high.

43 I.G. Field Johnson to R.I. Whitlaw [Greenway's office], 10 August 1894, MA, EC0016/GR1662/G507, item 6928; Colpitts, *Game in the Garden*, 77–80; Bossenmaier, *100-Plus Years*, 6; *Annual Report Saskatchewan* DA *1911*, 171.

44 Bossenmaier, *100-Plus Years*, 6–7, 10; *Conservation*, 2, no. 6 (July 1913): 2.

45 *Annual Report Saskatchewan* DA *1913*, 241; *Conservation* 5 (November 1918): 42; *Conservation* 4 (April 1920): 13; Colpitts, *Game in the Garden*, 84–5; "Sale of Game," 69.

46 Master, "Report of Committee on Fisheries, Game and Fur Bearing Animals," 254; Colpitts, *Game in the Garden*, 93; Goatcher, *Manitoba Legislation Affecting Wildlife*, 25.

47 *Annual Report Saskatchewan DA for Year Ended 30 April 1918*, 207; Bossenmaier, *100-Plus Years*, 10.

48 *Annual Report Saskatchewan DNR for Year Ended 30 April 1935* (typescript), 139; "Report on Game Conditions – Manitoba," 20 November 1930, MA, GR16000/G4537/15.1.2.A; *Winnipeg Tribune*, 26 December 1951.

49 "Sale of Game," 69; *Calgary Herald*, 11 September 1937.

50 John Patrick Gillese, "Water and Woodland Harvest," *Farm and Ranch Review*, November 1957; Lawton, "Game in Cold Storage," 71–3; *Winnipeg Tribune* 8 June 1949.

51 "Ways to Prepare Game," *Grain Growers' Guide*, 8 October 1924; "Where Game Trails End," *Country Guide*, September 1957; B.W. Cartwright, *News Flight* [DUC], no. 42-29 [1942] (copy in MA, GR43/G72).

CHAPTER SIX

1 Fromm, *The Anatomy of Human Destructiveness*, 136–40. Saskatchewan hunter David Carpenter argues that sport hunting, despite the appeal of its violence, is primarily a way of connecting with wild animals and natural environments and of forming a spiritual connection with the earth (Carpenter, *A Hunter's Confession*).

2 There is a significant literature on this subject. As only two examples, see Fine, "Rights of Men, Rites of Passage," 805–23; and Loo, "Of Moose and Men," 296–319.

3 Lefty (Plumas, MB), Young Co-operators page, *Western Producer*, 30 November 1950; Gottesman, "Native Hunting and the Migratory Birds Convention Act," 82; F.A. Twilley, "Killing for Fun," *Farm and Ranch Review*, February 1966; *Annual Report Saskatchewan DA for Year Ended 30 April 1920*, 106; *Winnipeg Tribune*, 16 October 1954.

4 William Lothian to My Dear Brother George, 7 February 1882, MA, MG8/B3, item 20; *Saskatchewan Federation of Fish and Game League Year Book 1943*, 15 (copy in SAB, R-E4531); Colpitts, *Game in the Garden*, 63–9.

5 D.G. Stuart, "Punnichy Saskatchewan 12th October 1937," SAB, R972, file 47.

6 *Annual Report Saskatchewan DA 1913*, 247–8; Jack Miner, "The Unfinished Treaty" [24 January 1927] (typescript), LAC, MG30 E441, vol. 12, file 1.

7 Bossenmaier, *100-Plus Years*, 7; "Report on Game Conditions – Manitoba," 20 November 1930, MA, GR1600/G4537/15.1.2.A; Smith to Taggart, 6 February 1935, SAB, NR1/1, file G5000G(1); *Regina Leader-Post*, 16 September 1929.

8 *Annual Report Alberta DA for Year Ended 31 March 1932*, 79; *Annual Report Alberta DA for Year Ended 31 March 1935*, 40.

9 *Fisheries Research in Saskatchewan*, 1; *Annual Report Saskatchewan DNR for Year Ended 31 March 1959*, p. W12; John Patrick Gillese, "Water and Woodland Harvest," *Farm and Ranch Review*, November 1957; *Annual Report Alberta DLF for Year Ended 31 March 1956*, 62.

10 Potter to Soper, 22 July 1935, LAC, RG109-D-2, vol. 5, file registry 'P'1922–46; *Conservation* 8, no. 6 (June 1919): 24; Doug Gilroy, "Prairie Wildlife No. 480," *Western Producer*, 2 May 1963.

11 Donihee, *The Evolution of Wildlife Law in Canada*, 7–10; Ritvo, *The Animal Estate*, 276–88.

12 Dunlap, "Sport Hunting and Conservation," 53; Loo, "Of Moose and Men," 306–8; Herman, *Hunting and the American Imagination*, 122–58; Ingram, "Beastly Measures," 230–1, 235. In 1885 Manitoba banned Sunday hunting, presumably to conform to sabbatarian prohibitions on playing sports on Sundays.

13 Seton, "Lobo, the King of the Currumpaw," in *Wild Animals I Have Known*, 41; "Sport," n.d. [c. 1920], LAC, MG29 D108, vol. 3, file 3/16. Seton employed the same theory in his writings about activities for Boy Scouts but omitted the killing and emphasized the chase (Wadland, *Ernest Thompson Seton*, 362–3).

14 Seton, *Life-Histories of Northern Animals*, vol. 1, 113; *Grain Growers' Guide*, 6 October 1920 (advertisement by Dominion Cartridge Co.).

15 "Alberta's Game Birds," *Good Roads Magazine*, September 1926; Dick Fairfax, "I Used to Be a Frying Pan Hunter" (published in two instalments), *Western Producer*, 8 and 15 March 1956; Otto Hohn, "Mallards by the Millions," *Farm and Ranch Review*, October 1962.

16 *Calgary Herald* (editorial), c. 1936 (copy in GA, A.S243).

17 *Annual Report Saskatchewan DA for Year Ending 30 April 1921*, 115; Frederic M. Baker, "Ethics and the Grizzly," *Country Guide*, February 1949.

18 Bossenmaier, *100-Plus Years*, 5–9; Goatcher, *Manitoba Legislation Affecting Wildlife*, 18–20, 50–2; *Winnipeg Tribune*, 2 June 1965; Untitled draft document, 1945, SAB, NR1/2, file 114.

19 *Winnipeg Tribune*, 1 August 1957; *Annual Report Saskatchewan DNR for Year Ended 31 March 1956*, 30–1.

20 Quoted in Colpitts, *Game in the Garden*, 94.

21 *Annual Report Saskatchewan DA 1914*, 223. Despite the provincial game commissioner's advice, the Saskatchewan government legalized the killing of cow moose in 1915 on the grounds that the buck law led to waste because cow moose that were shot by accident were left in the bush to cover up the crime. The game commissioner remained publicly opposed and seems to have organized opposition by the Saskatchewan Game Protective Association and others and succeeded in having the amendment reversed the next year (*Annual Report Saskatchewan DA 1915*, 242; *Annual Report Saskatchewan DA 1916*, 238–9; *Annual Report Saskatchewan DA 1918*, 235).

22 *Annual Report Manitoba* DMNR *for Year Ended 30 April 1932*, 174 (typescript).
23 *Annual Report* NWT DA *1904*, 108; *Calgary Herald*, 4 September 1909; *Annual Report Alberta* DA *1908*, 101; *Drumheller Mail*, 4 September 1919.
24 *Annual Report Saskatchewan* DA *1907*, 204; Francis Dickie, "When Grizzlies Roar," *Grain Growers' Guide*, 1 October 1926.
25 Howard Fredeen (Macrorie, SK), Nature Lovers page, *Western Producer*, 27 April 1939; *Claresholm Local Press*, 9 October 1947; F.A. Twilley, "Killing for Fun," *Farm and Ranch Review*, February 1966, 27.
26 *Annual Report Alberta* DA *1908*, 102; Dick Fairfax, "I Used to Be a Frying Pan Hunter" (published in two parts), *Western Producer*, 8 and 15 March 1956; "Alberta's Game Birds," *Good Roads*, September 1926.
27 Rowan to Lloyd, 2 February 1931, LAC, MG30 E441, vol. 15, file 14; *Annual Report Manitoba* DMNR *for Year Ended 31 March 1957*, 82; *Annual Report Manitoba* DMNR *for Year Ended 31 March 1959*, 80.
28 "Wild Goose Chase," *Saskatchewan Afield*, 1, no. 2 (September 1958): 10–11 (copy in SAB, R-E4531); *Winnipeg Tribune*, 15 June 1962.
29 On early game associations in the United States, see Tober, *Who Owns the Wildlife?*, 50–2. On game associations and conservation, see Reiger, *American Sportsmen and the Origins of Conservation*, 19–22. For a critique of Reiger, see Dunlap, "Sport Hunting and Conservation," 51–60.
30 *Macleod Gazette*, 7 March 1889; Edmonton *Bulletin*, 9 August 1894.
31 *Regina Leader-Post*, 16 September 1933; Loo, "Making a Modern Wilderness," 92–121. As historian George Colpitts shows, in southern Alberta it was important to protect local access to sport fishing from competing urban anglers (Colpitts, "Fish and Game Conservation in Alberta," 18–19).
32 *Saskatchewan Fish and League Year Book 1943*, 3 (copy in SAB R-E4531); "Questions and Answers about the Saskatchewan Fish and Game League," n.d. [c. 1940s], SAB, R-E4531; *Winnipeg Tribune*, 21 June 1963.
33 See, for example, claims by the Red Deer Rod and Gun Club in 1894 (*Regina Leader*, 16 August 1894) as well as those by the Alberta Fish and Game Association (*Annual Report of the Alberta* DA *1906*, 134) and by the Saskatchewan Fish and Game League ("Questions and Answers about the Saskatchewan Fish and Game League" [typescript], n.d. [c. 1940], SAB, R-E4531).
34 *Annual Report Saskatchewan* DA *1913*, 247–8. By 1924 Bradshaw had softened his stance somewhat. By then, there were eight game protective associations in the province with 700 members, all of whom were *ex officio* game guardians. He concluded that these organizations would make "a very real contribution" by promoting "a spirit of clean sportsmanship" (*Annual Report Saskatchewan* DA *for Year Ended 30 April 1924*, 302).
35 *The Nor'West Farmer*, 20 September 1905; *Calgary Herald*, 13 August 1909; *Saskatchewan Fish and Game League Year Book 1943*, 47 (copy in SAB R-E4531).

36 *Saskatchewan Afield* 1, no. 3 (1958): 2–3 (copy in SAB, R-E 4531); Goatcher, *Manitoba Legislation Affecting Wildlife*, 57; *Annual Report Saskatchewan DNR for Year Ended 31 March 1960*, 65.

37 *Annual Report Saskatchewan DA for Year Ended 30 April 1926*, 155; *Saskatchewan Fish and League Year Book 1943*, 7, 19 (copy in SAB R-E4531). On Alberta, see *Annual Report [Canada] Department of Fisheries for 1930–31*, 90; and *Annual Report Alberta DA 1906*, 134. On Manitoba, see Finding Aid, Winnipeg Game and Fish Association, MA, P3608.

38 Colpitts, "Fish and Game Conservation in Alberta," 23; Harris, "The Need and Value of Local Organizations," 145; *Conservation* 8, no. 9 (September 1919): 35.

39 *Annual Report Saskatchewan DA 1906*, 127; *Annual Report Saskatchewan DA for Year Ended 30 April 1921*, 99; *Saskatchewan Fish and Game League Year Book 1943*, 15 (copy in SAB R-E4531); Draft untitled document, 1945, SAB, NR1/2, file 114.

40 Hayden to Brownlee, 24 January 1928 (and attachments), PAA, Premiers Papers, file 59.

41 *Annual Report Alberta DLF for Year Ended 31 March 1944*, 75; *Annual Report Alberta DLF for Year Ended 31 March 1950*, 74.

42 Millar, "The Big Game of the Canadian Rockies," 118.

43 Colls to Paynter, 1948, LAC, RG109-D-2, vol. 5, file G.7.1.

44 *Annual Report Alberta DLF for Year Ended 31 March 1964*, 70; Soper, *Mammals of Alberta*, 329–30.

45 *Annual Report Saskatchewan DA 1916*, 225; *Annual Report Saskatchewan DA for Year Ended 30 April 1918*, 220; *Annual Report Saskatchewan DNR for Year Ended 31 March 1948*, 82; "Province of Alberta Predator Control Problem," n.d. [1949], PAA, 72.302, file 293.

46 *Annual Report Saskatchewan DNR for Year Ended 30 April 1946*, 74; *Annual Report Manitoba DMNR for Year Ended 31 March 1960*, 100.

47 *Winnipeg Tribune*, 5 June (editorial) and 1 July 1950 (editorial).

48 Taverner, "Scientific Advice for Wild Life Conservationists," 106; Walker to Soper, 31 March 1937, and Soper to Walker, 5 April 1937, LAC, RG109-D-2, vol. 5, file "W," 1934–47.

49 Ainley, *Restless Energy*, 270–2. By 1950 Saskatchewan had dropped bounties on hawks because "many better-type hawks were being destroyed" (*Regina Leader-Post*, 21 February 1950).

50 *Annual Report Saskatchewan DA 1909*, 181; *Grande Prairie Herald*, 12 April 1929.

51 *Annual Report Manitoba DAI 1915*, 112; *Annual Report Manitoba DAI 1919*, 61–2; *Annual Report: Manitoba DAI for Year Ended 30 April 1930*, 1. The amount paid in bounties in 1919 is equivalent to over $270,000 in 2016 currency, based on the Bank of Canada's inflation calculator.

52 Return on Agricultural Pests Competition [1929], PAA, 73.307, file 140; Poster [April 1929], PAA, Premiers Papers, file 15.
53 Farley, *Birds of the Battle River Region*, 47–8; "Declare War on the Crow" (Red Deer Board of Trade), *Blairmore Enterprise*, 30 May 1935.
54 Cartwright to Stephens, 10 January 1940, MA, GR1600/G4530/11.3.1. DUC sponsored research on predation of ducklings by northern pike and in 1939 proposed that lakes be cleared of pike by fishing or by dynamiting (see Main to McDiarmid, 5 October 1939, MA, GR1600/G4530, file 11.3.1; *Northern Pike (Jackfish) as Predator on Waterfowl and Muskrats: Report on Ecological Investigations* [Winnipeg: DUC], 1939, 9, 31 (copy in MA, EC0016/GR1661/G649); and Newsletter, 10 June 1940 (copy in MA, EC0016/GR43/G54/29). For an academic study partly and uncritically relying on this DUC evidence, see Pitman, "Waterfowl Predation in Canada by the Northern Pike," 64–7.
55 Cartwright to Godsmark, 16 February 1944, and Manitoba Game and Fish Association to Morton, 9 February 1944, MA, P3609, item 10; "Alberta Hunters Get Reward for Bagging Magpies," *Farm and Ranch Review*, May 1949.

56 *Saskatchewan Fish and Game Year Book 1943*, 37 (copy in SAB R-E4531).
57 *Annual Report Saskatchewan DNR for Year Ended 30 April 1937*, 58; *Annual Report Saskatchewan DNR for Year Ended 31 March 1948*, 79; *Annual Report Saskatchewan DNR for Year Ended 31 March 1949*, 96.
58 Rusconi to Bertrand, 15 March 1944, and Doake to Janzen, 25 April 1944, SAB, NR1/2, file 115.
59 *Drumheller Mail*, 24 May, 21 June, and 28 June 1928.
60 Webster to Taverner, 6 August 1931, LAC, RG45, vol. 294, 1–2.
61 Blondie Blue (Peebles, SK), Nature Lovers page, *Western Producer*, 15 July 1954; Robert Pharis, "Details of Magpie Extermination," *Country Guide*, October 1945.
62 *Country Guide*, August 1948, 30; Wood, *The Magpie Menace*, 34–5.
63 "Joy," Nature Lovers page, *Western Producer*, 31 May 1928.
64 *Annual Report Alberta DA 1929*, 35; *Annual Report Saskatchewan DNR for Year Ended 31 March 1949*, 96; Paynter to Lewis, 27 June 1949, LAC, RG109-D-2, vol. 5, file G7.1.
65 Bluebird, Nature Lovers page, *Western Producer*, 28 June 1928; Florence Bye, Nature Lovers page, *Western Producer*, 30 May 1929.
66 V.W. Jackson, "The Balance of Nature: The Value of Birds to Agriculture," *Grain Growers' Guide*, 11 June 1924; Kerry Wood, "The Crow: Friend or Enemy?," *Farm and Ranch Review*, March 1953.
67 *Annual Report Alberta DLF for Year Ended 31 March 1953*, 54.
68 Herbert to Halmrast, 16 January 1959, and Halmrast to Herbert, 21 January 1953, PAA, 72.302, file 1742.

69 "Big Game Monarchs Lure Army of Hunters," *Western Motordom*, October 1930.
70 Etter to Merkley, "Migratory Bird Situation," 22 January 1932, SAB, NR1/1, file G5000S.
71 Rowan, "Some Effects of Settlement on Wildlife in Alberta," 37–8.
72 Rowan to Lloyd, 23 February 1931, LAC, MG30 E441, vol. 15, file 14; Rowan to Soper, 15 November 1943, LAC, RG109, D-2, vol. 5, file "R." Rowan later elaborated on this in *The Ten Year Cycle: An Outstanding Problem of Canadian Conservation.* One hunter wrote in 1959 that based on his observations over many years, he thought that Rowan was correct (Gordon to Kuziak, 8 December 1959, SAB, R43, file 14).
73 Clipping, *Regina Star*, 7 September 1933, SAB, NR1/2, file 800C; *Regina Leader-Post*, 13 August 1934.
74 Jack Miner, "The Unfinished Treaty" [24 January 1927] (typescript), LAC, MG30 E441, vol. 12, file 1. Miner was not exaggerating. The daily duck limit in Manitoba at this time, for example, was 20 ducks per day from 15 September to 1 October and 40 per day from then to 30 November. If account was taken for Sundays when hunting was illegal, this would allow a total take of 2,500 birds if a hunter took the limit every day.
75 Murphy to Brownlee, 9 September 1931, PAA, Premiers Papers, file 10A.
76 *Calgary Herald*, 28 December 1935; *Regina Leader-Post*, 3 February 1932 and 13 August 1934.
77 In 1931 Saskatchewan reduced its season to one month and the possession limit (the number of ducks a hunter could have in his bag or in storage) to 30 birds. In 1932 Manitoba reduced its season to eight weeks with a seasonal limit (which did not govern birds in storage) to 150 birds (Goatcher, *Manitoba Legislation Affecting Wildlife*, 49).
78 E.A. Etter to Hoyes Lloyd, 14 June 1933, LAC, MG30 E441, vol. 5, file 17.
79 Jack Miner, "The Unfinished Treaty" [January 24, 1927] (typescript), LAC, MG30 E441, vol. 12 file 1; Soper to Cleveland Grant, 1 February 1940, LAC, RG109-D-2, vol. 5, file "G," 1925–46.
80 DUC, *Census 1938 & 1939 and Kee-man Record Book*. Winnipeg: DUC [1939], 30 (copy in MA, GR1600/G4530/11.3.1).
81 Ainley, *Restless Energy*, 256–7; Lawton to Lloyd, 23 February 1928, LAC, MG30 E441, vol. 10, file 7; *Regina Leader-Post*, 24 April 1937; Ralph Hedlin, "Crop Damage by Ducks," *Country Guide*, September 1953.
82 "Ducks Unlimited" (1937), PAA, Premiers Papers, file 663A; Lewis, *To Conserve a Heritage*, 33–9.
83 Huntington to du Pont, 7 April 1937 (copy), LAC, MG30 E441, vol. 5, file 6.
84 "Conservation: A National Issue," [text of radio broadcast] *News Flight*, no. 28, n.d. [c. 1938], DUC (copy in MA, EC0016/GR1661/649).

85 DUC, "Ducks and Duck Waters in the Canadian West," *Newsletter*, 10 May 1940 (copy in MA, EC0016/GR43/G54/29).
86 *Regina Leader-Post*, 24 April 1937.
87 Huntington to du Pont, 7 April 1937 (copy in LAC, MG30 E441, vol. 5, file 6). The organization has been flexible and has remade itself in step with broad social changes. While it remains supportive of sport hunting, DUC now also links its work in wetland conservation to the control of CO_2 emissions and the creation of carbon sinks. See, for example, http://www.ducks.ca/aboutduc/news/annual_report/pdf/ar2004.pdf.
88 *Annual Report Alberta DA 1938*, 48.
89 Soper to Walker, 20 April 1937, LAC, RG109-D-2, vol. 5, file "W," 1934–47.
90 Soper to Lloyd, 1 October 1943, LAC, RG109-D-2, vol. 5, file "T," 1935–46; Soper to Watson, 23 October 1946, RG109-D-2, vol. 5, file "W," 1934–47; Soper to Saunders, LAC, RG109-D-2, vol. 5, file "T." Charges of misrepresentation were also made in the United States. One 1938 report pointed out that Ducks Unlimited Inc. could not take all the credit for improving habitat in the United States given the federal government's rehabilitation of 4 million acres as waterfowl sanctuaries since 1934 (briefing memo attached by Hoyes Lloyd to file copy of Huntington to du Pont, 7 April 1937, LAC, MG30 E441, vol. 5, file 6).
91 The DUC "census" of duck populations on the prairies showed almost 41 million ducks in 1935, about 49 million in 1938, and about 59.6 million in 1939. DUC claimed that there had been 180 million in 1900 (*Newsflight* [DUC], no. 55 [1939]: 10).
92 L.B. Potter, "Let's Be Fair," *Western Producer*, 6 June 1940.
93 Gabrielson, "Conservation of Waterfowl," 58, 60–1. The book was edited by Francis Kortright, a Toronto conservationist.
94 Letter to the editor, *Winnipeg Tribune*, 8 January 1940 (copy in MA EC0016/GR43/G64). Tony Lascelles was the pen name of Hubert Unsworth Green, a former Banff park warden (http://www.archivesalberta.org/findingaids/whyte/Green/Green_main.html).
95 "Sport Fishing," *Keystone Province* 12, no. 2 (September–October 1941): 4 (copy in MA, EC0016/GR431G72).
96 *Annual Report Saskatchewan DA for Year Ended 30 April 1920*, 106; Jack Miner, "The Unfinished Treaty" [24 January 1927] (typescript), LAC, MG30 E441, vol. 12, file 1.
97 Kerry Wood, "The Farmer's Game Birds," *Country Guide*, September 1946; Kerry Wood, "The Deer Hunt," *Farm and Ranch Review*, September 1958; *Annual Report NWT DA 1903*, 124.
98 The number of deer licences sold in Saskatchewan grew tenfold, from 2,901 in 1945 to 29,634 in 1954 (*White Tailed Deer in Saskatchewan*, 1).

99 *White Tailed Deer in Saskatchewan*, 14.

100 *Winnipeg Tribune*, 22 November 1952.

101 F.A. Twilley, "Moose Hunt," *Farm and Ranch Review*, May 1958; W.J. Healy, "Buffalo Hunting in the 1840s," *Grain Growers' Guide*, 10 September 1924.

102 "Ye Editor Goes Bear Hunting," *Peace River Record*, 9 November 1922.

103 Jardine and Spary, "The Natures of Cultural History," 3–14; Brower, "'Take Only Photographs,'" n.p.; Dunaway, "Hunting with the Camera," 207–11.

104 Brower, "Trophy Shots," 17–18; "Hunting with a Camera," *Farm and Ranch Review*, November 1962; Baker, "Guest Editor's Introduction," 190. Historian Gregg Mitman further notes the value of the camera in biological fieldwork beginning in the 1940s for creating an objective and repeatable record of animal behaviour (Mitman, "When Nature *Is* the Zoo," 130–1).

105 Soper to Munro, 2 January 1946, LAC, RG109-D-2, vol. 5, file "M."

106 Doug Gilroy, "Prairie Wildlife No. 487," *Western Producer*, 20 June 1963; *Winnipeg Tribune*, 15 April 1933.

107 Williamson, "Game Preservation in the Dominion Parks," 140; *Calgary Herald* (editorial), 2 January 1936.

108 McGowan, "Hunting with a Camera," 22–5.

109 *Hunting with a Camera*, NFB (16 mm colour film), LAC, ISN 130427; Russell, "Are Wild Animals Really Wild?," 240.

110 In Alberta, for example, 99,000 hunting licences were issued to residents and non-residents in 2008, when the province's population was almost 3.6 million (*CTC Research and Evaluation, Sport Fishing and Game Hunting in Canada: An Assessment on the Potential International Tourism Opportunity*, 2012, 8, http://publications.gc.ca/collections/collection_2013/ic/Iu86-49-2012-eng.pdf). A relative decline of sport hunting on the basis of population is evident when it is recalled that 60,000 licences were issued in 1953 when the province's population was 1.1 million.

111 Bergman, "Obits for the Fallen Hunter," 818–30.

112 Robison and Ridenour, "Whither the Love of Hunting?," 418–36; *National Post*, 20 October 2004.

113 Dunlap, "Sport Hunting and Conservation," 58.

CHAPTER SEVEN

1 *Annual Report Saskatchewan DA for Year Ended 30 April 1926*, 161.

2 Crosby, *Ecological Imperialism*, 3–5, 288–93; Dunlap, "Remaking the Land," 304, 307. For this process in Mexico, see Melville, "Disease, Ecology, and the Environment."

3 Osborne, "Acclimatizing the World," 137; Ritvo, "Going Forth and Multiplying," 404, 410.

4 "Rock Pigeon," Texas Parks and Wildlife, at http://www.tpwd.state.tx.us/huntwild/wild/species/rockdove/.

5 Doughty, "Sparrows for America." See also Williams, "The Sparrow War."

6 *Report and Minutes of Evidence on Existing Natural Food Products [1887]*, 149; *Annual Report* NWT *DA 1903*, 205.

7 Mary Salinger (Dilke, SK), Nature Lovers page, 20 June 1929, and Honey Angel (Ormston, SK), Nature Lovers page, 5 January 1950, *Western Producer*. A manufacturer in Magrath, AB, developed the Catchall Sparrow Trap, which it was said could catch over 300 sparrows in less than twenty minutes. Details were not given, but the slogan "Make a catch before they hatch" suggested a type of nest trap (classified ad, *Western Producer*, 18 February 1937).

8 L.T. Trew, letter to the editor, *Western Producer*, 2 January 1930.

9 Cooke, "The Spread of the European Starling in North America (to 1928)," n.p. (online).

10 Nature Lovers page, *Western Producer*, 11 July 1929; Marion Nixon, "The Approach of Fall," *Regina Leader-Post*, 31 August 1954.

11 John Patrick Gillese, "The Starling Story," *Country Guide*, September 1952, 71–2; *Annual Report Alberta DA for Year Ended 31 March 1965*, 27. Alberta tested roost repellents in an effort to control them.

12 On the prairies the term "prairie chicken" usually referred to sharp-tailed grouse. Ornithologists classify the pinnated grouse as the true "prairie chicken" or what in Manitoba in the late 1800s was also called the "greater prairie chicken" (*Grouse and Introduced Game Birds of Saskatchewan*, 9–10). In Manitoba at the end of the 1890s sharp-tailed grouse were also called "old Manitoba prairie chicken." The more recently arrived pinnated grouse was called the "new Manitoba prairie hen" and the spruce grouse was called a "Canada grouse" (*Nor'West Farmer*, September 1898).

13 *Sharp-Tailed Grouse in Saskatchewan*, 5, 7, 20, 22–3.

14 McClung, *Clearing in the West*, 61; *Blairmore Enterprise* (editorial), 27 October 1910.

15 Rowan, *The Ten Year Cycle*, 11. Rowan thought the birds went to remote areas to die where their delicate bones disintegrated quickly.

16 *Annual Report Saskatchewan DA 1907*, 205; *Annual Report Saskatchewan DA 1911*, 171.

17 *Annual Report Manitoba DMNR for Year Ended 30 April 1933* (typescript), 129–30.

18 *Annual Report Saskatchewan DA 1915*, 229; *Annual Report Saskatchewan DA for Year Ended 20 April 1918*, 214. Poisoning as the explanation for sharp-tail depletion continued to find support at least until the 1920s (*Peace River Record*, 26 March 1926).

19 *Annual Report Saskatchewan DA 1909*, 181. One known example of migration at the high point of the cycle occurred in 1932 when enormous numbers of prairie sharp-tailed grouse moved as far as southern Ontario and

Quebec. They had vanished from there by 1934 (Rowan, *The Ten Year Cycle*, 11).

20 *Annual Report Alberta* DA *1915*, 167; *Annual Report Manitoba* DAI *1919*, 59, 61.

21 *Annual Report Saskatchewan* DA *for Year Ended 30 April 1922*, 340. Hungarian partridges (*Perdix perdix*) were also commonly called "gray" or "European partridge."

22 *Regina Leader-Post*, 16 September 1933.

23 *Cardston Globe* (editorial), 18 October 1919; Johnson to Chief Game Guardian, 26 January 1926, GA, M1327, file 29.

24 Hayden to Brownlee, 12 October 1927, and Lawton to Brownlee, 28 October 1927, PAA, Premiers Papers, file 59.

25 Hill to Costello, 19 December 1942, and Director to Costello, 21 December 1942, SAB, NR.1/2, file 117; various correspondence 1926–28, MA, P3611, item 9.

26 *Good Roads Magazine*, September 1923.

27 Minister to Patterson, 17 September 1934, MA, GR1600/G4537/15.1.1. The Saskatchewan Advisory Game Committee rejected the resolution, noting that scientific study showed that Hungarian partridges ate mice, weed seeds, and insects and by all accepted definitions were useful birds for both agriculture and sport hunting (Minutes, Advisory Game Committee, 27 June 1944, SAB, NR.1/2, file 115).

28 "Proceedings: Meeting of the Ecological Society of America at Toronto," 170–1; *Annual Report Saskatchewan* DA *for Year Ended 30 April 1922*, 341–2.

29 *Annual Report Saskatchewan* DA *for Year Ended 30 April 1925*, 287; *Annual Report Saskatchewan* DA *for Year Ended 30 April 1926*, 170.

30 Pratt to Potter, 20 July 1924, and Nixon to Potter, 17 March 1941, SAB, SHS50.

31 By 1955 Manitoba was using the ten-year cycle to set closed seasons on sharp-tails. Its planning was based on an assumption that years ending in a "2" were the population peaks (*Annual Report Manitoba* DMNR *for Year Ending 31 March 1955*, 64). On concern about decline of sharp-tails in the 1950s, see Doug Gilroy, Prairie Wildlife No. 77, *Western Producer*, 14 July 1955.

32 Rowan, *The Ten Year Cycle*, 13; *Annual Report Saskatchewan* DNR *for Year Ended 30 April 1946*, 72.

33 *Winnipeg Tribune*, 8 October 1953; *Sharp-Tailed Grouse in Saskatchewan*, 17–18; *Grouse and Introduced Game Birds of Saskatchewan*, 14.

34 *Regina Leader-Post*, 16 September 1933.

35 *Annual Report Manitoba* DMNR *for 15 July 1930 to April 30, 1931* (typescript), 72; Bossenmaier, *100-Plus Years*, 13; "Experimental Feed Plots," *The Keystone*, 1, no. 2 (September–October 1941), 10 (copy in MA, EC0016/GR432G72). The spread of wild turkeys in Manitoba and Alberta lies beyond the time frame of this book, but there are now small populations in both provinces.

36 *Annual Report Manitoba* DAI *for Year Ended 30 November 1913*, 38; *Annual Report Manitoba* DAI *for Year Ended 30 April 1926*, 5. The plantings in 1900 and 1925 were variously described as English, Chinese, and Mongolian pheasants. On introductions in the 1930s, see Bossenmaier, *100-Plus Years*, 13.

37 "Chinese Pheasants Settle Here," c. 1945, Manitoba Legislative Library, Clipping Collection M 8-9, p. 145; *Saskatchewan Fish and Game League Year Book 1943*, 7–9 (copy in SAB, R-E4531).

38 On the Moose Jaw project, see correspondence and clippings in SAB, NR1/1, G5000S and SAB, NR 1/2, P800C; and *Regina Leader-Post*, 10 and 24 June, 18 August, 8 September 1933, and 23 February 1934.

39 *Farm and Ranch Review*, October 1946; *Drumheller Mail*, 3 May and 2 August 1928.

40 *Farm and Ranch Review*, October 1946; List of Clubs, 16 July 1945, and Alberta Fish and Game Association to Huestis, 4 February 1945, GA, M1327, file 9; *Annual Report Alberta* DLM *for Year Ended 31 March 1946*, 65–6; *Fish, Fur and Feathers*, 224.

41 *Annual Report Alberta* DLF *for Year Ended 31 March 1950*, 75; *Annual Report Alberta* DLF *for Year Ended 31 March 1963*, 76; *Annual Report Alberta* DLM *for Year Ended 31 March 1948*, 76.

42 *Annual Report Manitoba* DMNR *for Year Ended 31 March 1950*, 60; *Annual Report Manitoba* DMNR *for Year Ended 31 March 1951*, 63; *Annual Report Manitoba* DMNR *for Year Ended 31 March 1956*, 70.

43 "Chukar Partridge," http://bna.birds.cornell.edu/bna/species/258/articles/introduction. The chukar partridge is the national bird of Pakistan.

44 Walker to Soper, 30 March 1939, LAC, RG109 D 2, vol. 5, file "W", Note to file, n.d. (1939), SAB, NR 1/2, file 112.

45 Dickson to Phelps, 28 June 1945, SAB, NR 1/2, file 115; "Notes on Meeting of Chukar Farm Advisory Committee," 18 March 1941, Hill to Forsyth, 2 May 1942, and Mosses to Forsyth, 16 August 1941, SAB, NR 1/2, file 112.

46 Various correspondence, chukar 1941–44, SAB, NR1/2, file 112; *Grouse and Introduced Game Birds of Saskatchewan*, 13.

47 Alberta unsuccessfully released 2,400 birds (raised at Brooks) in 1953 and 1954 (*Annual Report Alberta* DLF *for Year Ended 31 March 1961*, 91). On Manitoba, see, for example, *Annual Report Manitoba* DMNR *for Year Ended 30 April 1943* (typescript), 43.

48 *Annual Report Alberta* DLM *for Year Ended 31 March 1944*, 73.

49 "Experimental Feed Plots," *The Keystone* 1, no. 2 (September–October 1941), 10 (copy in MA, EC0016/GR431G72).

50 *Annual Report Manitoba* DMNR *for Year Ending 30 April 1946* (typescript), 57; "Keystone Conservation," Supplement No. 8, May 1947 (copy in MA, GR16000/G4527/36.2.1); *Annual Report Manitoba* DMNR *for Year Ended 31*

March 1949, 57. On the diet of upland game birds, see *Sharp-Tailed Grouse in Saskatchewan*, 8.

51 Moving native fish from one place to another within the region to supplement existing species has since been recognized as having sometimes introduced new genetic characteristics into existing resident populations.

52 *Regina Leader*, 12 March 1889; Barbour, "A Brief History of the Manitoba Fisheries," n.p. (online); *Annual Report [Canada] Department of Fisheries for the Fiscal Year Ended 30 June 1892*, Appendix 5, 188.

53 *Annual Report [Canada] Department of Fisheries 1887*, 307; *Annual Report [Canada] Department of Fisheries 1888*, Appendix 7, 229–30, and Appendix 5.

54 *Annual Report [Canada] Department of Fisheries 1888*, 219; *Annual Report [Canada] Department of Marine and Fisheries 1906*, xlvii. Fisheries authorities reviewed applications by residents for stocking lakes, and it was undertaken if the request was considered meritorious because of local need and the suitability of the lake.

55 *Red Deer Advocate*, 25 September 1908 and 3 April 1925; *Calgary Herald* (editorial), 3 August 1926.

56 *Western Motordom*, June 1930; *Drumheller Mail*, 5 July 1928; *Annual Report Alberta DLM for Year Ended 31 March 1934*, 97.

57 *Annual Report [Parks] DI 1899*, 6; *Annual Report [Parks] DI 1902–03*, 7.

58 *Annual Report [Parks] DI 1901–02*, 5.

59 Hayden to Brownlee, 21 July 1930, PAA, Premiers Papers, file 59; *Natural Resources of the Prairie Provinces*, 44.

60 *Annual Report Manitoba DMNR for Year Ended 30 April 1932* (typescript), 188.

61 *Annual Report Manitoba DMNR for Year Ending 30 April 1943* (typescript), 69–70; McDiarmid to Bentham, 11 July 1940, MA, GR1600/4519/36.3; *Annual Report Manitoba DMNR for Year Ended 31 March 1947*, 149–50; Stewart and Watkinson, *The Freshwater Fishes of Manitoba*, 19.

62 *Annual Report Saskatchewan DNR 1932–33* (typescript), 3–4; clippings July 1932 and 2 June 1933, SAB, NR1/2, file 800C; *Annual Report Saskatchewan DNR 1933–34* (typescript), 28.

63 Saskatchewan Fish and Game League, Annual Convention, 23–24 June 1941, SAB, NR.1/2, file 115; *Annual Report Saskatchewan DNR for Year Ended 31 March 1949*, 107–8.

64 *Annual Report Saskatchewan DNR for Year Ended 31 March 1949*, 111; *Annual Report Saskatchewan DNR for Year Ended 31 March 1953*, 82.

65 *Annual Report Alberta DLM for Year Ended 31 March 1932*, 63.

66 *Annual Report Alberta DLF for Year Ended 31 March 1950*, 100; *Annual Report Alberta DLF for Year Ended 31 March 1962*, 80; *Annual Report Alberta DLF for Year Ended 31 March 1957*, 49.

67 Memo for Premier Brownlee, 29 April 1931, and Hayden to Brownlee, 21 July 1930, PAA, Premiers Papers, file 59.

68 *Annual Report Alberta* DLM *for Year Ended 31 March 1942*, 61; Kooyman to Malaher, 23 February 1953 (conference reports), MA, GR1600/G4538/15.5.1.A; *Winnipeg Tribune*, 22 September 1958.

69 *Annual Report Alberta* DLF *for Year Ended 31 March 1957*, 49; *Winnipeg Tribune*, 9 March and 21 April 1954; *Annual Report Manitoba* DMNR *for Year Ended 31 March 1964*, 13. By the late 1950s studies showed that burbot were not "serious predators on valuable kinds of commercial fish," although "they may compete for food with desirable fish, and they do eat whitefish and Lake Trout eggs at certain times of the year" (*Annual Report Manitoba* DMNR *for Year Ended 31 March 1960*, 91).

70 *Annual Report Saskatchewan* DNR *for Year Ended 31 March 1959*, p. F19.

71 *Annual Report Saskatchewan* DNR *1936–37*, 52; *Annual Report Alberta* DLM *for Year Ended 31 March 1934*, 91.

72 *Annual Report Saskatchewan* DNR *for Year Ended 31 March 1960*, 104; *Annual Report Saskatchewan* DNR *for Year Ended 31 March 1964*, 15.

73 "Can We Hold the Line on Carp Invasion?," *Saskatchewan Afield* 1, no. 1 (1958), n.p. (copy in SAB, R-E4531); *Annual Report Saskatchewan* DNR *for Year Ended 31 March 1959*, p. F17.

74 *Annual Report Manitoba* DMNR *for Year Ended 31 March 1947*, 151. It was reported as a matter of interest that pumpkinseed fish (*Lepimus gibbosus*), "a new species of fish for Manitoba," were among the fish killed when the lake was poisoned.

75 *Annual Report Manitoba* DMNR *for Year Ended 31 March 1951*, 87. The fish killed in the 1951 experiment were yellow perch, rock bass, white sucker, tullibee, Johnny Darters, pearl dace, flathead minnow, and spot-tail minnow. Insects and crustaceans killed were not recorded.

76 *Annual Report Alberta* DLF *for Year Ended 31 March 1960*, 77; *Annual Report Saskatchewan* DNR *for Year Ended 31 March 1961*, 69; *Annual Report Saskatchewan* DNR *for Year Ended March 31, 1963*, 71; *Annual Report Saskatchewan* DNR *for Year Ended March 31, 1962*, 70.

77 *Edmonton Journal*, 26 November 2000.

78 Kooyman to Malaher, 23 February 1953 (conference reports), MA, GR1600/G4538/15.5.1.A.

79 Kerry Wood, "A Creel Full of Fish Facts," *Farm and Ranch Review*, August 1949; "The Pickerel Hole," *Country Guide*, May 1965.

80 "Sport Fish of Alberta," at http://www.albertasportfishing.com/sport_fish.php; Stewart and Watkinson, *The Freshwater Fishes of Manitoba*, 19.

81 *Calgary Herald*, 6 October 2002 and 29 June 2009. See also Mayhood, *Cumulative Human Impacts on Native Trout*.

82 *Annual Report Manitoba* DAI *for Year Ended 30 April 1926*, 4–5; *Winnipeg Tribune*, 30 May 1959; Goatcher, *Manitoba Legislation Affecting Wildlife*, 63, 68.

Concern about nutria grew from the case of Louisiana, where they had become invasive.

CHAPTER EIGHT

1 Seton, "Raggylug," in *Wild Animals I Have Known*, 81; Brody, *The Other Side of Eden*, esp. chaps 3, 4, and 6; Jacques, "The Homesteaders," in *Beside Still Waters*, 64.

2 "Ex Farmer" (letter to the editor), *Regina Leader*, 23 August 1894.

3 *Annual Report Alberta DA 1926*, 12. On similar community rabbit drives in southern Saskatchewan, see *Regina Leader-Post*, 14 February and 12 March 1934.

4 *Annual Report Manitoba DAI 1943–44*, 5; Renouf to Garson, 30 July 1943, MA, GR/EC0016/43/G57/75. Manitoba's 1944 Predator Control Act created a single framework for predator control throughout the province and expanded the range of animals on which bounties could be paid, including red foxes (Goatcher, *Manitoba Legislation Affecting Wildlife*, 93–6).

5 *Annual Report Alberta DLM for Year Ended 31 March 1946*, 71; *Annual Report Saskatchewan DNR for Year Ended 31 March 1949*, 87; *Annual Report Saskatchewan DNR for Year Ended 31 March 1950*, 99.

6 *Annual Report Saskatchewan DNR for Year Ended 31 March 1953*, 126.

7 Premier's Office to Douglas, 21 April 1937 (attachments), MA, GR/EC0016/1661/649; *Winnipeg Tribune*, 24 September 1963; *Farm and Ranch Review* (editorial), May 1956.

8 *Annual Report Saskatchewan DA for Year Ended 30 April 1919*, 67.

9 Clipping, 13 September 1932, SAB, NR 1/2, file P800C; *The U.F.A.*, 15 January 1932; *Peace River Record* (editorial), 7 August 1936; Smith to Taggart, 6 February 1935, SAB, NR1/1, G50000G(1).

10 Chalmers to Cowan, 12 October 1949, MA, GR1600/G4529/11.2.1; "Resolutions of the UFA Convention 1946," PAA, Premiers Papers, file 1513; *Annual Report Alberta DLM for the Year Ended 31 March 1948*, 774–5; "Notes for Discussion in Connection with the Dominion-Provincial Wildlife Conference, Ottawa, 27 and 28 February 1947," MA, GR1600/G4538/15.5.1A.

11 Norman Flanders, "Ducks Are a Farm Pest: So Let's Treat Them as Pests!," *Farm and Ranch Review*, October 1952.

12 Winters to McDiarmid, 26 June 1950, and Malaher to Stephens, 17 August 1950, MA, GR1600/G4529/11.2.1; Ralph Hedlin, "Crop Damage by Ducks," *Country Guide*, September 1953.

13 McDiarmid to Hawkins, 16 October 1946, and Chalmers to Stephens, 28 September 1949, MA, GR1600/G4529/11.2.1; *Winnipeg Tribune* (editorial), 12 December 1952.

14 *Annual Report Saskatchewan DNR for Year Ended 31 March 1953*, 93; *Annual Report Saskatchewan DNR for Year Ended 31 March 1960*, 85; *Annual Report Alberta DLF for Year Ended 31 March 1962*, 74.

15 *Annual Report Manitoba* DMNR *for Year Ended 31 March 1961*, 105.

16 Minutes of Council, Thursday July 1839 and Tuesday 22 January 1850, MA, MG2 B1, file "Minutes 1832–1861"; Minutes of Council, 29 December 1868, MA, MG2 B1, file "Minutes 1861–1869."

17 "Report of William Pearce," *Annual Report* DI *1886*, 20–1; "Report of William Pearce," *Annual Report* DI *1888*, 10; "Report of William Pearce," *Annual Report* DI *1887*, 14.

18 *Macleod Gazette* (editorial), 3 September 1891. See also *Macleod Gazette*, 26 January and 9 March 1894 and 11 May 1900; and *Regina Leader*, 12 July 1894 (reprinted in the *Lethbridge News*).

19 *Annual Report* NWT DA *1901*, 86; *Nor'West Farmer*, 5 February 1900 and 6 May 1901. On bounties in North America, see Zmyj, "A Fight to the Finish," 18–19.

20 *Annual Report Commissioner of the* NWMP *1893*, 108.

21 "Report of William Pearce," *Annual Report* DI *1896*, 31; *Annual Report* NWT DA *1898*, 85.

22 *Annual Report* NWT DA *1898*, 85; W.H. McKay, "Recollections as a Wolf Hunter," *Calgary Herald*, 15 and 22 January 1944.

23 *Annual Report* NWT DA *1899*, 78; *Macleod Gazette*, 26 May 1899.

24 Elofson, *Somebody Else's Money*, 64, 112–13, 161–2; *Annual Report* NWT DA *1900*, 82. Statistics are calculated from "Statement of Wolves Destroyed," *Annual Report* NWT DA *1904*, 69.

25 W.H. McKay, "Recollections as a Wolf Hunter," *Calgary Herald*, 15 and 22 January 1944; Wilson to Hamilton, 1 July 1943, SAB, SHS231, Clipping File, "Wolves."

26 On wolves in the Interlake, see *Winnipeg Tribune*, 29 March 1957; and in northern Alberta, *Peace River Record*, 19 February 1943.

27 *Regina Leader* (editorial), 27 October 1885; Mitchell, *In Western Canada before the War*, 23; *Annual Report* NWT DA *1901*, 86.

28 *Regina Leader*, 15 April 1890; *Lethbridge News*, 9 November 1892; *Annual Report* NWT DA *1901*, 108; *Annual Report Saskatchewan* DA *1905*, 62; *Annual Report Alberta* DA *1916*, 132.

29 *Peace River Record* (editorial), 13 October 1916.

30 *Nor'West Farmer*, 5 February 1900; *Annual Report Saskatchewan* DA *for Year Ended 30 April 1918*, 52.

31 Kitto, *The Peace River Country Canada*, 93; *Annual Report Alberta* DLM *for Year Ended 31 March 1949*, 60; Criddle, *The Habits and Economic Importance of Wolves*, 14, 16.

32 *Annual Report Saskatchewan* DA *for Year Ended 30 April 1921*, 321; *Annual Report Saskatchewan* DA *for Year Ended 30 April 1918*, 220; *Annual Report Saskatchewan* DA *1914*, 87. See also Loo, *States of Nature*, 166–8.

33 See, for example, Glenbow Archives photographs, "Coyotes" including NA 466-33 (1892) and NA611-1 (1912).

34 "Greyhounds and Coyotes," *Hunting and Fishing in Canada*, August 1938 (copy in SAB, RE2094); Walter Schedel (Pasweqin, SK), Nature Lovers page, *Western Producer*, 4 April 1929.

35 *Edmonton Journal*, 2 May 1929; *Annual Report Alberta DLM for Year Ended 31 March 1942*, 63.

36 *Annual Report Saskatchewan DNR for Year Ended 31 March 1948*, 79; *Annual Report Alberta DLF for Year Ended 31 March 1950*, 76–7; *Regina Leader-Post*, 15 December 1950.

37 Kerry Wood, "Organized Wild-Life Slaughter Disguised as 'Coyote Hunts,'" *Farm and Ranch Review*, January 1950; *Hints on Community Coyote Hunting*, n.p.

38 *Regina Leader-Post*, 8 November 1949.

39 *Regina Leader-Post*, 15 December 1950; *Winnipeg Tribune*, 19 and 22 November, 1954.

40 Kerry Wood, "Organized Wild-Life Slaughter Disguised as 'Coyote Hunts,'" *Farm and Ranch Review*, January 1950; Alfred Goff, letter to the editor, June 1950, and Edith Bowman, letter to the editor, February 1950, *Farm and Ranch Review*.

41 *Annual Report Saskatchewan DNR for Year Ended 31 March 1949*, 94; *Regina Leader-Post*, 15 December 1950.

42 *Annual Report Saskatchewan DNR for Year Ended 31 March 1949*, 94; *Annual Report Saskatchewan DNR for Year Ended 31 March 1950*, 90.

43 Province of Alberta, "Predator Control Problem," n.d. [1949], PAA, 72.302, file 293.

44 *Annual Report Saskatchewan DNR for the Year Ended 31 March 1950*, 98–9.

45 *Regina Leader-Post*, 12 September 1950. Part of the problem with estimating kills was that 1080 was not fast acting; after ingesting the poison, the coyote felt ill and went into hiding, where it died.

46 *Progress: A Survey of Saskatchewan Government Activity*, 13; *Regina Leader-Post*, 7 December 1950; *Annual Report Saskatchewan DNR for Year Ended 31 March 1952*, 105–6; *Annual Report Saskatchewan DNR for Year Ended 31 March 1956*, 27.

47 *Annual Report Manitoba DMNR for Year Ended 31 March 1950*, 60; *Winnipeg Tribune*, 22 December 1959; *Farm and Ranch Review*, October 1950.

48 C.R. Bowers, "Guilty?," *Country Guide*, July 1952; *Annual Report Alberta DA 1951*, 35; *Annual Report Alberta DA 1961*, 29.

49 *Annual Report Alberta DLF for Year Ended 31 March 1955*, 51.

50 "An Observer" (letter to the editor), *Western Producer*, 8 May 1952; *Regina Leader-Post*, 14 February 1951; *Western Producer*, 7 June 1951 (letter to the editor); *Farm and Ranch Review*, March 1958 (photo caption).

51 *Western Producer*, 8 May 1951 (letter to the editor).

52 Coyote control in the Red Deer district had successfully used a combina-

tion of aircraft, hounds, and hunters (*Regina Leader-Post*, 7 December 1950).

53 *Winnipeg Tribune*, 28 April 1952. On the 1942–3 mouse irruption, see Kerry Wood, "Meet Joe and Susie – Menaces to Unthreshed Grain," *Farm and Ranch Review*, January 1951.

54 R. Bowers, "Guilty?," *Country Guide*, July 1952; *Winnipeg Tribune*, 9 February 1952.

55 Kerry Wood, "Introducing a Sheepman Who Doesn't Hate Coyotes!," *Farm and Ranch Review*, September 1952; Kerry Wood, "The Yapping Yodler," *Country Guide*, February 1945; *Western Producer*, 8 May 1952 (letter to the editor).

56 *Winnipeg Tribune*, 9 January 1960; *Annual Report Manitoba DMNR for Year Ended 31 March 1959*, 73–4.

57 *Sharp-Tailed Grouse in Saskatchewan* [c. 1960], 9; Doug Gilroy, Prairie Wildlife No. 323, *Western Producer*, 28 April 1960.

58 Robert Bell, "Our Native Mammals of Economic Value" (Sommerville Lecture, 1895), LAC, MG29 B15, vol. 3, file 3.13.

59 For contemporary knowledge about them, see Criddle, "Some Canadian Rodents Injurious to Agriculture," 110–13; and V.W. Jackson, "Our Four Types of Gophers," *Grain Growers' Guide*, 17 June 1925.

60 *Annual Report Manitoba DA 1883*, 114; *Annual Report [Smith's Report] DI 1889*, 14–15; *Appendix to the Report of the Minister of Agriculture: Experimental Farms Reports 1889*, 148.

61 V.W. Jackson, "Our Four Types of Gophers," *Grain Growers' Guide*, 17 June 1925; *Annual Report NWT DA 1901*, 66.

62 Control of Columbia ground squirrels was discussed in almost every report of the Alberta Field Crops Commissioner from 1941 until 1965. Pocket gophers were also a concern in agricultural districts in Alberta during this period.

63 Seton, *Life-Histories of Northern Animals*, vol. 1, 380. On this and a corresponding decline in striped gophers, see Seton Journal, 7 June 1892, AMNA, E-3 vol. 4.

64 Soper, *Mammals of Alberta*, 134–6; *Managing Richardson's Ground Squirrels*, 1–4.

65 *Cardston Globe*, 23 March 1916. The female bears only one litter per year (http://research.uleth.ca/rgs/reproduction.cfm).

66 *Edmonton Bulletin*, 29 April 1895; *Nor'West Farmer*, 20 August 1900 (quoting *Sintaluta Times*); *Annual Report NWT DA 1901*, 66.

67 *Annual Report Alberta DA 1953*, 29. This method seems to have been popular for a time and was endorsed by one correspondent in Doug Gilroy's column (Doug Gilroy, Prairie Wildlife No. 117, *Western Producer*, 17 May 1956).

68 *Annual Report Saskatchewan DA for Year Ended 30 April 1922*, 234. For examples

of recipes, see *Farm and Ranch Review*, June 1907; and *Grain Growers' Guide*, 1 March 1922. Mrs James Tait's 1934 recipes are in GA, M106, file 6.

69 *Grain Growers' Guide*, 23 April 1913, 16 February 1921, and 15 March 1927.

70 *Annual Report Saskatchewan DA 1912*, 307; *Annual Report Saskatchewan DA 1915*, 220; "Poisoning Gophers," *Farm and Ranch Review*, June 1907; *Annual Report NWT DA 1901*, 66.

71 *Report and Minutes of Evidence on Existing Natural Food Products [1887]*, 130. On demands for help, see, for example, Jacobs to Department of Agriculture, 16 March 1915, LAC, RG17, vol. 1236, file 241600. Volume 576 of RG17 (Canada Department of Agriculture) consists entirely of correspondence regarding gopher control.

72 Secretary to Head Office, 23 July 1890, LAC, RG15-D-11-1; *Annual Report NWT DA 1898*, 37; *Annual Report Saskatchewan DA 1907*, 213.

73 *Regina Leader*, 3 January 1894 (letter to the editor); *Annual Report NWT DA 1898*, 86.

74 Seton Journal, 7 June 1892, AMNA, E-3, vol. 4; Crawford to Perley, 4 May 1888, LAC, RG17, vol. 576, file 64993; Perley to Macdonald, 30 April 1888, pp. 228016–18, LAC, MG26-A, REEL C-1791.

75 *Regina Leader*, 9 August 1894 (letter to the editor).

76 Goldsborough, "Man's War on Wildlife," 3, 5; "Memo [Jackson] Re Gopher Campaign 1919," 17 June 1919, MA, MG14/B45/3105; Hewitt, "The Conservation of Wild Life in Canada in 1917," 137.

77 Isern, "Gopher Tales," 193; *Annual Report Saskatchewan DA for Year Ended 30 April 1918*, 104; *Annual Report Saskatchewan DA for Year Ended 30 April 1920*, 56–7; *Annual Report Saskatchewan DA for Year Ended 30 April 1922*, 79, 234.

78 *Annual Report Alberta DA 1929*, 35; Tainish to Department of Agriculture, 12 March 1919, LAC, RG17, vol. 1336, file 266460; *House of Commons Debates*, 18 June 1920, col. 3793.

79 On contemporary events, see, for example, *Edmonton Journal*, 17 April 1990. On the 2002 Saskatoon "gopher derby," see *Calgary Herald*, 25 June 2002.

80 Isern, "Gopher Tales," 191, 197–8.

81 *Regina Leader-Post*, 6 March 1931; *Western Producer*, 13 March 1941.

82 Ralph Hedlin, "The Little Hunter," *Country Guide*, April 1957.

83 Wanda, Nature Lovers page, *Western Producer*, 13 June 1940. A parallel scene involving a weasel was described by another correspondent in 1928 (Raquel, Nature Lovers page, *Western Producer*, 8 March 1928).

84 Atkinson, *Manitoba Birds of Prey*, 3; *Annual Report Saskatchewan DA 1919–20*, 75, 78.

85 Farley, *Birds of the Battle River Region*, 25–6.

86 Walter Schowalter (Hayter, AB), Nature Lovers page, *Western Producer*, 28 July 1934.

87 Kerry Wood, "Most Live Hawks Are Good Hawks," *Country Guide*, March

1944, and editorial note, May 1944. See also Doug Gilroy, Prairie Wildlife No. 117, *Western Producer*, 17 May 1956. The same argument was occasionally also made with respect to the value of snakes for protecting crops from rodents. Horace Mitchell of the Saskatchewan museum, for example, argued in 1915 that snakes in Saskatchewan were "not only harmless but beneficial and should not be killed on sight as is everywhere and always customary" (*Annual Report Saskatchewan DA 1915*, 252).

88 Criddle, "Some Canadian Rodents Injurious to Agriculture," 113–15.

89 *Annual Report Alberta DLF for Year Ended 31 March 1951*, 48. On the rat-control program in Alberta, see McTavish and Zheng, "Rats in Alberta."

90 Kerry Wood, "The Predator Police Force," *Farm and Ranch Review*, March 1948; Kerry Wood, "Stacker of Mice," *Farm and Ranch Review*, 1961; "Badger Baiting Isn't Legal but We Still Badger Badgers," *Farm and Ranch Review*, March 1954.

91 *Annual Report Alberta DLF for Year Ended 31 March 1960*, 17, 92; *Annual Report Alberta DA for Year Ended March 31, 1965*, 27; Blackbirds Come for Dinner," *Farm and Ranch Review*, October 1962.

CHAPTER NINE

1 Spry, "The Transition from a Nomadic to a Settled Economy," 194.

2 Spry, "The Tragedy of the Loss of the Common in Western Canada," 204–5; Spry, "Aboriginal Resource Use in the Late Nineteenth Century," 85–7.

3 Friesen, "Grant Me Wherewith to Make My Living," note 5, 153, 141 55. The text of treaties 1–7 can be found in Morris, *The Treaties of Canada with the Indians*, Appendix, 299–375.

4 See Milloy, "'Our Country,'" 54–5; Ross and Sharit, "Forest Management in Alberta and Rights to Hunt, Trap and Fish," 652–4; Wetherell and Kmet, *Alberta's North*, 63–72; Tough, *"As Their Natural Resources Fail,"* 79–104.

5 Friesen, "Grant Me Wherewith to Make My Living," 145–7; Binnema and Niemi, "'Let the Line Be Drawn Now,'" 727.

6 Friesen, "Grant Me Wherewith to Make My Living," 148–50; *Annual Report DIA 1884*, liv–lv.

7 *Annual Report DIA 1891*, xix; *Annual Report [Canada] Department of Fisheries 1913–14* (Appendix 7), 227. On commercial fishing as wage employment, see *Annual Report [Canada] Department of Fisheries 1889* (Appendix 8), 238.

8 Friesen, "Grant Me Wherewith to Make My Living," 150; *Regina Leader*, 19 November 1889 and 14 October 1890.

9 *Annual Report Alberta DA 1908*, 110.

10 In Alberta, for example, anyone north of 55 degrees (about the latitude of the town of Athabasca) could kill game any time for food for themselves and their family. Travellers enjoyed the same right. The same rule applied in Manitoba north of latitude 54, about the latitude of The Pas (*Annual*

Report Alberta DA 1911, 141; Goatcher, *Manitoba Legislation Affecting Wildlife*, 41).

11 Bossenmaier, *100-Plus Years*, 5; Lawton to Deputy Superintendent General, 4 March 1919, LAC, RG10, vol. 6731, file 420-1, reel C-8093. In 1914 Alberta was granting trapping licences to First Nations people free of charge, and in 1919 Saskatchewan exempted them from having to have a trapping licence. These exemptions were not consistently granted and in some cases were withdrawn later. The same situation existed with respect to big game licences. In 1939 the only people in Alberta who could obtain a big game licence free of charge were First Nations – if they had the approval of the Indian agent (Wallace to House, 6 November 1939, PAA, 67.21, file 325).

12 Woods to Pedley, 8 January, and Pedley to Woods, 22 January 1908, PAA, 66.166, file 249; McLean to Lawton, 6 May 1912, and Scott to Lawton, 10 September 1914, LAC, RG10, vol. 6732, file 420-2A, reel C-8094.

13 Sennum to Oliver, 3 May 1909, and Secretary to Sennum, 31 May 1909, LAC, RG10, vol. 6732, file 420-2A, reel C-8094; *Annual Report DIA for Year Ended 31 March 1919*, 38.

14 Griesbach to Gilchrist, 10 September 1894, LAC, RG10, vol. 3581, file 878, pt A, reel C-10102; Bossenmaier, *100-Plus Years*, 8; *Annual Report DIA for Year Ended 31 March 1909*, xxvix.

15 *Annual Report DIA for Year Ended 31 March 1929*, 8–9.

16 Wetherell and Kmet, *Alberta's North*, 377; Scott circular letter, 30 May 1931, LAC, RG10, vol. 6731, file 420-1, reel C-08093.

17 Conn to Spargo, 11 January 1950, LAC, RG10, vol. 6734, file 420-2-1-1, reel C-8096.

18 Balazs, *A Short Analysis of the Transfer of Natural Resources to Alberta*, 20.

19 Hoadley to Murphy, 23 June 1932, and Spargo to Indian Affairs, 6 February 1935 (Resolution, 6 October 1933, attached), LAC, RG10, vol. 6733, file 420-2C, reel C-8094; Wetherell and Kmet, *Alberta's North*, 378–80.

20 Balazs, *Transfer of Natural Resources in 1930*, 111, 143; Donihee, *The Evolution of Wildlife Law in Canada*, 27. A similar approach in defining unoccupied land was seen in Manitoba (Goatcher, *Manitoba Legislation Affecting Wildlife*, 55).

21 *Globe and Mail*, 20 September 2003 and 28 April 2006.

22 *Annual Report DIA for Year Ended 31 March 1919*, 47–9; Tough, *"As Their Natural Resources Fail,"* 217–18, 295, 302.

23 *Annual Report DIA for Year Ended 31 March 1911*, 173; *Annual Report DIA for Year Ended 31 March 1915*, 73, 84; Monthly entries, November 1930–March 1931, Agent's Diary and Reports, Lesser Slave Lake Agency, LAC, RG10, vol. 11949, box 5, Shannon file no. 4, pt 2.

24 Grew to Allan, 11 March 1943, LAC, RG10, vol. 6733, file 420-2-2 2, reel 8095.

25 *Annual Report Manitoba* DMNR *for Year Ended 31 March 1957*, 79; *Annual Report Saskatchewan* DNR *for Year Ended 31 March 1963*, 118.

26 For a later expression of this view, see *Annual Report Saskatchewan* DA *[for] 1910*, 145.

27 Scott, "Relation of Indians to Wild Life Conservation," 27.

28 *Calgary Herald*, 11 April 1936.

29 *Annual Report Saskatchewan* DA *for Year Ended 30 April 1922*, 357.

30 *Annual Report* DIA *for Year Ended 31 March 1909*, xxviii–xxvix.

31 *Debates of the Senate*, 18 May 1887, 78; *Report and Minutes of Evidence on Existing Natural Food Products [1887]*, 18.

32 *Macleod Gazette*, 24 May 1887.

33 Ducharme, "Report on the Indian and Half Breed Question," 10 August 1939, SAB, R-126.2, no. 7; Gottesman, "Native Hunting and the Migratory Birds Convention Act," 85; Brightman, *Grateful Prey*, 117, 262.

34 Carter, *Lost Harvests*, 17–18.

35 *Annual Report* DIA *1888*, lxvii; Binnema and Niemi, "'Let the Line Be Drawn Now,'" 738.

36 *Macleod Gazette*, 11 October 1895; *Annual Report* NWT DA *1902*, 134; *Annual Report Alberta* DA *1908*, 109; *Annual Report* DIA *1886*, lii.

37 Colpitts, *Game in the Garden*, 88. On Aboriginal hunting practices and conservation, see also Binnema and Niemi, "'Let the Line Be Drawn Now,'" 730–2.

38 *Annual Report Saskatchewan* DA *1914*, 213; [T.C. Main], "Brief on Conditions in Northland" [1939], MA, EC0016/GR43/G64; *Winnipeg Tribune*, 29 June 1964.

39 *Annual Report Saskatchewan* DA *[for] 1911*, 175; *Annual Report Saskatchewan* DA *for Year Ended 30 April 1918*, 238.

40 *Annual Report Saskatchewan* DA *1910*, 145; *Annual Report Saskatchewan* DA *1915*, 220.

41 *Annual Report Saskatchewan* DA *for Year Ended 30 April 1921*, 112.

42 Brightman, *Grateful Prey*, 281–3; Colpitts, *Game in the Garden*, 101.

43 *Annual Report* DIA *for Year Ended 31 March 1929*, 8.

44 Spargo to Superintendent General of Indian Affairs, 6 February 1935, and Christianson to Secretary, 5 March 1935, LAC, RG10, vol. 6733, file 420-2C, reel C-8094.

45 "Report on Game Conditions – Manitoba," 20 November 1930, MA, GR1600/G4537/15.1.2.A; T.R.L. MacInness, "What Canada Is Doing for the Hunting Indians" (talk at North American Wild Life Conference, 3–7 February 1936), LAC, RG10, vol. 6731, file 420-1, reel C-8093.

46 See, for example, *Annual Report Alberta* DA *1929*, 32.

47 McLean to Schultz, 16 June 1893, MA, NG12/E1/6259–63; *Annual Report Manitoba* DMNR *for 15 July 1930 to 30 April 1931*, 69 (typescript).

48 *Annual Report DIA for Year Ended 31 March 1929*, 8. As the report noted, in "settled and organised localities," First Nations were discouraged "from dependence on the chase."

49 Balazs, *Transfer of Natural Resources in 1930*, 157.

50 Wetherell and Kmet, *Alberta's North*, 374–7.

51 Annual leases were commonest and provided tighter control by game officials. Saskatchewan issued ten-year leases in the late 1930s but replaced them with annual ones out of a fear that trappers would conserve animals until the end of the lease and then strip the district ("The History of the Northern Wildlife Conservation Programme," 18–19, SAB, R-E4287).

52 Shirley to Hill, 17 August 1942, SAB, NR 1/2, file 120; *Annual Report Manitoba DMNR for Year Ended 30 April 1941* (typescript), 50. See also *Annual Report Alberta DLM for Year Ended 31 March 1942*, 64.

53 "Leased Trapping Conserves Animal Families," *Calgary Herald*, c. 1935, GA, A.S243.

54 Memorandum for the Honourable Superintendent General, 23 February 1934, and Memorandum re Status of Hunting Indians, 5 March 1937, LAC, RG10, vol. 6731, file 420-1, reel C-8093.

55 Wetherell and Kmet, *Alberta's North*, 383; "Leased Trapping Conserves Animal Families," *Calgary Herald*, c. 1935, GA, A.S243; Malaher to Stevens, Re: *Keystone* article, 20 July 1949, MA, GR1600/G4527/36.2.1.

56 Wetherell and Kmet, *Alberta's North*, 384–5; "The History of the Northern Wildlife Conservation Programme," Fur Programme Report, January 1958, 18–19, SAB, R-E4287.

57 Ray, *I Have Lived Here Since the World Began*, 281.

58 Malaher to Stevens, Re: *Keystone* article, 20 July 1949, MA, GR1600/G4527/36.2.1; *Annual Report Manitoba DMNR for the Year Ended 30 April 1946* (typescript), 86.

59 Crerar to Tanner, 19 February 1939, LAC, RG10, vol. 6733, file 420-2-1, reel C-8095; Tunstead to Deake, 20 August 1943, SAB, NR 1/2, file 129.

60 Fender, "Public versus Private Ownership: Saskatchewan Fur Trapping," 237–40; *Progress: A Survey of Saskatchewan Government Activity*, 13, 43–4; "The History of the Northern Wildlife Conservation Programme," Fur Programme Report, January 1958, SAB, R-E4287; Quiring, *CCF Colonialism*, 104–5; *Annual Report Saskatchewan DNR for Year Ended 31 March 1948*, 72.

61 David Quiring argues that the CCF's northern development approaches were based on a socialist agenda to increase provincial wealth through centralized state exploitation of northern resources (including fur) and to assimilate and modernize Aboriginal northerners. The party's language sometimes suggested this, and indeed some of its policies were poorly thought out. But its policies were not as unique in practice as Quiring suggests. At least in terms of trapping, he would need to make

similar criticism of Manitoba's fur programs (Quiring, *CCF Colonialism in Northern Saskatchewan*, 13, 99–118).

62 See, for example, *Annual Report Alberta Fur Supervisor, 10 January 1949*, LAC, RG10, vol. 6734, file 420-2-1-3, reel 8096; Balazs, *Transfer of Natural Resources in 1930*, 21.

63 Wetherell and Kmet, *Alberta's North*, 382–7; Balazs, *Transfer of Natural Resources in 1930*, 1, 53, 134.

64 Balazs, *Transfer of Natural Resources in 1930*, 37; Clancy, "State Policy and the Native Trapper," 209.

CHAPTER TEN

1 Cohen to Attwood, 11 May 1931, MA, GR1600/G4531/14.3; *Natural Resources of the Prairie Provinces: A Brief Compilation* (1923), 55.

2 *Winnipeg Tribune*, 18 February 1950; National Parks Commissioner to Akseh Rauanheime, 11 August 1923, LAC, RG84-A-2-a, vol. 180, file U279; *Annual Report Alberta DLF for Year Ended 31 March 1959*, 67. In contrast, Ontario and Quebec trappers earned about $3.9 million on just over 1 million pelts, including a higher proportion of high-value furs such as beaver.

3 "Natural Resources and the Department of Natural Resources," n.d. (c. 1931), SAB, Pamphlet File, "Natural Resources, Saskatchewan."

4 Dick, review, 86–7; *Annual Report Manitoba DMNR for Year Ended 30 April 1937* (typescript), 66; *Annual Report Manitoba DMNR for Year Ended 30 April 1930, 1946* (typescript), 60.

5 Ray, *The Canadian Fur Trade in the Industrial Age*, 51, 61.

6 "A Letter about Furs," *Grain Growers' Guide*, 15 January 1927.

7 "Production of Raw Furs from the Wild," 22 October 1940, MA, GR1600/G4537/14.6.1; Letter to the editor of the *Brampton Times*, William Taylor, Woodlands [Manitoba], 10 February 1875 (typescript), 3, MA, MG8/B68.

8 *Annual Report Saskatchewan DA for Year Ended 30 April 1925*, 296; *Annual Report Manitoba DMNR for 11 Months Ended 31 March 1947*, 106; *Annual Report Manitoba DMNR for Year Ended 31 March 1958*, 78; Nadeau, *Fur Nation*, chaps 1, 2, and 4.

9 *Saturday Night* 36, no. 28 (1921): 36. Robert Brightman notes the irony of Aboriginal trappers having to confront criticism by Euro-Canadians about the ethics of trapping. As he phrases it, "many [Rock Cree] trappers are not strangers to moral paradox in the hunt; they participate in religions that both celebrate it and disguise it" (Brightman, *Grateful Prey*, 212).

10 *Regina Leader-Post*, 3 September 1934; Wetherell and Kmet, *Alberta's North*, 374.

11 Goatcher, *Manitoba Legislation Affecting Wildlife*, 52–3; "Production of Raw Furs from the Wild," 22 October 1940, MA, GR1600/G4537/14.6.1.

12 Bert Woodcock, "Steel Snare," *Country Guide*, November 1952.
13 Congdon, "Fur Bearing Animals in Canada," 109; Clipping, *Prince Albert Herald*, 5 December 1934 (copy in SAB, NR 1/2, file P800C(3)).
14 Pocahontas (Bratton, SK), Nature Lovers page, *Western Producer*, 24 April 1930; *Coleman Journal* (editorial), 19 September 1929.
15 Dick, review, 86–7.
16 Wanda, Nature Lovers page, 13 June, and Dot Wigglespear, Nature Lovers page, 18 July 1940, *Western Producer*.
17 The Canadian Association for Humane Trapping only received a federal charter as a non-profit charitable organization in 1954 (http://www.caht.ca/about/).
18 Allan to Fitzpatrick, 17 March 1944, LAC, RG10, vol. 6731, file 420-1-1-1 1, reel C-8093; *Western Producer* (letter to editor), 20 April 1950.
19 Bevington, "Frank Ralph Conibear," 386–7; Doug Gilmour, Prairie Wildlife No. 233, *Western Producer*, 31 July 1958.
20 Greenlay to Stevenson, 12 July 1954, MA, GR1600/G4538/15.6.1; Bossenmaier, *100-Plus Years*, 14.

21 *Saskatchewan Trappers Guide*, 11–12.
22 Patrol Report, Fort Chipewyan NWMP Detachment, 1 April 1913, LAC, RG18, vol. 440, file 167.
23 Conibear, "The Beaver's Lodge," 37.
24 Smith, *From the Land of Shadows*, 104–5, 108, 152–3.
25 Norman Carlisle, "Who Was This Cockney Lord of the Beaver Lands and What Was His Strange Tale?" (1959), (copy in SAB, Clipping File, "Grey Owl"); Grey Owl to My Young Friends, 20 December 1934, SAB, RE3553.
26 *Regina Leader*, 16 August 1887 (reprint from *Edmonton Bulletin*, 23 July 1887).
27 Ruttan to Secretary, 17 April 1899, LAC, RG10, vol. 6732, file 420-2, reel C-8094; Pugh to Soper, 7 March 1939, LAC, RG109-D-2, file "P."
28 J.L. Irwin, "The Beaver as a Conservation Engineer," *Farm and Ranch Review*, September 1946.
29 Smith, *From the Land of Shadows*, 103; *Annual Report Saskatchewan DA 1915*, 219.
30 *Annual Report Manitoba DAI 1916*, 111; *Annual Report Saskatchewan DA 1916*, 241.
31 Winkler to MacDonald (Durban, MB), 21 January 1919, MA, MG14/B45 item 2923; *Annual Report Saskatchewan DA 1911*, 173. On beaver policies, see also *Annual Report Saskatchewan DA 1914*, 228; and *Annual Report Alberta DA 1918*, 120. For examples of investigations of beaver damage in Manitoba, see MA, GR1600/G4529, file 2.2.1 (1951–53).
32 *Annual Report Saskatchewan DA for Year Ended 30 April 1918*, 223; *Annual Report Saskatchewan DA for Year Ended 30 April 1922*, 359.
33 *Annual Report Saskatchewan DNR for Year Ended 30 April 1939*, 2.
34 *Beaver in Saskatchewan*, 12.
35 Skead to Superintendent, 4 October 1949, LAC, RG10, vol. 6734, file 420-2

4LT-3, reel C-8096; "Brief on Conditions in Northland" (n.d.), MA, GR43/G64.

36 Reader to Malaher, 17 September 1945, MA, GR1600/G4529/2.2.1.

37 *Beaver in Saskatchewan*, 12; *Annual Report Saskatchewan DNR for Year Ended 31 March 1952*, 102; *Progress: A Survey of Saskatchewan Government Activity*, 13.

38 J.R. Reid to P.W. Doake, 29 October 1943, SAB, NR 1/2, file 111; Malaher to Stephens, 24 February 1949, MA, GR1600/G4529, file 2.2.1.

39 *Beaver in Saskatchewan*, 15–16; *Annual Report Alberta DLF for Year Ended 31 March 1952*, 49; *Annual Report Alberta DLF for Year Ended 31 March 1961*, 79.

40 Monthly Report, "N" Division NWMP for March 1909, LAC, RG18, vol. 1611, file 2.

41 Wetherell and Kmet, *Alberta's North*, 389; *Annual Report Manitoba DMNR for Year Ended 31 March 1962*, 61.

42 Mullen to English, 14 April 1939, PAA, 67.21, file 325; *Annual Report Alberta DLM for Year Ended 31 March 1944*, 86; *Annual Report Saskatchewan DNR for Year Ended 31 March 1963*, 68.

43 *Annual Report Manitoba DMNR for Year Ended 30 April 1936* (typescript), 63; *Annual Report Manitoba DMNR for Year Ended 31 March 1961*, 110; Clancy, "State Policy," 213.

44 "Don'ts for Fur Farmers," *Grain Growers' Guide*, 10 December 1924; *Annual Report Manitoba DMNR for Year Ended 30 April 1933* (typescript), 140.

45 *Annual Report Manitoba DMNR for the Year Ended 30 April 1934* (typescript), 105. See also "Stubborn Acres," *Grain Growers' Guide*, 6 February 1924.

46 "Production of Raw Fur from the Wild," 22 October 1940, MA, GR1600/G4537/14.6.1.

47 Stephens to Cunningham, 11 November 1942, and Kirk to Cunningham, 19 December 1944, MA, GR1600/G4530/12.1.2.1.

48 Fender, "Public versus Private Ownership," 228.

49 *Claresholm Review*, 17 July 1913.

50 *Annual Report Saskatchewan DA 1915*, 239.

51 *Annual Report Alberta DA 1913*, 78; *Annual Report Alberta DA 1914*, 192.

52 Report of Patrol to Fort McMurray, 29 January 1910, LAC, RG18, vol. 1643, file 125, Pt 1.

53 *Annual Report Alberta DA 1922*, 78; *Annual Report Alberta DA 1923*, 71; Gill Shark, "A Fox Is Where You Find Him," *Country Guide*, January 1949.

54 *Western Producer*, 21 July 1927.

55 *Annual Report [Fish and Game] Saskatchewan DNR for Year Ended 30 April 1940*, 3; Commissioner to Akseh Rauanheime, 11 August 1923, LAC, RG84-A-2-a, vol. 180, file U279.

56 Peter Macdonald, "A Saskatchewan Muskrat Farm," *Grain Growers' Guide*, 15 October 1924.

57 Ibid.

58 See, for example, correspondence in LAC, RG15, vol. 1140, file 4283438(3A).

59 Lawton to Harkin, 31 December 1923, and Bradshaw to Harkin, 23 May 1923, LAC, RG15, vol. 1139, file 4283438(2).

60 Extensive correspondence in 1924–27 on this is in LAC, RG15, vol. 1139, file 4283438(1B) and vol. 1140, file 4283438(3B).

61 Attwood to Conreliusen, 7 June 1935, MA, GR1600/G4538/15.6.1. The beaver sanctuaries set up around James Bay by the HBC and the federal and Quebec governments were a precedent. Beavers were transplanted into a marshy area, and trapping was controlled to let populations grow. The beavers carried out the engineering, and humans only had to suppress forest fires. On these projects, see Anderson, "7,000 Square Miles of Beaver Sanctuary," 16–18; Denmark, "James Bay Beaver Conservation," 38–43; and Loo, *States of Nature*, 103–9.

62 Allen to Tanner, 17 February 1939, LAC, RG10, vol. 6733, file 420-2-1, reel C-8095; Stowe, *The Last Great Frontiersman*, 12, 112–27.

63 "Experiment at Steeprock," 10–11; Denmark, "Conservation at Cumberland," 47–9.

64 *Annual Report Manitoba DMNR for the Year Ended 30 April 1937* (typescript), 58; Allen to Tanner, 17 February 1939, LAC, RG10, vol. 6733, file 420-2-1, reel C-8095; Garson to Underhill, 22 September 1947, MA, EC0016/GR1843/G703, file misc. A-W.

65 Trippensee, *Wildlife Management*, vol. 2, 23–32.

66 Allen to Jackson, 6 March 1941, LAC, RG10, vol. 6731, file 420-1-1-1, Pt 1, reel C-8093; Quiring, *CCF Colonialism in Northern Saskatchewan*, 109–11. By the early 1940s, three provincial projects totalling 2 million acres were under development at Cumberland House in Saskatchewan, including one organized by Jim Brady, the Métis leader and community organizer. The Department of Indian Affairs also established a project east of Cumberland House in the Sipanok area.

67 On the broader context for these techniques, see Trippensee, *Wildlife Management*, vol. 2, 37–43.

68 *Winnipeg Tribune*, 27 March 1954; "Draft" (article for Canada Year Book 1943), LAC, RG10, vol. 6713, file 420-1-1-1 1, reel C-8093.

69 Deputy Minister to Tisdale, 14 June 1945, MA, GR1600/G4532/14.5.1A; Russenholt to Stephens, 24 July 1939 (attachment), MA, GR 1600/G4530/11.3.1.

70 "Muskrat Important Unit in Keeping Nature's Balance," *Farm and Ranch Review*, October 1946.

71 *Annual Report Manitoba DMNR for Year Ended 30 April 1938* (typescript), 43; Cooney to Stevens, 6 July 1940, MA, GR1600/G4537/15.1.1.

72 "Draft" (article for Canada Year Book 1943), LAC, RG10, vol. 6713, file 420-1-1-1 1, reel C-8093.

73 *Annual Report Manitoba DMNR for Year Ended 31 March 1948*, 52.
74 "Outline of Game and Fisheries Activities Northern Manitoba [1946]," MA, GR1600/G4529/1.1.1.
75 Allan to Fitzpatrick, 17 March 1944, LAC, RG10, vol. 6731, file 420-1-1-1 1, reel C-8093; Manitoba Fur Advisory Committee, "Muskrat Development Projects," 24 February 1943, MA, GR1600/G4531/14.2.1.
76 "Report of Fur Advisory Committee, February 15–18, 1944," MA, GR1600/G4531/14.2.1. Saskatchewan used the same approach with respect to credit, creating the same problems as in Manitoba (Quiring, *CCF Colonialism*, 119–21).
77 Beresford to Stevens, 5 May and 4 August 1944 [Fisher River], MA, GR 1600/G4532/14.5.2 (file 1).
78 *Annual Report Manitoba DMNR for 11 Months Ended 31 March 1947*, 121.
79 Fur Advisory Board, Summary of Recommendations, n.d. [1943], MA, GR1600/G4531/14.2.1.
80 *Annual Report Manitoba DMNR for Year Ended 31 March 1956*, 74, 76.
81 Loney, "The Construction of Dependency," 57–78.
82 McMillan to Malaher, 25 July 1952, MA, GR1600/G4531/14.2.1; W.A. Ehrlich, L.E, Pratt, F.P. LeClaire, and J.A. Barr, *Report of Detailed Soil Survey of Pasquia Map Area in Northern Manitoba* (Winnipeg: Manitoba Soil Survey, Soils Report No. 11, 1960), 11–15; *Annual Report Manitoba DMNR for Year Ended 31 March 1960*, 108.

CHAPTER ELEVEN

1 *Annual Report [Parks] DI for Year Ended 31 March 1917*, 8.
2 Historian Alan MacEachern notes that this model did not always fit easily in other parts of the country (MacEachern, *Natural Selections*, 5–8).
3 Lothian, *A History of Canada's National Parks*, vol. 4, 52.
4 MacDonald, *Science and History at Elk Island*, 15–18.
5 For an early criticism of the reliance by prairie governments on federal initiatives, see Millar, "The Big Game of the Canadian Rockies," 116–17.
6 *Annual Report Saskatchewan DNR for 1931–32*, 12 (typescript); *An Act Respecting Provincial Parks*, chap. 60, 1930 (sec. 5), *Statutes of Alberta*; *Annual Report Alberta DLF [Parks] for 1953–54*, 92.
7 Doug Gilroy, Prairie Wildlife No. 324, *Western Producer*, 5 May 1960; *Winnipeg Tribune*, 2 September 1964 and 9 March 1965.
8 Waiser, *Saskatchewan's Playground*, 25–35.
9 Millar, "The Big Game of the Canadian Rockies," 110.
10 Bella, *Parks for Profit*, 24–7; Brown, "The Doctrine of Usefulness," 46–62.
11 *Annual Report DI 1886*, Whitcher's Report, 31 December 1886, 86–7, 91, 93.
12 Ibid., 86, 92, 93.
13 Ibid., 88, 90–1. Since there was confusion about the attributes of various

trout species, "which in these north-western wilds develop strange perplexities," Whitcher advised closed seasons on all trout rather than on particular species.

14 Foster, *Working for Wildlife*, 29.

15 A contrary view was expressed by Dr J.S. Lynch of Winnipeg in 1887 in a report for the CPR. He stressed that tourist development depended on sanitary conditions, public infrastructure, and the enforcement of standards of public morality and behaviour in the park. He recommended the protection of big game and the prohibition of hunting in the park, but gave no details and showed little commitment to wildlife conservation. Like Whitcher, he advised that the Stoney be banned from the park, not for reasons of wildlife protection, but for reasons of public health because there was "at present an epidemic of whooping cough at Banff" that originated with the Stoney ("Report Made for CPR on Banff by Dr J.S. Lynch, Winnipeg, July 1887," PAA, 70.347/55).

16 On tourism and depletion in the United States, see Nash, "The Exporting and Importing of Nature," 519–60.

17 *Annual Report [Parks] DI 1900–01*, 6; *Annual Report [Parks] DI 1902–03*, 6.

18 Binnema and Niemi, "'Let the Line Be Drawn Now,'" 725, 735. See also Matthew Wangler, "Canada's Rocky Mountain Parks," 55–7. On the removal of Aboriginal peoples from national parks in the United States as an assertion of a particular view of wilderness and nature, see Spence, *Dispossessing the Wilderness*. On the Montreal Lake Cree and Prince Albert National Park, see Waiser, *Saskatchewan's Playground*, 40–1.

19 *Annual Report [Parks] DI 1905–06*, 16.

20 *Annual Report [Parks] DI 1901–02*, 5, 7; *Annual Report [Parks] DI 1900*, 4–5. On these early measures to enable the tourist gaze, see also *Annual Report [Parks] DI 1904–05*, 10; and *Annual Report [Parks] DI for Year Ending 31 March 1910*, 4.

21 Jardine and Spary, "The Natures of Cultural History," 12; *Blairmore Enterprise*, 12 April 1946.

22 *Annual Report [Parks] DI for Year Ended 31 March 1917*, 6.

23 *Annual Report [Parks] DI for Year Ended 31 March 1912*, 12.

24 Taylor, *Jasper*, 118; *Annual Report [Deputy Minister] DI for Year Ended 31 March 1916*, 32; *Annual Report [Parks] DI for Year Ended 31 March 1917*, 7; *Annual Report [Parks] DI for Year Ended 31 March 1919*, 6.

25 *Annual Report [Parks] DI for Year Ended 31 March 1921*, 47.

26 *Annual Report [Parks] DI for the Year Ended 31 March 1917*, 7, 53–4.

27 *Annual Report [Parks] DI for Year Ended 31 March 1923*, 79. This was done at most of the western parks. After the Second World War, DDT was added to the petroleum or, later, was put directly into the water. On land, DDT was spread by fogging and spraying. Adding DDT to water severely affected

fish, and in 1949 tests were conducted in Saskatchewan to calculate the concentrations of DDT needed to kill mosquito larva and not fish (*Annual Report Saskatchewan DNR for Year Ended 31 March 1950*, 77).

28 Taylor, *Jasper*, 115.

29 Taylor, "Legislating Nature," 128.

30 Hart, *Harkin*, 85–6, 109–14; *Annual Report [Parks] DI for Year Ended 31 March 1916*, 9.

31 *Annual Report [Parks] DI for Year Ended 31 March 1917*, 7–8; *Annual Report [Parks] DI for the Year Ended 31 March 1922*, 104.

32 While detailed analysis of their iconography is lacking, postcards, photographs, and skins and heads of wild animals were part of the tourist experience, serving as triggers for memory. Before the First World War, the Brewster Trading Co. and Norman Luxton's Morley Trading Co. in Banff and the Exshaw Trading Co. at Exshaw traded in animal souvenirs as well as in Aboriginal artifacts. This demonstrated the common linkage in popular culture of Aboriginal people with wilderness (Colpitts, *Game in the Garden*, 153).

33 Taylor, "Legislating Nature," 127, 133.

34 Smith, *From the Land of Shadows*, 91.

35 Waiser, *Saskatchewan's Playground*, 77–8.

36 Quoted in Hart, *J.B. Harkin*, 453.

37 Sandlos, "'Not Wanted in the Boundary,'" 189–221.

38 *Annual Report [Parks] DI for Year Ended 31 March 1920*, 6; *Annual Report Alberta DA 1911*, 142; Pincher Creek Board of Trade letterhead and map on stationary (example in PAA, Premier's Papers, file 369, 1933); *Good Roads*, August 1922, 7. On the value of the overflow, see Colpitts, *Game in the Garden*, 70–5, 119–21.

39 *Annual Report [Parks] DI for Year Ended 31 March 1919*, 18; Williamson, "Game Preservation in the Dominion Parks," 137; "Extract from the Superintendent's Conference, Copy for file," n.d. [c. 1927], LAC, RG84, vol. 115, file 212, reel T-12893.

40 Taylor, *Jasper*, 143–5.

41 Ibid., 146–51; McGowan, "Hunting with a Camera," 22–5.

42 *Annual Report [Parks] DI for Year Ended 31 March 1917*, 52; *Annual Report [Parks] DI for Year Ended 31 March 1920*, 27; *Annual Report [Parks] DI for Year Ended 31 March 1921*, 47.

43 Memo to Cory, 25 August 1922, LAC, RG84, vol. 120, file U266, reel T-12898.

44 Hart, *Harkin*, 295–7; Harkin circular letter, 31 January 1925, LAC, MG30 E441, vol. 14, file 17.

45 Harkin circular letter, 31 January 1925, LAC, MG30 E441, vol. 14, file 17.

46 Hart, *Harkin*, 300, 460–71; Taylor, *Jasper*, 119–21.

47 Hart, *Harkin*, 448.

48 Duncan to Lloyd, 13 April 1930, and Lloyd to Duncan, 22 May 1930, LAC, MG30 E3441, vol. 5, file 5.

49 Roe to Rowatt, 14 February 1933, LAC, RG84, vol. 168, file RM266 (Pt I), reel T-10432.

50 "Comments on Predation," attached to Pratt to Kerr, 16 April 1938, SAB, NR 1/2, file 131.

51 MacEachern, "Changing Ecologies," 363. In the United States, predator control ended in national parks in the early 1930s, and in 1940 the parks began "seeking to preserve all species in a natural balance." This policy was entrenched after the Second World War (Dunlap, "Value for Varmits," 158, 161).

52 Dunlap, "Ecology, Nature and Canadian National Parks Policy," 143–4; Taylor, *Jasper*, 129.

53 Hart, *Harkin*, 475; Taylor, *Jasper*, 126–7. One of the few people who objected to killing cougars was the naturalist Dan McGowan, who argued that they should be left alone to allow a self-regulating balance of nature to develop in the park.

54 Taylor, *Jasper*, 128, 132–5, 141. Fire suppression continued at Jasper until the 1980s.

55 Fisher to Aberhart, 15 July 1939 (italics in original), and Mullen to Aberhart, 31 July 1939, PAA, Premiers Papers, file 663B.

56 *Prince Albert Herald*, 25 February 1935, copy in SAB, NR 1/2, file P800C(3); Waiser, *Saskatchewan's Playground*, 113.

57 Smart to Superintendent, 11 February 1944, and Heaslip to Smart, 18 February 1944, LAC, RG84, vol. 168, file RM266, Pt I, reel T-10432. For similar complaints at Riding Mountain, see Hebner to Minister of the Interior, 1 May 1928, LAC, RG84, vol. 168, file RM266, reel T-10432.

58 *Regina Leader-Post*, 15 December 1950.

59 *Report and Minutes of Evidence on Existing Natural Food Products [1887]*, 5, 18–19, 63–5; *Nor'West Farmer*, April 1898; Colpitts, *Game in the Garden*, 53, 56; Brower, *Lost Tracks*, 131–3; Seton Journal, 7 September 1892, AMNH, E-3, vol. 4.

60 Mair, *The American Bison*, 106–7.

61 Lothian, *A History of Canada's National Parks*, vol. 4, 24–31; *Annual Report [Parks] DI for Year Ended 31 March 1908*, 11.

62 Brower, *Lost Tracks*, 2, 5; Isenberg "The Returns of the Bison," 180.

63 Macdonald, *Science and History at Elk Island*, 16–17; *Regina Leader*, 22 November 1907; *House of Commons Debates*, 15 February 1916, 845.

64 *Annual Report [Parks] DI for Year Ended 31 March 1922*, 105.

65 *Annual Report [Parks]DI for Year Ended 31 March 1908*, 12; Brower, *Lost Tracks*, 96.

66 *Annual Report [Parks] DI for Year Ended 31 March 1913*, 19.

67 Macdonald, *Science and History at Elk Island*, 20, 50–1.
68 *Claresholm Review*, 23 January 1912; *Edmonton Journal*, 1 November 1930; Brower, *Lost Tracks*, 89.
69 Brower, *Lost Tracks*, 44, 131–42.
70 Hart, *Harkin*, 152; Graham, *Canada's Wild Buffalo*, 12; Williamson, "Game Preservation in the Dominion Parks," 132.
71 "Report of William Ogilvie, DLS, Exploratory Survey of Part of the Lewis … and Mackenzie Rivers," *Annual Report DI 1889*, 91–2.
72 Hart, *Harkin*, 151–3; Sandlos, *Hunters at the Margin*, 23–45. Wood Buffalo became famous in 1954 when it was discovered that it was the nesting site of whooping cranes.
73 Wetherell and Kmet, *Alberta's North*, 370.
74 *Annual Report [Parks] DI for Year Ended 31 March 1919*, 6; Brower, *Lost Tracks*, 42–3.
75 Brower, *Lost Tracks*, 101–5.
76 *Annual Report [Parks] DI for Year Ended 31 March 1920*, 6–7. Plans for culling were announced the next year as well (*Annual Report [Parks] DI for Year Ended 31 March 1921*, 6).
77 Hart, *Harkin*, 285–6; Lothian, *A History of Canada's National Parks*, vol. 4, 33.
78 *Annual Report [Parks] DI for Year Ended 31 March 1923*, 74–5.
79 Lothian, *A History of Canada's National Parks*, vol. 4, 33.
80 Ibid., vol. 1, 66
81 *Annual Report [Parks] DI for the Year Ended 31 March 1923*, 75.
82 Hart, *Harkin*, 290–2.
83 Ibid., 293; Thomas Childs, "Buffalo: Massacre and Salvage," *Western Producer*, 20 September 1956; Soper to Farley, 18 April 1933, LAC, RG109-D-2, vol. 5, file F. Perhaps to make his position on the matter clear for posterity, Hoyes Lloyd kept the original correspondence protesting the shipments. These letters later became part of the papers he donated to Library and Archives Canada (LAC, MG30 E441, vol. 3, file 13, "Buffalo 1924–1930").
84 Macdonald, *Science and History at Elk Island*, 23, 39; Hart, *Harkin*, 292.
85 Sandlos, *Hunters at the Margin*, 59; Brower, *Lost Tracks*, 58–60.
86 "A Holiday in Buffalo Park," *Farm and Ranch Review*, 20 June 1918.
87 Grant MacEwan, "Tragedy Continues to Stalk the Buffalo," *Farm and Ranch Review*, November 1964.
88 Lothian, *A History of Canada's National Parks*, vol. 4, 37; *Farm and Ranch Review*, July 1947.
89 Lothian, *A History of Canada's National Parks*, vol. 4, 38.
90 Loo, *States of Nature*, 141–8; Sandlos, *Hunters at the Margin*, 89–105.
91 Williamson, "Game Preservation in the Dominion Parks," 137–8.
92 Seton, *Life-Histories of Northern Animals*, vol. 1, 47; Millar, "The Big Game of the Canadian Rockies," 104.

93 *Annual Report Manitoba DAI 1911*, 34–5; Brook, "Historical Review of Elk-Agriculture Conflicts," 76. Saskatchewan banned elk hunting in 1918. It was the last province to do so, which meant that killing elk was then banned everywhere in Canada (*Annual Report Saskatchewan DA for Year Ended 30 April 1918*, 216).

94 Hart, *Harkin*, 356–8.

95 Tony Lascelles, "Mammals of Manitoba" [Part 2], *Winnipeg Free Press* (magazine), 9 May 1936; "Riding Mountain National Park" (tourist pamphlet), n.d. (c. 1940s), copy in MA, GR1661/277/1050.

96 *Annual Report Alberta DA 1908*, 103; Macdonald, *Science and History at Elk Island*, 15.

97 Hart, *Harkin*, 172–5.

98 *Annual Report Alberta DA 1923*, 69; *Annual Report Alberta DLF for Year Ended 31 March 1950*, 81. The Pembina-Brazeau Elk Preserve was established in 1927 to protect this herd. It operated until 1950 (*Fish, Fur and Feathers*, 71).

99 Taylor, *Jasper*, 118, 141; *Annual Report Saskatchewan DNR for Year Ended 30 April 1938*, 4; *Annual Report Saskatchewan DNR for Year Ended 30 April 1939 (Game and Fur Report)*, 10.

100 *Annual Report Alberta DLM for the Year Ended 31 March 1946*, 70.

101 *Annual Report Manitoba DMNR for Year Ending 30 April 1934* (typescript), 97; Minutes, Manitoba Game Commission, 27 April 1934, MA, GR1600/G4537, file 15.1.1.

102 Brook, "Historical Review of Elk-Agriculture Conflicts," 76; "Notes for Discussion in Connection with Dominion-Provincial Wildlife Conference, Ottawa, February 27 and 28th 1947," MA, GR1600/G4538/15.5.1.A.

103 *Annual Report Manitoba DMNR for Year Ended 31 March 1951*, 62; "Notes for Discussion in Connection with Dominion-Provincial Wildlife Conference, Ottawa, February 27 and 28th 1947," MA, GR1600/G4538/15.5.1.A; *Annual Report Manitoba DMNR for Year Ended 31 March 1948*, 50.

104 Brook, "Historical Review of Elk-Agriculture Conflicts," 76. Similar elk reduction was undertaken at other parks, including Prince Albert National Park (Waiser, *Saskatchewan's Playground*, 117–19).

105 *Annual Report Alberta DA 1908*, 103; Seton, *Life-Histories of Northern Animals*, vol. 1, 219–21, 237.

106 *Annual Report Saskatchewan DA 1907*, 204; *Annual Report Saskatchewan DA 1908*, 176–8.

107 Williamson, "Game Preservation in the Dominion Parks," 129; Campbell to Cory, 21 July 1910, LAC, RG84-A-1a, vol. 628, file 578206, reel T-8104. On lobbying for a permanent preserve, see, for example, *Annual Report Alberta DA 1911*, 142; and *Annual Report Saskatchewan DA 1914*, 212.

108 Gibson, "Three Tiny Prairie Parks," 61; Hart, *Harkin*, 156–8; H.H.C Anderson, "Charlie's Pets," *Grain Growers' Guide*, 1 September 1926; *Annual Report Saskatchewan DA 1916*, 236.

109 Hart, *Harkin*, 156–8; Gibson, "Three Tiny Prairie Parks," 61; *Annual Report [Parks]* DI *for Year Ended 31 March 1922*, 105.

110 Gibson, "Three Tiny Prairie Parks," 61; Graham to Lloyd, 19 January 1921, LAC, MG30 E441, vol. 3, file 15.

111 Gibson, "Three Tiny Prairie Parks," 62. The Department of the Interior also undertook a few stopgap measures such as feeding wild pronghorns in the winter of 1922 when they were threatened by starvation because of deep snow.

112 Gibson, "Three Tiny Prairie Parks," 64; Aubrey Fullerton, "Antelope Stage a Comeback," *Country Guide*, October 1945; *Annual Report [Parks]* DI *for Year Ended 31 March 1923*, 75.

113 *Annual Report Alberta* DA *1934*, 46.

114 Clippings, *Calgary Herald*, 14, 17, and 21 December 1934, and not dated, GA, A.S243.

115 MacDonald, *Science and History at Elk Island*, 41, 53.

116 Ibid., 41; Hedlin, "The Buffalo as a Northern Resource," 27.

117 Blais, "Preserving the Balance of Nature," 5.

118 Grant MacEwan, "Andy Russell on the Grizzly Trail," *Farm and Ranch Review*, July–August 1965; *Winnipeg Tribune*, 2 September 1964 and 9 March 1965. On parallel debates in the United States, see Spaulding, "Yosemite and Ansel Adams," 634–7.

CHAPTER TWELVE

1 [Frank Brazier], "The Well-Tempered Bird Watcher," c. 1955, SAB, R-1676, file 3.

2 Stearns, "The Royal Society and the Company," 8–13; Baillie, "Naturalist on Hudson Bay," 36–9; Houston and Houston, "Samuel Hearne, Naturalist," 27; Semenchuk, *The Atlas of Breeding Birds of Alberta*, 9–11.

3 Burman, *The Present Status of Natural Science*, 3–5. On Macoun's career with the Geological Survey, see Waiser, *The Field Naturalist.*

4 *Annual Report Saskatchewan* DA *1913*, 247–8.

5 H.M. Speechley, "Common Prairie Birds" (Part I), *Grain Growers' Guide*, 21 April 1915.

6 *Annual Report Saskatchewan* DA *for Year Ended 30 April 1921*, 88.

7 Star Bright, Nature Lovers page, *Western Producer*, 26 July 1928.

8 Shepherd to Potter, 23 October 1941, SAB, SHS 50.

9 On academic respectability and wildlife art, see the discussion about Rungius and the Glenbow Museum in Kaye, *Hiding the Audience*, 117–25.

10 Berger, *Science, God, and Nature in Victorian Canada*, 3; Kohler, *All Creatures*, 114.

11 Strickland, "Entomology in the Early Days of Alberta," 11.The Rothschild flea collection is now in the British Museum. Baron Rothschild's daughter,

Miriam, followed in her father's footsteps, becoming a world famous Siphonaptera specialist.

12 Journal entry, 2 August 1884, AMNA, E-3, vol. 1, Pt 2, and journal entries, 23 January and 30 March 1892, AMNH, E-3, vol. 2; File note, 1888, LAC, MG29 D108, vol. 16, file 16.5.

13 "Ornithological Section" (c. 1941), MA, P5070, item 1; R.D. Colquette, "Freeholders of the Wild," *Country Guide*, November 1946.

14 Issues of the program's publication, also called *Camera Trails*, from 22 April 1937 to 17 June 1937 can be found in SAB, R-942.

15 "The First Twenty-Five Years of the Regina Natural History Society, 1933 to 1958," SAB, R-942, file I.15; *Edmonton Journal*, 19 October 1945.

16 SAB, R-E3033. In 1976 several of her Nature Notes columns were published in a single volume titled *A Second Look*.

17 Clarence Tillenius, "Footnote to 'Through Field and Wood,'" *Country Guide*, September 1962.

18 Grant MacEwan, "The Naturalist from Red Deer, Alberta," *Farm and Ranch Review*, July 1957; Kerry Wood, "The Menace of the .22 and What Can Be Done about It," *Farm and Ranch Review*, May 1950; Kerry Wood, "Nature's Jubilee," *Farm and Ranch Review*, September 1955.

19 Kerry Wood, "Pet Peeves …," *Farm and Ranch Review*, August 1960; Kerry Wood, "Does Anyone Like Nature?," *Farm and Ranch Review*, August 1957.

20 "Doug Gilroy (1915–2003)," http://econet.ca/sk_enviro_champions/gilroy.html (accessed 14 March 2014); Doug Gilroy, Prairie Wildlife No. 25, *Western Producer*, 15 July 1954.

21 Doug Gilroy, Prairie Wildlife No. 23, 5 August 1954, and Prairie Wildlife No. 67, 5 May 1955, *Western Producer*.

22 Doug Gilroy, Prairie Wildlife No. 214 and No. 215, *Western Producer*, 13 and 20 March 1958.

23 Doug Gilroy, Prairie Wildlife No. 27, 29 July 1954, and Prairie Wildlife No. 397, 28 September 1961, *Western Producer*.

24 Doug Gilroy, Prairie Wildlife No. 28, *Western Producer*, 5 August 1954. On the Federation of Ontario Naturalists, see Doug Gilroy, Prairie Wildlife No. 293, 1 October 1959, and Prairie Wildlife No. 591, 10 July 1963, *Western Producer*.

25 Nature Alberta (formerly the Federation of Alberta Naturalists) continues to use the term "natural history" to indicate the study of wildlife through observation. It calls a student of natural history a "naturalist" (http://naturealberta.ca/about-us/what-we-do/ [accessed 5 March 2014]).

26 *Nor'West Farmer*, 20 November and 20 December 1902.

27 Drouin and Bensaude-Vincent, "Nature for the People," 421–3; Berger, *Science, God, and Nature in Victorian Canada*, xi–xiii, 5–6; Jardine and Spary, "The Natures of Cultural History," 3–13.

28 Burman, *The Present Status of Natural Science*, 1–2.

29 Robert Bell, *Evolution with Prevision*. Baccalaureate address, eighth annual commencement, Thomas Clarkson Memorial School of Technology, Potsdam, New York, 1907, 1–2 (reprinted from *Clarkson Bulletin*, 4, no. 3 (July 1907) (copy in LAC, MG29 B15 file 3-52).

30 Friesen, "The Manitoba Historical Society" (online, not paginated).

31 See, for example, Browne, "Biogeography and Empire," 319. The quote from Marx is in Guntau, "The Emergence of Geology as a Scientific Discipline," 283.

32 Zeller, *Inventing Canada*, 6.

33 Stewart, "The First Half Century" (online, not paginated).

34 Burman, *The Present Status of Natural Science*, 1–2; *Regina Leader*, 4 September 1884.

35 "President's Inaugural Address," *Transactions of the Manitoba Historical and Scientific Society* (1889), reprinted in Levere and Jarrell, *A Curious Field Book*, 144–8.

36 Friesen, "The Manitoba Historical Society" (online, not paginated); Stewart, "The First Half Century" (online, not paginated).

37 Toogood, "Modern Observations," 348–51.

38 H.M. Speechly, "The Natural History Society of Manitoba 1920 to 1940," MA, P5070, item 1.

39 Ibid.

40 "Appendix E: Territorial Natural History Society," *Annual Report of the NWT DA 1904*, 157–8, 162.

41 *Regina Leader*, 30 October 1902; *Nor'West Farmer*, 20 November 1902.

42 "Appendix E: Territorial Natural History Society," *Annual Report NWT DA 1904*, 170.

43 Alberta Natural History Society, PAA, 67.293.

44 The founders were Fred Bradshaw, former Saskatchewan game commissioner; E.H.M. Knowles, a Regina barrister interested in wildlife photography; and J.H. Taylor, a life insurance representative and, at first, an avid duck and goose hunter whose hunting "whetted an interest in ornithology that ended his career as a hunter." Another of the founders was H.C. Andrews, principal of the Saskatchewan Teachers College in Moose Jaw and later in Regina, who specialized in the science curriculum for public schools ("The First Twenty-Five Years of the Regina Natural History Society 1933 to 1958," SAB, R-942, file I.15).

45 "The First Twenty-Five Years of the Regina Natural History Society 1933 to 1958," SAB, R-942, file I.15.

46 *Western Producer*, 21 February 1946.

47 *Annual Report Saskatchewan DNR for Year Ended 31 March 1952*, 136; "The First Twenty-Five Years of the Regina Natural History Society 1933 to 1958,"

R-942, file I.15; *A Guide to Saskatchewan Mammals,* in SAB, R-1337. On Isabel Priestly, see C. Stuart Houston, "The Birds of the Yorkton District, Saskatchewan," *Canadian Field Naturalist* 63, no. 6 (November–December 1949): 216–17.

48 *Regina Leader-Post,* 25 October 1954 (two articles).

49 Frank Brazier, "The Well Tempered Bird Watcher," SAB, R-1676, file 3.

50 Ina Watson, Nature Lovers page, *Western Producer,* 18 April 1934; "A Ray of Sunlight," Nature Lovers page, *Western Producer,* 10 May 1928.

51 *Western Producer,* 3 July 1930; Dan McGowan, "Gulls," *Grain Growers' Guide,* 30 July 1924. For a similar appreciation of this bird, see R.W. Sutton, "Franklin's Gull," *Country Guide,* May 1946.

52 Hughes to Potter, n.d. [1920], SAB, SHS50; Leonard Burke (Aberley, SK), Nature Lovers page, *Western Producer,* 19 July 1928.

53 Pocahontas (Bratton, SK), Nature Lovers page, *Western Producer,* 24 April 1930; *Winnipeg Tribune,* 15 April 1933.

54 Various entries, 1885, AMNH, E-3, vol. 1, Pt 2. Showing the limited number of people qualified to comment, the list was sent to Robert Bell and John Macoun in Ottawa, to five individuals in Manitoba (some of whom could not comment), and to George Guernsly, a birdwatcher at Fort Qu'Appelle (1885, Entries passim, E-3 vol. 1, Pt 2, AMNH).

55 Farley to Soper, 24 January 1938, RG109-D-2, vol. 5, file "K" (1932–47). While such networks were significant among birdwatchers, E.A. Strickland, professor of entomology at the University of Alberta, noted the same pattern among entomologists. Indeed, he claimed that entomologists could not be successful without the help of collectors and amateur entomologists throughout the region (Strickland, "Entomology in the Early Days of Alberta," 6).

56 See, for example, "Watching the Winter Birds," *Country Guide,* December 1957; and Elizabeth Huggett, "You Can Have Birds Galore If You Put Out Feed," *Farm and Ranch Review,* February 1954.

57 "Bird Watching Is Fun on Our Prairie Farms," *Farm and Ranch Review,* July 1950; *Regina Leader-Post,* September 1959; *Winnipeg Tribune,* 7 January 1964.

58 With the transfer of natural resources in 1930, most of these sanctuaries were transferred to the provinces. In the 1930s many of them were seriously damaged by drought and the system was subsequently reorganized.

59 *Calgary Herald,* 16 March 1929; Bullick, *Calgary Parks and Pathways,* 91; Elphinstone, *Inglewood Bird Sanctuary,* 24–6.

60 *Calgary Herald,* 19 January 1935; Elizabeth Garbutt, "Sanctuary," *Alberta Poetry Year Book 1934–35,* 12; Elphinstone, *Inglewood Bird Sanctuary,* 25.

61 Walker to Brownlee, 24 October 1932, PAA, Premiers Papers, file 15.

62 Walker to Soper, 9 September 1936, and Walker to Soper, 2 November 1936, LAC, RG109-D2, vol. 5, file "W."

63 Walker to Soper, 29 March 1935 and 10 December 1934, LAC, RG109-D-2, vol. 5, file "w."

64 Soper to Walker, 15 December 1934 and 2 April 1935, LAC, RG109-D-2, vol. 5, file "w."

65 *Calgary Herald*, 11 January 1936; *Annual Report Alberta DA 1938*, 48; Walker to Aberhart, 23 September 1940, PAA, Premiers Papers, file 771B; Bullick, *Calgary Parks*, 88, 91; Elphinstone, *Inglewood Bird Sanctuary*, 27–9.

66 Doug Gilroy, Prairie Wildlife No. 179, *Western Producer*, 18 July 1957; *Regina Leader-Post*, 13 December 1958; "The First Twenty-Five Years of the Regina Natural History Society 1933 to 1958," SAB, R-942, file I.15; *Globe and Mail*, 2 August 2003; *Annual Report Saskatchewan DNR for the Year Ended 31 March 1958*, 113.

67 *Regina Leader-Post*, 24 October 1942; *Newsflight*, no. 29, Ducks Unlimited (Canada), n.d. (c. 1938), (copy in MA, EC0016/GR1661/649).

68 Ralph Stueck, "Jonathan the Talking Goose" (two items with same title), SAB, R.E3294.

69 Anna Tillenius, "Sanctuary for Honkers," *Country Guide*, April 1957; *Annual Report Manitoba DMNR for Year Ended 31 March 1957*.

70 R.D. Colquette, "He Kept Wild Birds as a Hobby," *Grain Growers' Guide*, 1 August 1927.

71 *Newsflight*, no. 29, Ducks Unlimited (Canada), n.d. (c. 1938), (copy in MA, EC0016/GR1661/649).

72 *Winnipeg Tribune*, 24 November 1962.

73 *Globe and Mail*, 20 July 2004; Bossenmaier, *100-Plus Years*, 13; Scott, "Bring Back the Bluebirds," 30–1.

74 *Annual Report Saskatchewan DNR for Year Ended 31 March 1958*, 32, 102; *Annual Report Saskatchewan DNR for Year Ended 31 March 1963*, 38, 43.

75 Nature Alberta, http://naturealberta.ca/about-us/what-we-do/ (accessed 5 March 2014); Emel and Wolch, "Witnessing the Animal Moment," 11–12. The six clubs that came together in 1970 in Alberta were the Edmonton Natural History Club (which grew out of the University of Alberta Botany Club, formed in 1937); the Edmonton Bird Club (established 1949); the Calgary Bird Club (established 1955); and the Bow Valley Naturalists, the Lethbridge Natural History Society, and the Cold Lake Bird Club, all established in 1967 (*Fish, Fur and Feathers*, 313–14).

CHAPTER THIRTEEN

1 Baker, "Guest Editor's Introduction," 190.

2 Berger, "Why Look at Animals?," 26; Anderson, "Culture and Nature at the Adelaide Zoo," 276, 283.

3 Bennett, *The Birth of the Museum*, 1–13.

4 It will be recalled that in 1898 Dr George established the North-West

Entomological Society, which later became the Territorial Natural History Society and then the Alberta Natural History Society.

5 Wonders, "A Sportsman's Eden Part II" (online, not paginated). See also Lampard, "Henry George, MD, 1864–1932," 12–13.

6 *Regina Morning Leader*, 12 November 1908.

7 *Regina Leader*, 15 March 1913; *Annual Report Saskatchewan DA 1913*, 253, 262.

8 *Annual Report Saskatchewan DA 1916*, 245, 248. The creation of a military and natural history museum was announced in 1919 to commemorate Saskatchewan's war effort. While nothing came of the project, an architectural competition resulted in preliminary drawings by the firm Hobbs and Hyde of Montreal for a building in which military exhibits would have had twice the space as natural history ones (*Regina Leader*, 30 January, 9 April, and 20 August 1919).

9 McTavish, "Learning to See in New Brunswick," 558–9.

10 *Annual Report Saskatchewan DA 1913*, 253, 262. On group work in American museums at this time, see Kohler, *All Creatures*, 77–9.

11 *Annual Report Saskatchewan DA 1915*, 241; *Annual Report Saskatchewan DA for Year Ended 30 April 1921*, 126.

12 *Annual Report Saskatchewan DA 1913*, 253, 263 (italics in original); *Annual Report Saskatchewan DA for Year Ended 30 April 1920*, 113.

13 Farber, "The Development of Taxidermy and the History of Ornithology," 550–66; Desmond, "Displaying Death, Animating Life," 160–4.

14 *Annual Report Saskatchewan DA 1913*, 263, 265; *Annual Report Saskatchewan DA 1914*, 237; *Annual Report Saskatchewan DA for Year Ended 30 April 1918*, 212.

15 *Annual Report Saskatchewan DA for Year Ended 30 April 1918*, 212.

16 *Annual Report Saskatchewan DA 1913*, 263, 265. On the history of the study of bird eggs (called "oology") in Alberta, see Semenchuk, *The Atlas of Breeding Birds of Alberta*, 12. For an earlier example of the debate about killing birds for collecting, see Atkinson, *Insectivorous Birds of Manitoba*, 1. A parallel 1915 view is Taverner, *Suggestions for Ornithological Work in Canada*, 25–6.

17 "A History of the Royal Saskatchewan Museum," 35 (typescript, 1994), SAB, R-E4121; "The First Twenty-Five Years of the Regina Natural History Society 1933 to 1958," SAB, R-942, file I.1; *Annual Report Saskatchewan DNR for Year Ended 31 March 1950*, 113; *Annual Report Saskatchewan DNR for Year Ended 31 March 1952*, 113–17. After 1950 the museum expanded its activity in human history through archaeological research and display. In 1950 it began work on a film, *Indian Kitchen Middens*, to encourage public interest in Aboriginal pre-contact history.

18 *Annual Report Saskatchewan DA 1914*, 238; *Annual Report Saskatchewan DNR for Year Ended 30 April 1937*, 72 (typescript); *Annual Report Saskatchewan DA 1913*, 265; *Annual Report Saskatchewan DA for Year Ended 30 April 1920*, 79–80.

19 "The First Twenty-Five Years of the Regina Natural History Society 1933 to 1958," SAB, R-942, file I.15; *Annual Report Saskatchewan* DNR *for Year Ended 31 March 1953*, 103.

20 Young, "The Saskatchewan Museum of Natural History," 265. Fred Bard was director until 1970. Like "his mentor," Fred Bradshaw, he came from the natural history tradition. He started working at the museum in 1925 at the age of seventeen and worked with Dewey Soper on duck population counts in the 1930s. As seen earlier, like Bradshaw, "he made the [whooping] crane a personal project" and he promoted their preservation and carried out scientific work on them (Dunlap, "Organisation and Wildlife Preservation," 205).

21 *Annual Report Saskatchewan* DNR *for Year Ended 31 March 1950*, 113–16; *Annual Report Saskatchewan* DNR *for Year Ended 31 March 1953*, 103–4.

22 "The First Twenty-Five Years of the Regina Natural History Society 1933 to 1958," 26, SAB, R-942, file I.15; Young, "The Saskatchewan Museum of Natural History, 276.

23 Young, "The Saskatchewan Museum of Natural History," 265, 270.

24 Ibid., 270, 276.

25 *Annual Report Saskatchewan* DNR *for Year Ended 31 March 1958*, 113; *Annual Report Saskatchewan* DNR *for Year Ended 31 March 1962*, 100.

26 Young, "The Saskatchewan Museum of Natural History, 276; *Annual Report Saskatchewan* DNR *for Year Ended 31 March 1957*, 113–14.

27 *Annual Report Saskatchewan* DNR *for the Year Ended 31 March 1959*, p. CO-13; *Annual Report Saskatchewan* DNR *for the Year Ended 31 March 1960*, 135. For analysis of the way exhibit techniques shaped museum research activity in the United States in an earlier period, see Kohler, *All Creatures*, 110.

28 *Annual Report Saskatchewan* DNR *for the Year Ended 31 March 1960*, 142; *Annual Report Saskatchewan* DNR *for the Year Ended 31 March 1958*, 109; Doug Gilroy, Prairie Wildlife No. 182, *Western Producer*, 5 August 1957.

29 Among the private museums, the Lund Wildlife Exhibit, located in a wartime building near downtown Prince Albert, Saskatchewan, was assembled by taxidermist and collector G.E. Lund ("Lund's Wildlife Exhibit," *Forest and Outdoors*, February 1951; "The Lund Exhibit Is a Work of Art in Still Life," *Western Producer Magazine*, 17 July 1952).

30 Cartwright to Cunningham, 11 March 1935, MA, GR1600/G4529, file 8.2.1. Harrold's obituary is in the *Canadian Field Naturalist* 43 (September 1929): 132–3.

31 Cowan to Russell, 9 May 1944, MA, GR1600/G4529, file 8.2.1; "The Manitoba Museum: Its History, Development, Present Status and Future Needs," UMA, MSS24, file "Manitoba Museum 1946–80."

32 *Annual Report* DI *1886*, Whitcher's Report, 31 December 1886, 87; *Annual Report [Parks]* DI, *1890*, 6; Wonders, "A Sportsman's Eden Part II" (online version, not paginated).

33 *Annual Report [Parks]* DI, *1902–03*, 7.
34 *Annual Report [Parks]* DI *for Year Ended 31 March 1908*, 8, 20; *Annual Report [Parks]* DI *for Year Ended 31 March 1921*, 29.
35 *Handbook of the Rocky Mountain Park Museum* (Ottawa: Department of the Interior, 1914).
36 Horowitz, "Animal and Man in the New York Zoological Park," 426–38.
37 *Debates, House of Commons*, 16 July 1903, col. 6790; *The Last of the Buffalo: Comprising a History of the Buffalo Herd of the Flathead Reservation and An Account of the Great Roundup* (1908), (SAB, pamphlet file "Bison American"); *Annual Report [Parks]* DI *for the Year Ending 31 March 1908*, 7.
38 *Annual Report [Deputy Minister]* DI, *1897*, 4, 35–6; *Annual Report [Parks]* DI, *1898*, 346.
39 *Annual Report [Parks]* DI, *1899*, 4; *Annual Report [Parks]* DI *1900*, 4.
40 *Annual Report [Parks]* DI, *1904–05*, 9; *Annual Report [Parks]* DI *for Year Ended 31 March 1912*, 12; *Annual Report [Parks]* DI *for Year Ended 31 March 1913*, 26; *Annual Report [Parks]* DI *for Year Ended 31 March 1911*, 21; *Annual Report [Parks]* DI *for Year Ended 31 March 1920*, 15.

41 *Annual Report [Parks]* DI *for Year Ended 31 March 1923*, 79; *Annual Report [Parks]* DI *for Year Ended 31 March 1920*, 15.
42 *Annual Report [Parks]* DI, *1904–05*, 9–10; *Annual Report [Parks]* DI *for Year Ended 31 March 1910*, 8; *Annual Report [Parks]* DI *for Year Ended 31 March 1920*, 16; *Annual Report [Parks] for the Year Ending 31 March 1911*, 20.
43 *Annual Report [Parks]* DI, *1900–01*, 5; *Annual Report [Parks]* DI, *1905–06*, 7.
44 *Annual Report [Parks]* DI *for Year Ended 31 March 1908*, 9; *Annual Report [Parks]* DI *for Year Ended 31 March 1910*, 8; *Annual Report [Parks]* DI *for Year Ended 31 March 1912*, 24.
45 *Annual Report [Parks]* DI *for Year Ended 31 March 1913*, 25; *Annual Report [Parks]* DI *for Year Ended 31 March 1920*, 15; *Annual Report [Parks]* DI *for Year Ended 31 March 1923*, 75, 79.
46 *Annual Report [Parks]* DI *for Year Ended 31 March 1911*, 21; *Annual Report [Parks]* DI *for Year Ended 31 March 1913*, 25.
47 Willis, "Looking at the Zoo," 678.
48 *Annual Report [Parks]* DI *for Year Ended 31 March 1923*, 79; Ritvo, *The Animal Estate*, 220; *Annual Report [Parks]* DI *for Year Ended 31 March 1910*, 9; *Annual Report [Parks]* DI *for Year Ended 31 March 1911*, 21; *Annual Report [Parks]* DI, *1902–03*, 5. For an interesting exploration of cuteness in popular culture and its social attributes, see Genosko, "Natures and Cultures of Cuteness."
49 *Annual Report [Parks]* DI, *1901–02*, 5; *Annual Report [Parks]* DI *for Year Ended 31 March 1910*, 5, 9.
50 *Annual Report [Parks]* DI *for Year Ended 31 March 1908*, 7.
51 *Annual Report [Parks]* DI, *1901–02*, 4; *Annual Report [Parks]* DI *for Year Ended 31 March 1910*, 9.

52 *Annual Report [Parks]* DI, *1902–03*, 5; *Annual Report [Parks]* DI *for Year Ended 31 March 1913*, 26.

53 *Annual Report [Parks]* DI, *1900–01*, 5; *Annual Report [Parks]* DI *for Year Ended 31 March 1911*, 7; *Annual Report [Parks]* DI *for Year Ended 31 March 1912*, 24.

54 *Annual Report [Parks]* DI, *1902–03*, 5.

55 *Annual Report [Parks]* DI, *1906–07*, 6; *Annual Report [Parks]* DI *for Year Ended 31 March 1911*, 21.

56 *Annual Report [Parks]* DI, *1904–05*, 9–10; *Annual Report [Parks]* DI, *1906–07*, 5–6; *Annual Report [Parks]* DI *for Year Ended 31 March 1912*, 23.

57 *Annual Report [Parks]* DI *for the Year Ended 31 March 1911*, 20.

58 *Annual Report [Parks]* DI *for Year Ended 31 March 1908*, 5; Wonders, "A Sportsman's Eden Part II" (not paginated); "A Walk through Banff History," http://www.itaylorresearch.com/walk_banff.html (accessed 1 April 2014).

59 Seton, "The Pacing Mustang," in *Wild Animals I Have Known*, 160; Laughing Devil, Nature Lovers page, *Western Producer*, 15 March 1928; Beresford Richards, "To a Little Wild Duck Captive Born," *Alberta Poetry Year Book 1934–35*, 24; Gill Shark, "Don't Tame Pets from the Wild," *Country Guide*, September 1949.

60 Clipping, 18 January 1929, CCA, PRDF, series III, box 48, file 1927–31, Calgary Zoo, Botanical Garden, and Prehistoric Park Newspaper Ledger.

61 Seton Journal, 7 September 1892, AMNH, E-3, vol. 4.

62 River Park closed in 1941 ("Closed Canadian Parks," http://cec.chebucto.org/ClosPark/River.html (accessed April 15, 2014); A Brief History of the Assiniboine Park Zoo," http://www.assiniboinepark zoo.ca/about/the-zoo.php (accessed 3 April 2014).

63 *Annual Report Manitoba* DAI *1905*, 478–9; *Annual Report Manitoba* DAI *1906*, 526; *Annual Report Manitoba* DAI *for 11 Months Ended 30 November 1912*, 233.

64 Champion to Reader, 15 October 1927, CCA, PDRF, series III, box 33, file 1927; "Spring List of 1926 of Surplus Zoological Specimens," CCA, PRDF, series III, box 33, file 1926.

65 Champion to Reader, 15 October 1927, CCA, PDRF, series III, box 33, file 27.

66 Grey Owl to Cottingham, 14 October 1931, and Smart (A/Superintendent) to Campbell, 2 November 1931, SAB, R-E3863.

67 Grey Owl to Cottingham, 14 October 1931 (italics in original), SAB, R-E3863.

68 "A Brief History of the Assiniboine Park Zoo," http://www.assiniboinepark zoo.ca/about/the-zoo.php (accessed 3 April 2014); *Winnipeg Tribune*, 30 January 1951.

69 *Winnipeg Tribune*, 8 October 1959; "A Brief History of the Assiniboine Park Zoo," http://www.assiniboineparkzoo.ca/about/the-zoo.php (accessed 3 April 2014).

70 The Borden Park zoo was closed and replaced with the city-run Storyland Valley Zoo in 1959.

71 "City Parks Department Annual Report 1920," CCA, RG1, vol. 265, file 1861; Extracts from Parks Department Annual Reports, CCA, PDRF, series III, box 34, file 1954; Reader to Superintendent (Banff), 4 January and 28 February 1922, and Reader to Rawlings (Vancouver), 22 April 1922, CCA, PDRF, box 33, file 1922.

72 Extracts from annual reports, CCA, PRDF, series III, box 34, file 1954. Some local supplies dried up; by 1929 the national parks branch no longer supplied mule deer to zoos because, as Commissioner Harkin told Reader in 1929, they became quite tame in the open but when confined "they start to fight and knock themselves about," often injuring themselves or dying. Nor did they "travel well, often dying before reaching their destination" (Harkin to Reader, 21 December 1929, PRDF, series III, box 33, file 1929).

73 Extracts from annual reports, CCA, PRDF, series III, box 34, file 1954.

74 City Commissioner to Reader, 11 July 1931, CCA, PRDF, series III, box 33, file 1931.

75 *Calgary Herald*, 19 September 1931; Calgary Zoological Society to City Commissioners, 25 March 1929, CCA, CCDF, box 199, file 1342; "Annual Report of the Calgary Zoological Society 1932," CCA, RG1, vol. 275, file 1915.

76 Extracts from annual reports, CCA, PRDF, series III, box 34, file 1954; "Calgary Parks Department Annual Report 1936," CCA, RG1, vol. 278, file 1933; "Annual Report Calgary Zoological Society 1932," CCA, RG1, vol. 275, file 1915; "Foodstuffs used at St George's Island Zoo, January 1–June 1, 1930," CCA, PRDF, series III, box 33, file 1930.

77 In 1929, for example, an elephant or a pair of Bengal tigers cost $2,500, while a wildebeest or a Chapman's zebra cost around $550 (Chapman's Zoological Animal Corporation Price List, January 1929, CCA, PRDF, series III, box 33, file 1929).

78 *Calgary Herald*, 19 September 1931; Specimens in Zoo 1933, CCA, PRDF, series III, box 34, file 1954; "Annual Report Parks Department 1935," CCA, RG1, vol. 277, file 1929; Jennings (Banff) to Patrick, 22 February 1938, CCA, PRDF, series III, box 34, file 1938; "Annual Report Parks Department 1940," Inventory, CCA, RG1, vol. 281, file 1950.

79 Baines to Reader, 31 March 1931, and Reader to Chritchley and Calhoun, 16 June 1931, CCA, PRDF, series III, box 33, folder 1931; Potter to Soper, 24 April 1939, LAC, RG109-D-2, vol. 5, file "P" 1922–46.

80 "Reasons for a Proposed Educational Building for St George's Island Park" (attached to Baines to Reader, 13 March 1939), CCA, CPRF, series 3, box 34, file 1938.

81 "Calgary Zoological Society Annual Report, 28 February 1947," CCA, PRDF, series III, box 36, file 1947.

82 "Annual Report, Parks Department," CCA, RG1, vol. 288, file 1994; Curator's Report 1959, CCA, PRDF, series III, box 34, file 1954.
83 Harvie to Forsland, 7 August 1953 and passim, CCA, PRDF, series III, box 37, file 1953; *The Honker*, no. 10 (Summer 1956), CCA, PRDF, series III, box 37, file 1956.
84 Calgary Zoological Society to G.R. Bidlake (Yukon), 13 February 1953, and T. Kjar to Harvie, 11 August 1953, CCA, PRDF, series III, box 3, file 1953.
85 Clipping, *Montreal Star*, 15 August 1953, CCA, PRDF, series III, box 37, file 1954; "Annual Report Parks Department 1957," CCA, RG1, vol. 293, file 2022; Curator's Report, 1959, CCA, PRDF, series III, box 34, file 1954; Baines to Boothman ("Annual Report, 1963"), 10 January 1964, CCA, PRDF, series III, box 35, file 1964.
86 "Calgary Zoological Society Annual Report, February 28, 1947," CCA, PRDF, series III, box 34, file 1947; Publicity Committee Release, 30 April 1962, CCA, PRDF, series III, box 35, file 1962.
87 *Calgary Zoo News* 1, no. 1 (Spring 1963), CCA, PRDF, series III, box 42, file 1963; "Calgary Zoological Society Annual Report February 28, 1947," CCA, PRDF, series III, box 34, file 1947; Publicity Committee Release, 30 April 1962, CCA, PRDF, series III, box 35, file 1962.
88 "Parks Department Annual Report 1959," CCA, RG1, vol. 295, file 2033; Baines to Boothman ("Annual Report 1963"), 10 January 1964, CCA, PRDF, series III, box 35, file 1964.
89 *Calgary Zoo News* 2, no. 1 (Winter 1964), CCA, PRDF, series III, box 42, file 1964.
90 "Parks Department Annual Report 1954," CCA, RG1, vol. 292, file 1954; Willis, "Looking at the Zoo," 682.
91 "Annual Report Parks Department 1956," CCA, RG1, vol. 293, file 2022; "Annual Report Parks Department 1957," CCA, RG1, vol. 294, file 2025.
92 *Monthly News Letter*, June–July 1962, CCA, PRDF, series III, box 42, file 1962.
93 "Annual Report Parks Department 1962," CCA, PRDF, series III, box 35, file 1962; *Calgary Zoo's News* 2, no. 1 (Winter 1964), CCA, PRDF series III, box 42, file 1964.
94 "Zoo Planning" (1961), CCA, PRDF, series 3, box 43, file 1961.
95 Calgary Zoological Society to G.R. Bidlake (Yukon), 13 February 1953, CCA, PRDF, series III, box 3, file 1953; "Annual Report Parks Department 1956," CCA, RG1, vol. 293, file 2022; "Annual Report Parks Department 1955," CCA, RG1, vol. 293, file 2019.
96 *Monthly News Letter*, June–July 1962, CCA, PRDF, series III, box 42, file 1962.
97 Di Cintio "The Zookeeper," 41. As an example of the continuation of this trend, the Calgary Zoo in 2013 was working on captive breeding of whooping cranes.
98 Al Oeming closed the park in 1998 citing changing public attitudes and

rising hostility from animal rights activists (*Globe and Mail*, 9 April 2014). Another zoo was created in Saskatoon around this time. The city purchased the Golden Gate Animal Farm in 1964 and kept the animals at the Forestry Farm just outside the city. It then opened an Animal Park at the Forestry Farm in 1972 (DZ Design, *Forestry Farm Conceptual Development Plan*, Saskatoon: n.d. [1975], 2–3).

99 Colpitts, *Game in the Garden*, 122–3; *Regina Leader-Post*, 25 March 1929; *Saskatchewan Fish and Game League Year Book 1948*, 71 (copy in SAB R-E4531).

100 Colpitts, *Game in the Garden*, 122–3; *Regina Leader-Post*, 25 March 1929; *Moose Jaw Wild Animal Park and Kiddie's Zoo* [guidebook], n.p, n.d. [1963], 41 and passim.

101 "Bluebird," Nature Lovers page, *Western Producer*, 27 September 1934; "Tarry" (Kayville, SK), Nature Lovers page, *Western Producer*, 28 October 1948.

102 *Annual Report Saskatchewan DNR for Year Ended 31 March 1953*, 119; *Moose Jaw Wild Animal Park and Kiddie's Zoo* [guidebook], n.p., n.d., [1963], 41 and passim. The park was closed by the 1990s and the site was abandoned.

103 Secord, "The Crisis of Nature," 542; Mitman, "When Nature *Is* the Zoo," 117–20. On the contemporary appeal of conservation as a goal of zoos, see Fennell, *Tourism and Animal Ethics*, 78–84.

CONCLUSION

1 H.S.F., "Lo the Poor Fish," *Country Guide*, May 1947.

2 Forkey, *Canadians and the Natural Environment*, 40–1.

3 "Questions and Answers about the Saskatchewan Fish and Game League" (typescript) (c. 1950), SAB, R-E4531.

4 *Saskatchewan Federation of Fish and Game League Year Book 1943*, 11 (copy in SAB, R-E 4531).

5 This is in contrast to the western United States, where the federal government was a more active force. On this, see Warren, *The Hunter's Game*.

6 Dunlap, *Saving America's Wildlife*, 6.

7 *Annual Report Saskatchewan DA for Year Ended 30 April 1918*, 207.

8 Ainley, *Restless Energy*, 253.

9 *Saturday Night* 65, no. 1 (1949): 96.

10 Forkey, *Canadians and the Natural Environment*, 104–6.

11 Pembina Institute, *Toward a Sustainable Economy*, 1–2.

12 Fennell, *Tourism and Animal Ethics*, 29–30, 32.

13 Hamilton, *Requiem for a Species*, 135.

14 On predators in the re-wilding movement, see Chapron et al., "Recovery of Large Carnivores," 1517–19.

Bibliography

ARCHIVAL COLLECTIONS

American Museum of Natural History (AMNH)
E-3, Ernest Thompson Seton Journals

City of Calgary Archives (CCA)
CCA, City Clerk's and Department Fonds
CCA, Parks and Recreation Department Fonds
CCA, RG1, City of Calgary Council Fonds

Glenbow Archives (GA)
GA, A.S243, Leslie Sara Fonds
GA, M106, Tait Family Fonds
GA, M1327, Austin de B. Winter Fonds

Library and Archives Canada (LAC)
LAC, MG26-A, Macdonald Fonds
LAC, MG29 B15, Robert Bell Papers
LAC, MG29 D108, Ernest Thompson Seton Papers
LAC, MG30 E441, Hoyes Lloyd Papers
LAC, RG10, Records of the Department of Indian affairs
LAC, RG109 – D-2, Canadian Wildlife Service Papers, Correspondence of J. Dewey Soper
LAC, RG15, Records of the Department of the Interior
LAC, RG17, Records of Canada Department of Agriculture
LAC, RG18, Records of the Royal Canadian Mounted Police
LAC, RG45, Records of the Geological Survey of Canada
LAC, RG84, Records of Parks Canada
LAC, RG13, Records of the Department of Justice

Manitoba Archives (MA)
MA, EC0016/GR1661, Premier Bracken Papers
MA, EC0016/GR1662, Premier Thomas Greenway Papers
MA, EC0016/GR1843, Premier Garson Papers

MA, EC0016/GR43, Premier Bracken Papers
MA, GR1600, Manitoba Department of Mines and Natural Resources Papers
MA, MG12/E1, Schultz Papers
MA, MG14, Winkler Papers
MA, MG2 B1, District of Assiniboia Council Minutes
MA, MG8/B3, William Lothian Papers
MA, MG8/B68, William Taylor Papers
MA, P2993/9, E.S. Russenholt Papers
MA, P3608–3617, Winnipeg Game and Fish Association Papers (various accessions)
MA, P5070, Manitoba Naturalist Society Papers

Provincial Archives of Alberta (PAA)
PAA, 66.166, Department of the Attorney General Papers
PAA, 67.21, Department of Agriculture Papers
PAA, 67.293, Alberta Natural History Society Papers
PAA, 70.347, Barford Family Fonds
PAA, 72.302, Department of Agriculture Papers
PAA, 73.30, Department of Agriculture Papers
PAA, Premiers Papers 1927–46

Queens University Archives (QUA)
QUA, Collection 2026a, Charles Mair Fonds

Saskatchewan Archives Board (Regina) (SAB)
SAB, R-126.2, P.E. Lavoie Papers
SAB, SHS82, "Various Materials on Bison"
SAB, Clipping File "Bison, American"
SAB, Clipping File "Wolves"
SAB, NR1/1, Saskatchewan Department of Natural Resources Papers
SAB, NR1/2, Saskatchewan Department of Natural Resources Papers
SAB, Pamphlet File "Bison American"
SAB, Pamphlet File "Natural Resources, Big Game"
SAB, Pamphlet File "Natural Resources, Saskatchewan"
SAB, Pamphlet File "Recreation"
SAB R-972, D.G. Stuart Papers
SAB, R-1337, William Alexander Marjerrison Papers
SAB, R-1456, R.D. Symons Papers
SAB, R-1676, Saskatchewan Natural History Society Papers
SAB, R-192.4, Papers of the Department of Provincial Secretary, Correspondence of Lt Governor
SAB, R-283, Saskatchewan Sessional Papers

SAB, R-43, P.H. Gordon Papers
SAB, R-942, Regina Natural History Society Papers
SAB, R-E 2786, Grey Owl
SAB, R-E 3033, Elizabeth Roley Cruickshank Papers
SAB, R-E 3553, Grey Owl Letters
SAB, R-E 3863, Grey Owl Papers
SAB, R-E 4531, Saskatchewan Fish and Game League Papers
SAB, R-E2094, Grey Owl Papers
SAB, R-E3294, Ralph Stueck Papers
SAB, R-E4121, William Riddlell Papers
SAB, R-E4287, Fur Advisory Committee Papers
SAB, R-E4531, Saskatchewan Fish and Game League Papers
SAB, SHS50, Lawrence Potter Papers

University of Manitoba Archives (UMA)
UMA, MSS24, Winnipeg Tribune Collection

Other
Manitoba Legislative Library, Clipping Collection, M 8–9

GOVERNMENT PRINTED SERIAL RECORDS (actual titles vary)*

Alberta. Annual Report Department of Agriculture 1905–65
Alberta. Annual Report Department of Lands and Forests 1950–65
Alberta. Annual Report Department of Lands and Mines 1931–49
Canada. Annual Report Department of Fisheries [various titles] 1885–1930–31
Canada. Annual Report Department of Indian Affairs 1873–1929
Canada. Annual Report Department of the Interior 1876–1931
Canada. Annual Report of the North-West Mounted Police [after 1921, Royal Canadian Mounted Police] 1879–1929
Canada. *Debates of the Senate* 1878–88
Canada. *House of Commons Debates* 1877–1920
Manitoba. Annual Report Department of Agriculture and Immigration 1882–83, 1902–16, 1919–21, 1925–49
Manitoba. Annual Report Department of Mines and Natural Resources 1931–65

* There is much variation in formal titles and the periods covered by government annual reports. Titles are condensed here to create a consistent format without sacrificing clarity.

North-West Territories. Annual Report Department of Agriculture 1898–1904
Saskatchewan. Annual Report Department of Agriculture [Game Reports] 1905–27
Saskatchewan. Annual Report Department of Natural Resources [Game and Fisheries] 1931–65

NEWSPAPERS AND MAGAZINES (outside dates, major citations only)

Calgary Herald, 1902–57
Conservation, 1912–20
Country Guide, 1908–66
Edmonton Journal, 1921–45
Farm and Ranch Review, 1905–66
Grain Growers' Guide, 1913–29
Macleod Gazette, 1887–1900
Nor'West Farmer, 1897–1905

Regina Leader, 1883–1921
Regina Leader-Post, 1929–54
Saturday Night, 1887–1965
Western Producer, 1926–65
Winnipeg Tribune (researched using subject files in UMA, MSS24, Winnipeg Tribune Collection)

BOOKS AND ARTICLES

Abrahamson, Kurt. "Landscapes Lost and Gained: On Changes in Semiotic Resources." *Human Ecology Review* 6, no. 2 (1999): 51–61.

Adams, Charles. "The Relation of General Ecology to Human Ecology." *Ecology* 16, no. 3 (July 1935): 316–35.

Ainley, Marianne Gostonyi. *Restless Energy: A Biography of William Rowan 1891–1957*. Montreal: Véhicule Press, 1993.

Alberta Poetry Year Book 1934–35. Edmonton: Canadian Authors' Association, Edmonton Branch, 1934.

Anderson, Alan, and Brenda Niskala. "Finnish Settlements in Saskatchewan: Their Development and Perpetuation." *Finnish Diaspora I: Canada, South America, Africa, Australia and Sweden*, edited by Michael Karni, 155–82. Toronto: Multicultural History Society of Ontario, 1981.

Anderson, H. Allen. *The Chief: Ernest Thompson Seton and the Changing West*. College Station: Texas A&M University Press, 1986.

Anderson, J.W. "7,000 Square Miles of Beaver Sanctuary." *The Beaver*, Outfit 265 (June 1934): 16–18.

Anderson, Kay. "Culture and Nature at the Adelaide Zoo: At the Frontiers of

'Human' Geography." *Transactions of the Institute of British Geographers*, new series 20, no. 3 (1995): 275–94.
Atkinson, George E. *Insectivorous Birds of Manitoba.* Historical and Scientific Society of Manitoba Transaction No. 80 (14 November 1901). Winnipeg: Stovel Co. Printers, 1903.
– *Manitoba Birds of Prey and the Small Mammals Destroyed by Them.* Winnipeg: Historical and Scientific Society of Manitoba, 1899.
– *A Review-History of the Passenger Pigeon of Manitoba.* The Historical and Scientific Society of Manitoba, Transaction No. 68 (February 1905). Winnipeg: Manitoba Free Press Co, 1905.
Atwood, Margaret. *Survival: A Thematic Guide to Canadian Literature.* Toronto: Anansi, 1972.
Baillie, James L., Jr. "Naturalist on Hudson Bay." *The Beaver*, Outfit 277 (December 1946): 36–9.
Baker, Steve. "Guest Editor's Introduction: Animals, Representation, and Reality." *Society and Animals* 9, no. 3 (2001): 189–201.
Balazs, Dawn. *A Short Analysis of the Transfer of Natural Resources to Alberta in 1930.* Edmonton: Treaty and Aboriginal Rights Research of the Indian Association of Alberta, 1975 (typescript).
– *Transfer of Natural Resources in 1930 and Registered Traplines.* Edmonton: Treaty and Aboriginal Rights Research, Indian Association of Alberta, 1976.
Balcombe, Jonathan. *Second Nature: The Inner Lives of Animals.* New York: Palgrave Macmillan, 2010.
Barbour, Alexander. "A Brief History of the Manitoba Fisheries." *Manitoba Historical Society Transactions*, series 3, 1955–56, http://www.mhs.mb.ca/docs/transactions/3/manitobafisheries.shtml.
Beaudoin, Alwynne. "What They Saw: The Climatic and Environmental Context for Euro-Canadian Settlement in Alberta." *Prairie Forum* 24, no. 1 (Spring 1999): 1–40.
Beaver in Saskatchewan. Conservation Bulletin No. 1. Regina: Department of Natural Resources, n.d. [c. 1955].
Bella, Leslie. *Parks for Profit.* Montreal: Harvest House, 1987.
Bennett, Tony. *The Birth of the Museum: History, Theory, Politics.* London and New York: Routledge, 1995.
Berger, Carl. *Science, God, and Nature in Victorian Canada.* The 1982 Joanne Goodman Lectures. Toronto: University of Toronto Press, 1983.
Berger, John. "Why Look at Animals?" In John Berger, *About Looking*, 1–26. New York: Pantheon Books, 1980.
Bergman, Charles. "Obits for the Fallen Hunter: Reading the Decline – and Death? – of Hunting in America." *American Literary History* 17, no. 4 (2005): 818–30.
Bevington, Angie. "Frank Ralph Conibear, 1896–." *Arctic* 36, no. 4 (1983): 386–7.

Binnema, Theodore (Ted), and Melanie Niemi. "'Let the Line Be Drawn Now': Wilderness, Conservation and the Exclusion of Aboriginal People from Banff National Park in Canada." *Environmental History* 11, no. 4 (October 2006): 724–50.

Blais, Gabrielle. "Preserving the Balance of Nature: Wildlife Conservation in Canada's National Parks." *The Archivist* 12, no. 2 (March–April 1985): 3–5.

Bodsworth, Fred. *Last of the Curlews.* Toronto: McClelland and Stewart, [1954] 1963.

Bossenmaier, Eugene F. *100-Plus Years of Wildlife Protection and Development in Manitoba.* Winnipeg: Manitoba Department of Renewable Resources and Transportation Services, 1978.

Braz, Albert. "Wither the White Man: Charles Mair's 'Lament for the Bison.'" *Canadian Poetry* vol. 49, http://canadianpoetry.org/volumes/vol49/braz.html

Brightman, Robert. *Grateful Prey: Rock Cree Human–Animal Relationships.* 2nd rev. edn. Regina: Great Plains Research Center, [1973] 2002.

Brink, Jack. *Imagining Head-Smashed-In: Aboriginal Buffalo Hunting on the Northern Plains.* Edmonton: AU Press, 2008.

Brody, Hugh. *The Other Side of Eden: Hunters, Farmers and the Shaping of the World.* Vancouver and Toronto: Douglas and McIntyre, 2000.

Brook, Ryan K. "Historical Review of Elk-Agriculture Conflicts in and around Riding Mountain National Park, Manitoba, Canada." *Human–Wildlife Conflicts* 3, no. 1 (2009): 72–87.

Brower, Jennifer. *Lost Tracks: National Buffalo Park 1909–1939.* Edmonton: AU Press, 2008.

Brower, Matthew. "'Take Only Photographs': Animal Photography's Construction of Nature Love." *Invisible Culture: An Electronic Journal for Visual Culture,* no. 9 (2005): n.p., http://www.rochester.edu/in_visible_culture/Issue_9/brower.html.

– "Trophy Shots: Early North American Photographs of Nonhuman Animals and the Display of Masculine Prowess." *Society and Animals* 13, no. 1 (2005): 13–30.

Brown, Robert Craig. "The Doctrine of Usefulness: Natural Resource and National Park Policy in Canada 1887–1914." In *The Canadian National Parks: Today and Tomorrow,* edited by J.G. Nelson and R.C. Space, 46–62. Calgary: National and Provincial Parks Association of Canada, 1968.

Browne, Janet. "Biogeography and Empire." In *Cultures of Natural History,* edited by N. Jardine, J.A. Secord, and E.C. Spary, 305–21. Cambridge: Cambridge University Press, 1996.

Bullick, Terry. *Calgary Parks and Pathways.* Calgary: Rocky Mountain Books, 1991.

Burman, W.A. *The Present Status of Natural Science in Manitoba and the Northwest.* Historical and Scientific Society of Manitoba. Winnipeg: Manitoba Free Press Co. 1900.

Burnett, J. Alexander. *A Passion for Wildlife: The History of the Canadian Wildlife Service.* Vancouver: UBC Press, 2003.

Butler, William Frances. *The Great Lone Land: A Narrative of Travel and Adventure in the North-West of America.* Rutland, VT, and Tokyo: Charles Tuttle Co. Publishers, 1968 (facsimile of 1872 edition).

Carpenter, David. *A Hunter's Confession.* Vancouver: Greystone Books, D&M Publishers, 2010.

Carter, Sarah. *Lost Harvests: Prairie Indian Reserve Farmers and Government Policy.* Montreal and Kingston: McGill-Queen's University Press, 1990.

Chambers, Ernest J., ed. *Canada's Fertile Northland: A Glimpse of the Enormous Resources of Part of the Unexplored Regions of the Dominion.* Ottawa: Government Printing Bureau, 1907.

Chapron, Guillaume, et al. "Recovery of Large Carnivores in Europe's Modern Human-Dominated Landscapes." *Science* 346, no. 6216 (19 December 2014): 1517–19.

Clancy, Peter. "Game Policy in the Northwest Territories: The Shaping of Economic Position." Paper presented at the Annual Meeting, Canadian Political Science Association, 6–8 June 1983.

– "State Policy and the Native Trapper: Post-War Policy toward Fur in the Northwest Territories. In *Aboriginal Land Use in Canada: Historical and Legal Aspects,* edited by Kerry Abel and Jean Friesen, 191–217. Winnipeg: University of Manitoba Press, 1991.

Coldwell, John Adams. *Sir Charles God Damn: The Life of Sir Charles G.D. Roberts.* Toronto: University of Toronto Press, 1986.

Coleman, Jon T. "Two by Two: Bringing Animals into American History." *Reviews in American History* 33, no. 4 (2005): 481–92.

Colpitts, George. "Fish and Game Conservation in Alberta." *Alberta History* 42, no. 1 (Autumn 1994): 16–26.

– *Game in the Garden: A Human History of Wildlife in Western Canada to 1940.* Vancouver: UBC Press, 2002.

– "Wildlife Promotion, Western Canadian Boosterism and the Conservation Movement, 1890–1914." *American Review of Canadian Studies* 28, nos 1–2 (Spring–Summer 1998): 103–30.

Commission of Conservation Canada. *National Conference on Conservation of Game, Fur-Bearing Animals and Other Wild Life. Under the Direction of the Commission for Conservation in Co-operation with the Advisory Board on Wild Life Protection, February 18 and 19, 1919.* Ottawa: King's Printer, 1919.

Congdon, F.T. "Fur Bearing Animals in Canada, and How to Prevent Their Extinction." *Commission of Conservation Canada, Report of the First Annual Meeting Held at Ottawa, January 18th to 21st, 1910,* 107–14. Ottawa: Mortimer Co., 1910.

Conibear, Frank. "The Beaver's Lodge." *The Beaver,* Outfit 282 (September 1951), 36–7.

Connor, Ralph. *The Man from Glengarry.* Toronto: Westminster Co., 1901.
Conrad, Norman. *Reading the Entrails: An Alberta Ecohistory.* Calgary: University of Calgary Press, 1999.
Cooke, May Thacher. "The Spread of the European Starling in North America (to 1928)." Washington: United States Department of Agriculture, Circular No. 40, 1928.
Cottingham, John. "A Brute to the Brutes: Descartes Treatment of Animals." *Philosophy* 53 (October 1978): 551–9.
Cotton, Theodore. "From Game Refuges to Ecosystem Assessments: Wildlife Management in the Western National Forests." *Journal of the West* 38, no. 4 (1999): 36–44.
Criddle, Norman. *The Birds of a Manitoba Garden* (pamphlet), n.p., n.d. [c. 1920].
– *The Habits and Economic Importance of Wolves in Canada.* Ottawa: Department of Agriculture, Bulletin No. 13, new series, 1925.
– "Some Canadian Rodents Injurious to Agriculture." *Agricultural Gazette* 2, no. 2 (1915): 110–15.

Cronon, William. "Cutting Loose or Running Aground?," *Journal of Historical Geography* 20, no. 1 (1994): 22–43.
Crosby, Alfred. *Ecological Imperialism.* Cambridge: Cambridge University Press, 1986.
Crozier, Lorna. *A Saving Grace: The Collected Poems of Mrs. Bentley.* Toronto: McClelland and Stewart, 1996.
D.G. Wetherell and Associates. *Agriculture in Western Canada: A Historical Assessment* (typescript). Report for National Museum of Science and Technology, 1994.
Demeritt, David. "Ecology, Objectivity and Critique in Writings on Nature and Human Societies." *Journal of Historical Geography* 20, no. 1 (1994): 22–37.
– "The Nature of Metaphors in Cultural Geography and Environmental History." *Progress in Human Geography* 18, no. 2 (1994): 163–85.
Denmark, Donald. "Conservation at Cumberland." *The Beaver*, Outfit 270, no. 4 (March 1940): 47–9.
– "James Bay Beaver Conservation." *The Beaver*, Outfit 279 (September 1948): 38–43.
Denny, C.E. *The Riders of the Plains: A Reminiscence of the Early and Exciting Days in the North West.* Calgary: Herald Company, 1905.
den Otter, A.A. "'The Wilderness Will Rejoice and Blossom Like the Crocus': Bishop David Anderson's Perceptions of Wilderness and Civilization in Rupert's Land." *Historical Papers 2001: Canadian Society of Church History.* [Langley, BC: Canadian Society of Church History] (2001): 81–100.
Desmond, Jane. "Displaying Death, Animating Life: Changing Fictions of 'Liveness' from Taxidermy to Animatronics." In *Representing Animals*, edited by Nigel Rothfels, 158–79. Bloomington and Indianapolis: Indiana University Press, 2002.

Di Cintio, Marcello. "The Zookeeper." *Alberta Views* 13, no. 6 (July–August 2010): 38–43.
Dick, L.R. Review. *Ecology* 4, no. 1 (January 1923): 86–7.
Dombrowski, Daniel A. "Bears, Zoos, and Wilderness: The Poverty of Social Constructionism." *Society and Animals* 10, no. 2 (2002): 195–202.
Donihee, John. *The Evolution of Wildlife Law in Canada.* Canadian Institute of Resources Law, Occasional Paper No. 9. Calgary: University of Calgary, 2000.
Dorsey, Kirk. "Scientists, Citizens and Statesmen: U.S.-Canadian Wildlife Protection Treaties in the Progressive Era." *Diplomatic History* 19, no. 3 (Summer 1995): 407–29.
Doughty, Robin. "Sparrows for America: A Case of Mistaken Identity." *Journal of Popular Culture* 14, no. 2 (1980): 212–28.
Dowson, Eric. *Country Places and Other Poems.* Nanaimo, BC: n.p., n.d.
Doyle, James. "Science, Literature and Revolution: The Life and Writings of Dyson Carter." *left history* 5, no. 2 (Fall 1997): 7–29.
Drouin, Jean Marc, and Bernadette Bensaude-Vincent. "Nature for the People." In *Cultures of Natural History*, edited by N. Jardine, J.A. Secord, and E.C. Spary, 408–25. Cambridge: Cambridge University Press, 1996.
Ducks Unlimited (Canada). *Census 1938 & 1939 and Kee-man Record Book.* Winnipeg: Ducks Unlimited (Canada) [1939].
Dunaway, Finnis. "Hunting with the Camera: Nature Photography, Manliness, and Modern Memory 1890–1910." *Journal of American Studies* 34, no. 2 (August 2000): 207–30.
Dunlap, Thomas. "Ecology, Nature and Canadian National Parks Policy: Wolves, Elk and Bison as a Case Study." In *To See Ourselves: Ecology and Culture in Canada*, edited by Rowland Lorimer, Michael M'Gonigle, Jean-Pierre Revéret, and Sally Ross, 139–47. Montreal: Association of Canadian Studies, 1991.
– "Organisation and Wildlife Preservation: The Case of the Whooping Crane in North America." *Social Studies of Science* 21, no. 2 (May 1991): 197–221.
– "The Realistic Animal Story: Ernest Thompson Seton, Charles Roberts, and Darwinism." *Forest and Conservation History* 36 (April 1992): 56–62.
– "Remaking the Land: The Acclimatization Movement and Anglo Ideas of Nature." *Journal of World History* 8, no. 2 (1997): 303–19.
– *Saving America's Wildlife.* Princeton: Princeton University Press, 1988.
– "Sport Hunting and Conservation 1880–1920." *Environmental Review* 12, no. 1 (Spring 1985): 51–60.
– "That Kaibab Myth." *Journal of Forest History* 32, no. 2 (April 1985): 60–8.
– "Value for Varmits: Predator Control and Environmental Ideas 1920–1939." *Pacific Historical Review* 53, no. 2 (May 1984): 141–61.
Egerton, Frank. "Changing Concepts of the Balance of Nature." *Quarterly Review of Biology* 48, no. 2 (June 1973): 322–50.
– "The History of Ecology: Achievement and Opportunities, Part One." *Journal of the History of Biology* 16 (1983): 259–311.

– "The History of Ecology: Achievement and Opportunities, Part Two." *Journal of the History of Biology* 18 (1985): 103–43.

Ehrenreich, Barbara. *Living with a Wild God: A Nonbeliever's Search for the Truth about Everything*. New York and Boston: Twelve, 2014.

Elofson, Warren. "Grassland Management in Southern Alberta: The Frontier Legacy." *Agricultural History* 86 (Fall 2012): 143–68.

– *Somebody Else's Money: The Walrond Ranch Story 1883–1907*. Calgary: University of Calgary Press, 2009.

Elphinstone, Dave. *Inglewood Bird Sanctuary: A Place for All Seasons*. Calgary: Rocky Mountain Books, 1990.

Emel, Jody, and Jennifer Wolch. "Witnessing the Animal Moment." In *Animal Geographies: Place, Politics and Identity in the Nature-Culture Borderlands*, edited by Jody Emel and Jennifer Wolch, 1–24. London: Verso, 1998.

Epp, Henry. "Henry Kelsey's Journals and Correspondence." In *Three Hundred Prairie Years: Henry Kelsey's "Inland Country of Good Report*, 196–235. Regina: Canadian Plains Research Center, 1993.

"Experiment at Steeprock." *The Beaver*, Outfit 270, no. 1 (June 1939): 10–11.

Farber, Paul Lawrence. "The Development of Taxidermy and the History of Ornithology." *Isis* 68, no. 4 (December 1977): 550–66.

Farley, Frank. *Birds of the Battle River Region: With Notes on Their Present Status, Migration, Food Habits and Economic Value*. Edmonton: Institute of Applied Arts, 1932.

Fender, Ann Harper. "Public versus Private Ownership: Saskatchewan Fur Trapping and Trading Legislation in the 1940s." In *Selected Papers of the Seventh North American Fur Trade Conference, Halifax, Nova Scotia, 1995*, edited by JoAnne Fiske, Susan Sleeper-Smith, and William Wicker, 223–43. East Lansing: Michigan State University Press, 1998.

Fennell, David. *Tourism and Animal Ethics*. London and New York: Routledge, 2012.

Fiamengo, Janice. "Looking at Animals; Encountering Mystery: The Wild Animal Stories of Ernest Thompson Seton and Charles G.D. Roberts." *Journal of Canadian Studies* 44, no. l (2010): 36–59.

Fine, Lisa M. "Rights of Men, Rites of Passage: Hunting and Masculinity at Reo Motors of Lansing, Michigan, 1945–1975." *Journal of Social History* 33, no. 4 (2000): 805–23.

"Fisheries of Manitoba." In *Commission of Conservation Canada: Report on "Lands, Fisheries and Game, and Minerals" for the Year 1911*, 164–75. Ottawa: Mortimer Co., 1911.

Fisheries Research in Saskatchewan. Regina: Department of Natural Resources, Conservation Bulletin No. 3, n.d. (c. 1960).

Fish, Fur and Feathers: Fish and Wildlife Conservation in Alberta 1905–2005. Edmonton: Federation of Alberta Naturalists, 2005.

Flores, Dan. "Bison Ecology and Bison Diplomacy: The Southern Plains from 1800 to 1850." *Journal of American History* 78 (September 1991): 465–85.

Forkey, Neil S. *Canadians and the Natural Environment in the Twenty-First Century*. Toronto: University of Toronto Press, 2012.

Foster, Janet. *Working for Wildlife: The Beginning of Preservation in Canada*. 2nd edn. Toronto: University of Toronto Press, 1998.

Fowke, Vernon. *Canadian Agricultural Policy*. Toronto: University of Toronto Press, 1946.

Francis, R. Douglas. *Images of the West: Changing Perception of the Prairies 1690–1960*. Saskatoon: Western Producer Prairie Books, 1989.

Francis, R. Douglas, Richard Jones, and Donald B. Smith, eds. *Origins: Canadian History to Confederation*. 3rd edn. Toronto: Harcourt Brace and Co., 1996.

Francis, R. Douglas, and Chris Kitzan. *The Prairie West as Promised Land*. Calgary: University of Calgary Press, 2007.

Friesen, Gerald. "The Manitoba Historical Society: A Centennial History." *Manitoba History*, no. 4 (1982), www.mhs.mb.ca/docs/mb_history/04/centennial.shtml.

Friesen, Jean. "Grant Me Wherewith to Make My Living." In *Aboriginal Resource Use in Canada: Historical and Legal Aspects*, edited by Kerry Abel and Jean Friesen, 141–55. Winnipeg: University of Manitoba Press, 1991.

Fromm, Eric. *The Anatomy of Human Destructiveness*. New York: Holt, Rinehart and Winston, 1973.

Gabrielson, Ira. "Conservation of Waterfowl." In *The Ducks, Geese and Swans of North America*, edited by Francis Kortright, 55–61. Harrisburg, PA: The Stackpole Company, and Washington: Wildlife Management Institute, 1942.

Genosko, Gary. "Natures and Cultures of Cuteness." *Invisible Culture: An Electronic Journal for Visual Culture*, http://hdl.handle.net/1802/3753 (accessed 22 January 2007).

Gibson, Jim. "Three Tiny Prairie Parks Saved the Pronghorn." *Canadian Geographic* 107, no. 6 (December 1987/January 1988): 59–65.

Gillese, John Patrick. "The Wildlife Journals of William Rowan." *The Beaver*, Outfit 289 (Autumn 1958): 46–53.

Gillespie, Greg. "'I Was Much Pleased with Our Sport among the Buffalo': Big Game Hunters, Travel Writers and Cultural Imperialism in the British American West, 1847–72." *Canadian Historical Review* 83, no. 4 (December 2002): 555–84.

Glavin, Terry. *Waiting for the Macaws and Other Stories from the Age of Extinctions*. Toronto: Viking Canada, 2006.

Goatcher, Melody L. *Manitoba Legislation Affecting Wildlife from 1876–1972*. 2nd printing. Winnipeg: Manitoba Department of Renewable Resources and Transportation Services, Fisheries and Wildlife Branch, March 1977.

Goldsborough, Gordon. "Man's War on Wildlife." *Manitoba Chapter Wildlife Society Newsletter* (April 2010): 2–13.

Gottesman, Dan. "Native Hunting and the Migratory Birds Convention Act: Historical, Political and Ideological Perspectives." *Journal of Canadian Studies* 18, no. 3 (Fall 1983): 67–89.

Grady, Wayne. *Chasing the Chinook: On the Trail of Canadian Words and Culture.* Toronto: Viking, 1998.

Graham, Maxwell. *Canada's Wild Buffalo: Observations on the Wood Buffalo Park.* Ottawa: King's Printer, 1923.

Gray, Philip Howard. "The Early Animal Behaviorists: Prolegomenon to Ethology." *Isis* 59, no. 4 (Winter 1968): 372–83.

Griesbach, W.A. "Sheriden Lawrence: A Tribute." *Alberta Historical Review* 4 (1956): 20–2.

Grouse and Introduced Game Birds of Saskatchewan. Regina: Museum of Natural History, Popular Series No. 5, 1963 (reprinted 1974).

Guntau, M. "The Emergence of Geology as a Scientific Discipline." *History of Science* 16, no. 4 (1978): 280–90.

Hamilton, Clive. *Requiem for a Species.* London: Earthscan, 2010.

Hampton, Bruce. "Shark of the Plains: Early Western Encounters with Wolves." *Montana* 46, no. 1 (1996): 2–13.

Hanner, John. "Government Response to the Buffalo Hide Trade, 1871–1883." *Journal of Law and Economics* 24 (October 1981): 239–71.

Harris, S. "The Need and Value of Local Organizations." In *Commission of Conservation Canada, National Conference on Conservation of Game, Fur-Bearing Animals and Other Wild Life. Under the Direction of the Commission of Conservation in Co-operation with the Advisory Board on Wild Life Protection, February 18 and 19, 1919,* 141–5. Ottawa: King's Printer, 1919.

Harrison, Dick. *Unnamed Country: The Struggle for a Canadian Prairie Fiction.* Edmonton: University of Alberta Press, 1977.

Hart, E.J. (Ted). *J.B. Harkin: Father of Canada's National Parks.* Edmonton: University of Alberta Press, 2010.

Haskell, William. "Protection of Migratory Birds." In *Commission of Conservation Canada, Report of the Fifth Meeting Held at Ottawa, January 20–21, 1914,* 66–73. Toronto: Bryant Press, 1914.

Hayes, Cecilia, and Jeff Gale. *Social Learning in Animals: The Roots of Culture.* San Diego: Academic Press, 1996.

Hays, Samuel. *Conservation and the Gospel of Efficiency: The Progressive Conservation Movement 1890–1920.* Cambridge: Harvard University Press, 1959.

Hedlin, Ralph. "The Buffalo as a Northern Resource." *The Beaver,* Outfit 291 (Summer 1960): 24–8.

Herman, Daniel. *Hunting and the American Imagination.* Washington: Smithsonian Institution Press, 2011.

Hewitt, Gordon. "Conservation of Birds and Mammals in Canada." In *Commission for Conservation Canada, Committee on Fisheries, Game and Fur Bearing Animals, Conservation of Fish, Birds and Game. Proceedings of a Meeting of the Committee, November 1 and 2, 1915*, 141–51. Toronto: Methodist Book and Publishing House, 1916.

– *The Conservation of the Wild Life of Canada*. New York: Charles Scribner's Sons, 1921.

– "The Conservation of Wild Life in Canada in 1917: A Review." In *Commission of Conservation Canada, Report of the Ninth Meeting Held at Ottawa, November 27–28, 1917*, 118–39. Ottawa: 1918.

Hints on Community Coyote Hunting. Edmonton: Alberta Department of Agriculture, Livestock Branch, January 1950.

Holliday, N.J. "Norman Criddle: Pioneer Entomologist of the Prairies." *Manitoba History*, no. 51 (February 2006): online version.

Horowitz, Barbara Natterson, and Kathryn Bowers. *Zoobiquity: What Animals Can Teach Us about Being Human*. N.p.: Doubleday Canada, 2012.

Horowitz, Helen. "Animal and Man in the New York Zoological Park." *New York History* 5, no. 4 (1975): 126–55.

Houston, C. Stuart, and Marc J. Bechard. "Decline of the Ferruginous Hawk in Saskatchewan." *American Birds*, March–April 1984, 166–70.

Houston, Stuart, and Mary Houston. "Samuel Hearne, Naturalist." *The Beaver* 67, no. 4 (August–September 1987): 23–7.

Hunting with a Camera. National Film Board (16 mm. colour film, 17 minutes), Producer: Stephen Greenlees. Copy at LAC: ISN 130427.

Ingold, Tim. "Review of Adrian Franklin, *Animals and Modern Culture: A Sociology of Human–Animal Relations in Modernity*." *Society and Animals* 9, no. 2 (2001): 183–8.

Ingram, Darcy. "Beastly Measures: Animal Welfare, Civil Society, and State Policy in Victorian Canada." *Journal of Canadian Studies* 41, no. 1 (Winter 2013): 221–52.

Isenberg, Andrew C. "The Moral Ecology of Wildlife." In *Representing Animals*, edited by Nigel Rothfels, 48–63. Bloomington: Indiana University Press, 2002.

– "The Returns of the Bison: Nostalgia, Profit and Preservation." *Environmental History* 2, no. 2 (April 1997): 179–96.

Isern, Thomas. "Gopher Tales: A Study in Western Canadian Pest Control." *Agricultural History Review* 36, no. 2 (1988): 188–98.

Jacques, Edna. *Beside Still Waters*. Toronto: Thomas Allen, 1939.

Jardine, Nicolas, and Emma Spary. "The Natures of Cultural History." In *Cultures of Natural History*, edited by N. Jardine, J.A. Secord, and E.C. Spary, 3–14. Cambridge: Cambridge University Press, 1996.

Jasen, Patricia. "Native People and the Tourist Industry in Nineteenth Century Ontario." *Journal of Canadian Studies* 28, no. 4 (1993–94): 5–27.

Jones, Karen. "*Never Cry Wolf*: Science, Sentiment, and the Literary Rehabilitation of *Canis Lupus*." *Canadian Historical Review* 84, no. 1 (March 2003): 65–93.

Kalmbach, E.R. *The Crow in Its Relation to Agriculture: Is it a Farm Pest?* Washington: US Department of Agriculture, Farmers' Bulletin 1102, 1920.

Karr, Clarence. "Robert Stead's Search for an Agrarian Ideal." *Prairie Forum* 14, no. 1 (Spring 1989): 37–57.

Kaye, Frances W. *Hiding the Audience: Viewing Arts and Arts Institutions on the Prairies*. Edmonton: University of Alberta Press, 2003.

Keller, Betty. *Black Wolf: The Life of Ernest Thompson Seton*. Vancouver: Douglas and McIntyre, 1984.

Kingsland, Sharon E. *The Evolution of American Ecology, 1890–2000*. Baltimore: Johns Hopkins University Press, 2005.

Kitto, F.H. *The Peace River Country Canada: Its Resources and Opportunities*.3rd rev. edn. Ottawa: Department of the Interior, 1930.

Kohler, Robert G. *All Creatures: Naturalists, Collectors, and Biodiversity 1850–1950*. Princeton and Oxford: Princeton University Press, 2006.

586 Kohlstedt, Sally. *Teaching Children Science: Hands on Nature Study in North America 1890–1930*. Chicago: University of Chicago Press, 2010.

Kollin, Susan. "The Wild, Wild North: Nature Writing, Nationalist Ecologies and Alaska." *American Literary History* 12, nos 1/2 (Spring/Summer 2000): 41–78.

Kricher, John. *The Balance of Nature: Ecology's Enduring Myth*. Princeton and Oxford: Princeton University Press, 2009.

Lampard, Robert. "Henry George, MD, 1864–1932." *History Now* [newsletter of the Historical Society of Alberta] 2 (April 2008): 10–13.

Lawton, Benjamin. "Game in Cold Storage." *National Conference on Conservation of Game and Fur-Bearing Animals and Other Wildlife, February 18 and 19, 1919*, 71–3. Ottawa: King's Printer, 1919.

Leopold, Aldo. *A Sand Country Almanac*. New York: Ballantine Books, [1949] 1970.

Levere, H., and Richard A. Jarrell, eds. *A Curious Field Book: Science and Society in Canadian History*. Toronto: Oxford University Press, 1974.

Lewis, Margaret. *To Conserve a Heritage*. Edmonton: Alberta Fish and Game Association, 1979.

Loney, Martin. "The Construction of Dependency: The Grand Rapids Hydro Project." *Canadian Journal of Native Studies* 7, no. 1 (1987): 57–78.

Loo, Tina. "Making a Modern Wilderness: Conserving Wildlife in Twentieth Century Canada." *Canadian Historical Review* 82 (March 2001): 92–121.

– "Of Moose and Men: Hunting for Masculinities in British Columbia, 1880–1939." *Western Historical Quarterly* 32, no. 3 (Autumn 2001): 296–319.

– *States of Nature: Conserving Canada's Wildlife in the Twentieth Century*. Vancouver: UBC Press, 2006.

Lothian, W.F. *A History of Canada's National Parks.* Ottawa: Parks Canada, 1979.
MacDonald, Graham. *Science and History at Elk Island: Conservation Work in a Canadian National Park 1906–1994.* Microfiche Report Series 525. Ottawa: Canadian Heritage Parks Service, 1994.
MacEachern, Alan. "Changing Ecologies: Preservation in Four National Parks." In *Canadian Environmental History: Essential Readings,* edited by David Freeland Duke, 361–86. Toronto: Canadian Scholars' Press, 2006.
– "The Conservation Movement." In *Canada, Confederation to Present,* edited by Chris Hackett and Bob Hesketh. CD-ROM. Edmonton: Multimedia, 2003.
– *Natural Selections: National Parks in Atlantic Canada 1935–1970.* Montreal and Kingston: McGill-Queen's University Press, 2001.
– "Voices Crying in the Wilderness: Recent Works in Canadian Environmental History." *Acadiensis* 31, no. 2 (Spring 2002): 215–26.
MacKenzie, John A. *The Empire of Nature: Hunting, Conservation and British Imperialism.* Manchester: Manchester University Press, 1988.
MacLachlan, Ian. "The Historical Development of Cattle Production in Canada." Online at https://www.uleth.ca/dspace/bitstream/handle/10133/303/Historical_cattle_Canada.pdf?sequence=3.
Mair, Charles. *The American Bison: Its Habits, Methods of Capture and Economic Use in the North-West, with Reference to Its Threatened Extinction and Possible Preservation* (first published in *Transactions of the Royal Society of Canada* 8, section II, 1890). Montreal: Dawson Brothers, Publishers, 1891.
Mallory, Enid L. "Canada's Wildlife." *Canadian Geographical Journal* 80, no. 3 (1970): 98–109.
Managing Richardson's Ground Squirrels. Edmonton: Alberta Agriculture, AGDEX 684 – 2 July 2012.
Mandelbaum, David. *The Plains Cree: An Ethnographic, Historical and Comparative Study.* Regina: Canadian Plains Research Center, [1940] 1979.
Mandeville, François, *This Is What They Say: A Story Cycle Dictated in Northern Alberta in 1928, Edited and Translated from Chipewyan by Ron Scollon.* Vancouver: Douglas and McIntyre, and Seattle: University of Washington Press, 2009.
Martin, Calvin. *Keepers of the Game.* Berkeley: University of California Press, 1978.
Master, Oliver. "Report of Committee on Fisheries, Game and Fur Bearing Animals." In *Commission of Conservation, Canada, Report of the Eighth Annual Meeting Held at Ottawa January 16–17, 1917,* 249–56. Montreal: Federated Press, 1917.
Mayhood, David. *Cumulative Human Impacts on Native Trout Stock in Relation to Petro-Canada's Sullivan Field Development Project.* Calgary: Alberta Wilderness Association, 2008, at http://www.pekisko.ca/docs/sullivan/D_00350472_mayhood.pdf.
McCandless, Robert. *Yukon Wildlife: A Social History.* Edmonton: University of Alberta Press, 1985.

McClung, Nellie. *Clearing in the West.* Toronto: Thomas Allen, 1935.

McCormack, P.L. "Transportation and Settlement: Problems in the Expansion of the Frontier of Saskatchewan and Assiniboia in 1904." *Prairie Forum* 5, no. 1 (1980): 1–18.

McGowan, Dan. *A Guide to Reading: Nature Study in Canada (1) Animals and Birds,* and *A Guide to Reading: Nature Study in Canada (2) Trees and Shrubs.* Ottawa: Canadian Legion Educational Services, 1944.

– "Hunting with a Camera." *The Beaver,* Outfit 279 (June 1948): 22–5.

McTavish, Lianne. "Learning to See in New Brunswick, 1862–1929." *Canadian Historical Review* 87, no. 4 (December 2006): 553–81.

McTavish, Lianne, and Jingjing Zheng. "Rats in Alberta: Looking at Pest-Control Posters from the 1950s." *Canadian Historical Review* 92, no. 3 (September 2011): 515–46.

Melville, Elinor G.K. "Disease, Ecology, and the Environment." In *The Oxford History of Mexico,* edited by Michael C. Meyer and William H. Beezley, 213–43. New York: Oxford University Press, 2000.

Merchant, Carolyn. *The Death of Nature: Women, Ecology and the Scientific Revolution.* New York: Harper and Row, 1980.

Merriman, R. Owen. *Attracting Birds with Food and Water.* Ottawa: Department of the Interior, 1923.

Midgley, Mary. "Against Humanism." *Eurozine,* 3 November 2010 (first published in *New Humanist,* May 2010).

Mighetto, Lisa. "Science, Sentiment, and Anxiety: American Nature Writing at the Turn of the Century." *Pacific Historical Review* 54, no. 1 (February 1985): 33–50.

"Migratory Birds Convention." In *Commission of Conservation Canada, National Conference on Conservation of Game, Fur-Bearing Animals and Other Wildlife,* 99–121. Ottawa: King's Printer, 1919.

Millar, W.N. "The Big Game of the Canadian Rockies: A Practical Method for Its Preservation." In Canadian Commission of Conservation, Committee on Fisheries, Game and Fur-Bearing Animals, *Conservation of Fish, Birds and Game: Proceedings of a Meeting of the Committee, November 1 and 2, 1915,* 100–24. Toronto: Methodist Book and Publishing House, 1916.

Milloy, John. "'Our Country': The Significance of the Buffalo Resources for a Plains Cree Sense of Territory." In *Aboriginal Resource Use in Canada: Historical and Legal Aspects,* edited by Kerry Abel and Jean Friesen, 51–70. Winnipeg, University of Manitoba Press, 1991.

Milne, Jack. *Trading for Milady's Furs: In the Service of the Hudson's Bay Company 1923–1943.* Saskatoon: Western Producer Prairie Books, 1975.

Mitchell, Elizabeth B. *In Western Canada before the War: Impressions of Early Century Prairie Communities.* Saskatoon: Western Producer Prairie Books, [1915] 1981.

Mitchener, A.V. *Field Crop Insects and Their Control in the Prairie Provinces.* Winnipeg: Line Elevators Farm Service, Bulletin No. 8, 1956.

Mitman, Gregg. "When Nature *Is* the Zoo: Vision and Power in the Art and Science of Natural History." *Osiris*, 2nd series (1996): 117–43.

Mitman, Gregg, and Richard W. Burkhardt, "Struggling for Identity: The Study of Animal Behavior in America, 1930–1945." In *The Expansion of American Biology*, edited by Keith R. Benson, Ronald Rainger, and Jane Maienschein, 164–94. New Brunswick, NJ: Rutgers University Press, 1991.

Moodie, D.W., and Barry Kaye. "Taming and Domesticating the Native Animals of Rupert's Land." *The Beaver*, Outfit 307, no. 3 (Winter 1976): 10–19.

Morell, Virginia. "Minds of Their Own." *National Geographic Magazine*, March 2008, at http://ngm.nationalgeographic.com/2008/013/animal-minds/virginia-morell-text.

Morris, Alexander. *The Treaties of Canada with the Indians.* Toronto: Coles Publishing Co., 1971 (facsimile reprint of 1880 edition).

Moyles, R.G., ed. *Challenge of the Homestead: Peace River Letters of Clyde and Myrle Campbell 1919–1924.* Calgary: Alberta Records Publication Board, 1988.

Nadeau, Chantal. *Fur Nation: From Beaver to Brigitte Bardot.* London and New York: Routledge, 2001.

Nash, Charles W. *The Birds of Ontario in Relation to Agriculture.* Toronto: Ontario Department of Agriculture, 1898.

Nash, Roderick. "The Exporting and Importing of Nature: Nature Appreciation as a Commodity, 1850–1980." *Perspectives in American History* 12 (1979): 519–60.

– *Wilderness and the American Mind.* New Haven: Yale University Press, 1982.

Natural Resources of the Prairie Provinces: A Brief Compilation Respecting the Development of Manitoba, Saskatchewan and Alberta. Ottawa: Department of the Interior, 1923 and 1925 editions.

Nelson, J.G. *The Last Refuge.* Montreal: Harvest House, 1973.

Nelson, Joseph S., and Paetz, Martin J. *The Fishes of Alberta.* Edmonton: University of Alberta Press, 1992.

North American Banders' Study Guide, The. Point Reyes Station, California: North American Banding Council, 2012.

"North and West Homesteading at Rosthern." In *Mennonite Memories: Settling in Western Canada*, edited by Lawrence Kippenstein and Julius Toews, 163–86. Winnipeg: Centennial Publications, 1977.

Olson, Roberta J.M., and Kathleen Husler. "Petropolis: A Social History of Urban Animal Companions." *Visual Studies* 18, no. 3 (2003): 133–43.

Ondrack, Jack. *Big Game Hunting in Alberta.* Edmonton: Wildlife Publishing, 1985.

Osborne, Michael. "Acclimatizing the World: A History of the Paradigmatic Colonial Science." *Osiris*, 2nd series, 15 (2000): 135–51.

Pembina Institute, *Toward a Sustainable Economy: Grounding and Integrating Themes in Environmental Economics.* Drayton Valley, AB: Pembina Institute, 1998.

Pitman, C.R.S. "Waterfowl Predation in Canada by the Northern Pike (or Jackfish) Esox Indus L." *Bulletin of the British Ornithologists' Club* 82 (1962): 64–7.

Potyondi, Barry. "Loss and Substitution: The Ecology of Production in Southwestern Saskatchewan 1860–1930." *Journal of the Canadian Historical Association 1994*, 213–35.

Preece, Rod, and David Fraser. "The Status of Animals in Biblical and Christian Thought: A Study in Colliding Values." *Society and Animals* 8, no. 3 (2000): 245–63.

"Proceedings. Meeting of the Ecological Society of America at Toronto Meeting of December 28, 1921." *Ecology*, 3.2 (1922):166–76.

Progress. A Survey of Saskatchewan Government Activity. 4th edition. Regina: Bureau of Publications, Government of Saskatchewan, 1955.

"Provincial Revenue from Fisheries and Game." *Commission of Conservation Canada, Report on Lands, Fisheries, and Game and Minerals for the Year 1911.* Ottawa: Mortimer Co., 1911.

Quiring, David M. *CCF Colonialism in Northern Saskatchewan: Battling Parish Priests, Bootleggers and Fur Sharks.* Vancouver: UBC Press, 2004.

Ray, Arthur J. *The Canadian Fur Trade in the Industrial Age.* Toronto: University of Toronto Press, 1990.

– *I Have Lived Here Since the World Began: An Illustrated History of Canada's Native People.* Toronto: Lester Publishing and Key Porter Books, 1996.

– *Indians in the Fur Trade: Their Role as Hunters, Trappers and Middlemen in the Lands Southwest of Hudson Bay 1660–1870.* Toronto: University of Toronto Press, 1974.

– "Periodic Shortages, Native Welfare, and the Hudson's Bay Company 1670–1930." In *The Subarctic Fur Trade: Native Social and Economic Adaptations*, edited by Shepard Krech III, 22–41. Vancouver: UBC Press, 1984.

– "Some Conservation Schemes of the Hudson's Bay Company 1821–1850: An Examination of the Problems of Resource Management in the Fur Trade." *Journal of Historical Geography* 1, no. 1 (1975): 49–68.

Reiger, John. *American Sportsmen and the Origins of Conservation.* New York: Winchester Press, 1975.

Report and Minutes of Evidence of the Select Committee of the Senate on the Existing Natural Food Products of the North-West Territories, and the Best Means of Conserving and Increasing Them. Journals of the Senate of Canada, vol. 21, Appendix 1, 1887.

Ritvo, Harriet. *The Animal Estate: The English and Other Creatures in the Victorian Age.* Cambridge: Harvard University Press, 1987.

– "Going Forth and Multiplying: Animal Acclimatization and Invasion." *Environmental History* 17, no. 2 (April 2012): 404–14.

– "History and Animal Studies." *Society and Animals* 10, no. 4 (2002): 403–6.

Robison, Kristofer K., and Daniel Ridenour. "Whither the Love of Hunting?

Explaining the Decline of a Major Form of Rural Recreation as a Consequence of the Rise of Virtual Entertainment and Urbanism." *Human Dimensions of Wildlife: An International Journal* 17, no. 6 (2012): 418–36.

Roe, F.G. "The Red River Hunt." *Transactions of the Royal Society of Canada.* Section II, 1935, 171–218.

Ross, Monique M., and Cheryl Y. Sharit. "Forest Management in Alberta and Rights to Hunt, Trap and Fish under Treaty 8." *Alberta Law Review* 36, no. 3 (1998): 645–91.

Rothman, Hal. *The Culture of Tourism: Selling the Past to the Present in the American Southwest.* Albuquerque: University of New Mexico Press, 2003.

Rowan, William. "Some Effects of Settlement on Wildlife in Alberta." In *Transactions of the Canadian Conservation Association Third Annual Meeting, Quebec, 1952.* The Association, 1952.

– *The Ten Year Cycle: An Outstanding Problem of Canadian Conservation.* Edmonton: University of Alberta, Department of Extension Publications, 1948.

Rowles, Edith. "Bannock, Beans and Bacon: An Investigation of Pioneer Diet." *Saskatchewan History* 5, no. 1 (Winter 1952): 1–14.

Roy, Gabrielle. *Enchantment and Sorrow: The Autobiography of Gabrielle Roy.* Translated by Patricia Claxton. Toronto: Lester and Orpen Dennys, 1987.

Royal Commission on Canada's Economic Prospects. *Progress and Prospects of Canadian Agriculture.* Ottawa: Royal Commission on Canada's Economic Prospects, 1957.

Russell, Andy. "Are Wild Animals Really Wild?" *Reader's Digest,* May 1960, 234–40.

Russell, Nerissa."The Wild Side of Animal Domestication." *Society and Animals* 10, no. 3 (2002): 285–302.

"Sale of Game." In *National Conference on Conservation of Game, Fur-Bearing Animals and Other Wild Life. Under the Direction of the Commission of Conservation in Co-operation with the Advisory Board on Wild Life Protection, February 18 and 19, 1919,* 69–71. Ottawa: King's Printer, 1919.

Salisbury, Joyce E. *The Beast Within: Animals in the Middle Ages.* New York and London: Routledge, 1994.

Sandlos, John. "From within Fur and Feathers: Animals in Canadian Literature." *Topia* 4 (2001): 73–91.

– *Hunters at the Margin: Native People and Wildlife Conservation in the Northwest Territories.* Vancouver: UBC Press, 2007.

– "'Not Wanted in the Boundary': The Expulsion of the Keeseekoowenin Ojibway Band from Riding Mountain National Park." *Canadian Historical Review* 89, no. 2 (June 2008): 189–221.

Saskatchewan Poetry Book, 1952–53, The, Regina: Saskatchewan Poetry Society, 1952.

Saskatchewan Trappers Guide. Regina: Department of Natural Resources and Department of Northern Saskatchewan, n.d. (c. 1975).

Scott, Colin. "Knowledge Construction among Cree Hunters: Metaphors and Literal Understanding." *Journal de la Société des Americanistes,* 75, no. 75 (1989): 193–208.

Scott, Duncan Campbell. "Relation of Indians to Wild Life Conservation." In *Commission of Conservation Canada, National Conference on Conservation of Game, Für-Bearing Animals and Other Wild Life. Under the Direction of the Commission of Conservation. In Co-operation with the Advisory Board on Wild Life Protection. February 18 and 19, 1919,* 19–39. Ottawa: King's Printer, 1919.

Scott, Lorne. "Bring Back the Bluebirds." *The Beaver,* Outfit 306, no. 4 (Spring 1976): 30–1.

Secord, James A. "The Crisis of Nature." In *Cultures of Natural History,* edited by N. Jardine, J. A. Secord, and E.C. Spary, 447–59. Cambridge: Cambridge University Press, 1996.

Semenchuk, Glen P., ed. *The Atlas of Breeding Birds of Alberta.* Edmonton: Federation of Alberta Naturalists, 1992.

Seton, Ernest Thompson. *The Arctic Prairies.* New York: Charles Scribner's Sons, 1912.

– *Life-Histories of Northern Animals: An Account of the Mammals of Manitoba.* New York: Charles Scribner's Sons, 1909.

– *Wild Animals I Have Known.* Facsimile reprint of 1898 edition with Afterword by David Arnason. Toronto: McClelland and Stewart, 1991.

Sharp-Tailed Grouse in Saskatchewan. Regina: Saskatchewan Department of Natural Resources, n.d. [c. 1960], Conservation Bulletin No. 4.

Smith, Donald B. *From the Land of Shadows.* Saskatoon: Western Producer Prairie Books, 1990.

Soper, Dewey. *The Mammals of Alberta.* Edmonton: Hanly Press, 1964.

Spaulding, Jonathan. "Yosemite and Ansel Adams: Art, Commerce and Western Tourism." *Pacific Historical Review* 65, no. 4 (1996): 615–39.

Spence, Mark David. *Dispossessing the Wilderness: Indian Removal and the Making of the National Parks.* New York: Oxford University Press, 1999.

Spry, Irene. "Aboriginal Resource Use in the Nineteenth Century in the Great Plains of Modern Canada." In *Aboriginal Resource Use in Canada: Historical and Legal Aspects,* edited by Kerry Abel and Jean Friesen, 81–92. Winnipeg: University of Manitoba Press, 1991.

– ed. *Papers of the Palliser Expedition 1857–1860.* Toronto: Champlain Society, 1968.

– "The Tragedy of the Loss of the Common in Western Canada." In *As Long as the Sun Shines and Water Flows: A Reader in Canadian Native Studies,* edited by Ian A.G. Getty and Antoine Lusier, 203–28. Vancouver: UBC Press, 1983.

– "The Transition from a Nomadic to a Settled Economy in Western Canada 1856–96." *Transactions of the Royal Society of Canada,* vol. 6 (June 1968), section II, 187–201.

Stead, Robert J.C. *Songs of the Prairie.* New York: Platt and Peck Co., 1912.

Stearns, R.P. "The Royal Society and the Company." *The Beaver,* Outfit 276 (June 1945): 8–13.

Steinberg, Ted. "Down to Earth: Nature, Agency and Power in History." Forum Essay. *American Historical Review* 107, no. 3 (2002): http://www.historycooperative.org/journals/ahr/107.3/.

Stewart, David. "The First Half Century: A Sketch of the Early Years of the Historical and Scientific Society of Manitoba." *Manitoba Pageant* 24, no. 3 (Spring 1979): www.mhs.mb.ca/docs/pageant/24/firsthalfcentury.shtml.

Stewart, Kenneth, and Douglas Watkinson. *The Freshwater Fishes of Manitoba.* Winnipeg: University of Manitoba Press, 2004.

Stowe, Leland. *The Last Great Frontiersman: The Remarkable Adventures of Tom Lamb.* Toronto: Stoddard, 1982.

Strickland, E.H. "Entomology in the Early Days of Alberta." *Proceedings of the First Annual Meeting of the Entomological Society of Alberta 1953,* 5–12.

Suchet, Sandie. "'Totally Wild?' Colonising Discourses, Indigenous Knowledge and Managing Wildlife." *Australian Geographer* 33, no. 2 (2002): 141–57.

Taverner, Percy Algeron. *Bird Houses and Their Occupants.* Offprint from the *Ottawa Field Naturalist* 32, no. 7 (January 1919).

– "A Discussion of the Origin of Migration." *The Auk* 21, no. 3 (July 1904): 322–33.

– *The Hawks of the Canadian Prairie Provinces in Their Relation to Agriculture.* Ottawa: Canadian Department of Mines, Geological Survey Museum Bulletin, no. 28 (1918).

– "Scientific Advice for Wild Life Conservationists." *Canadian Field Naturalist* 40, no. 5 (May 1926): 106.

– *Suggestions for Ornithological Work in Canada* (offprint, *The Ottawa Field Naturalist* 29 [April–May 1915]), 25–6.

Taylor, C.J. *Jasper: A History of the Place and Its People.* Markham, ON: Fifth House, 2009.

– "Legislating Nature: The National Parks Act of 1930." In *To See Ourselves: Ecology and Culture in Canada,* edited by Rowland Lorimer, Michael M'Gonigle, Jean-Pierre Revéret, and Sally Ross, 125–37. Montreal: Association of Canadian Studies, 1991.

Thomas, Keith. *Man and the Natural World: Changing Attitudes in England 1500–1800.* Toronto: Penguin Books, 1984.

Thomas, Lewis G. "Alberta Perspectives, 1905." *Alberta History* 28 (1980): 1–5.

Thorpe, W.H. *Learning and Instinct in Animals.* Cambridge: Harvard University Press, 1963.

Tober, James A. *Who Owns the Wildlife? The Political Economy of Conservation in Nineteenth Century America.* Westport, CT: Greenwood Books, 1981.

Toogood, Mark. "Modern Observations: New Ornithology and the Science of Ourselves, 1920–1940." *Journal of Historical Geography* 37 (2011): 348–57.

Tough, Frank. *"As Their Natural Resources Fail": Native People and the Economic History of Northern Manitoba 1870–1930.* Vancouver: UBC Press, 1996.

Trippensee, Reuben Edwin. *Wildlife Management.* Vol. 2: *Fur Bearers, Waterfowl and Fish.* New York: McGraw Hill, 1953.

Voisey, Paul, ed. *A Preacher's Frontier: The Castor, Alberta Letters of Rev. Martin W. Holdom 1909–1912.* Calgary: Alberta Records Publications Board (Historical Society of Alberta), 1996.

Wadland, John Henry. *Ernest Thompson Seton: Man and Nature and the Progressive Era 1880–1915.* New York: Arno Press, 1978.

Waiser, W.A. *The Field Naturalist: John Macoun, the Geological Survey, and Natural Science.* Toronto: University of Toronto Press, 1989.

– *Saskatchewan's Playground: A History of Prince Albert National Park.* Saskatoon: Fifth House Publishers, 1989.

Wangler, Matthew. "Canada's Rocky Mountain Parks: Rationality, Romanticism and Modern Canada." In *The Prairie West as Promised Land,* edited by R. Douglas Francis and Chris Kitzan, 53–76. Calgary: University of Calgary Press, 2007.

Warren, Louis S. *The Hunter's Game: Poachers and Conservationists in Twentieth Century America.* New Haven: Yale University Press, 1997.

Weir, T.R. "Settlement in Southwest Manitoba 1870–1891." *Papers Read Before the Historical and Scientific Society of Manitoba,* series 3 (1960–61): 54–64.

Wetherell, Donald G. "Making Tradition: The Calgary Stampede, 1912–1939." In *Icon, Brand, Myth: The Calgary Stampede,* edited by Max Foran, 21–45. Edmonton: Athabasca University Press, 2008.

Wetherell, Donald G., and Irene R.A. Kmet. *Alberta's North: A History 1890–1950.* Edmonton: University of Alberta Press, 2000.

– *Useful Pleasures: The Shaping of Leisure in Alberta 1896–1945.* Regina: Canadian Plains Research Centre, 1990.

White Tailed Deer in Saskatchewan. Conservation Bulletin No. 2. Regina: Saskatchewan Department of Natural Resources, n.d. [c. 1956].

Williams, David. "The Sparrow War." *American History* 37, no. 2 (June 2002): 38–42.

Williamson, F.H.H. "Game Preservation in the Dominion Parks." In *Commission of Conservation Canada, Committee on Fisheries, Game and Fur Bearing Animals: Conservation of Fish, Birds and Game: Proceedings at a Meeting of the Committee, November 1 and 2, 1915,* 125–40. Toronto: Methodist Book and Publishing House, 1916.

Willis, Susan. "Looking at the Zoo." *South Atlantic Quarterly* 98, no. 4 (1999): 669–85.

Wonders, Karen. "Hunting Narratives of the Age of Empire: A Gender Reading of Their Iconography." *Environment and History* 11, no. 3 (August 2005): 269–91.

– "A Sportsman's Eden Part II." *The Beaver* 79, no. 6 (December 1999–January 2000): 30–8.

Wood, Kerry. *The Magpie Menace.* Red Deer: Advocate Press, 1949.
Wood, William. *Draft of a Plan for Beginning Animal Sanctuaries in Labrador.* "To Be Submitted to the Fourth Annual Meeting of the Conservation Commission of the Dominion of Canada in 1913" [privately printed].
Woodsworth, James. *Thirty Years in the Canadian North-West.* Toronto: McClelland, Goodchild and Stewart, 1917.
Worster, Donald. *Nature's Economy: A History of Ecological Ideas.* 2nd edn. Cambridge: Cambridge University Press, 1994.
– "Transformations of the Earth: Toward an Agroecological Perspective in History." *Journal of American History* 76, no. 4 (March 1990): 1087–106.
Young, R.G. "The Saskatchewan Museum of Natural History." *Canadian Geographical Journal* 51, no. 1 (July 1955): 264–76.
Zeller, Suzanne. *Inventing Canada: Early Victorian Science and the Idea of a Transcontinental Nation.* Montreal and Kingston: McGill-Queen's University Press, 2009.
Zmyi, Peter M. "A Fight to the Finish. The Extermination of the Gray Wolf in Wyoming, 1890–1910." *Montana* 46, no. 1 (1996): 14–25.

Index